MICHIGAN

0 10   50
Scale of Miles
UMMZ—1957—WLB

MICHIGAN FLORA

MICHIGAN FLORA

A guide to the identification and occurrence of the native and naturalized seed-plants of the state.

Part III

DICOTS (Pyrolaceae–Compositae)

Edward G. Voss

Cranbrook Institute of Science
Bulletin 61   *and*
University of Michigan Herbarium

1996

Series designed by William A. Bostick
Composition by Huron Valley Graphics, Inc.
Printed and bound by Edwards Brothers, Inc.
Edited by Marion M. Sisneros

To the honor of CLARENCE ROBERT HANES (1874–1956)
and FLORENCE NUTTEN HANES (1886–1966),
whose generous planning made possible
the completion of *Michigan Flora* and
publication of its three volumes.

Preface

A FLORA is an inventory or census of a basic resource. Like soil surveys and topographic maps, good ones (I am happy to say) frequently prove useful a great many years after completion. These documents, unlike textbooks, reviews of research, or even encyclopedias, seldom undergo revision every few years—despite the fact that some of their data do indeed become obsolete from the moment of going to press.

Fifty years ago, as a high school student in Ohio, I discovered in the Toledo Public Library a very large book by Charles C. Deam, *Flora of Indiana*. It was an impressive work not only for its bulk but also for its meticulous detail, critical evaluation of dubious records, thoroughness, and utility for identification. For the list price of $3.50, I saved my allowance and bought my own (now well worn) copy—one of the best investments I ever made! Similarly, I discovered a little volume, slim and concise but nevertheless very practical and useful for my summers in Michigan, H. A. Gleason's *Plants of Michigan*. So there was indeed a more orderly and precise route to identification than trying to match pictures! With Deam and Gleason as role models (we might now say), is it any wonder that 10 years later I was starting a full flora of Michigan that, contrary to all expectations, has taken 40 years to complete?

Deam examined nearly 85,000 specimens for his flora, of which well over half were in his personal herbarium. While we have had no single collector of Deam's stature in Michigan, we have had several whose collections each approached or exceeded 10,000 in the state. No time has been spent on an accurate count, but to say that the present flora (all three parts) is based on examination of over 250,000 specimens would be about right. My rough estimate is that the number of Michigan specimens in herbaria in the state has more than doubled over the past 40 years—so that it has not been practicable to examine critically every single one when there is an excess of common species from previously known localities. Even so, the sheer quantity of available material has contributed to delay in completion of this work.

The present volume covers all the families not included in Parts I (1972) and II (1985), i.e., it includes the "sympetalous dicots": those with united petals forming a cup-, trumpet-, funnel-, or wheel-shaped corolla. It concludes with a key to all families (or smaller special groups) in the entire Flora. The introductory material in the previous parts need not all be repeated here; nevertheless, readers may benefit from reminders on a few points.

I do stress that this is a local flora for one state, and it tries to avoid repetitive information which can be found elsewhere about distribution as a whole; standard descriptions of families, genera, and species; variants occurring elsewhere; and other data largely of extra-territorial origin. On the other hand, it does emphasize particular information about ranges, localities, collectors, and records within its geographical scope—in hopes that anyone compiling data for a larger area will be able to make use of what is presented for *this* portion of the territory. Michigan does encompass enough terrain to make concise generalizations difficult or impossible on matters of blooming time, habitat, and overall abundance. Detroit is closer to the East Coast of the United States than it is to the farthest point in Michigan. Some readers will doubtless wish that I had done more to indicate all taxa that have threatened or endangered status. With some exceptions, I have avoided providing what is inherently such ephemeral information. Michigan's Endangered Species Act requires biennial updating of the list; plants are regularly being added or removed as a result of new information or interpretations. (The assembly of data for this volume, in fact, will surely result in alterations in the official lists.)

The three published parts of *Michigan Flora* present a total of 2465 distribution maps, and that could be used as a count of the "number of species" in the state by those who demand such figures. However, a few of these maps represent hybrids well known by binomials; many other hybrids are mentioned in the text without being mapped. Similarly, a few waifs of dubious status are mentioned in the text without being mapped. About 60 species are known to have been collected (or recognized) in Michigan since the maps for Parts I and II were published (no census is ever stable!). Furthermore, the demand to know "How many species?" cannot be met without agreement not only on hybrids and the status of waifs or garden escapes, but also on taxonomic judgments regarding the distinctiveness of species. On this point there is always some understandable and healthy difference of opinion among experts in numerous groups (not alone *Crataegus*, *Viola*, and *Aster*).

The percentage of alien species is even harder to tabulate, especially since labels on many collections do not make clear whether the collector thought the specimen represented an indigenous occurrence. Some species are apparently represented by both indigenous and naturalized origins (e.g., *Erysimum cheiranthoides*, *Achillea millefolium*, and *Prunella vulgaris*). Others are native not far from Michigan, but occurrence labeled as "along railroad," e.g., could be interpreted to mean either a prairie relic or a waif spread by passing cars.

A large portion of this volume is devoted to the Composite Family, in which many species are of questionable nativity in the state. Among other families included are the heaths, gentians, milkweeds, mints, snapdragon allies, and honeysuckles. I suspect that plants with showy flowers and well known among the populace at large are better represented in this volume than in the previous two.

It is my intention, in collaboration with my colleague in the University Herbarium, A. A. Reznicek, to prepare a compact one-volume manual with updated maps. All herbaria, especially those outside of Michigan, cannot again be examined for additions, so persons aware of state or county records are encouraged to supply specimens or advise where they may be sought.

Acknowledgments

As this work comes to its conclusion, I must—as for the previous two volumes—acknowledge an immense debt of gratitude to those who have gone before: authors of floras and revisions, annotators of specimens in herbaria, and of course those who did the basic work, namely *collect* specimens. As always, my approach has been to check what I find in herbaria and literature against each other and against my own field experience, using whatever proves to be most practical for distinguishing our taxa, for indicating their habitats, and for documenting their distribution in the state. Naturally, it is impossible for me to agree with everyone, and certainly anyone is free to disagree with what I have done. I have generally tried to give clues to alternative viewpoints and to suggest sources for more detailed information. Furthermore, for a work as long in preparation as this, it is impractical to go back and consult specimens across the state and beyond every time a new revision appears in the literature, in order to see if other characters really do work on all Michigan material. During the last rush to get this final volume pulled together and into press, scant attention was paid to refinements in the treatments of groups long since accommodated in text and keys.

Many recent collectors active in the state were mentioned under Acknowledgments in Part II. About that time, Robert W. Smith began his very thorough collecting in Lenawee County, for which most mapped records are supported by his specimens. Don Henson (in the Upper Peninsula) and Russ and Deb Garlitz (especially in the northern Lower Peninsula) have continued their assiduous collecting. James Wells and colleagues concentrated in recent years on Macomb County and on islands in Lakes Michigan and Huron, of which the Charity Islands have been most productive for mapping purposes. Calhoun and Crawford counties are among the areas much better known as a result of collections made by personnel of the Michigan Natural Features Inventory. Documentation by environmental consultants, especially W. W. Brodowicz, has produced good records of a number of rarities. Peter Fritsch profitably concentrated during several seasons in Hillsdale County. Berrien County receives intense attention from many persons at Andrews University and the Morton Arboretum.

The principal herbaria in which Michigan specimens in all or most families have been examined for this volume, through visits or by means of loans, are listed below. Unless otherwise indicated, each is in Michigan. Symbols in parentheses are the standard ones as listed in *Index Herbariorum* (Holmgren, Holmgren, & Barnett 1990).

Albion College, Albion (ALBC)
Alma College, Alma (ALMA)
Andrews University, Berrien Springs (AUB)
Butler University, Indianapolis, Indiana (BUT)
Central Michigan University, Mount Pleasant (CMC)
Cornell University, Ithaca, New York (CU & BH)
Cranbrook Institute of Science, Bloomfield Hills (BLH)
Harvard University, Cambridge, Massachusetts (GH & A)
Isle Royale National Park, Houghton (IRP)
Miami University, Oxford, Ohio (MU)
Michigan State University, East Lansing (MSC)
Missouri Botanical Garden, St. Louis, Missouri (MO)
Morton Arboretum, Lisle, Illinois (MOR)
Northern Michigan University, Marquette (NM)
University of Michigan, Ann Arbor (MICH)
University of Michigan Biological Station, Pellston (UMBS)
University of Notre Dame, Notre Dame, Indiana (ND)
Wayne State University, Detroit (WUD)
Western Michigan University, Kalamazoo (WMU)

Selected specimens requested on loan from two institutions helped with several rare or unusual species:

New York Botanical Garden, Bronx, New York (NY)
U. S. National Herbarium, Washington, D. C. (US)

I am deeply indebted to the curators and support staff of all these institutions for their prompt and generous assistance with often massive loans and their hospitality when I have visited them, generally for very long working hours. A few specimens were recorded while on loan to MICH from various herbaria for research projects of other staff or students, and several small local herbaria were examined (including some dicots) when Part I was in preparation but not again. Time and circumstances unfortunately have not allowed checking *every* possible herbarium containing Michigan specimens.

The collections of the Ford Forestry Center of Michigan Technological University (MCTF) were presented in 1988 to Michigan State University and are now cited as part of the latter herbarium. Through the kindness of William R. Overlease, I have been able to examine all Part III families from his private herbarium (chiefly from Benzie Co.) and to retain selected specimens for the University of Michigan; the remainder are being presented by him to Butler University. The collections of Anne M. Richards (mostly Marquette Co.) have been given to Northern Michigan University, and the collections of Erna R. Eisendrath (mostly Charlevoix Co.) are now largely incorporated into the herbarium of the University of Michigan Biological Station.

The annotations of specialists on herbarium specimens are always helpful, whether provided on their initiative during their research or on my invitation for the present Flora. Early in this project, E. E. Sherff looked at many specimens of *Bidens* for me, and T. G. Yuncker did likewise with *Cuscuta*. Also in the 1950s, Robert W. Long agreed to look at numerous *Helianthus*, some of which were further examined by C. B. Heiser. Most recently, Randall J. Bayer checked some hundreds of *Antennaria*. To all these, I am most grateful.

A number of authorities on certain families or genera have reviewed portions of my manuscript—some of them many years ago—and I have taken their advice whenever I could do so: T. M. Barkley, Lincoln Constance, Tom S. Cooperrider, William G. D'Arcy, John V. Freudenstein, Charles B. Heiser, Rogers McVaugh, James S. Pringle, Earl E. Sherff, Sylvia M. Taylor, Edward E. Terrell, John W. Thieret, Arthur O. Tucker, S. P. Vander Kloet, James R. Wells. None of these should be blamed for any perversity on my part in following an alternative opinion, but all deserve thanks for improvements and several also for determining troublesome specimens submitted to them. Several students, alumni, and colleagues have noted problems in draft keys, especially the ones to families. Tony Reznicek has helped as always with countless identifications, literature references, helpful diagnostic characters, and general advice and encouragement.

Many students and other part-time assistants over the years have helped in recording data from specimen labels and organizing the records. Not least among their achievements has been dispelling some of my computer illiteracy (since 1991, all records have gone into a computer database as well as the usual files). Particularly essential was the help of Derek Artz, Joann Constantinides, and Nancy (White) Paul in finishing on schedule the record-keeping and assembling of maps, illustrations, and index in the final months of Part III preparation.

While the Michigan Flora Project has been part of the research program of the University Herbarium, the special expenses (since 1969) of travel, hourly assistance, preparation of maps and illustrations, computerization, and finally publication have been covered in large part by the Clarence R. and Florence N. Hanes Fund, to whose trustees all users of this Flora owe a great deal—as do I—for their aid and encouragement, thus making possible a substantially more comprehensive flora than was first envisioned. For Part III, especially valuable help with computerization and travel has come from a bequest to the Herbarium along with an endowment from E. Genevieve Gillette. Throughout preparation of Part III, the continued support of the director of the University of Michigan Herbarium, William R. Anderson, has made possible the concentrated and sustained effort required by a project of this magnitude.

ILLUSTRATIONS

The illustrations have been selected to show representative diversity among the families covered in this volume. The color photos give special (but by no means exclusive) attention to rare, unusual, showy, Michigan-related, or otherwise interesting species and to those for which line drawings in available sources were unsatisfactory. These photos are all from 35 mm transparencies of Michigan plants. The source (county) of each is indicated, along with the last name of the photographer. Doubtless more showy photographs than some of these exist somewhere, but I did not care to conduct (or judge) a nationwide search for the most spectacular photos. Nevertheless, I am grateful to those talented friends who responded to my invitation to submit slides, especially of certain species particularly needed: Frederick W. Case, Jr. (Saginaw, Michigan), Don Henson (Manistique, Michigan), Steven Jessup (Berkeley, California), and James R. Wells (Hendersonville, Tennessee; formerly Cranbrook Institute of Science).

The line drawings were mostly not made from Michigan plants. They have been borrowed from previously published sources, which it is a pleasure to acknowledge here. The skilled photographic copying of these was the work of David Bay, photographer for the Department of Biology.

Figure 1, the map of Michigan counties, is used through the courtesy of the University of Michigan Museum of Zoology.

Figures 9, 10, 11, 12, 13, and 14, by Ronald A. With, and 145 and 148, by Leslie A. Garay, are from *Shrubs of Ontario*, by James H. Soper and Margaret L. Heimburger, published by the Royal Ontario Museum (1982), and are used by permission.

Figures 8, 137, and 156, by Thomas Cobbe, are reprinted from Cecil Billington's *Shrubs of Michigan* (Bull. 20, revised 1949); Figures 122, 158, 167, 175, 180, 210, 214, 239, 256, 257, and 261, by Ruth Powell Brede, are reprinted from Helen Smith's *Michigan Wildflowers* (Bull. 42, revised 1966); both published by Cranbrook Institute of Science.

Figures 20, 21, and 117, by Sarah Phelps, and 22, by Janice Glimn Lacy, are from Otis' *Michigan Trees*, revised by B. V. Barnes and W. H. Wagner, published by the University of Michigan Press (1981), and are used by permission of the Press.

Figures 24, 27, 28, and 29, by Marion Platek, are from *The Gentians of Canada Alaska and Greenland*, by John M. Gillett (Publ. 1180 of Agriculture Canada, 1963); Figure 132, by Ilgvars Steins, is from *The Plantains of Canada*, by I. John Bassett (Monogr. 7 of Agriculture Canada, 1973); Figures 246, 247, 248, 249, 250, 251, 253, and 255, by Ilgvars Steins, are from *The Thistles of Canada*, by R. J. Moore and C. Frankton (Monogr. 10 of Agriculture Canada, 1974); all are reproduced with the permission of the Minister of Supply and Services Canada, 1995.

Figures 6, 7, 16, 18, 32, 46, 93, 119, 121, 157, 166, and 169 were prepared for the generic flora of the southeastern United States under the direction of Carroll E. Wood, who generously made them available; figures 32, 46, 119, 166, and 169 are used with permission of the Journal of the Arnold Arboretum, © 1977, 1986, 1989, and 1991 President and Fellows of Harvard College; most of the others also appeared earlier in the same journal.

Figures 47, 50, 52, 55, 96, 104, 109, 115, 123, 124, 126, 147, and 151, from *The Dicotyledoneae of Ohio, Part 2: Linaceae through Campanulaceae*, by Tom S. Cooperrider, are reprinted by permission. Copyright 1995 by the Ohio State University Press. All rights reserved.

Figures 112 and 118, by F. Schuyler Mathews, are from his *Field Book of American Wild Flowers* (rev. ed., 1927) © G. P. Putnam's Sons, and are used by permission of The Putnam Publishing Group.

Figures 35, 36, 37, 38, and 39 are from T. G. Yuncker's "Revision of the North American and West Indian Species of Cuscuta" (Illinois Biol. Monogr. 6 (2–3), published in 1921).

Figures 15, 49, 76, 83, 84, 92, 105, 133, 136, 160, 178, 179, 182, 188, 190, 191, 192, 195, 196, 198, 199, 204, 220, 234, and 245, by F. Schuyler Mathews, are from W. J. Beal's *Michigan Weeds* (Bull. 267. 1911); Figures 44 and 80 are from Darlington, Bessey, and Megee's *Some Important Michigan Weeds* (Spec. Bull. 304. 1940); both published by the Michigan Agricultural Experiment Station.

Figures 30, 33, 67, 72, 75, 77, 86, 91, 106, 134, and 135, by Regina O. Hughes, are from Clyde Reed's *Selected Weeds of the United States* (Handb. 366. 1970), published by the U. S. Department of Agriculture.

Figures 26, 120, and 150 are from W. J. Hooker's *Flora Boreali-Americana*, London, 1833–1838.

Figures 129, 189, 266, and 267 were drawn for me by Edward M. Barrows and previously appeared in *The Michigan Botanist* 6: 46–48 (1967). The embellishment with Figure 74 is the work of Karin Douthit.

All of the remaining line illustrations are reproduced by permission of the New York Botanical Garden from H. A. Gleason's *New Britton and Brown Illustrated Flora of the Northeastern United States and Adjacent Canada* (Vol. 3, 1952).

On February 22, 1956, the executive board of the Horace H. Rackham School of Graduate Studies of the University of Michigan approved the Michigan Flora Project for five years of funding through the Faculty Research Fund. It is a pleasure to be able to draft this preface and submit the final portion of manuscript just 40 years later!

E. G. V.
University of Michigan Herbarium

Ann Arbor
February 22, 1996

Contents

MICHIGAN FLORA

Introductory Section

Introduction

The need for knowledge about Michigan's plant life has never been greater. Environmental assessment, outdoor recreation, nature photography, natural area protection, concern for habitat of animals—not to mention humans—all involve fundamental information about the diversity of plants, on which the other forms of life ultimately depend.

The purpose of *Michigan Flora* remains as it began—to help all interested persons to expand their knowledge about the botanical riches of this particular portion of the world: how to identify the plants, where they are known to grow (in terms of both geography and habitat), and what still needs to be found out about them. Indirectly, at least, it should also help to dispel some myths and misinformation. Would that every writer who mentions plants might consult a reliable flora before disseminating drivel about their distribution, names, and habits! Almost every hiker, homeowner, or student sooner or later encounters a plant (through allergy, stomach-ache, or simple curiosity) that remains a mystery until it is named and something learned of its nature.

THE BASIS OF THIS FLORA

Michigan Flora is essentially a distillation and integration (or reconciliation) of nearly two centuries of botanical collecting, exploration, and publication in this region. Thousands upon thousands of pressed and dried specimens rest in herbaria. Countless published records are distributed through an immense diversity of literature (some of it relatively obscure); if these are documented by preserved specimens, one can determine whether they are indeed correct by modern standards of classification and nomenclature. However, most specimen records have never been reported upon in print. And of course, there are doubtless many species growing in one corner of the state or another that remain to be discovered.

Thomas Nuttall, in 1810, was the first person with a professional interest in natural history to study plants in what is now the state of Michigan. He encountered in his travels many species that he described as new to science, including three from the area of Mackinac Island. These three, each illustrating, incidentally, a different overall distribution pattern in North America, have been displayed as the frontispieces of the three parts of *Michigan Flora*. Dr. Dennis Cooley began collecting in Michigan, mostly in Macomb and Oakland counties, when he moved to the state in 1827; his specimens (now at Michigan State University) are the first by a resident collector. When Michigan became a state in 1837, the first department of government created by the legislature was a geological survey, additionally charged with surveying the plant and animal life of the state. McVaugh (1970) has discussed the work of the "First Survey," under the direction of Douglass Houghton. Many of its collections still exist, and indeed some species found by its collectors 1837–1840 remain the only ones of their kind known from Michigan. Information on the lives and travels of these and other early collectors in the state is in Voss (1978). Brief comments on some of the more recent collectors are in the first two parts of this Flora and continue in the Acknowledgments in this one.

The distribution maps presented here are based solely on extant specimens collected from the early 19th century onward. Every dot on a map is supported by one or more specimens that I have examined and for which full label data are on file in the University of Michigan Herbarium. The specimens themselves may be in any of a large number of herbaria. No matter where the specimens are located, if the data are on file the herbarium sheet bears a rubber-stamp "Noted, Michigan Flora Project" with the year of recording. Records filed since April of 1991, including *all* specimens in families from Solanaceae through Compositae, are in a computer database in addition to the paper file slips that cover all three parts of *Michigan Flora*. Especially for common species and in circumstances when time was limited, not every specimen in a herbarium was necessarily checked and recorded. However, approximately a quarter-million specimens have been fully recorded over the past 40 years, and the files will explain the basis for every map dot: when collections were made in a county (or island group), exactly where, by whom, and how often—insofar as this information appeared on the labels or could be added from other sources. Also in the files are the herbarium location of each collection and the dated initials of specialists who have annotated them. For introduced species, I have often tried in the text to indicate the earliest records from the state; and for species that appear to be extirpated, the latest records.

While every map dot is backed up by one or more specimens with identification checked, there is no way to be certain of accurate labeling as to source. A few collections for which there is some strong reason to be suspicious are merely mentioned in the text and not mapped. Some others are mapped, but queried in the text (see "Credibility of Records," below). With these cautions, those who need to know the status of a species in Michigan will find the details in the maps and text. If the range of a species is merely said in a manual or other source to include "Michigan," does that mean it is found throughout the state? Or was it found as a waif only in the 1890s? "Read all about it" in the present volume.

Another source of confusion is not specimens incorrectly identified but ones cited in manuals or monographs on the basis of misinterpretation of labels, such as taking a collector's address to be the locality for a specimen. For example, collections made in the 1830s and 1840s and part of the herbarium of Dennis Cooley, with labels explicitly attributing them to Massachusetts or Illinois (and sometimes even to other collectors), have been cited from Macomb County, Michigan, in reputable monographs simply because Cooley's home was in that county and his address was stamped on his labels.

As in previous volumes, in addition to the 83 counties in Michigan, seven islands or island groups in the Great Lakes are mapped separately as they are distinct enough geographically and/or phytogeographically from the mainland counties to which they are politically attached. These are shown on the map in Fig. 1:

Charity Islands (Arenac County)
Beaver Islands (Charlevoix County) [The entire group.]
Drummond Island (Chippewa County) [Including a few small adjacent islands but not islands of the St. Mary's River.]
Isle Royale (Keweenaw County) [The entire archipelago comprising the National Park.]
Fox Islands (Leelanau County)
Manitou Islands (Leelanau County)
Mackinac, Round, and Bois Blanc Islands (Mackinac County) [But all other islands in the Straits of Mackinac are included with the mainland.]

Absence of a dot from a county (or island) means only that no specimen has been seen in the herbaria examined (as listed in the Acknowledgments). If the species is known from neighboring areas, it is likely that it does occur in the apparently missing county, at least if appropriate habitat is there. Common weeds and trees, as well as garden escapes, tend to be neglected by collectors and thus to be under-represented on the maps, while species considered to be rare or otherwise interesting are usually more thoroughly documented. Some localities, such as lakes, straddle a county line or refer to more than one county. If a collector failed to specify a county on a label there may be no way to assign it. Records from such localities are not mapped unless there is no other record from either of two counties; then the dot is placed on the county line.

Just as geographic information is summarized on the maps from existing specimens, so too is habitat information summarized in general terms from the labels of Michigan specimens—adjusted with some common sense, personal experience, and original literature for this region (statements of

1. Map of Michigan, showing counties and major islands in the Great
   Lakes.

habitat for other regions often do *not* apply here). Too many labels use-lessly give a range of habitats apparently visited by the collector on the same day, without specifying the precise one from which the specimen on the sheet actually came. Collectors conform to no uniform set of habitat descriptions, so I can give only an impression of representative kinds of places where each species grows, not a catalog of all possible sites according to any rigid definitions. In the concise histories often given for the apparent early spread of exotic species in the state, I have sometimes listed several counties, usually in alphabetical order (which carries no significance); however, if the sequence is not alphabetical, the context should suggest whether it is geographical (e.g., north to south) or (usually) chronological. Remember that especially when collectors were relatively sparse, the history of collections may document the history of collectors more thoroughly than the distribution of plants (see example of *Xanthium strumarium*, under genus 31 in the Compositae).

TAXONOMY AND NOMENCLATURE

Little needs to be said on these topics that was not covered in Parts I and II. I have tried, as always, to be nomenclaturally correct and to avoid perpetuating names that are bound to fall into disuse as conviction spreads that they are incorrect under the *International Code of Botanical Nomenclature* (Greuter et al. 1994). The basic principle of *nomenclature* is to use the oldest available name or epithet at the rank employed for a taxon and applicable within the circumscription adopted for that taxon. Of course, the *taxonomic* questions of rank and circumscription are often more difficult to deal with, but those questions have to be settled before the Code can be applied in determining the correct name. Many name changes are inevitable as ideas of rank and circumscription change with increasing knowledge. If two or more taxa are merged, the oldest name prevails and others are not used. If a taxon is divided, only one of the elements continues to bear the former name. If a variety is elevated to the rank of species, or vice versa, the name inevitably changes (at least in number of words).

For example: Whether the plant long known as *Aster ptarmicoides* is truly an *Aster*, or belongs in *Solidago*, or deserves to be in a third genus, *Oligoneuron*, is a question of taxonomy, on which reasonable botanists may differ even when confronted by the same set of facts. By way of contrast, nomenclatural rules sort out the possible names: *Doellingeria ptarmicoides*, published by Nees in 1832, is illegitimate as it was a renaming of *Chrysopsis alba*, published by Nuttall in 1818; this renaming was improper since there was no obstacle to transfer of the earlier epithet *alba* to *Doellingeria* if that is where Nees wanted to classify the species. However, existence of the names *Solidago alba*, published by Miller in 1768, and of *Aster alba*, published by Sprengel in 1826, both for different plants, prevents later transfer of Nuttall's *alba* to those genera. (Different species may not bear the same binomial!) Torrey and Gray, in 1841, were the first properly to use the name *Aster ptarmicoides* for this plant and so the epithet is attributed to them (not to Nees); Boivin, in 1972, transferred this epithet to *Solidago*, as needed to reflect taxonomic opinion that the species was really a goldenrod and not an aster. In 1993, Nesom accepted a narrower view of generic limits and since there was no prior use of Nuttall's epithet in *Oligoneuron*, published the new combination *Oligoneuron album*. (There are yet other names for this species in the literature, but let's keep this example simple!)

From the beginning, I have been most concerned with establishing *what* species are documented from Michigan, *where* in the state they have been found (as to both geography and habitat), and *how* they can most reliably be distinguished from one another. I have been less concerned with adjusting family delineations or generic assignments and such taxonomic issues — important as they are. The latest opinion does not always prove to be the most durable, and for convenience of the average user I have been slow to adopt every "up-to-date" realignment. Surely there will be many more proposed before the Flora goes out of print. Users can always adjust as fits their needs, whether to be thoroughly up to date or to be more consistent with widespread previous usage. Citation of synonyms helps in making the necessary correlations, and I have given synonyms here when adopting names that are likely to be unfamiliar. Often I provide a few words, without going into complex bibliographic detail, that may help explain the *reasons* why a different name (or citation of different authors for it) is used here. Sometimes it is a taxonomic reason such as a different generic assignment or defining a taxon broadly or narrowly (lumping vs. splitting). Sometimes a name is nomenclaturally incorrect under the Code and may not be used for any plant or must be used for a different plant than the one to which it had been incorrectly applied.

For circumscription of families, I follow Cronquist's *Integrated System* (1981). However, as this Flora was begun in 1956, before full explication of that or any other modern system, the *sequence* of families in this volume is necessarily still the antiquated Englerian one. (Recent research suggests that even the "Cronquist system" is moving toward obsolescence.)

Common names are governed by no code and far too often seem to have been coined rather than actually to reflect common use. I cite them sparingly, and when they are not taxonomically correct use a hyphen or a single word (e.g., bogbean is not a legume) or quotation marks (e.g., the ericaceous "laurel" is not in the Lauraceae). So-called common names that are not in fact in common use do not promote communication, many plants do not have such names at all, and the only reliable key to tracking down further information is by means of scientific names, which are reasonably uniform and certainly universal across language and other cultural barriers.

CREDIBILITY OF RECORDS

Some collectors are, unfortunately, rather careless in writing their labels, and I have had to ignore several with contradictory or incomplete data that could not be resolved from additional notes, itineraries, or other sources. For example, a number of specimens from "Pratt Lake, Michigan" (where there are four such lakes in as many different counties) had to be temporarily set aside until finally the discovery of two key species in the lot led me to assign them to Ogemaw County—the only one containing a Pratt Lake that would have both these species in the vicinity. (There were no notebooks of the collector available and no one with a knowledge of his travels.)

Problems relating to collections of O. A. Farwell are well known to, and lamented by, most botanists who have had to deal with his copious collections. There is no doubt that he had a sharp eye for all sorts of variants (and a compulsion to name them). However, there is frequently cause to doubt the accuracy of labels from the Keweenaw Peninsula, especially in the 1880s. But time has proved him too often correct in documenting odd range extensions for us summarily to discount *every* unreasonable-looking label. Specimens totally out of line phenologically with label dates do lead us to question locality as well. Collection numbers inserted, with a fraction, in his sequence are likewise often suspect (see also note in Mich. Fl. i: 285). Plants hundreds of miles beyond their otherwise known range are often questioned in the text, at least if not again seen for over a century, although some such disjuncts are known to this day in Farwell's territory.

Quite a number of prairie species are among Farwell's mysteries; perhaps there really was a good prairie element in Keweenaw County, one which has mostly not been found since. A few of the species are indeed found elsewhere in the Upper Peninsula—but some have never been found again near the Keweenaw. Should these be discounted? Not a prairie plant, but another case in point, is *Krigia biflora*, which does occur widely in both peninsulas of the state. All Upper Peninsula collections bear glandular hairs at the base of the involucre and the summit of the pedicel, while all Lower Peninsula collections are glabrous. There is one exception to that documented distribution: The only Keweenaw County specimen, gathered by Farwell in 1908, is perfectly glabrous—casting some justified doubt, perhaps, on its alleged origin. Through the kindness of the Cranbrook Institute of Science, Farwell's field notes (such as they are) and his copiously annotated old Britton and Brown floras have been readily available to me and these have sometimes been very helpful for confirming or casting doubt upon a dubious specimen. Usually, however, these sources and the scraps of specimen labels that Farwell wrote (see McVaugh, Cain, & Hagenah 1953) are in full accord with one another, even for material about which one must feel very uneasy.

REFERENCES

Since publication of Part II, with its list of useful references, several important volumes dealing with whole floras or at least multiple families have appeared and have been frequently consulted. These include fine new editions of Swink and Wilhelm's *Plants of the Chicago Region* and the Gleason and Cronquist *Manual of Vascular Plants*, an excellent *Flora of the Great Plains* by the Great Plains Flora Association, a thorough revision of the Kartesz *Synonymized Checklist*, and an outstanding installment on the Ohio flora by my college classmate Tom Cooperrider (an example of convergent evolution?). A new version of Gleason's *Plants of Michigan* is nearing publication and will prove valuable for those who want a compact key to common and conspicuous species in the state. These and several other works particularly helpful in identifying local plants covered in the present volume or in interpreting past reports for Michigan are cited below (and not repeatedly in the text), as are other standard references for style and conventions. Information on the flora of the Canadian province that forms the entire eastern and northern border of Michigan is in the catalogs by Morton and Venn.

Many of the countless publications dealing with single families or genera are listed at appropriate places in the text, including some with information on the lives and uses of various species, especially in this part of the continent. References cited in the text simply by author and date, if not included below, are given in full at the beginning of the relevant genus or family.

Some serials, especially those published by scientific societies, can be confusing in regard to date of publication (i.e., actual distribution). The year in which a meeting or conference was held, the year in which a preface was dated, and even the year stated on the published document may one or all of them differ from the true date of publication. In accord with widespread practice among botanical bibliographers, when there is conflict, the actual date of publication is followed, in full citations, by quoting the stated date [in brackets].

Bailey Hortorium. 1976. Hortus Third A Concise Dictionary of Plants Cultivated in the United States and Canada. Macmillan, New York. xiv + 1290 pp.

Barnes, Burton V., & Warren H. Wagner, Jr. 1981. Michigan Trees. Univ. Mich. Press, Ann Arbor. 383 pp.

Beal, W. J. 1905 ["1904"]. Michigan Flora. Rep. Michigan Acad. 5: 1–147.

Brummitt, R. K., & C. E. Powell (eds.). 1992. Authors of Plant Names. Royal Botanic Gardens, Kew. 732 pp.

Cole, Emma J. 1901. Grand Rapids Flora. A. Van Dort, Grand Rapids. 170 pp.

Cooperrider, Tom S. 1995. The Dicotyledoneae of Ohio Part 2. Linaceae through Campanulaceae. Ohio State Univ. Press, Columbus. 656 pp.

Cronquist, Arthur. 1981. An Integrated System of Classification of Flowering Plants. Columbia Univ. Press, New York. 1262 pp.

Deam, Charles C. 1940. Flora of Indiana. Dep. Conservation, Indianapolis. 1236 pp.

Fernald, Merritt Lyndon. 1950. Gray's Manual of Botany. Ed. 8. Am. Book Co., New York. lxiv + 1632 pp. [Some corrections in later printings, including 1987 facsimile reprint of 1970 printing, by Dioscorides Press, Portland, Oregon.]

Gleason, Henry A. 1952. The New Britton and Brown Illustrated Flora of the Northeastern United States and Adjacent Canada. New York Bot. Gard., New York. 3 vol. [Minor corrections in later printings.]

Gleason, Henry A., & Arthur Cronquist. 1991. Manual of Vascular Plants of Northeastern United States and Adjacent Canada. 2nd ed. New York Bot. Gard., New York. lxxv + 910 pp.

Gleason, Henry Allan. [in press] Plants of Michigan. Revised and expanded, by Richard K. Rabeler. George Wahr Publ., Ann Arbor.

Great Plains Flora Association (T. M. Barkley, ed.). 1986. Flora of the Great Plains. Univ. Press of Kansas, Lawrence. 1392 pp.

Greuter, W., J. McNeill, et al. 1994. International Code of Botanical Nomenclature (Tokyo Code) Adopted by the Fifteenth International Botanical Congress, Yokohama, August–September 1993. Regnum Vegetabile 131. xviii + 389 pp.

Hanes, Clarence R., & Florence N. Hanes. 1947. Flora of Kalamazoo County, Michigan. Vascular Plants. [Authors,] Schoolcraft, Mich. 295 pp.

Holmgren, Patricia K., Noel H. Holmgren, & Lisa C. Barnett (eds.). 1990. Index Herbariorum Part I: The Herbaria of the World. 8th ed. Regnum Vegetabile 120. 693 pp.

Hultén, Eric. 1971. The Circumpolar Plants. II Dicotyledons. Sv. Vetakad. Handl. IV. 13(1). 463 pp.

Kartesz, John T. 1994. A Synonymized Checklist of the Vascular Flora of the United States, Canada, and Greenland. 2nd ed. Timber Press, Portland, Oregon. 2 vol.

Komarov, V. L. (ed.), et al. 1968– . Flora of the U.S.S.R. Israel Program for Scientific Translations, Jerusalem. Vol. 1–21 + 24 [of 30]. [Supplemented by S. K. Czerepanov. 1973. Additamenta et Corrigenda ad "Floram URSS" (tomi I–XXX). Nauka, Leningrad. 667 pp.]

McVaugh, Rogers, Stanley A. Cain, & Dale J. Hagenah. 1953. Farwelliana: An Account of the Life and Botanical Work of Oliver Atkins Farwell, 1867–1944. Cranbrook Inst. Sci. Bull. 34. 101 pp.

Morton, J. K., & Joan M. Venn. 1984. The Flora of Manitoulin Island. 2nd ed. Univ. Waterloo Biol. Ser. 28. 181 + 106 pp.

Morton, J. K., & Joan M. Venn. 1990. A Checklist of the Flora of Ontario Vascular Plants. Univ. Waterloo Biol. Ser. 34. x + 218 pp.

McVaugh, Rogers. 1970. Botanical Results of the Michigan Geological Survey under the Direction of Douglass Houghton, 1837–1840. Michigan Bot. 9: 213–243.

Schwarten, Lazella, & Harold William Rickett. 1958. Abbreviations of Titles of Serials Cited by Botanists. Bull. Torrey Bot. Club 85: 277–300.

Smith, Helen V. 1966. Michigan Wildflowers. Cranbrook Inst. Sci. Bull. 42, revised. 468 pp.

Smith, Norman F. 1995. Trees of Michigan and the Upper Great Lakes. 6th ed. Thunder Bay Press, Lansing. 178 pp.

Steyermark, Julian A. [1963]. Flora of Missouri. Iowa St. Univ. Press, Ames. lxxxiii + 1725 pp.

Swink, Floyd, & Gerould Wilhelm. 1994. Plants of the Chicago Region. 4th ed. Indiana Acad. Sci., Indianapolis. 921 pp.

Tutin, T. G., et al. (eds.). 1965–1980. Flora Europaea. Cambridge Univ. Press, Cambridge. 5 vol.

Voss, Edward G. 1972. Michigan Flora Part I Gymnosperms and Monocots. Cranbrook Inst. Sci. Bull. 55 & Univ. Michigan Herb. 488 pp.

Voss, Edward G. 1978. Botanical Beachcombers and Explorers: Pioneers of the 19th Century in the Upper Great Lakes. Contr. Univ. Michigan Herb. 13. 100 pp.

Voss, Edward G. 1985. Michigan Flora Part II Dicots (Saururaceae–Cornaceae). Cranbrook Inst. Sci. Bull. 59 & Univ. Michigan Herb. 724 pp.

Taxonomic Section

PLATE 1

C. **Ledum groenlandicum**
Voss   Mackinac Co

F. **Gaultheria procumbens**
Case   Saginaw Co.

A. **Pterospora andromedea**
Case   Alpena Co.

G. **Dodecatheon meadia**
Voss   Menominee Co.

E. **Epigaea repens**
Voss   Emmet Co.

D. **Vaccinium myrtilloides**
Voss   Luce Co.

B. **Kalmia polifolia**
Case   Otsego Co.

B. Vaccinium membranaceum
Voss   Luce Co.

D. Lysimachia thyrsiflora
Case   Wexford Co.

PLATE 2

A. Vaccinium ovalifolium
Voss   Luce Co.

C. Primula mistassinica
Voss   Emmet Co.

G. Gentiana alba
Voss   Washtenaw Co.

E. Sabatia angularis
Wells   Berrien Co.

**F. Gentiana
   puberulenta**
Case   Allegan Co.

B. Calystegia sepium
Voss   Emmet Co.

A. Asclepias tuberosa
Voss   Cheboygan Co.

C. Cuscuta gronovii
Voss   Washtenaw Co.

F. Leucophysalis grandiflora
Voss   Emmet Co.

D. Cynoglossum boreale
Voss   Emmet Co.

PLATE 3

G. Chaenorrhinum minus
Voss   Emmet Co.

E. Lycopus americanus
Voss   Luce Co.

F. Epifagus virginiana
Voss   Emmet Co.

C. **Gratiola aurea**
Henson   Gogebic Co.

E. **Conopholis americana**
Wells   Oakland Co.

B. **Chelone obliqua**
Voss   Washtenaw Co.

PLATE 4

A. **Besseya bullii**
Jessup   Jackson Co.

D. **Mimulus glabratus**
var. **michiganensis**
Voss   Cheboygan Co.

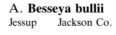
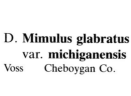

D. **Plantago cordata**
Voss   Hillsdale Co.

PLATE 5

A. **Utricularia resupinata**
Voss   Luce Co.

F. **Viburnum opulus**
Voss   Emmet Co.

B. **Ruellia humilis**
Wells   St. Joseph Co.

E. **Lonicera canadensis**
Voss   Emmet Co.

C. **Littorella uniflora**
Voss   Iron Co.

A. Lobelia cardinalis
Case   Saginaw Co.

E. Bidens coronatus
Voss   Washtenaw Co.

B. Agoseris glauca
Voss   Otsego Co.

PLATE 6

D. Bidens cernuus
Wells   Oakland Co.

G. Adenocaulon bicolor
Voss   Luce Co.

C. Megalodonta beckii
Voss   Cheboygan Co.

F. Matricaria discoidea
Voss   Emmet Co.

A. **Tragopogon dubius × T. pratensis**
Voss   Cheboygan Co.

B. **Silphium terebinthinaceum**
Voss   Delta Co.

PLATE 7

F. **Solidago simplex** var. **gillmanii**
Voss   Chippewa Co.

D. **Solidago altissima**
Voss   Emmet Co.

E. **Solidago hispida**
Voss   Luce Co.

C. **Solidago houghtonii**
Voss   Emmet Co.

A. **Helianthus mollis**
Wells   Monroe Co.

PLATE 8

B. **Aster ciliolatus**
Voss   Cheboygan Co.

F. **Petasites frigidus**
Voss   Emmet Co.

C. **Cirsium pitcheri**
Voss   Presque Isle Co.

E. **Petasites sagittatus**
Voss   Menominee Co.

D. **Liatris scariosa**
Voss   Otsego Co.

USING THIS FLORA: KEYS AND STYLE

The keys and descriptive comments in Part III follow the general style of the previous two. Now that the Flora is completed, keys to all the families in the entire work can be provided, and they appear at the end of the present volume, followed by the glossary and index. The user with a plant to identify and for which the family is not known must necessarily start with the general keys, which sometimes lead to a portion of a family (e.g., an unusual genus). Volume and page numbers are provided, to which one can then turn to continue the process of identification.

Keys must always be read carefully, and both halves of a couplet read, before one decides which of two choices better fits the specimen at hand. Such comparison will also provide a contrast in use of words, which sometimes have subtle differences in application in different contexts. If there is a problem, sometimes both choices have to be followed up in hope that eventually one will lead clearly to the correct answer. Always compare comments in the text, illustrations if any, habitat and distributional information, and any other available data before concluding that keying has been successful. Consult a work with more complete illustrations or descriptions, if doubt remains. Or consult a flora for a larger area or a different nearby area, such as the Gleason and Cronquist *Manual* or the *Flora of the Great Plains*. It is always possible that if an unknown plant doesn't fit the keys, it could be new to the state (or the keys could be defective!).

When possible, more than one character is mentioned in each couplet, with whatever is usually the most useful or reliable character mentioned first. Occasionally an exception is stated for a character ["(except in species 9)"] to suggest that the stated exception be checked in case of doubt over making a choice on the basis of other features presented. Such exceptions are generally species quite easily recognized on other characters or ones so unlikely to be seen (such as old waifs) that increasing the complexity of the key to accommodate them appeared unwise.

No special typography distinguishes the names of species not considered indigenous in the local flora. However, all species are considered indigenous in at least part of Michigan unless there is a statement to the contrary for the species or for the entire genus. Records of particular interest (the rarest species, ones previously unreported from the state, unusual forms, doubtful records, etc.) are often documented within parentheses with collector's name and collection number (if any) in italics; the year of collection, if cited, following "in" and not italicized; and the herbaria where specimens have been seen indicated by the standard symbols as listed in the Acknowledgments (p. xiii). Thus, persons questioning, or wishing to see, any of these records will have some context for their search. Square brackets in locality data indicate information (usually the county) which was not in the source (label or literature), but which has been added. Parentheses are similarly used for type localities (abbreviated "TL"). Whenever the type locality for a name mentioned is in Michigan, that is indicated by the abbreviation or by a fuller discussion.

Measurements of plant parts are based primarily on dry specimens—of which a far more diverse sample than fresh ones is available while keys are being written. The commonest range is stated, with further extremes at one or both ends indicated in parentheses or (if taken from literature rather than specimens) in square brackets. No precise statistical significance may be assumed from these extreme measurements. The symbol ±, meaning "more or less," is frequently used as a space-saving way of saying "rather," "somewhat," or "to a greater or lesser extent." "Ca." ("about") is sometimes used when measurements have been few or difficult to determine.

A good hand lens (magnifying about 10× or 12×) is usually necessary for seeing details, and good light is essential. A dissecting microscope is best, freeing the hands for manipulation. Checking characters of pappus and other features in the head of a composite (the largest family in this volume) requires patience. A rule graduated in millimeters or, even better, half-millimeters is essential, for it never pays to guess at measurements and ratios.

Illustrations

The scale of enlargement or reduction of a whole plant or inflorescence (or part of one) is given first in the legends, followed by indication of any details and their scale. The scales are only approximate and usually are those stated (if any) in the sources from which the drawings were taken, as credited in the Acknowledgments. On each page of figures, they are numbered in the same sequence as the species appear in the text. Sometimes space considerations have required that a species be shown "out of order" on a different page, however, and it is then numbered in the sequence for that page. Separate figure numbers are assigned to details only when they are on a different page from the main drawing or if they have been taken from a different source (requiring separate acknowledgment). All figures and color illustrations are cited after the names of the species in the text; occasionally one may also be cited in keys or comments, when critical points can be significantly clarified by the illustration.

Abbreviations and Symbols

Abbreviations for titles of periodicals basically follow Schwarten and Rickett (1958), except that names of states are not abbreviated. Citations for authors of scientific names follow Brummitt and Powell (1992), except that abbreviations that would replace only two letters by a period are not adopted, and a few unique, long-familiar, concise abbreviations for joint authors (e.g., T. & G. for Torrey & A. Gray) are used when there can be no confusion. Herbarium symbols (which in some cases, but not all, are abbreviations or acronyms) are listed on p. xiii. Generic names are abbreviated to the first letter when it is clear from the context what name is meant.

ca.	about (Latin: circa)
cf.	compare (Latin: confer)
cm	centimeter (see scale below, the numbered metric units)
dm	decimeter (= 10 centimeters; see scale below)
e.g.	for example (Latin: exempli gratia)
f.	form (forma)
i.e.	that is (Latin: id est)
m	meter (= 10 decimeters, 100 centimeters, 1000 millimeters)
mm	millimeter (= 0.1 centimeter; see scale below)
sens. lat.	in a broad sense (Latin: sensu lato)
sens. str.	in a narrow sense (Latin: sensu stricto)
sp.	species (singular)
spp.	species (plural)
ssp.	subspecies (singular)
TL:	type locality: the locality from which a type specimen came
Tp.	Township
var.	variety (varietas)
±	more or less
×	in figure legends, scale of enlargement or reduction; otherwise the sign of a hybrid: before a specific epithet, × indicates that the binomial applies to a hybrid; between two binomials, it is a formula indicating a hybrid (sometimes putative) of that parentage

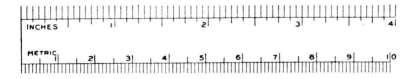

Magnoliopsida (Dicots)

PYROLACEAE   Shinleaf or Wintergreen Family

This family has frequently been included (as a subfamily) in the Ericaceae. There are good arguments each way. As in the Ericaceae, the anthers open by terminal (basal) pores rather than longitudinal slits; in most species, these pores are at the end of a short cylindrical tubular prolongation of the anther. I am indebted to John V. Freudenstein for helping to clarify this family for me.

REFERENCE

Haber, Erich, & James E. Cruise. 1974. Generic Limits in the Pyroloideae (Ericaceae). Canad. Jour. Bot. 52: 877–883.

KEY TO THE GENERA

1. Style ± strongly declined, at least 4 mm long; inflorescence a ± symmetrical raceme; plant nearly or quite scapose, with leaves crowded at the base ....1. **Pyrola**
1. Style straight (essentially none visible in *Chimaphila*), short or long; inflorescence various (usually 1-flowered, corymbose, or 1-sided); leaves various
  2. Inflorescence corymbose or umbellate; leaves evergreen, at least twice as long as broad, sharply (though often rather remotely) toothed; filaments strongly widened and pubescent (or at least ciliolate) toward the base; peduncle without bracts; style so short as to be concealed by the stigma ....2. **Chimaphila**
  2. Inflorescence racemose or 1-flowered; leaves deciduous, less than twice as long as broad, entire to crenulate or minutely denticulate; filaments neither expanded nor pubescent; peduncles (or scape) usually with 1 or more bracts; style not concealed by the stigma
    3. Inflorescence 1-flowered, the corolla 15–20 (22) mm broad, flat (petals widely spreading); anthers prolonged into a short cylindrical tube below the pore; valves of capsule glabrous; style (not including prominent stigma lobes) 3–5 mm long .
       ....3. **Moneses**
    3. Inflorescence racemose, the corolla 3–7 mm broad, ± campanulate (petals close about reproductive parts); anthers not prolonged into tubes; valves of capsule with cobwebby fibers on the margins when dehiscing; style various (but stigma only very shallowly lobed)
      4. Raceme 1-sided; style 2.5–6 (6.5) mm long, protruding at maturity from the corolla; sepal margins finely toothed or erose ....4. **Orthilia**

4. Raceme symmetrical; style ca. 1.5 mm or less long, scarcely if at all protruding beyond the corolla; sepal margins entire ....1. **Pyrola (minor)**

1. **Pyrola** Shinleaf; Pyrola

Orthilia secunda has traditionally been included in this genus, as *P. secunda* L., but recent authors are here followed in keeping it distinct.

REFERENCES

Haber, Erich. 1972. Priority of the Binomial Pyrola chlorantha. Rhodora 74: 396–397.
Haber, Erich. 1983. Morphological Variability and Flavonol Chemistry of the Pyrola asarifolia Complex (Ericaceae) in North America. Syst. Bot. 8: 277–298.

KEY TO THE SPECIES

1. Style straight, ca. 1.5 mm or less long; flower clearly with radial symmetry ...1. **P. minor**
1. Style ± strongly declined, (4) 5–9 (9.5) mm long at anthesis and on fruit; flower slightly bilateral
  2. Leaf blades thin (deciduous), dull above, elliptic-oblong, distinctly longer than their petioles; calyx lobes 1–1.8 (2.5) mm long, deltoid; bracts of the raceme narrowly cuneate, less than 12 mm broad; anthers hardly if at all constricted beneath the pores; petals white ....2. **P. elliptica**
  2. Leaf blades thick, ± coriaceous and evergreen, ± shiny above (or distinctly shorter than their petioles), nearly or quite orbicular (occasionally broadly elliptic); calyx lobes various (if less than 1.8 mm long, the leaf blades usually shorter than the petioles); bracts of the raceme mostly over 1 mm broad (except sometimes in *P. chlorantha*); anthers abruptly contracted to distinct short tubes beneath the pores (fig. 2); petals white to greenish or pink
    3. Calyx lobes 1–1.5 (1.7) mm long and mostly a little broader; leaf blades dull, all or mostly shorter than their petioles (or leaves absent); petals greenish white; major bracts at base of stem (among the leaves) less than 6 mm long ....3. **P. chlorantha**
    3. Calyx lobes (1.3) 1.5–3.2 (4) mm long and equally broad or (usually) distinctly longer than broad; leaf blades shiny above, shorter to longer than the petioles; petals white to pink or reddish; major bracts at base of stem over 6 mm (often over 10 mm) long
      4. Petals white (very rarely pinkish); calyx lobes ± ovate-oblong, not overlapping at base, distinctly longer than broad; leaf blades rounded to nearly truncate at the base (± tapered-decurrent onto the very narrowly winged petiole) ....4. **P. rotundifolia**
      4. Petals pink to reddish (rarely white); calyx lobes ± triangular, mostly slightly overlapping at the very base [use lens!]; leaf blades rounded to subcordate or cordate at the base (only shortly if at all decurrent) ....5. **P. asarifolia**

1. **P. minor** L.

Map 1. Very local under conifers (cedar, jack pine) and at edges of alder or spruce-fir thickets, usually very near Lake Superior. A circumpolar species, so similar around the world in northern latitudes that hardly any attempt has been made to recognize infraspecific taxa.

Very distinctive when the flowers are examined, but rather easily overlooked. However, it seems to be genuinely rare in the state. The leaves are thin in texture, rather like those of *Orthilia secunda*, with minute callus-teeth, but the blades more often obtuse or rounded at the apex.

2. **P. elliptica** Nutt.

Map 2. Usually in moist to dry chiefly deciduous (but sometimes coniferous) woods; often on banks of streams or ponds and along rivers. One of the relatively few herbaceous species blooming in mid-summer in deciduous forests, where the white flowers are rather conspicuous.

The bracts at the base of the pedicels are less than 0.8 mm broad (occasionally as broad as 1 mm at the base), whereas in *P. rotundifolia* and *P. asarifolia*, which are sometimes confused with it, these bracts are 1–3.2 mm broad (at least at middle and lower pedicels).

3. **P. chlorantha** Sw.   Fig. 2

Map 3. Moist to rather dry coniferous or mixed woods, often in moss or litter of conifer needles.

Sometimes known as *P. virens* Schreber. The leaf blades are only rarely as long as 3–4 cm, whereas they are longer in the preceding species and, often, in the next two. The bracts in the raceme may overlap in size those of *P. elliptica* on one hand or those of the next two on the other; but they tend to be more often minutely toothed or papillose on the margins and to be partly adnate to the pedicels.

4. **P. rotundifolia** L.

Map 4. Usually associated with deciduous trees in the southern part of the state, but northward often under red and jack pine, or in cedar swamps and peatlands (fens).

P. rotundifolia usually reaches the peak of blooming in the second or third week of July in our area (sometimes as early as the end of June in southern Michigan); it is our latest-blooming species of the genus, with flowers sometimes lingering on until September near Lake Superior.

1. Pyrola minor   2. Pyrola elliptica   3. Pyrola chlorantha

As treated here, this is a circumpolar species, the American plants mostly placed in var. *americana* (Sweet) Fernald, distinguished from the typical, largely Old World, variety by longer calyx lobes, petals, and anthers. The large bracts in the inflorescence will readily distinguish this species from *P. elliptica*, also white-flowered, should there be doubt on other characters. Some authors treat *P. americana* Sweet as a distinct species, and others treat it as a subspecies of *P. asarifolia*—while still others treat *P. asarifolia* as a subspecies of *P. rotundifolia*!

5. **P. asarifolia** Michaux

Map 5. Cedar swamps and other moist forests; peatlands, "marl bogs," and springy places; interdunal hollows and borders of shore thickets. Very local in the southern part of the state.

This is our earliest-blooming species, usually at its peak by the first week of July. In the Lake Superior area, and elsewhere when near cool shores (as on islands), it may bloom later, and if such plants are the rare white-flowered form they are rather easily mistaken for *P. rotundifolia* (which blooms still later in the same sites).

The calyx lobes in this species, while quite variable in shape, are usually more noticeably (though minutely) erose- or papillose-margined than in our other species. We have two varieties, not sharply distinguished: var. *asarifolia*, with subcordate to cordate leaf blades and calyx lobes distinctly longer than broad; and var. *purpurea* (Bunge) Fernald, with leaf blades rounded to truncate at the base and calyx lobes relatively short (often little if at all longer than broad and occasionally with a more ovate, less sharply triangular shape). (In general, var. *purpurea* averages smaller in all its characters than var. *asarifolia*.) All eastern North American plants are simply placed by Haber (1983) in ssp. *asarifolia*.

2. **Chimaphila**   Wintergreen

KEY TO THE SPECIES

1. Leaves broadest below the middle, marked with white or pale green along the midrib and main lateral veins ....1. **C. maculata**
1. Leaves broadest beyond the middle, dark green throughout ....2. **C. umbellata**

1. **C. maculata** (L.) Pursh   Fig. 3   Spotted Wintergreen

Map 6. Deciduous forests of several kinds, often with some conifers, but especially under oaks on sandy soils, as on wooded dunes.

2. **C. umbellata** (L.) W. P. C. Barton   Pipsissewa; Prince's-pine

Map 7. In a great diversity of dry to boggy woods, especially with pine and oak, persisting after some disturbance; dry, rocky openings and thickets. Very local in the southernmost part of the state, widespread in the north.

Eastern North American plants of this circumpolar species have been referred to var. *cisatlantica* S. F. Blake; var. *occidentalis* (Rydb.) S. F. Blake of western North America has been reported from the Upper Peninsula, but any differences seem inconsequential here (see Michigan Bot. 20: 60. 1981).

The attractive pink flowers nod in an inflorescence held above the crowded glossy-green leaves—a very attractive plant blooming in mid to late summer. The fruit, unlike the flowers, is held erect.

3. Moneses

1. **M. uniflora** (L.) A. Gray   Fig. 4   One-flowered Shinleaf
Map 8. Cedar swamps and other moist coniferous and mixed forests. The only Oakland County collection was made in 1888 and those from St. Clair County, 1892–1894; otherwise, strictly northern.

This very uniform circumpolar species is the only one in the genus. The flowers are rather frequently 4-merous (rather than 5-merous), and very rarely the stamens may be converted to additional petals, forming a "double" rose-like flower (see Michigan Bot. 16: 136. 1977). Plants are usually less than 10 cm tall, but occasionally become as tall as 14 cm, when the fruit is erect. The flowers nod shyly.

4. Pyrola rotundifolia

5. Pyrola asarifolia

6. Chimaphila maculata

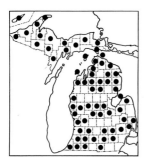
7. Chimaphila umbellata

8. Moneses uniflora

9. Orthilia secunda

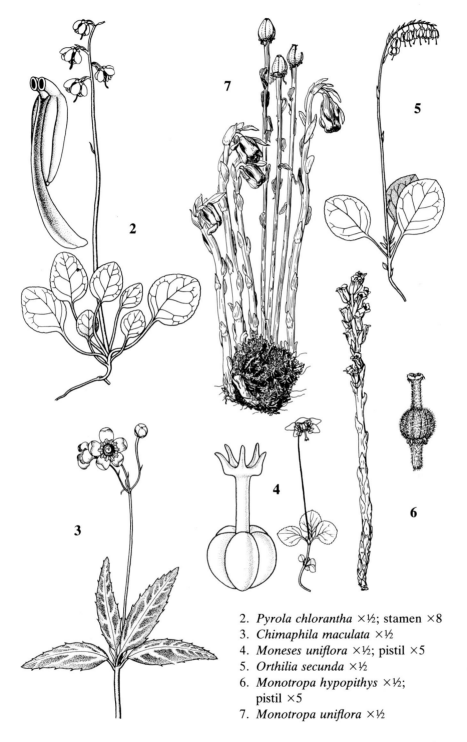

2. *Pyrola chlorantha* ×½; stamen ×8
3. *Chimaphila maculata* ×½
4. *Moneses uniflora* ×½; pistil ×5
5. *Orthilia secunda* ×½
6. *Monotropa hypopithys* ×½;
   pistil ×5
7. *Monotropa uniflora* ×½

4. **Orthilia**

This genus has long been included in *Pyrola* by most authors, but besides its distinctive 1-sided raceme and long straight styles, it differs in chromosome number, separate pollen grains (rather than coherent in tetrads), and other characters.

1. **O. secunda** (L.) House   Fig. 5   One-sided Shinleaf
    Map 9. Moist woods, coniferous and mixed; borders of thickets and dunes; sometimes in deciduous forests and other drier places. A circumpolar species, including var. *obtusata* (Turcz.) House, generally considered to inhabit regions farther to the north; but plants with smaller and obtuse to rounded leaf blades from some Upper and northern Lower Peninsula bogs and swamps have been referred to this variety.

The axis of the inflorescence is turned ± horizontal at flowering time, so the flowers hang like little bells. However, the axis soon becomes fully erect and the ripe fruit remains pendent (not always so clearly 1-sided).

MONOTROPACEAE   Indian-pipe Family

The Monotropaceae are often included in the Pyrolaceae, or as a subfamily of the Ericaceae (if the Pyrolaceae are also included in the Ericaceae); they differ in their "saprophytic" habit and also, generally, in having anthers opening by slits rather than apical pores, the pollen grains separate. All species in the family lack chlorophyll and as far as known are dependent for nourishment on a fungus which forms mycorrhizae with both the "saprophyte" and a forest tree. The stems, bearing reduced scale-like leaves, arise from a tight ball of mycorrhizal roots.

REFERENCES

Copeland, Herbert F. 1941. Further Studies on Monotropoideae. Madroño 6: 97–119.
Wallace, Gary D. 1975. Studies of the Monotropoideae (Ericaceae): Taxonomy and Distribution. Wasmann Jour. Biol. 33: 1–88.

KEY TO THE GENERA

1. Corolla of partly united petals, 5–7 mm long; calyx of 5 normal sepals, regular; plant ± densely glandular-pubescent, clammy; inflorescence (axis) erect from the beginning . . . . .1. **Pterospora**
1. Corolla of separate petals, 8–19 mm long; calyx of 0–4 bract-like sepals, irregular; plant glabrous or ± finely pubescent (not glandular); inflorescence bent or nodding when young (becoming erect) . . . . .2. **Monotropa**

1. Pterospora

REFERENCE

Bakshi, Trilochan S. 1959. Ecology and Morphology of Pterospora andromedea. Bot. Gaz. 120: 203–217.

1. **P. andromedea** Nutt.   Plate 1-A   Pine-drops
Map 10. Quite local and not to be relied upon to appear every year; nearly always in habitats with conifers (especially pines but also hemlock, spruce, fir, white-cedar), in dryish (or rocky) soil, often with common juniper and sometimes aspen or birch.

Bakshi ascribed this species to "moist soil" in western North America, discounting records from dry sites in the east, but never seeing as large colonies as we sometimes find in the Great Lakes region (nor fully investigating the mycorrhizal system). The species is disjunct between the Great Lakes region and the Black Hills and mountains to the west; also scattered eastward to Quebec and New England, but becoming very rare and local (see Michigan Bot. 20: 69–70. 1981). John Ball found it in 1860 to grow "in great quantities in many places through the islands" of Manitoulin (Ontario) and Drummond. Often said to be parasitic on the roots of pine, but no evidence of such a connection is known and if it is to be called a parasite, the "host" would presumably be the fungus which participates in a relationship more aptly termed symbiotic, involving also mycorrhizae with pine or other conifer.

The corolla is white and flask-shaped, resembling that of *Andromeda* (whence the name). The sepals and vegetative parts of the plant are reddish or maroon, very clammy-viscid. Individuals are usually twice (or more) as tall as the average *Monotropa*.

2. Monotropa

REFERENCE

Campbell, Ella O. 1971. Notes on the Fungal Association of Two Monotropa Species in Michigan. Michigan Bot. 10: 63–67.

KEY TO THE SPECIES

1. Flowers few to many; plant ± pubescent, at least in inflorescence, yellowish to ± orange or even reddish when fresh, becoming at most a dark brown long after flowering ....1. **M. hypopithys**
1. Flower solitary; plant glabrous (except for petals and filaments), pure white when fresh (rarely pinkish), soon turning black (except for anthers) when bruised, old, or dry ....2. **M. uniflora**

1. **M. hypopithys** L.　　Fig. 6　　　　　　　　Pinesap; False Beech-drops
Map 11. Coniferous or deciduous woods or even conifer swamps.

Specimens in fruit might be mistaken for small ones of *Pterospora andromedea*; besides lacking the clammy pubescence of that species, the style is 2.4–5 mm long, hirsute, and noticeable on the fruit (fig. 6). In *Pterospora*, the style is ca. 2 mm long or less, glabrous, and nearly or quite concealed by the expanding fruit.

This species is sometimes segregated into a different genus from the next, but the generic name *Hypopitys* is illegitimate and cannot be used (see Wood in Jour. Arnold Arb. 42: 68. 1961).

2. **M. uniflora** L.　　Fig. 7　　　　　　　　　　　　　　Indian-pipe
Map 12. Usually in slightly more moist woods than the preceding species, occasionally even in deep sphagnum moss in a bog.

Plants of decided pink color [f. *rosea* Fosberg] are rare, but have been found in Barry, Cheboygan, and Luce counties.

ERICACEAE　　　　　　　　　　　　　　　　　　　Heath Family

Following Cronquist, the family is here treated as including subfamily Vaccinioideae (*Vaccinium* and *Gaylussacia* of our flora), sometimes recognized as a distinct family on the basis of inferior ovary and fleshy fruit. There is, however, growing evidence that Pyrolaceae and Monotropaceae are better included as additional subfamilies in the Ericaceae. There is no doubt about either the distinctness of these three groups or their close relationship; the rank at which they are recognized "is purely a matter of taste" (Cronquist 1981). Judd and Kron (1993) "consider the circumscription of Ericaceae to be more than a matter of 'taste' and the segregation of Pyrolaceae and Monotropaceae to be a distortion of genealogical relationships." The keys and maps here apply equally well no matter what the family circumscription.

10. Pterospora　　　　　11. Monotropa hypopithys　　12. Monotropa uniflora
　　andromedea

8. *Kalmia angustifolia* in fruit, ×⅔
9. *Chamaedaphne calyculata* in fruit, in flower ×½
10. *Gaylussacia baccata* in fruit, in flower ×½

In all of our genera except four (*Kalmia, Calluna, Ledum,* and *Vaccinium* subg. *Oxycoccus*) the corolla has a tubular and usually flask- or urn-shaped appearance, ± swollen and then constricted just below the apex (rarely nearly cylindrical, as in *Epigaea*). Leaves of all our species are evergreen except in *Gaylussacia* and some species of *Vaccinium.* (Presence of leaves on twigs of a previous year is convenient evidence that they are evergreen.) A key to genera based solely on vegetative characters would be difficult, but the Ericaceae are for the most part easily recognized in the field—and so abundant and important in many habitats—that a generic key not based primarily on flowers or fruits is offered here.

Lyonia ligustrina (L.) DC. (maleberry) does not range north of southern Ohio in our longitude. It has been reported from Michigan, but the only specimen found so labeled by the collector (or anyone else!), from Newaygo County, is actually *Nyssa sylvatica* Marshall (sour-gum).

REFERENCES

Judd, Walter S., & Kathleen A. Kron. 1993. Circumscription of Ericaceae (Ericales) as Determined by Preliminary Cladistic Analyses Based on Morphological, Anatomical, and Embryological Features. Brittonia 45: 99–114.

Wood, Carroll E., Jr. 1961. The Genera of Ericaceae in the Southeastern United States. Jour. Arnold Arb. 42: 10–80.

KEY TO THE GENERA

1. Leaves opposite or whorled; flowers 8–20 mm broad or leaves less than 1.5 mm broad
   2. Flowers less than 4 mm broad; sepals 4, much exceeding the corolla; leaves closely appressed, less than 1.5 mm wide, auricled at the base . . . . .1. **Calluna**
   2. Flowers 8–20 mm broad; sepals 5, much shorter than the corolla; leaves ± spreading, at least the wider ones (2) 5–23 mm broad, not auricled . . . . .2. **Kalmia**
1. Leaves all alternate; flowers in most species less than 8 mm broad; leaves all or mostly over 1.5 mm broad
   3. Leaf blades with very densely pubescent lower surfaces (tomentose or minutely puberulent), mostly 2.5–15 (20) times as long as broad, the margins strongly revolute
      4. Flowers with urn-shaped pink (rarely white) corolla and conspicuous calyx (± 1.5– 2.5 mm long); leaves very minutely but densely white- (rarely grayish-) puberulent beneath . . . . .3. **Andromeda**
      4. Flowers with spreading, separate, white petals and minute calyx (not over 0.5 mm); leaves densely tomentose beneath with brownish or white woolly hairs . . . . .4. **Ledum**
   3. Leaf blades glabrous to somewhat pubescent beneath (if ± densely pubescent, the hairs straight and leaf margins flat), less than 3 (rarely 4) times as long as broad, usually with flat margins
      5. Leaf blades (evergreen) with scurfy appearance (especially beneath) from round orange-brown (to whitish) scales, otherwise glabrous; fruit a dry capsule; flowers white, subtended by reduced leaves in a ± elongate raceme-like inflorescence; plant a bushy-branched shrub . . . . .5. **Chamaedaphne**
      5. Leaf blades without scurfy scales (if leaves resin-dotted, then deciduous and ± pubescent); fruit ± fleshy (at least within); flowers and habit various

6. Plant an upright, bushy-branched, stiff-stemmed shrub; leaves deciduous; ovary inferior; fruit a blue or black berry
  7. Leaves with shiny orange resinous dots (especially beneath), entire, usually ± pubescent beneath (at least on midrib); fruit containing 10 hard nutlets . . . . .6. **Gaylussacia**
  7. Leaves without resinous dots, entire to finely toothed, glabrous to pubescent; fruit containing many fine seeds . . . . .7. **Vaccinium** (couplet 4)
6. Plant with primary stems lax, creeping or prostrate (sometimes with short erect shoots, and in *Gaultheria procumbens* the primary stem a shallow or surficial rhizome with erect usually unbranched shoots); leaves evergreen; ovary inferior or superior; fruit red or white and berry-like (except in *Epigaea*)
  8. Larger leaf blades (3.5) 4–7.5 (9) cm long, cordate or subcordate at base, on distinct petioles at least a fourth as long; corolla with nearly parallel-sided tube and flaring lobes; anthers opening by longitudinal slits (plants functionally dioecious); fruit a nearly globose capsule filled with fleshy white placental material and subtended by both sepals and bracts .8. **Epigaea**
  8. Larger leaf blades less than 4.5 (5) cm long, rounded or tapered to petioles less than a fourth as long; corolla urn-shaped or cleft to the base with reflexed petals; anthers opening by terminal pores (flowers mostly perfect); fruit fleshy throughout, subtended by sepals (or bracts) or (in most species) the ovary inferior and the fruit incorporating the calyx
   9. Leaf blades with scattered dark elongate glands or bristles beneath; flowers 4-merous
    10. Leaves rounded or even minutely notched (though with a gland) at apex; berries in terminal clusters, red; stems short-puberulent with incurved hairs or glabrous; plant very rare (only from Isle Royale), without wintergreen odor . . . . .7. **Vaccinium** (**vitis-idaea**)
    10. Leaves abruptly acute at gland-tipped apex; berries solitary in axils, white; stems with coarse appressed bristly hairs; plant common, with wintergreen flavor and odor . . . . .9. **Gaultheria** (**hispidula**)
   9. Leaf blades without bristles or glands beneath; flowers 4-merous or 5-merous
    11. Corolla 4-merous, cleft to the base, the petals strongly reflexed; leaf blades whitened beneath, less than 1.2 (1.4) cm long; calyx not fleshy in the mature fruit, but ovary inferior . . . . .7. **Vaccinium** (couplet 3)
    11. Corolla 5-merous, sympetalous, urn-shaped; leaf blades pale or dark green but not whitened beneath, mostly 1–4.5 (5) cm long; calyx fleshy and incorporated in the mature fruit or the ovary clearly superior
     12. Plant with wintergreen odor and flavor; leaf margins with obscure, remote teeth (bearing rusty brown bristles when young); leaves borne only at summit of erect shoots from surficial or shallow rhizome; flowers July–August . . . . .9. **Gaultheria** (**procumbens**)
     12. Plant without wintergreen odor or flavor; leaf margins strictly entire; leaves all along branches and shoots; flowers May (–early June) . . . . .10. **Arctostaphylos**

1. Calluna

A number of species of *Erica* (heath) are cultivated. Few are hardy this far north, and none are known to be escaped here. They would key to *Calluna*, which they resemble in 4-merous flowers and tiny linear leaves, but the latter are mostly whorled and not auriculate; the sepals are shorter than the petals.

REFERENCES

Gimingham, C. H. 1989. Heather and Heathlands. Jour. Linn. Soc. Bot. 101: 263–268.
Munson, Richard H. 1984. Heaths and Heathers Cultivated in North America (Ericaceae). Baileya 22: 101–133.

1. **C. vulgaris** (L.) Hull   Scotch Heather

Map 13. Introduced into cultivation from Europe, where it is a heathland dominant on acid soils where there has been human disturbance. Many cultivars are grown. Locally escaped in North America. Collected in 1961 (*Hagenah 4743*, BLH, MICH) along a roadside in eastern Mackinac County (Marquette Tp.), where it seems to have spread into second-growth mixed woods. Reported as naturalized in Marquette County 40 years earlier. Apparently planted in a borrow pit along highway US-10 in Midland County.

2. **Kalmia**   "Laurel"

These showy plants have the largest (and most open, saucer-shaped) flowers of any of our native Ericaceae. The corolla (except in albinos) is pink, and as it opens each anther is held in a little pocket (visible externally on the buds). The filament is therefore arched backward, under tension. When a visiting insect trips the filament, the anther springs loose from its pocket and the insect receives a shower of pollen.

Several species are in cultivation in North America, including our two although the most frequent is the mountain-laurel, *K. latifolia* L., which ranges well south and east of Michigan. *Kalmia* foliage is considered a stock-poisoning agent, but no human is likely to eat enough of it to cause trouble. These plants should not be confused with the true laurel (in the Lauraceae), the source of bay leaves (see Mich. Fl. ii: 236). The generic name commemorates Pehr Kalm, a Finnish student of Linnaeus, whose specimens brought back from travels in eastern North America 1748–1751 helped stimulate Linnaeus to complete the final draft of his *Species Plantarum*.

13. Calluna vulgaris

14. Kalmia polifolia

15. Kalmia angustifolia

REFERENCES

Ebinger, John E. 1974. A Systematic Study of the Genus Kalmia (Ericaceae). Rhodora 76: 315–398.

Holmes, Mary L. 1956. Kalmia, the American Laurels. Baileya 4: 89–94.

Southall, Russell M., & James W. Hardin. 1974. A Taxonomic Revision of Kalmia (Ericaceae). Jour. Elisha Mitchell Sci. Soc. 90: 1–23.

KEY TO THE SPECIES

1. Leaves opposite, sessile or subsessile, strongly whitened (by dense minute puberulence) beneath; inflorescences terminal; calyx and pedicels eglandular . . . . .1. **K. polifolia**
1. Leaves mostly in whorls of 3, distinctly petioled, green beneath; inflorescences axillary, the branches terminating in leafy shoots; calyx and pedicels glandular . . . . .2. **K. angustifolia**

1. **K. polifolia** Wangenh.   Plate 1-B   Pale- or Bog-laurel
    Map 14. Primarily in bogs but also in some fens and interdunal peaty swales; does not do well when succession produces too much shade.
    The young twigs are 2-edged, rather like those of *Hypericum kalmianum*.
    Polifolia was an old generic name taken up by Linnaeus as a species epithet in *Andromeda* (for a species more northern than ours). The epithet in *Kalmia* is here treated the same (although its origin is not so explicit); if it were to be considered merely a Latin adjective ("leaves like *Polium*" — a mint) it should be spelled *poliifolia* (the Code makes such spelling corrections obligatory for adjectival epithets, but they cannot be made for generic names).

2. **K. angustifolia** L.   Fig. 8   Sheep-laurel; Lambkill
    Map 15. Locally abundant and every bit as handsome as the previous species when in bloom, but more restricted in range in the state. Bogs and borders of peatlands; moist coniferous woods or swamps, including low areas in jack pine plains and oak or aspen woods.
    Usually reaches the peak of blooming in late June or the first half of July—a full month or more after *K. polifolia*. The corolla is generally a little smaller than in the preceding species: less than 12 mm broad, whereas in our *K. polifolia* it runs 13–20 mm.

3. Andromeda

1. **A. glaucophylla** Link   Bog-rosemary
    Map 16. Bogs and, especially, fens, often on floating mats — even to the edge of the water; relic in conifer swamps, and occasionally in other wet places such as interdunal hollows or rock crevices and pools on Lake Superior (Isle Royale). Very rare and local in southeastern Michigan and, indeed, long considered endangered (now probably extirpated) in Ohio.

Occasionally treated as a variety or subspecies of the circumpolar, far-northern *A. polifolia* L., a dwarf shrub with leaves glabrous beneath and the pedicels erect (especially in fruit) rather than nodding. Reports of typical *A. polifolia* from Michigan, if not based on outright misidentifications, are presumably derived from earlier reports by authors who did not distinguish the two taxa.

The leaves are typically a conspicuous pale blue-green above, at least when young (darker when mature), not glossy as in *Kalmia polifolia*, and they generally have an "alligator" pattern of lateral veins (often ± impressed). They are usually much larger than those of the seasoning herb, rosemary [*Rosmarinus officinalis* L., in the mint family], but appear similar, especially when dried, in the linear shape with strongly revolute margins. *Andromeda glaucophylla* leaves are usually white with minute pubescence beneath, but are occasionally a grayish green.

4. Ledum

REFERENCE

Kron, Kathleen A., & Walter Judd. 1990. Phylogenetic Relationships within the Rhodoreae (Ericaceae) with Specific Comments on the Placement of Ledum. Syst. Bot. 15: 57–68.

1. **L. groenlandicum** Oeder   Plate 1-C   Labrador-tea
Map 17. Typically in acid bogs and conifer swamps, but more widespread northward, in interdunal swales, in mossy conifer woods, on shaded sandy bluffs along Lake Superior, in rock crevices on Isle Royale.

This taxon is sometimes treated as a subspecies of *L. palustre* L.—making the latter a circumpolar species rather than strictly an Old World one. And some recent authors have included *Ledum* in the genus *Rhododendron*.

The leaves have long been used to make a tea, in the Great Lakes region as well as elsewhere, at least in times of emergency. Usually reaches the peak of blooming around early to mid June — later than *Kalmia polifolia*, with which it often grows; but close to Lake Superior, as on Whitefish Point, flowering may peak around the first of July in late seasons, and it is a truly glorious sight to see the two species blooming together in great masses. The woolly pubescence on the undersides of the leaves is white when young but turns rusty brown by the second year on the evergreen leaves (or on herbarium specimens). The style persists on the capsule, which is narrowly elliptic and splits from the base along 5 sutures.

5. Chamaedaphne

Although I have seen listing of "*Chamaedaphne* sp." in ecological tables, that represents unwarranted caution, for there is only one familiar species, circumpolar in range (although absent from westernmost Europe).

At times the genus has borne the name *Cassandra*, but *Chamaedaphne* Moench has been officially conserved against an earlier homonym and hence has priority.

REFERENCE

Lems, Kornelius. 1956. Ecological Study of the Peat Bogs of Eastern North America III. Notes on the Behavior of Chamaedaphne calyculata. Canad. Jour. Bot. 34: 197–207.

1. **C. calyculata** (L.) Moench   Fig. 9   Leatherleaf
   Map 18. Doubtless our most abundant "bog shrub," often forming very large clones, with lower stems buried in sphagnum, peat, or other substrate (and apparently ± resistant to fire). Can be expected in every open bog and many shrubby fens, wet or dryish, even persisting long after drying or burning of the peat or succession to swamp; small plants also grow in rock crevices and pool margins along Lake Superior, as at Isle Royale.

6. Gaylussacia
   This generic name was published in 1819 and honors the distinguished French chemist Joseph Louis Gay-Lussac (1778–1850), whose famous "law" regarding gas volumes had been presented 11 years earlier and who also investigated the chemistry of plants. He received help from and became a close friend of Baron Alexander von Humboldt, whose explorations of Latin America with A. Bonpland 1799–1804 resulted in the recognition of this genus and countless other new taxa. Several additional species of *Gaylussacia* grow in the southeastern United States, and others in South America.

1. **G. baccata** (Wangenh.) K. Koch   Fig. 10   Huckleberry;
    Crackleberry
   Map 19. Like so many Ericaceae, thrives in acid situations, from old dunes and plains with oak, pine, and birch to wet bogs with such plants as *Chamaedaphne* and *Larix*.

16. Andromeda
    glaucophylla

17. Ledum groenlandicum

18. Chamaedaphne
    calyculata

The nearly black (very rarely glaucous) fruit ripens distinctly later than the common blueberries, with which it may grow; it is just as flavorful, but many people find the crunchy berries (actually drupes with 10 little pits) much less pleasant to chew upon. The corollas vary from yellow to a rich reddish, but rarely they or the fruit may be white. The leaves, which turn a handsome red in the fall, can be rubbed between the fingers or on a piece of paper to see the characteristic resinous substance which, besides the fruit, readily distinguishes our true huckleberry from the blueberries.

7. **Vaccinium**   Blueberries and Cranberries

Vaccinium stamineum L. (deerberry), with conspicuously exserted stamens, has been reported from Michigan on the basis of a collection made at Ann Arbor (no further details) in 1903 by a student of C. A. Davis. There has been no confirmation of this species' occurring in the area, and it was ignored in lists for Washtenaw County. The specimen was probably from a plant in cultivation.

Blueberries are one of the most recent fruits to be domesticated and cultivated, although fruit from wild plants has long been harvested by native peoples and settlers. A wild blueberry pie, or a generous handful in the pancake batter or muffin mix, will satisfy one as to the bounties of nature. While wild stands of lowbush blueberries are sometimes harvested, cultivars derived from *V. corymbosum* are the favored commercial source. Michigan is the nation's top producer of commercial blueberries, accounting for half the country's total. The primary growing region is the five southernmost counties along Lake Michigan (Berrien to Muskegon).

The common name "huckleberry" is widely applied to various blueberries, but that name should be reserved for the true huckleberry (*Gaylussacia*), with the seeds enclosed in hard stony pits.

REFERENCES

Hall, Ivan V., Lewis E. Aalders, Nancy L. Nickerson, & Sam P. Vander Kloet. 1979. The Biological Flora of Canada 1. Vaccinium angustifolium Ait., Sweet Lowbush Blueberry. Canad. Field-Nat. 93: 415–430.

Vander Kloet, S. P. 1978a. Systematics, Distribution, and Nomenclature of the Polymorphic Vaccinium angustifolium. Rhodora 80: 358–376.

Vander Kloet, S. P. 1978b. The Taxonomic Status of Vaccinium pallidum, the Hillside Blueberries Including Vaccinium vacillans. Canad. Jour. Bot. 56: 1559–1574.

Vander Kloet, S. P. 1980. The Taxonomy of the Highbush Blueberry, Vaccinium corymbosum. Canad. Jour. Bot. 58: 1187–1201.

Vander Kloet, S. P., & I. V. Hall. 1981. The Biological Flora of Canada 2. Vaccinium myrtilloides Michx., Velvety-leaf Blueberry. Canad. Field-Nat. 95: 329–345.

Vander Kloet, S. P. 1983a. The Taxonomy of Vaccinium § Oxycoccus. Rhodora 85: 1–43.

Vander Kloet, S. P. 1983b. The Taxonomy of Vaccinium § Cyanococcus: A Summation. Canad. Jour. Bot. 61: 256–266.

Vander Kloet, S. P. 1988. The Genus Vaccinium in North America. Agr. Canada Publ. 1828. 201 pp.

Young, Steven B. 1970. On the Taxonomy and Distribution of Vaccinium uliginosum. Rhodora 72: 439–459.

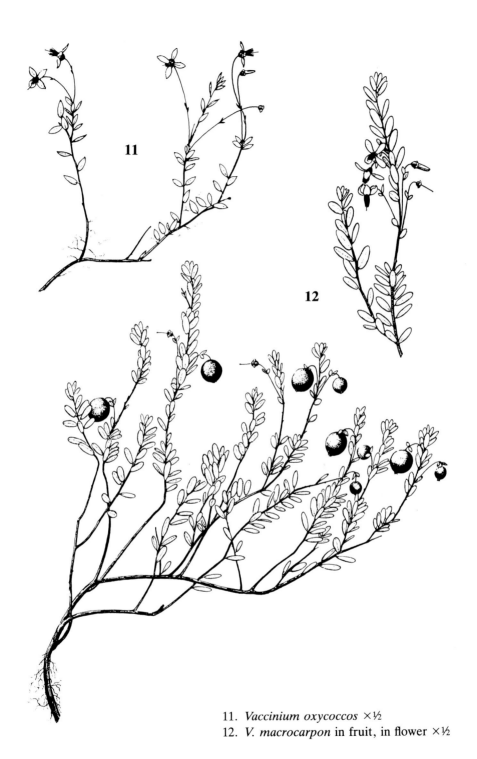

11. *Vaccinium oxycoccos* ×½
12. *V. macrocarpon* in fruit, in flower ×½

KEY TO THE SPECIES

1. Stems creeping, prostrate, or trailing, ± lax; leaves evergreen, all or mostly less than 8 mm wide; flowers 4-merous; mature fruit a red berry
   2. Flowers (and fruit) ± crowded, distinctly longer than their stalks; corolla cleft about halfway, ± campanulate, with ascending or spreading lobes; leaf blades light green beneath with scattered minute dark elongate glandular projections . . . .
   . . . . .1. **V. vitis-idaea**
   2. Flowers (and fruit) distinctly shorter than their very slender elongate stalks; corolla cleft to the base, the lobes (petals) strongly reflexed; leaf blades whitened beneath, smooth and glabrous
      3. Leaves mostly acute in outline at the apex; bracts on pedicels red, less than 0.8 (1) mm broad . . . . .2. **V. oxycoccos**
      3. Leaves mostly obtuse or rounded (or even minutely notched) in outline at apex; bracts on pedicels green (rarely red), (0.7) 0.8–2.7 (3) mm broad . . . .
      . . . . .3. **V. macrocarpon**
1. Stems ± erect, stiffly bushy; leaves deciduous, of various width; flowers 5-merous [reported 4-merous in *V. uliginosum*]; mature fruit a blue to black berry (may be reddish when immature)
   4. Flowers and fruit 1 (−2) in axils of leaves (or axillary buds); calyx lobes rounded or obsolete, deciduous, leaving at most a ring on summit of the fruit; anthers with a pair of conspicuous awns or horns laterally; twigs smooth or wrinkled; plants restricted to Upper Peninsula
      5. Leaves all less than 1.5 cm broad (usually ± 1 cm); plant not over 5 dm tall; new branchlets essentially terete (or obscurely angled); pedicels 1–5 (7) mm long; berries less than 10 mm in diameter
         6. Leaf blades toothed, membranous, bright green; flowers solitary in axils of foliage leaves, on current year's growth; bud scales 2 . . . . .4. **V. cespitosum**
         6. Leaf blades entire, coriaceous, blue-green; flowers 1–2 from buds on previous year's stem; bud scales more than 2 . . . . .5. **V. uliginosum**
      5. Leaves (at least the larger ones) over 1.5 cm broad; plant over 5 dm tall; new branchlets ± sharply ridged, angled, or 2-edged; pedicels (3) 4–12 mm long; berries (8) 10–18 mm in diameter
         7. Leaf blades entire (or obscurely toothed on lower half), nearly obtuse to rounded at apex; mature fruit blue-glaucous . . . . .6. **V. ovalifolium**
         7. Leaf blades toothed their whole length, clearly acute; mature fruit purple-black
         . . . . .7. **V. membranaceum**

19. Gaylussacia baccata

20. Vaccinium vitis-idaea

21. Vaccinium oxycoccos

4. Flowers and fruit in terminal or lateral racemes or crowded clusters; calyx lobes acute (to obtuse), persistent as a small toothed crown on the fruit; anthers without horns; twigs warty or wrinkled
  8. Plants ca. 1 m tall or higher, not rhizomatous (but often in alder-like clumps); fully mature leaf blades all or mostly (3.5) 4–5.5 (7) cm long; corollas (6) 6.5–8.5 (9.5) mm long . . . . 8. **V. corymbosum**
  8. Plants less than 1 m tall (usually 0.5 m or less), rhizomatous (usually forming large colonies); fully mature leaf blades (1.2) 1.5–3.5 (rarely 4) cm long; corollas (4) 4.5–6 (6.5) mm long
    9. Leaves beneath and young twigs ± pubescent; margins of leaves entire . . . . . .
    . . . . .9. **V. myrtilloides**
    9. Leaves and twigs glabrous or nearly so (pubescence, if any, on margins and main veins, very rarely scattered across undersides of leaves, and in lines on twigs); margins of leaves finely serrate or sometimes (in *V. pallidum*) entire
      10. Leaf blades entire or at least partly or irregularly serrate (both conditions sometimes on the same plant), broadly elliptic, mostly about twice as long as broad or shorter, pale beneath . . . . .10. **V. pallidum**
      10. Leaf blades finely but regularly and closely serrulate with bristle-tipped teeth their whole length, narrowly elliptic, mostly 2–3 times as long as broad, pale or green beneath . . . . .11. **V. angustifolium**

1. **V. vitis-idaea** L.   Mountain-cranberry; Lingonberry; Cowberry

Map 20. A circumpolar arctic-subarctic species, ranging from the tundra south in North America to New England and the Lake Superior region. For over a century, the only known Michigan material was collected in 1868 (*A. E. Foote*, MICH, OS) on Smithwick Island. Later searching for the species there and elsewhere at Isle Royale long proved fruitless, despite the fact that it is widespread along the Ontario shore of Lake Superior on rock ledges bordering the lake and inland in the boreal forest. Finally, in 1994, E. Judziewicz rediscovered the species in Michigan, in sphagnum under *Thuja* on Passage Island (*10932*, MICH).

The low mats of this species perhaps resemble a diminutive *Arctostaphylos*. The fruit is tart but tasty and where common is made into preserves or an excellent juice much like that of the true cranberries. North American plants have been referred to var. *minus* Lodd., which also occurs in Eurasia.

2. **V. oxycoccos** L.   Fig. 11   Small Cranberry

Map 21. Bog mats (rarely in fens or cedar swamps except on sphagnum hummocks), persisting under spruce and tamarack.

Another circumpolar species. Sometimes placed in a segregate genus, *Oxycoccus*, in which case the species has been known as *O. microcarpus* Rupr. (sens. lat.) or *O. palustris* Pers. (sens. str., for the tetraploid plants [*O. quadripetalus* is a name not validly published by Gilibert]). But the near tautonym *Oxycoccus oxycoccos* (L.) MacMillan seems to be correct if one recognizes that genus (Taxon 36: 126–128. 1987).

Inattentive collectors often mix this species with the next and even with *Gaultheria hispidula*; the number of mixed collections in herbaria is rather large. The acute aspect of the leaf tips is often enhanced by more revolute leaf margins toward the apex, resulting also in a somewhat ovate blade. In both cranberries, the stamens are 8, each anther terminating in 2 long slender tubules; the cone-like mass of stamens and strongly reflexed petals combine to give the flower the appearance of a miniature shooting-star (*Dodecatheon*). See also comments under the next species.

3. **V. macrocarpon** Aiton   Fig. 12   Large Cranberry
Map 22. Characteristic of bog mats (sometimes fens), persisting in some shade; interdunal swales and hollows. Often forms dense stands ringing the open water of a bog pond—a treacherous place to gather its fruit!

This is the species of cranberry usually grown commercially. Cranberry-growing has been attempted in Michigan but never proved to be a longterm success except west of Whitefish Point, especially about 1880–1932. Gathering of wild cranberries (either species) in the fall (just before the first hard frost) is, however, a pleasant occupation for those who do not mind risking wet feet (or more). The fresh ripe berries are often dark purplish rather than red.

Unlike the preceding, this is a strictly North American species. It is sometimes segregated as *Oxycoccus macrocarpus* (Aiton) Pers., but the cranberries are probably best placed in *Vaccinium* sect. *Oxycoccus*.

The leaf blades tend to be more elliptic-oblong or even slightly spatulate in comparison with those of the preceding species. The leaves, flowers, and fruits average a little larger. Occasionally the 2 (usually ± opposite) bracts on the pedicels are ambiguous, e.g., relatively narrow and reddish in *V. macrocarpon*, resembling large ones of *V. oxycoccos*. Additional points of distinction are these: (1) The stems in *V. macrocarpon* are usually prolonged well beyond the flowers or fruit, while in *V. oxycoccos* the flowers and fruit usually appear to be terminal or nearly so, without a prolonged leafy shoot. (2) *V. oxycoccos* blooms about a month earlier, generally in June or sometimes as late as mid July in the Upper Peninsula; *V. macrocarpon* blooms in July, or sometimes as early as mid June in southern Michigan. (3) *V. macrocarpon* is diploid (2n = 24), while in our area *V. oxycoccos* is tetraploid (2n = 48). (Vander Kloet includes the diploid *V. microcarpum* (Rupr.) Schmalh., not in this region, in *V. oxycoccos*.)

4. **V. cespitosum** Michaux   Dwarf Bilberry
Map 23. A native of North America (often erroneously attributed also to Eurasia), ranging from Labrador to Alaska, south (in eastern North America) to the Great Lakes region. In our area, grows very locally on fairly dry (or only seasonally wet) soils, in openings among aspen and other hardwoods; much lower in stature than even *V. angustifolium*, with which it often grows.

The bilberries all have tasty fruit, which lacks the persistent 5-pointed crown of sepals that characterizes the true blueberries.

5. **V. uliginosum** L.   Alpine Bilberry

Map 24. A circumpolar arctic-alpine species, at its southern border in the Great Lakes region at Isle Royale, including especially Passage Island and some others, where it grows in crevices and margins of rock pools.

This is a polymorphic species of intergrading races and several chromosome numbers, discussed by Young (1970). The flowers are usually said to be mostly 4-merous, but all of our material (insofar as flowering) has 5-merous flowers. The fruit of both this and the preceding species is blue-glaucous, as in *V. ovalifolium* and *V. myrtilloides*.

6. **V. ovalifolium** Sm.   Plate 2-A   Oval-leaved Bilberry

Map 25. Widespread in the West from Alaska to Oregon and Montana, disjunct to the Great Lakes region (Keweenaw Peninsula to Algoma District of Ontario) and again around the mouth of the St. Lawrence (see Michigan Bot. 20: 70–71. 1981). While often found with *V. membranaceum* on wooded dunes and rocks, this species seems more often than that one to be in deep shade in forests of hemlock, beech and maple, or mixed conifers.

The leaves are usually a pale, light blue-green, especially beneath. The fruit is of good quality, though not so large nor so tasty as in the next species. In both, the berry separates only reluctantly—with a slight tear—from the pedicel, meaning that the wounded fruit has to be eaten (or frozen) promptly before spoiling. In this species, like *V. cespitosum* and *V. membranaceum*, the buds have 2 valvate scales, while our other species of *Vaccinium* have more than 2 imbricate bud scales.

7. **V. membranaceum** Torrey   Plate 2-B   Tall Bilberry

Map 26. An understory shrub, often on wooded dunes and rocks, under hardwoods and/or conifers, and in bordering thickets. This is another western American species, disjunct to the south side of Lake Superior (only in Michigan and recently found in Pukaskwa National Park, Ontario, but not in the St. Lawrence region; see Michigan Bot. 20: 63–65. 1981). There are fine colonies of both this species and the preceding (which often grow together) along Lake Superior near the western edge of Muskellunge Lake State Park in Luce County, at Mt. Bohemia in Keweenaw County, and many other places.

The dark fruit is very large (to 18 mm), juicy and tasty — really too big to use conveniently in blueberry pancakes, but choice in pies or eaten by the handful when picked. The leaves are not at all glaucous, but often a rather yellow-green; I have seen then as long as 6.5 or even 8 cm.

The name was proposed by David Douglas on the basis of his travels in the Pacific Northwest, but it was not validly published by him, or by Hooker, as made clear as long ago as 1897 by Britton. Nevertheless, the name is still often attributed to incorrect authors (Douglas, Hooker, or Britton) rather than to Torrey, who validated it in 1874 in a report on the botany of the Wilkes Expedition.

8. **V. corymbosum** L. Highbush Blueberry

Map 27. Rarely in dry oak or oak-hickory woods, but usually in ± wet sandy or peaty places: low woods, high shrub zones and borders of bogs, often with tamarack and poison sumac; boggy thickets.

Herbarium specimens for which the collector neglected to note the stature of the plant on the label can be difficult to identify, especially when the leaves are immature, for the habit of the plant and the size of the mature leaves are the best distinguishing features. As treated here in a broad sense (following Vander Kloet 1980), *V. corymbosum* is a polymorphic complex, consisting (in the north) largely of tetraploids originally derived from hybridization of certain lowbush southern diploid taxa. Diploid plants with pubescent leaves and twigs, the fruit black, have often been recognized as *V. atrococcum* (A. Gray) A. Heller. This complex varies not only in chromosome number, pubescence, and fruit color, but also in leaf color (usually green, sometimes pale beneath) and margin (either entire or finely toothed). The leaves may mature at the same time as the flowers or distinctly later. Hybridization with lowbush taxa (producing "half-high" shrubs) has been suspected.

Specimens with flowers but immature leaves may be distinguished from *V. myrtilloides* if the leaves are glabrous and/or serrate; from *V. an-*

22. Vaccinium
    macrocarpon

23. Vaccinium cespitosum

24. Vaccinium uliginosum

25. Vaccinium ovalifolium

26. Vaccinium
    membranaceum

27. Vaccinium
    corymbosum

gustifolium, if the leaves are entire or strongly pubescent; from *V. pallidum*, if they are green beneath. On the other hand, *some* specimens of *V. corymbosum* may not be distinguishable on these grounds, and little but flower size can be used to separate them from the lowbush blueberries. Collections from Lake 16 in Cheboygan County have been identified by Vander Kloet as *V. corymbosum* × *angustifolium* and "show considerable introgression" with the latter. These are so far north of the otherwise documented range of *V. corymbosum* that there could be some doubt about them; an Emmet County collection from near Brutus with corollas (6) 6.5 mm long may be of similar status.

9. **V. myrtilloides** Michaux   Plate 1-D   Velvetleaf or
                                                           Canada Blueberry
  Map 28. Thrives on dry sandy or rocky open or ± shaded ground, often with *V. angustifolium*, but more frequently than that species also found in moist places, including bogs and wet woods.
  Quite distinct in its entire, ± densely pubescent leaves. The fruit is usually blue with a glaucous bloom, but may be purple-black; the white-fruited f. *chiococcum* (W. Deane) Fernald has been collected in Cheboygan, Houghton, and Marquette counties and doubtless occurs elsewhere. (The dried fruit has turned blue on all these specimens.) This species and *V. pallidum* are diploids (2n = 24).

10. **V. pallidum** Aiton   Hillside or Dryland Blueberry
  Map 29. Usually in dry sandy woods dominated by various oak or hickory species, rarely in moister areas.
  Including *V. vacillans* Torrey (see Vander Kloet 1978b). The inflorescence is a bit more openly racemose than the dense clusters of our other lowbush blueberries. Very rarely the leaves may be narrower than the usual broadly elliptic to ovate shape, and such plants may be difficult to distinguish from *V. angustifolium* if the leaf margins are serrulate or from small-leaved specimens of *V. corymbosum* if the stature is not known (or is unusually tall). Fortunately, *V. pallidum* is restricted to the southern part of the Lower Peninsula, except for a 1995 collection from Iron County (*Henson 4000*, MICH).

11. **V. angustifolium** Aiton   Low Sweet Blueberry
  Map 30. Often abundant (and much sought by berry pickers) in dry, sandy or rocky, open or ± shaded ground with oak, pines, and/or aspen; but also in low places, including peatlands with *Chamaedaphne* and other shrubs, though less frequently than *V. myrtilloides* in such places; thrives after clearing or burning.
  A polymorphic tetraploid complex (see Vander Kloet 1978a), once widely known as *V. pensilvanicum* Lam., an illegitimate name. Plants with green leaves but blue-glaucous fruit have been segregated from the narrow-

est interpretion of *V. angustifolium* as var. *laevifolium* House or *V. lamarckii* Camp. Plants with glaucous leaves, on the other hand, but dark blue-black fruit have been segregated as var. *nigrum* (A. W. Wood) Dole or *V. brittonii* E. P. Bicknell; these may now be called *V. angustifolium* f. *nigrum* (A. W. Wood) B. Boivin if one wishes to name this variant. The two phases often grow together and are usually quite distinct, but they cross-pollinate freely and do not breed true, according to Vander Kloet, who also believes that the interpretation of "typical" *V. angustifolium* as a diploid is mistaken.

8. Epigaea

REFERENCE

Clay, Keith. 1983. Myrmecochory in the Trailing Arbutus (Epigaea repens L.). Bull. Torrey Bot. Club 110: 166–169.

1. **E. repens** L.   Plate 1-E   Trailing-arbutus
Map 31. Typically in dry, usually sandy woodland and forest with oak, aspen, and pines, but also sometimes associated with beech, maple, and/or hemlock and in swamps of cedar or other conifers.

The exquisitely fragrant flowers, in dense clusters, often nearly hidden on the ground beneath the leaves of this and other plants, open early in the spring—by mid April or mid May depending on season and latitude, with some lingering till the first of June in late seasons. The corollas are white to pink, the pink shade intensifying with age. A double-flowered form has been collected in Charlevoix County (*Fessenden* in 1929, ALBC). The slender trailing stems are easily gathered, leading to scarcity of the plant from over-picking in some areas; these prostrate stems, however, arise ultimately from a stout woody tap-rooted crown. The stem, petioles, and usually leaf blades (especially on margins and veins beneath) are ± shaggy with stiff rusty-brown hairs. This species is dioecious. The seeds are dis-

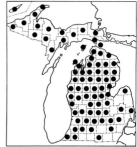

28. Vaccinium myrtilloides   29. Vaccinium pallidum   30. Vaccinium angustifolium

persed by ants attracted to the fleshy placental material when the capsule opens.

9. **Gaultheria**

Both our species are flavored with wintergreen in all parts (even the flowers) and are aromatic when bruised. *G. hispidula* was long maintained in a separate genus, *Chiogenes*. The fruit is composed largely of the fleshy calyx, the lobes of which are visible at the apex; small bractlets at the base of the fruit might, however, easily be mistaken for a calyx.

KEY TO THE SPECIES

1. Leaf blades less than 1 cm long, with scattered dark bristles beneath; flowers 4-merous, ca. 2–3 mm long, open May–June; fruit pure white, ripe July–August; stem prostrate, with appressed dark bristles . . . . .1. **G. hispidula**
1. Leaf blades (1.5) 2–4.5 (5) cm long, without bristles beneath (though on margins of young leaves); flowers 5-merous, 6–9.5 mm long, open late July–August; fruit red, ripe late in the fall (but over-wintering); stem with ± erect shoots bearing incurved whitish hairs or glabrate . . . . .2. **G. procumbens**

1. **G. hispidula** (L.) Bigelow   Creeping-snowberry
Map 32. Forming flat mats in moist woods, thickets, and swamps of cedar, spruce, tamarack, or other conifers; often on mossy logs and hummocks.
The foliage is sometimes confused with that of the cranberries, but differs in the wintergreen flavor, strigose stems, more acute leaf tips, green undersides of the leaves, and bristle-like glands or hairs on the undersides (and often margins) of the leaves.

2. **G. procumbens** L.   Plate 1-F; fig. 13   Teaberry; Wintergreen
Map 33. Thrives in dry, usually ± sandy woodland with oak, pine, paper birch, aspen, red maple, bracken, blueberry; but also frequent in low woods, even conifer swamps; less often with beech, maple, hemlock. Generally thought to be a species which does well after fire.

31. Epigaea repens

32. Gaultheria hispidula

33. Gaultheria procumbens

The wintergreen flavor and aroma are stronger in leaves and fruits than in the preceding species and make this a favorite nibble for many persons in the outdoors. The tender young leaves are most palatable, and the old ones can be used for tea. Commercial wintergreen oil, however, if not wholly synthetic, is now more likely to be extracted from *Betula lenta* than from *Gaultheria*. Frederick Stearns, one of Michigan's pioneer pharmaceutical manufacturers, reported in the 1850s that "large quantities" of wintergreen berries "are annually offered for sale in Detroit, and are simply eaten as a relish."

10. **Arctostaphylos**

We have only a single species, a circumpolar one. There are numerous others in the western part of North America, known generally as manzanita and ranging in habit from low shrubs to small trees, but often with the shreddy outer bark and smooth reddish or mahogany stems that can be seen on older plants of our species.

REFERENCES

Rosatti, Thomas J. 1982. Trichome Variation and the Ecology of Arctostaphylos in Michigan. Michigan Bot. 21: 171–180.

Rosatti, Thomas J. 1987. Field and Garden Studies of Arctostaphylos uva-ursi (Ericaceae) in North America. Syst. Bot. 12: 61–77.

1.   **A. uva-ursi** (L.) Sprengel   Fig. 14   Bearberry; Kinnikinick

Map 34. One of the few Ericaceae not characteristic of acid habitats. Often forms large mats on sand dunes as well as jack pine and oak plains and on limestone pavements and gravel ridges (or other rock in the western Upper Peninsula).

The generic name and specific epithet both mean "bearberry," in Greek and Latin, respectively. Whether the species forms any part of the diet of bears is, however, doubtful. The raw fruit is not appealing to humans although it is said to be palatable when cooked. American Indians used the leaves as a substitute for tobacco, as noted in 1820 by Henry Rowe Schoolcraft in the Upper Peninsula.

The leaf blades are mostly 1–2.5 cm long, often slightly obovate. Several varieties or subspecies have been named, based largely on pubescence characters of young stems (long and short hairs, or none; and these glandular or not, in various combinations). Exhaustive field and greenhouse studies by Rosatti have documented nearly continuous variation (and sufficient inconstancy in characters of cuttings or seedlings as well as in ecology and geography) so as not to support taxonomic recognition of variants.

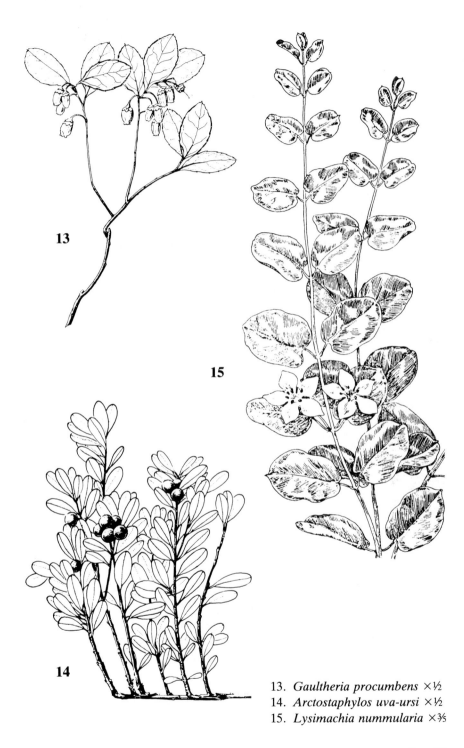

13. *Gaultheria procumbens* ×½
14. *Arctostaphylos uva-ursi* ×½
15. *Lysimachia nummularia* ×⅗

PRIMULACEAE   Primrose Family

This family includes the true primroses, not to be confused with the "evening-primroses," which belong to an entirely different family (Onagraceae) and which should never be called merely "primroses."

In modern systems of classification, the order to which the Primulaceae belong and the order including the Ericaceae are generally classified far from the rest of the sympetalous families. Both Cronquist and Takhtajan, for example, include the Primulales and Ericales as relatively advanced orders in the subclass Dilleniidae. (All the remaining sympetalous families—the rest of Part III of this Flora—are in the subclass Asteridae.) Families in both of these orders generally have 5 carpels rather than 2 as is common in the other sympetalous families. In our flora, the Primulaceae are unique in the sympetalous corolla with adnate stamens of the same number as the petals (corolla lobes) and *opposite* them (e.g., fig. 18). The flowers of *Primula* and some other genera are heterostylous.

REFERENCES

Channell, R. B., & C. E. Wood, Jr. 1959. The Genera of the Primulales of the Southeastern United States. Jour. Arnold Arb. 40: 268–288.

Iltis, Hugh H., & Winslow M. Shaughnessy. 1960. Preliminary Reports on the Flora of Wisconsin No. 43. Primulaceae—Primrose Family. Trans. Wisconsin Acad. 49: 113–135.

KEY TO THE GENERA

1. Principal leaves all basal; inflorescence an umbel
  2. Leaves all or mostly 12–30 cm long; plant ca. 25–50 (or more) cm tall; flowers ca. 2–3 cm long, including stamens in a cone-like assemblage and reflexed corolla lobes (plate 1-G) . . . . .1. **Dodecatheon**
  2. Leaves (or blades if a petiole is present) less than 6 (8) cm long; plants often shorter; flowers less than 1 (2) cm long, with corolla lobes spreading but not reflexed
    3. Corolla white, shorter than the calyx; scapes and bracts at base of umbel minutely stellate-pubescent; style minute, less than 0.3 mm long . . . . .2. **Androsace**
    3. Corolla pale to deep pink-magenta (rarely white) with yellow "eye," yellow, or purple, exceeding the calyx; scapes and bracts glabrous or pubescence not stellate; style ca. 1–3.5 mm long (or even longer in a rare escape) . . . . .3. **Primula**
1. Principal leaves all or partly cauline; inflorescence various, but not a stalked umbel
  4. Corolla yellow to red; leaves opposite or in several whorls
    5. Leaves sessile, not over twice as long as broad; corolla red; capsule circumscissile . . . . .4. **Anagallis**
    5. Leaves petioled or over twice as long as broad (or both); corolla yellow; capsule dehiscing longitudinally . . . . .5. **Lysimachia**
  4. Corolla white; leaves alternate or in a single whorl
    6. Leaves in a single whorl; flowers usually 7-merous, solitary on 1–3 (4) pedicels . . . . .6. **Trientalis**
    6. Leaves alternate; flowers 5-merous, in a raceme
      7. Flowers in axillary and terminal racemes, the corolla lobes ca. 1 mm long or less; stem and leaves glabrous; ovary half-inferior . . . . .7. **Samolus**

7. Flowers in a dense terminal raceme, the corolla lobes ca. 3.5–5 mm long; stem
   and leaves finely pubescent; ovary wholly superior . . . . . . .5. **Lysimachia (clethroides)**

1. Dodecatheon

REFERENCE

Ingram, John. 1963. Notes on the Cultivated Primulaceae 2. Dodecatheon. Baileya 11: 69–90.

1. **D. meadia** L.   Plate 1-G   Shooting-star
   Map 35. This is a handsome prairie and woodland species, barely reaching
north into Michigan, where it is very rare. Known from a fen in Berrien
County and a prairie-like strip between highway and railroad in Menominee
County; collected on Belle Isle in 1896 (no habitat stated and conceivably
planted).
   The pink color of the corolla varies considerably in intensity. Persons who
rely on picture-book taxonomy often misidentify cranberry as shooting-star,
for both have pink flowers with reflexed lobes and a cone of yellow anthers.
However, in the cranberries the petals are only 4 (rather than 5) and the
flowers are at best half the size of those in *Dodecatheon*—apart from the
totally different habit of the plants.

2. Androsace

1. **A. occidentalis** Pursh   Fig. 16   Rock-jasmine
   Map 36. A tiny annual of early spring (blooming in April), near the
eastern edge of its range in Michigan. It is a species of dry open ground,
superficially similar to some of the early Cruciferae of similar habitat, but
with very inconspicuous flowers. Collected in 1899 and 1931 at Niles.

3. Primula   Primrose

REFERENCES

Hausen, B. M., H. W. Schmalle, D. Marshall, & R. H. Thomson. 1983. 5,8-Dihydroxyflavone
   (Primetin) the Contact Sensitizer of Primula mistassinica Michaux. Arch. Dermatol. Res.
   275: 365–370. [Corrected 277: 157. 1985; the earlier report cited by the authors was surely
   not based on this species.]
Soper, James H., Edward G. Voss, & Kenneth E. Guire. 1965. Distribution of Primula
   mistassinica in the Great Lakes Region. Michigan Bot. 4: 83–86 [reprinted Quart. Jour.
   Am. Primrose Soc. 30: 109–112. 1972].
Vogelmann, Hubert W. 1955. A Biosystematic Study of Primula mistassinica Michx. Ph. D.
   thesis, Univ. Michigan. 218 pp.
Vogelmann, H. W. 1960. Chromosome Numbers in Some American Farinose Primulas with
   Comments on their Taxonomy. Rhodora 62: 31–42.

KEY TO THE SPECIES

1. Scape and leaves glabrous or farinose; flowers less than 1 cm long, the corolla pale to deep magenta with yellow "eye"; leaves tapered at base but without distinct petiole; style ca. 1–3.5 mm long; plant native in calcareous habitats . . . .1. **P. mistassinica**
1. Scapes and leaves (especially beneath) pubescent; flowers ca. 1.5–2.5 cm long, the corolla yellow or purplish; leaves mostly contracted to a distinct winged petiole; style much exceeding 3.5 mm (ca. 5–10 mm); plant rarely escaped from cultivation
. . . . .2. **P. veris**

1. **P. mistassinica** Michaux   Plate 2-C   Bird's-eye Primrose
   Map 37. This attractive little plant is the only true primrose native in the region, although larger species of the genus are popular in cultivation. Ranging from Labrador to western Yukon, as it comes south it seems concentrated near the shores of the Great Lakes, where it is often abundant in damp calcareous meadows and interdunal flats. Otherwise, especially inland, it is local in marly bogs (fens), on calcareous banks and sandstone cliffs, and in other cool damp places. Wherever one finds butterwort (*Pinguicula vulgaris*) one is almost certain to find this *Primula* closely associated, although the latter is more widespread. Other characteristic associates include such calciphiles as *Selaginella selaginoides, Carex crawei, C. garberi, Eleocharis quinquefolia* [*E. pauciflora*], *Rhynchospora capillacea*, and *Parnassia parviflora*.

Rarely (as at Miner's Falls in Alger County) the corolla is pure white, but normally it ranges from pale to deep pink-magenta. The yellow "eye" is rarely lacking. A form with yellow corolla was collected on Thunder Bay Island (*Penskar & Ludwig* in 1981, MICH). The flowers are seldom over 1 cm broad, very fragrant, and usually at their peak in early to mid May. As in other species with heterostylous adaptations to cross-pollination, some plants have long styles, with stamens inserted lower on the corolla tube, while other plants have shorter styles with stamens inserted higher on the tube. Flowering plants tend to be short (often less than 10 cm), but the scape elongates somewhat later in the season and a few plants may attain as

34. Arctostaphylos   35. Dodecatheon meadia   36. Androsace occidentalis
    uva-ursi

much as 35 cm in height. An occasional flower is seen as late as July or August. The umbels may be reduced to a single flower or have as many as 8 or even 11 flowers.

There is also great variability in the amount of "mealiness" on plants. Some are densely farinose especially on the inside of the calyx and the undersides of the leaves. Others lack any of the whitish or yellowish "meal." Farinose plants with alleged seed differences as well were named—with some hesitancy—as *P. intercedens* by Fernald (TL: shores of Lake Huron & Lake Michigan [Straits of Mackinac area]). Almost all authors, however, have been willing to accept Vogelmann's conclusion that in the absence of any consistent morphological or cytological differences, *P. intercedens* should not be recognized as a good species.

P. mistassinica was named for Lake Mistassini, in northern Quebec, near where André Michaux found it on September 5, 1792. Winter was already setting in and snow was in the air. Michaux, en route to Hudson Bay, had to turn back as his Indian guides warned of the danger of pushing their canoes any farther north. (Michaux could not have seen hosts of primroses with golden-eyed fragrant blossoms shaking in the wind, as one imaginative writer, obviously unacquainted with the published account, described the occasion!) In the herbarium at Paris, I have seen Michaux's ample collection from the far point of his journey: 10 little primroses in fruit (with a few withered corollas) mounted on a sheet labeled in his hand "Rivièrre des Goélands." (For the locality, see Rousseau in Mém. Jard. Bot. Montréal 3: 30. 1948.)

Hausen et al. have analyzed Michigan material of *P. mistassinica* and isolated a substance capable of producing a contact dermatitis, although no problems with this species are actually known.

2. **P. veris** L. Cowslip

Map 38. A European species widely cultivated, spreading to grassy roadsides near the north end of the Mackinac Bridge and to be expected elsewhere (*D. Garlitz 553*; *Voss 16211*, MICH).

4. **Anagallis**

1. **A. arvensis** L. Common or Scarlet Pimpernel

Map 39. A species of Eurasia, weedy there as well, but considered introduced (in part escaped from cultivation) in North America. Except for a 1901 collection by Farwell from Detroit, all Michigan specimens have been gathered since 1935, most often along roadsides, lawns, and gardens, including the saline median of the I-94 expressway; but also along streams and doubtless in other disturbed sites.

The corolla is usually a distinctive brick-red. Blue and white forms are known but have not been collected in Michigan. Besides the characters given in the key, hairy filaments and leaves ± gland-dotted (especially beneath) will help to identify this distinctive plant.

This species contains glycosides which are reputed to be poisonous to livestock when ingested in large quantities. On the other hand, the species is also reported to have been used medicinally.

5. **Lysimachia** Loosestrife

The common name is the same as for the quite dissimilar, purple-flowered genus *Lythrum* (Lythraceae).

Some species have in the past been segregated into additional genera, but these are now not generally recognized (*Nummularia, Steironema, Naumburgia*).

REFERENCES

Coffey, Vincent J., & Samuel B. Jones, Jr. 1980. Biosystematics of Lysimachia section Seleucia (Primulaceae). Brittonia 32: 309–322.

Cooperrider, Tom S., & Bruce L. Brockett. 1974. The Nature and Status of Lysimachia ×producta (Primulaceae). Brittonia 26: 119–128.

Cooperrider, Tom S., & Bruce L. Brockett. 1976. The Nature and Status of Lysimachia ×producta (Primulaceae)—II. Brittonia 28: 76–80.

Ingram, John. 1960. Notes on the Cultivated Primulaceae 1. Lysimachia. Baileya 8: 85–97.

Ray, James Davis, Jr. 1956. The Genus Lysimachia in the New World. Illinois Biol. Monogr. 24 (3–4). 160 pp. [Some corrections of Michigan citations in Michigan Bot. 6: 20–21. 1967.]

KEY TO THE SPECIES

1. Leaf blades orbicular or nearly so, on short, distinct, glabrous petioles; plant prostrate, creeping . . . . .1. **L. nummularia**
1. Leaf blades at least twice as long as broad, sessile or on indistinct, ciliate, or pubescent petioles; plant erect
  2. Flowers in ± dense terminal or axillary racemes or panicles
    3. Leaves alternate, pubescent; flowers white . . . . .2. **L. clethroides**
    3. Leaves opposite or whorled, glabrous or pubescent; flowers yellow
      4. Inflorescence a panicle or flowers in axils of upper leaves; stem and leaves (at least beneath) soft-pubescent; leaves usually at least partly whorled

37. Primula mistassinica   38. Primula veris   39. Anagallis arvensis

5. Flowers on slender pedicels in axils of upper leaves; sepals green throughout; corolla glandular-margined ....3. **L. punctata**
5. Flowers in panicles; sepals dark-margined; corolla not glandular ......4. **L. vulgaris**
   4. Inflorescence a raceme; stem and leaves glabrous (or with a little loose pubescence, as along midrib beneath); leaves all opposite (often with axillary tufts)
      6. Raceme terminal (rarely a subsidiary raceme or even lateral leafy branches with racemes); corolla lobes (4.5) 5.5–7.5 (8.5) mm long ....5. **L. terrestris**
      6. Racemes on axillary peduncles; corolla lobes (2.5) 3–5 (5.5) mm long ......
      ....6. **L. thyrsiflora**
2. Flowers solitary on axillary peduncles
  7. Leaf blades ± rounded to subcordate at base, on long (ca. 1–3 cm) strongly ciliate petioles; flowers 1.7–2.5 (3.3) cm broad ....7. **L. ciliata**
  7. Leaf blades ± tapered, sessile or on short glabrous to ciliate petioles; flowers 1–1.8 (2.5) cm broad
     8. Leaves linear, less than 5 (6) mm wide, the lateral veins obscure at best, the margins smooth and ± revolute ....8. **L. quadriflora**
     8. Leaves lanceolate to broadly elliptic, the principal ones (8) 9–28 (34) mm wide, mostly with distinct pinnate venation, the margins minutely scabrous, papillose, or pubescent (rarely smooth and glabrous in *L. quadrifolia*), flat
       9. Calyx, corolla, and leaves ± streaked or dotted with red or black; anthers ca. 1 mm long or less; corolla lobes essentially entire ....9. **L. quadrifolia**
       9. Calyx, corolla, and leaves without streaks or dots; anthers ca. 2–3 mm long; corolla lobes coarsely toothed (erose) and/or mucronate at apex
         10. Sepals with 3–5 distinct nerves; leaf blades ovate to lanceolate, green beneath (hardly paler than above); stem ca. 4–7 mm in diameter at base ....
         ....10. **L. hybrida**
         10. Sepals apparently nerveless or very weakly nerved; leaf blades lanceolate to lance-elliptic, usually paler beneath; stem ca. 1–2.5 mm in diameter at base
         ....11. **L. lanceolata**

1. **L. nummularia** L.   Fig. 15   Moneywort

Map 40. A native of Europe and Asia Minor, established as a weed escaped from cultivation elsewhere in the world. Its large yellow axillary flowers make it an attractive plant and its creeping habit withstands mowing, so it thrives in some lawns, cemeteries, and borders of gardens; also found in swamp forests, on stream and river banks as well as damp shores, and in ditches and meadows.

40. Lysimachia   41. Lysimachia clethroides   42. Lysimachia punctata
    nummularia

This species is sometimes misidentified (when without flowers) as some native plant, such as *Chrysosplenium* or *Mimulus*, from which it can easily be distinguished vegetatively by the orange to black glandular dots visible especially as tiny swellings on the undersides of the leaves.

2. **L. clethroides** Duby — White or Gooseneck Loosestrife

Map 41. A native of eastern Asia, occasionally escaped from cultivation. Collected in 1986 (*Smith 1780*, MICH) as spreading to a roadside bank from a garden in Adrian.

3. **L. punctata** L. — Garden Loosestrife

Map 42. A stout, handsome, large-flowered European species, rarely escaped from cultivation in our area, where it has been collected along roadsides and in dumping grounds.

4. **L. vulgaris** L. — Garden Loosestrife

Map 43. Another cultivated species from the Old World. Collected in Wayne County in 1912 and in sand along Ogontz Bay in Delta County in 1987 (*Henson 2318A*, MICH).

A coarse plant, bushy in aspect, with the flowers in a large leafy panicle; the dark borders on the calyx lobes are distinctive.

5. **L. terrestris** (L.) BSP. — Swamp-candles

Map 44. Marshes, fens, wet shores, damp hollows and thickets, ditches, even occasionally tamarack swamps.

The leaves vary from nearly linear to broadly lance-elliptic. However, hybrids with *L. quadrifolia* are also wide-leaved, but with some leaves in whorls of 3 or 4 and a tendency to have some axillary inflorescences. This hybrid has been named *L. ×producta* (A. Gray) Fernald, on the basis of specimens collected in sandy damp woods in Branch County, Michigan, by the First Survey in 1837 (see Michigan Bot. 6: 20. 1967). It was also found by the Haneses in a tamarack swamp in Kalamazoo County, and by Billington in Oakland County in low, open, sandy ground near Royal Oak. Hybrids with *L. thyrsiflora* have wider leaves than usual for *L. terrestris*, all opposite, and denser inflorescences with a tendency for some to occur in the axils. This hybrid [*L. ×commixta* Fernald] has been collected in Alger County by Gleason. Wounds to the apex of a plant may result in the growth of lateral branches with terminal racemes, and such plants should not be confused with hybrids.

Late in the season, plants may produce red bulblets in the leaf axils. These segmented structures, which may be as long as 1 cm, are actually why this taxon bears the epithet *terrestris*, when *palustris* would seem more appropriate for an inhabitant of wet places. Linnaeus, in first naming it from Philadelphia, thought that in such bulblets he had a terrestrial (rather than arboreal) mistletoe, and only long afterward was it realized that the name applied in fact to part of a loosestrife.

6. **L. thyrsiflora** L.   Plate 2-D   Tufted Loosestrife
    Map 45. Like the preceding species, in almost any sort of wet place: streamsides and lake margins, bogs and fens, ditches and damp thickets, wet spots in swamps (cedar, tamarack, black ash).
    This species is easily recognized by the dense tufts of yellow flowers on long peduncles in the axils of mid-cauline leaves. There is often a little loose pubescence on the leaves along the midrib beneath, in the inflorescences, and on the stem. The top of the plant is often deformed by disease.

7. **L. ciliata** L.   Fringed Loosestrife
    Map 46. Low ground and thickets, usually ± shaded, including streamsides and floodplains, ditches, marsh edges, swamp forests (coniferous or deciduous), meadows, and shores; sometimes in drier places and almost weedy in habit.

8. **L. quadriflora** Sims
    Map 47. Fens, wet prairies, and calcareous marshy shores.

9. **L. quadrifolia** L.   Fig. 17   Whorled Loosestrife
    Map 48. Usually in rather dry forests and woodlands with oak, hickory, sassafras, aspen, and/or occasionally pine; rarely in swampy ground.
    The leaves are usually 4 in a whorl, but they may be as many as 6 or, very rarely, only 2.

10. **L. hybrida** Michaux
    Map 49. A plant of moist ground, collected in 1919 near Fish Hawk Lake in Gogebic County (*Darlington*, MICH, det. Coffey). A 1904 collection (*Pepoon 226*, MSC) from near Magician Lake, Cass and Van Buren counties appears to be the same, and Farwell (*8024*, BLH, MSC) found the species in 1927 at Eloise (Wayne County).

43. Lysimachia vulgaris   44. Lysimachia terrestris   45. Lysimachia thyrsiflora

Some authors have included this taxon in *L. lanceolata*; the two are not very easy to distinguish.

11. **L. lanceolata** Walter

Map 50. In both moist and dry situations, such as meadows and shores; often in oak woods; sometimes in ± disturbed sites.

6. Trientalis

REFERENCE

Anderson, Roger C., & Michael H. Beare. 1983. Breeding System and Pollination Ecology of Trientalis borealis (Primulaceae). Am. Jour. Bot. 70: 409–415.

1. **T. borealis** Raf.   Star-flower

Map 51. This is essentially a northern plant, ranging south rather locally in the mountains to northern Georgia. With us, typical of damp and boggy coniferous woods and thickets, but grows in almost any kind of coniferous and mixed woods, ranging from jack pine and oak, or beech and maple, to tamarack or cedar swamp, as well as with aspens; persisting on eroded sand banks and in cutover woods; advancing into bogs (though not thriving on open mats).

The attractive white corollas, usually with 7 petals united only at the very base, are open in the late spring and they drop intact — like fallen stars. Small bees pollinate the flowers and there is extensive vegetative spread by tubers. (See Anderson & Beare for work in Michigan.)

7. Samolus

1. **S. parviflorus** Raf.   Fig. 18   Water-pimpernel; Brookweed

Map 52. Creek beds, mudflats, sandbars, river banks; salt marshes, ditches, and wet hollows.

Sometimes treated as a variety or subspecies of the Eurasian *S. valerandi* L. and also long known as *S. floribundus* Kunth.

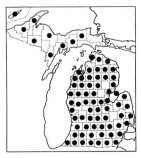
46. Lysimachia ciliata

47. Lysimachia quadriflora

48. Lysimachia quadrifolia

STYRACACEAE   Storax Family

1. Halesia

REFERENCE

Reveal, James L., & Margaret J. Seldin. 1976. On the Identity of Halesia carolina L. (Styracaceae). Taxon 25: 123–140.

1. **H. carolina** L.   Silver-bell

Map 53. A southern species, barely ranging north to southernmost Illinois and Ohio (two counties each). Often cultivated northward, and collected, in full bloom, as escaped in a wooded ravine in Oakland County (*Thompson* in 1946, BLH).

A shrub or small tree with alternate, simple, serrate leaves, easily recognized by its spring-blooming, 4-merous, white, bell-shaped, pendent flowers; inferior ovary ripening into a large 4-winged, dry, indehiscent fruit; and young twigs stellate-pubescent, older ones with chambered pith. Some authors have followed Reveal and Seldin in considering this name to have been long misapplied, leaving *H. tetraptera* J. Ellis as the correct name for this species.

OLEACEAE   Olive Family

The edible olive, *Olea europaea* L., is surely the most widely known and used member of this family, but it is found in Michigan only on grocery-store shelves, in salads, and in similar habitats. Shrubs of the familiar yellow-flowered *Forsythia* spread vegetatively from cultivation if not kept pruned back, but are barely established (no more than *Philadelphus* or *Rhodotypos* of similar habit but quite different families). Many of the planted forsythias are sterile hybrids.

49. Lysimachia hybrida   50. Lysimachia lanceolata   51. Trientalis borealis

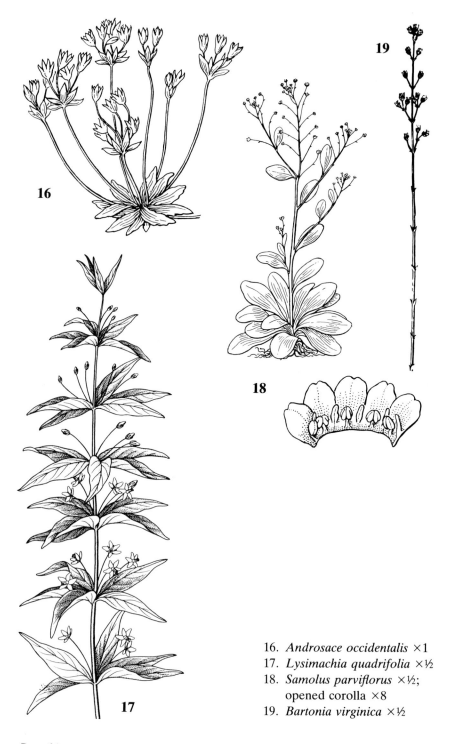

16. *Androsace occidentalis* ×1
17. *Lysimachia quadrifolia* ×½
18. *Samolus parviflorus* ×½;
    opened corolla ×8
19. *Bartonia virginica* ×½

REFERENCES

Hardin, James W. 1974. Studies of the Southeastern United States Flora. IV. Oleaceae. Sida 5: 274–285.

Wilson, Kenneth A., & Carroll E. Wood, Jr. 1959. The Genera of Oleaceae in the Southeastern United States. Jour. Arnold Arb. 40: 369–384.

KEY TO THE GENERA

1. Leaves pinnately compound (very rarely simple); corolla none; fruit a samara; plant a tree .... 1. **Fraxinus**
1. Leaves simple; corolla conspicuous; fruit a capsule or berry-like drupe; plant a shrub
    2. Flowers pink to purple (rarely white), numerous in a dense panicle at least 7 cm long; fruit a capsule; leaf blades ovate, with truncate or subcordate base and acute to acuminate apex .... 2. **Syringa**
    2. Flowers white, in panicles less than 7 cm long; fruit a berry-like drupe; leaf blades mostly elliptic-oblong, scarcely if at all more acute at apex than at base ....
    .... 3. **Ligustrum**

1. Fraxinus   Ash

The first juvenile leaves on seedlings are simple, and rare mutant trees may bear simple leaves. In all of our species, the flowers appear before the leaves, on twigs of the previous year. Since the inconspicuous flowers are often unisexual (plants dioecious or polygamous), vegetative features are stressed in the key. Mites often distort the inflorescences, producing persistent galls. Leaflet number, as given in some keys, is seldom very useful, although leaves with 11 leaflets occur frequently in the first two species and not in the others. The fruit is conspicuously winged, a samara much like that of the maples but not in pairs.

The wood of ash—white ash in particular—is known for its hard, elastic, shock-resisting properties. Oars and paddles, tennis racquets, ball bats, tool handles, hockey sticks, barrels, and snowshoes illustrate its many uses. Black and red ash have more often been used in basket-making, especially by skilled American Indians. Furthermore, ash makes an excellent, quiet-burning firewood.

Fraxinus excelsior L., the European ash, sometimes planted, has a broadly winged linear-oblong fruit like those of *F. quadrangulata* and *F. nigra*, and likewise lacks a calyx; it has sessile leaflets glabrous even at the base, broad terminal buds, and terete green twigs (contrasting strongly with black buds).

REFERENCES

McCormac, James S., James K. Bissell, & Stanley J. Stine, Jr. 1995. The Status of Fraxinus tomentosa (Oleaceae) in Ohio with Notes on Its Occurrence in Michigan and Pennsylvania. Castanea 60: 70–78.

Miller, Gertrude N. 1955. The Genus Fraxinus, the Ashes, in North America, North of Mexico. Cornell Univ. Agr. Exp. Sta. Mem. 335. 64 pp.

Taylor, Sylvia May Obenauf. 1972. Ecological and Genetic Isolation of Fraxinus americana and Fraxinus pennsylvanica. Ph. D. thesis, Univ. Michigan. 161 pp.

Wagner, W. H., Jr., Sylvia Taylor, Gerald Grieve, Ronald O. Kapp, & Keith Stewart. 1988. Simple-leaved Ashes (Fraxinus: Oleaceae) in Michigan. Michigan Bot. 27: 119–134.

KEY TO THE SPECIES

1. Branchlets ± sharply 4-angled or even slightly winged, with flattish sides . . . . .1. **F. quadrangulata**
1. Branchlets (at least internodes) essentially terete
  2. Lateral leaflets definitely sessile, 6–10; flowers without a calyx; samara strongly flattened, scarcely if at all thickened over the seeds, ca. (4.5) 6–8 (10) mm broad across the middle of the seed, winged to the base . . . . .2. **F. nigra**
  2. Lateral leaflets (at least the lower ones) petiolulate (with at least a short though sometimes winged stalk), 4–8; flowers with a calyx (persistent at base of fruit); samara distinctly thickened over the mature seed, ca. 1.3–7 mm broad across the middle of the seed, very narrowly if at all winged at the base
    3. Leaflets whitish beneath and minutely papillose (or densely pitted in appearance) [use 30× or stronger lens], the lateral ones on petiolules often 5–10 mm long; leaves and young twigs glabrous or nearly so (sometimes with hairs along main veins of leaflets beneath); samaras with distended portion over the seed ca. 4–6.5 (7) times as long as wide (measured at middle), winged only along apical half or less of the seed . . . . .3. **F. americana**
    3. Leaflets green beneath, not papillose, the lateral ones on petiolules rarely as long as 5 mm (usually shorter and often winged to the base—but much longer in the rare *F. profunda*); leaves and young twigs densely pubescent to glabrous; samaras with distended portion ca. (6.5) 7.5–12 times as long as wide, often winged (very narrowly) to below the middle of the seed
      4. Calyx at base of fruit ca. (2.2) 3–5 mm long; wing of samara ca. (6.5) 7–10 mm broad; leaflets pubescent beneath . . . . .4. **F. profunda**
      4. Calyx at base of fruit ca. 0.5–2 (2.5) mm long; wing of samara (4) 4.5–6.5 (8.5) mm broad; leaflets glabrous or pubescent beneath . . . . .5. **F. pennsylvanica**

1. **F. quadrangulata** Michaux   Blue Ash

Map 54. Deciduous woods, usually on floodplains, occasionally on uplands.

As in the next species, the calyx is lacking (or early deciduous) and the fruit is strongly flattened, but the lateral leaflets have short petiolules.

52. Samolus parviflorus   53. Halesia carolina   54. Fraxinus
                                                                                                quadrangulata

2. **F. nigra** Marshall   Fig. 20   Black Ash
   Map 55. Usually on mucky or peaty soils in swamps, such as river
floodplains; sometimes a dominant tree, or merely scattered (as often in
cedar swamps).
   There is a patch of dense pubescence at the base of each leaflet. The
terminal bud is ± prolonged, usually a bit longer than broad, whereas in
our other species, except for *F. pennsylvanica*, the terminal bud is at least
as broad as long.

3. **F. americana** L.   Fig. 21   White Ash
   Map 56. Usually in well drained upland deciduous forests, especially
beech-maple.
   The distinctive papillae (giving the whitish color) can often be best seen
on the underside of a leaflet where it has been folded; they should not be
confused with the larger glandular-looking dots often seen in the islets
formed by the ultimate leaf veins in *Fraxinus*. Papillae do not occur on
young seedlings. A form with the leaves mostly simple was described from
Kalkaska County: f. *barrii* W. H. Wagner.

4. **F. profunda** (Bush) Bush   Pumpkin Ash
   Map 57. Not known from Michigan until 1992, when discovered by
McCormac and Reznicek at sites in Hillsdale County not far north of the
Ohio border in deciduous swamps with *Acer saccharinum, Fraxinus
pennsylvanica, Populus heterophylla,* and *Quercus bicolor.* Two years later,
found by Brodowicz in Berrien County, likewise in seasonally inundated
swamp forests.
   Long known as *F. tomentosa* F. Michaux, an illegitimate name.

5. **F. pennsylvanica** Marshall   Fig. 22   Red or Green Ash
   Map 58. Swamps and shores, often in areas that are quite wet, at least
seasonally; occasionally invading peatlands and rarely in upland deciduous
forests.

55. Fraxinus nigra

56. Fraxinus americana

57. Fraxinus profunda

The widely planted "Marshall ash" is *F. pennsylvanica*. Of all the plants in our flora with names including this epithet, no other was spelled originally (and hence has now to be spelled) with the double *n*.

Plants with leaves and new twigs glabrous have been called green ash, but are not now considered of sufficient taxonomic significance to warrant recognition as a variety or subspecies. Pubescent plants (red ash) are rather easily recognized, but glabrous ones may sometimes be confused with *F. americana*, especially when one does not have sufficient magnification to see the papillose lower surface of the leaflets in the latter. The distinctive slender petiolules in *F. americana* usually contrast well with the very short petiolules in *F. pennsylvanica*, in which furthermore they are often narrowly winged so as to give the leaflets an almost sessile aspect. The leaflets may be paler green beneath than above, but are only very rarely as whitish as in typical *F. americana*. Waxy flakes or strips peeling from year-old twigs of *F. americana* contrast with the usually smooth (if not pubescent) twigs of *F. pennsylvanica*. Natural hybrids between these two species are apparently very rare, according to Taylor.

2. Syringa

Several species, with hundreds of cultivars and hybrids, are in cultivation. The common lilac is the most popular of all and spreads to form thickets at old homesites, including farms. Very rarely it seems to grow from seed. Lilacs have doubtless spread from cultivation in many more counties than those mapped, from which specimens have been collected and labeled to suggest more than mere persistence where once planted. The generic name is — unfortunately—sometimes used as a common name for *Philadelphus* (mock-orange).

REFERENCE

LaRue, Carl D. 1948. The Lilacs of Mackinac Island. Am. Midl. Nat. 39: 505–508.

58. Fraxinus pennsylvanica   59. Syringa vulgaris   60. Ligustrum vulgare

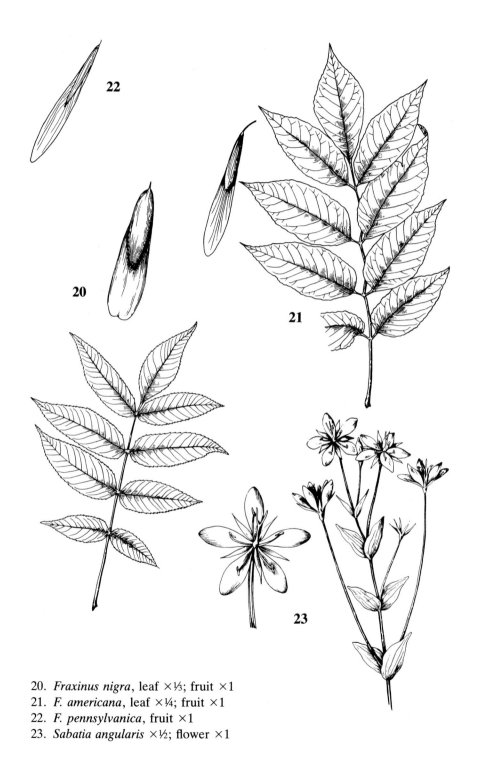

20. *Fraxinus nigra*, leaf ×⅓; fruit ×1
21. *F. americana*, leaf ×¼; fruit ×1
22. *F. pennsylvanica*, fruit ×1
23. *Sabatia angularis* ×½; flower ×1

1. **S. vulgaris** L.   Common Lilac
    Map 59. Most lilacs are Asian in origin, but this species is apparently
from southeastern Europe. Spreads to roadsides, fencerows, clearings and
fields, shores, usually but not always near evidence of old habitation.
    The leaves of *S. vulgaris* have cordate-ovate blades, while in *S.* ×*persica*
L., Persian lilac, the blades are lanceolate, tapered at the base.
    Although the largest lilac in the state (on Mackinac Island) is said to
have a trunk 13 inches in diameter at standard breast height, some of the
large specimens for which Mackinac Island is famous were measured by
LaRue in 1947, with diameters from 14 to 23 inches (measured closer to the
ground to avoid branched portion of the trunk).

3. Ligustrum   Privet
    Several species of this genus are commonly cultivated, especially for
hedges, and they doubtless escape far more often than the maps suggest.
Numerous cultivars have been named, as has a hybrid of the last two
species below.

KEY TO THE SPECIES

1. Corolla tube equalling or shorter than the lobes; leaves glabrous; twigs and pedicels
   glabrescent to finely puberulent (hairs all or mostly less than 0.2 mm long) ......
   ....1. **L. vulgare**
1. Corolla tube about twice as long as lobes, or longer; leaves glabrous or pubescent;
   twigs and pedicels glabrous or pubescent with many hairs over 0.2 mm long
  2. Twigs, pedicels, calyx, and midrib of leaves beneath pubescent with many of the
     hairs over 0.2 mm long; leaves promptly deciduous ....2. **L. obtusifolium**
  2. Twigs, pedicels, calyx, and leaves glabrous or nearly so; leaves persistent well into
     winter ....3. **L. ovalifolium**

1. **L. vulgare** L.   Common Privet
    Map 60. A native of the Mediterranean region but widely cultivated in
this area, escaping to disturbed ground and forests, dry or damp.

2. **L. obtusifolium** Siebold & Zucc.
    Map 61. A Japanese species, occasionally escaping to fields, thickets,
shores, woodlands, and doubtless other habitats.

3. **L. ovalifolium** Hassk.   California Privet
    Map 62. Another Japanese species, collected out of cultivation on the
bank of a filled area along the River Raisin floodplain (*Smith 1649* in 1986,
MICH).

LOGANIACEAE   Logania Family

1. Buddleja
Sometimes this genus is segregated into a separate family, Buddlejaceae.
A proposal to alter the spellings (in the list of conserved names) to
Buddleia and Buddleiaceae, which have been used by some authors, was
strongly rejected at the 1975 International Botanical Congress.

1. B. davidii Franchet   Butterfly-bush
Map 63. A native of China, familiar in cultivation and hardy farther
north than other species in the genus. Collected on a beach at Mackinac
Island in 1934 (*Potzger 4611*, BUT, ND). To be expected elsewhere as an
escape.
This is a shrub with opposite, crenulate-serrate, nearly sessile leaves.
The undersides of the leaves and branches of the inflorescence are covered
with stellate hairs. The 4-merous flowers are ± purple, with an orange
center, in a dense elongate terminal inflorescence, and are very attractive
to butterflies.

GENTIANACEAE   Gentian Family

The Menyanthaceae are now generally recognized as a separate family
on the basis of various anatomical characters as well as alternate, petiolate
leaves. Our only representative in Michigan is *Menyanthes*, with trifoliolate
leaves. All of our Gentianaceae have the leaves (or most of them) simple,
opposite, entire, and essentially sessile. From the Caryophyllaceae and
Guttiferae, with similar foliage, they can be readily distinguished by the
sympetalous corolla, solitary style (sometimes cleft apically) or none, pari-
etal placentation (it is free-central in Caryophyllaceae, axile in Guttiferae),
and 2 carpels. Most of our species (not genera) of Gentianaceae have the
flowers blue or bluish, a color absent in Caryophyllaceae and Guttiferae.
All "gentians" are protected from exploitation by the 1943 wildflower
amendment to Michigan's "Christmas Tree Law," but whether this means
all members of the family or merely of the genus *Gentiana* (as understood
in 1943?) is not clear.

REFERENCES

Gillett, John M. 1963. The Gentians of Canada Alaska and Greenland. Canada Dep. Agr.
    Publ. 1180. 99 pp.
Mason, Charles T., Jr., & Hugh H. Iltis. 1966 ["1965"]. Preliminary Reports on the Flora of
    Wisconsin No. 53. Gentianaceae and Menyanthaceae—Gentian and Buckbean Families.
    Trans. Wisconsin Acad. 54: 295–329.
Wood, Carroll E., Jr., & Richard E. Weaver, Jr. 1982. The Genera of Gentianaceae in the
    Southeastern United States. Jour. Arnold Arb. 63: 441–487.

KEY TO THE GENERA

1. Leaves all reduced to tiny narrow scales, less than 3 (5) mm long; stigmas decurrent on obscure style to top of the ovary ....1. **Bartonia**
1. Leaves all or mostly well developed; stigma distinct, not decurrent
  2. Cauline leaves in whorls of 4; corolla lobes with a conspicuous circular fringed nectar gland just below the middle; plant normally ca. 1 m or more tall .....2. **Frasera**
  2. Cauline leaves opposite; corolla with nectaries none or basal and without fringe; plant less than l m tall
    3. Corolla bright pink (rarely albino); style slender, ca. one fourth as long as the ovary or longer; stigmas linear or ovate
      4. Lobes of corolla 9–17 mm long, much longer than the tube, with yellow spot at base; stigmas linear (nearly or quite equalling the undivided portion of the style); calyx lobes mostly 1–2.3 mm broad ....3. **Sabatia**
      4. Lobes of corolla 2–5.5 mm long, much shorter than the tube, pink (or whitish) at base; stigmas ± ovate, less than 1 mm long, less than half as long as the style; calyx lobes less than 1 mm broad ....4. **Centaurium**
    3. Corolla blue, purplish, greenish, or white; style none or obsolete (or stubby, less than a fifth as long as ovary, with a large fan-like stigma, in *Gentianopsis*)
      5. Petals 4, each with a spur ca. 2.5–4 (5) mm long projecting at the base, greenish, often tinged with purplish, entire (not fringed) ....5. **Halenia**
      5. Petals 5, without spur, or 4 and fringed, blue, purplish, or white
        6. Pedicels mostly all distinctly longer than the flowers; flowers 4-merous, the corolla lobes fringed, spreading in sunshine; seeds covered with conspicuous papillae ....6. **Gentianopsis**
        6. Pedicels none or much shorter than the flowers; flowers 5-merous, the corolla lobes entire, erect; seeds smooth
          7. Flowers (1) 1.3–2.3 cm long, on short but distinct pedicels; sinuses of calyx without membranous connective between the lobes; corolla lacking plaits [folds] between the lobes; seeds round ....7. **Gentianella**
          7. Flowers 2.5–4.3 cm long, sessile or subsessile in mostly involucrate clusters; sinuses of calyx with a membranous connective between the lobes; corolla with distinct plaits between the lobes; seeds flattened and winged .....8. **Gentiana**

61. Ligustrum obtusifolium

62. Ligustrum ovalifolium

63. Buddleja davidii

1. Bartonia   Screwstem; Bartonia

These often pale, essentially leafless plants obviously derive their nutrition from other plants, but whether or not with the aid of a fungus seems to be unknown.

REFERENCES

Gillett, John M. 1959. A Revision of Bartonia and Obolaria (Gentianaceae). Rhodora 61: 43–62.
Henson, Don. 1985. Bartonia paniculata, New to Michigan. Michigan Bot. 24: 19–20.
Reznicek, A. A., & R. Emerson Whiting. 1976. Bartonia (Gentianaceae) in Ontario. Canad. Field-Nat. 90: 67–69.

KEY TO THE SPECIES

1. Mid-cauline leaves alternate; anthers ca. 0.3–0.5 mm long . . . . .1. **B. paniculata**
1. Mid-cauline leaves opposite or subopposite; anthers ca. 0.5–0.9 (1.1) mm long . . .
. . . . .2. **B. virginica**

1. B. paniculata (Michaux) Muhl.

Map 64. First collected in Michigan in 1983 in a large patterned fen in Luce County (*Henson 1560*, MICH). In 1995, discovered in western Chippewa County in two sandy intermittent wetlands in Tahquamenon Falls State Park (*Penskar et al. 1003, 1004*, MICH) and in open sphagnum hummocks at the edge of a tamarack swamp in Allegan County (*Reznicek et al. 10141*, MICH). This very inconspicuous species will undoubtedly turn up elsewhere in the region, where it is quite disjunct from its usual Atlantic Coastal Plain range and even from its Ontario occurrences.

The capsule dehisces completely through the obscure style. A variant of this species (usually treated as a subspecies) with anthers as long as in the next species occurs along the north Atlantic coast.

2. B. virginica (L.) BSP.   Fig. 19

Map 65. Moist sandy ditches, shores, and hollows (including borrow pits); depressions in woods (of oak, etc.); low mixed woods, mossy swamps (hardwoods, tamarack); often in sphagnum; occasionally in drier ground.

Plants vary in stature from as short as 5 cm, with a single terminal flower, to as tall as 30 or even 40 cm, with as many as 40–50 flowers in a narrow compound cyme. Rarely a depauperate individual in a population may have all the leaves alternate, but such plants can be distinguished from the preceding species by their slightly larger anthers and capsules dehiscing only below the obscure style.

2. Frasera

This genus is often included in the widespread genus *Swertia*, from which it is not very clearly distinguished. The correct name is then *Swertia caroliniensis* (Walter) Kuntze.

REFERENCE

Pringle, James S. 1990. Taxonomic Notes on Western American Gentianaceae. Sida 14: 179–187.

1. **F. caroliniensis** Walter   Fig. 24   American Columbo
   Map 66. Dry (oak, hickory, sassafras) or sometimes moist woods and openings.

3. Sabatia

This genus of attractive plants is centered in the southeastern United States, where all the species occur; only one ranges into Michigan. The stems (in our species) are ± sharply 4-angled or even very narrowly winged, as they are in *Centaurium* and some other genera.

REFERENCE

Wilbur, Robert L. 1955. A Revision of the North American Genus Sabatia (Gentianaceae). Rhodora 57: 1–33; 43–71; 78–104.

1. **S. angularis** (L.) Pursh   Plate 2-E; Fig. 23   Rose-pink;
                                                                        Rose Gentian
   Map 67. Moist sandy shores, depressions in dunes, marshy ground.
   A few plants with pure white corollas (except for the yellow spot) have been found among locally abundant pink-flowered plants in Berrien County [f. *albiflora* House].

4. Centaurium   Centaury

Both species treated here are Eurasian natives, sporadically established, perhaps originally as escapes from cultivation. The attractive flowers are sensitive to light, and mostly remain closed except during the middle of sunny days.

64. Bartonia paniculata

65. Bartonia virginica

66. Frasera caroliniensis

KEY TO THE SPECIES

1. Flowers sessile or nearly so, the calyx immediately subtended by bracts; stems usually arising from a basal rosette; mature corolla lobes ca. (4) 4.5–5.5 mm long
....1. **C. erythraea**
1. Flowers all or mostly on short pedicels ca. (1) 2–4.5 mm long; stems without a basal rosette; mature corolla lobes ca. 2–3.5 mm long ....2. **C. pulchellum**

1. **C. erythraea** Rafn
Map 68. Damp disturbed and open ground, including old fields, meadows, shores, roadside ditches.

This species has often been called *C. umbellatum* Gilib., a name not validly published, and *C. minus* Moench, a confused name, at least in part applying to a different species.

2. **C. pulchellum** (Sw.) Druce
Map 69. In habitats similar to those of the preceding, including roadsides and lawns.

Often a shorter and more delicate plant than *C. erythraea*, up to 20 cm tall, and usually very bushy in aspect.

5. **Halenia**

1. **H. deflexa** (Sm.) Griseb.   Figs. 25, 26   Spurred Gentian
Map 70. Cedar swamps; woods and thickets of spruce, fir, cedar, less often pine and aspen; frequently in springy, mossy places and often especially common at the borders of coniferous woods (as along shores); occasionally in fens and borrow pits.

6. **Gentianopsis**   Fringed Gentian
Long included in *Gentiana*, and by Gillett in *Gentianella*, but differing in several characters and recognized as a distinct genus by an increasing number of authors. Tiny plants bearing a single flower may sometimes be

67. Sabatia angularis

68. Centaurium erythraea

69. Centaurium pulchellum

found, especially very late in the season (October), often after blooming of normal-sized plants—which themselves are quite variable in shape and stature.

REFERENCES

Iltis, Hugh H. 1965. The Genus Gentianopsis (Gentianaceae): Transfers and Phytogeographic Comments. Sida 2: 129–153.
Pringle, James S. 1967. The Fringed Gentians—Observations on their Life Histories in Ontario. Wood Duck 21: 14–17.

KEY TO THE SPECIES

1. Principal mid-cauline leaves (those on main stem at or just below the lower branches) ovate to ovate-lanceolate, less than 4 times as long as broad or over 1 cm broad (or usually both) . . . . .1. **G. crinita**
1. Principal mid-cauline leaves linear to linear-lanceolate, (5) 6–21 times as long as broad and less than 1 cm at their widest . . . . .2. **G. procera**

1. **G. crinita** (Froel.) Ma

Map 71. Meadows, marshes, and damp shores; wet slopes and banks; ditches, moist old fields and roadsides; usually on calcareous sands and gravels.

Depauperate plants may be confused with *G. procera*, for the species are not as clearly separable as one might wish (and are considered to hybridize in some regions). Indeed, most of the northernmost records mapped (Alpena, Arenac, Clare, Gladwin, Grand Traverse, Lake, Montmorency, and Roscommon counties) are based on specimens with rather small, ambiguous and unusually shaped leaves, though not as linear as expected (especially on small individuals) in *G. procera*.

2. **G. procera** (Holm) Ma

Map 72. A calciphile like the preceding species; sandy, gravelly, rocky, and marly shores, wet meadows, crevices in limestone (or dolomite) pavements; often makes a magnificent blue display when in bloom in interdunal

70. Halenia deflexa   71. Gentianopsis crinita   72. Gentianopsis procera

hollows and calcareous flats along the northern shores of Lakes Michigan and Huron.

Sometimes treated as a variety or subspecies of *G. crinita*. There has been some attempt recently to resurrect an old name, which I am not fully convinced applies to this species, in the combination *G. virgata* (Raf.) Holub.

Floral differences often cited between this species and the preceding do not hold up. The corolla in our material of both may be as long as 7–7.5 cm, and both may have stipitate ovaries. There does seem to be a tendency for the cilia (fringe) at the end of the corolla lobes in *G. procera* to be reduced to little more than irregular teeth. Robust plants may have leaves as broad at the base as in *G. crinita*, but they are relatively much longer. The small leaves or bracts below the long pedicels are sometimes ovate (shaped as in the larger leaves of *G. crinita*), while the principal leaves are much longer and linear-lanceolate.

7. Gentianella

Long included in *Gentiana* in the broad sense, this is a large genus worldwide, well represented in regions of high altitudes and high latitudes. Our only species is restricted to temperate eastern North America. One other, a circumpolar species, might be found in the Lake Superior region of the state, for it grows on the north side of that lake in Ontario, as well as from Labrador to Alaska in North America, south to Mexico in the western mountains. This is *G. amarella* (L.) Börner, which differs most obviously in having a long beard arising at the base of each corolla lobe. I have seen one specimen (ex herb. F. C. Gates) labeled "Northern Michigan" but suspiciously lacking collector, date, or exact locality. There are no published reports from Michigan in the botanical literature.

REFERENCE

Gillett, John M. 1957. A Revision of the North American Species of Gentianella Moench. Ann. Missouri Bot. Gard. 44: 195–269. [Includes the species now placed in *Gentianopsis*.]

1. **G. quinquefolia** (L.) Small   Stiff Gentian
Map 73. Creek and river banks, marshy meadows; bluffs and wooded hillsides; usually in ± calcareous sites.

The corolla is blue or purplish; as in other blue gentians, whitish flowers are known but these seem not to have been noted in Michigan. (The white-flowered form of *G. amarella* is frequent in populations on the north side of Lake Superior.) Plants from the western part of the range of this species, including those from Michigan to Iowa, represent ssp. *occidentalis* (A. Gray) J. M. Gillett. The tips of the corolla lobes are mucronate, while no mucro or tooth occurs on the lobes in *G. amarella*.

24. *Frasera caroliniensis* ×½
25. *Halenia deflexa* ×½
26. *H. deflexa*, flower ×2

8. Gentiana
Gentian

The preceding two genera were long included in *Gentiana*. The species of this genus, as now defined, which grow in Michigan mostly have flowers with corollas "closed" or nearly so, requiring some struggle by the bees that enter to pollinate them. Only the very rare *G. puberulenta* has fully open flowers. Individuals with white flowers may be expected in all the species in which the corolla is ordinarily blue or purplish. Interspecific hybrids occur rarely, and are discussed by Pringle (1965a, with a key, & 1967).

REFERENCES

Costelloe, Barbara H. 1988. Pollination Ecology of Gentiana andrewsii. Ohio Jour. Sci. 88: 132–138.
Pringle, James S. 1965a ["1964"]. Preliminary Reports on the Flora of Wisconsin. No. 52. Gentiana Hybrids in Wisconsin. Trans. Wisconsin Acad. 53: 273–281.
Pringle, James S. 1965b. The White Gentian of the Prairies. Michigan Bot. 4: 43–47.
Pringle, James S. 1966. Gentiana puberulenta sp. nov., a Known but Unnamed Species of the North American Prairies. Rhodora 68: 209–214.
Pringle, James S. 1967. Taxonomy of Gentiana, section Pneumonanthae, in Eastern North America. Brittonia 19: 1–32. [Includes all our species.]
Pringle, James S. 1968. The Status and Distribution of Gentiana linearis and G. rubricaulis in the Upper Great Lakes Region. Michigan Bot. 7: 99–111.

KEY TO THE SPECIES

1. Plaits between the corolla lobes distinctly longer than the lobes, toothed or erose (fig. 27); calyx lobes minutely ciliate . . . . .1. **G. andrewsii**
1. Plaits distinctly shorter than the corolla lobes, entire or toothed (fig. 28); calyx lobes smooth, papillose, or ciliate
   2. Stem puberulent, especially on the angles; calyx lobes linear, broader at the very base; corolla open at anthesis . . . . .2. **G. puberulenta**
   2. Stem glabrous; calyx lobes broadest above the base (or linear in *G. linearis* and *G. rubricaulis*); corolla nearly closed at anthesis
      3. Calyx lobes distinctly keeled basally on the underside and in one species also ciliolate; apex of plaits between the corolla lobes ± strongly erose or at least bifid; plants only in the southern half of the Lower Peninsula
         4. Leaf blades less than 2 cm broad; calyx lobes (and at least the involucral leaves) densely ciliolate; flowers mostly blue . . . . .3. **G. saponaria**
         4. Leaf blades (at least the upper) (1.7) 2.7–5 cm broad; calyx lobes and leaves eciliate; flowers white, marked with greenish . . . . .4. **G. alba**
      3. Calyx lobes without a keel or cilia (at most minutely papillose or scabrous); apex of plaits between the corolla lobes broadly triangular, terminating in a single off-center tooth (sometimes with a much smaller subsidiary tooth on each side); plants only in the Upper Peninsula and north half of the Lower Peninsula (though farther south in Ontario)
         5. Leaves deep green, linear or nearly so, not over 1 cm broad; calyx lobes green throughout, ± exposed by the spreading, narrow involucral leaves . . . . .5. **G. linearis**
         5. Leaves ± pale green, lanceolate to ovate, at least the larger ones over 1 cm broad; calyx lobes green at most toward apex, usually well concealed by the broad, enveloping involucral leaves . . . . .6. **G. rubricaulis**

1. **G. andrewsii** Griseb.　　Fig. 27　　　　Closed or Bottle Gentian
Map 74. Marshy or at least moist ground: meadows and wet prairies; shores, thickets, and ditches; river banks, floodplains, swamp forests.

Each corolla lobe terminates in a small tooth or mucro. White-flowered individuals [f. *albiflora* Britton] occur occasionally with the normal sky-blue ones, and pale blue flowers also occur uncommonly. White- or blue-flowered plants may be easily distinguished from *G. saponaria*, which also has ciliolate calyx lobes, by the broader leaves of *G. andrewsii* (as broad as in *G. alba*) as well as the very short tips of the corolla lobes. Hybrids with *G. alba* [*G. ×pallidocyanea* J. S. Pringle] have been identified by Pringle from Genesee and Wayne counties. The hybrid with *G. puberulenta* [*G. ×billingtonii* Farwell] was originally described from Squirrel Island, on the Ontario side of the St. Clair River, and may turn up in the state, as might the rare hybrid with *G. rubricaulis* [*G. ×grandilacustris* J. S. Pringle], described from Minnesota.

Pollination is by bumblebees, "the only insects observed to have the strength and learning capability necessary to open the closed corolla consistently" (Costelloe).

2. **G. puberulenta** J. S. Pringle　　Plate 2-F　　　　Prairie Gentian
Map 75. Sandy, seasonally wet to dry areas. There are very few recent records from the state.

Long known as *G. puberula* Michaux, a name which belongs in the synonymy of *G. saponaria* (Pringle 1966). The flowers are ordinarily a very deep rich blue.

3. **G. saponaria** L.　　　　　　　　Soapwort Gentian
Map 76. This moist prairie and oak savana species was collected at Niles in 1867 (*Geo. Ames*, MICH, MO) but apparently has not been seen in Michigan since. Rare but still found as close to Michigan as the nearby Indiana Dunes area and the Oak Openings near Toledo, Ohio.

73. Gentianella
   quinquefolia

74. Gentiana andrewsii

75. Gentiana puberulenta

4. **G. alba** Nutt.   Plate 2-G   White Gentian
Map 77. Dry or moist prairies and oak woodland. Now nearly extirpated in Michigan.

White-flowered plants of other species have frequently been misidentified as this normally white-flowered one, which has long been known as *G. flavida* A. Gray (Pringle 1965b).

5. **G. linearis** Froel.   Bog Gentian
Map 78. Meadows, shores, river margins, mossy thickets, and adjacent somewhat disturbed areas. This is primarily a boreal species of eastern North America, but ranges down in the mountains to the southern Appalachians.

The small calyx lobes (linear or nearly so) are fully green and appear quite different from those of the next species. A form with the corolla violet rather than blue is known only from Michigan.

6. **G. rubricaulis** Schwein.   Fig. 28   Red-stemmed Gentian
Map 79. Widespread in many moist, usually alkaline, sites: shores, interdunal hollows, meadows; bogs and fens; river and stream margins; sandy and marly excavations; alder thickets, coniferous swamps, depressions in pine plains; rock crevices and pool margins on Lake Superior.

The distribution of this northern species centers in the Great Lakes region, and it is usually omitted from books that deal with common wildflowers. Both blue- and white-flowered forms have often been misidentified as *G. andrewsii*, which is readily distinguished by its long fimbriate plaits (exceeding the corolla lobes) and prominent, ciliolate calyx lobes.

This has often been treated as part of the preceding species, as var. *latifolia* A. Gray [the correct name would be var. *lanceolata* A. Gray] or ssp. *rubricaulis* (Schwein.) J. M. Gillett but, as Pringle (1967, 1968) has noted, it is amply and consistently distinct. The flowers are more often a muddy or purplish blue than in *G. linearis*, which is usually bright blue.

76. Gentiana saponaria

77. Gentiana alba

78. Gentiana linearis

Although long included, often as a subfamily, in the Gentianaceae, this family is now generally recognized as distinct — indeed, in a different order. We have a single species in Michigan, although a second might be expected in soft waters of the Upper Peninsula. This would be *Nymphoides cordata* (Elliott) Fernald, floating-heart, which grows as near as Sudbury and elsewhere in the northern Lake Huron region of Ontario. As the name indicates, the cordate leaf blade suggests that of a small *Nymphaea*; the elongate stem (resembling a petiole) also bears pedicels of the white emergent flowers and usually a cluster of tuberous adventitious roots.

REFERENCES

Gillett, John M. 1963. The Gentians of Canada Alaska and Greenland. Canada Dep. Agr. Publ. 1180. 99 pp.
Mason, Charles T., Jr., & Hugh H. Iltis. 1966 ["1965"]. Preliminary Reports on the Flora of Wisconsin No. 53. Gentianaceae and Menyanthaceae—Gentian and Buckbean Families. Trans. Wisconsin Acad. 54: 295–329.
Wood, Carroll E., Jr. 1983. The Genera of Menyanthaceae in the Southeastern United States. Jour. Arnold Arb. 64: 431–445.

1. **Menyanthes**

1. **M. trifoliata** L.   Fig. 29   Buckbean; Bogbean
Map 80. Fens (but sometimes in deep sphagnum), cedar swamps (often in the wettest places, such as hollows and trails), wet alder or tamarack swamps, old interdunal swales.

This circumpolar species is the only one in the genus. The 5-merous heterostylous flowers have white corollas with prominent hairs on the lobes. They bloom in spring. The large trifoliolate leaves are pale, the leaflets with entire margins, and are easily recognized.

79. Gentiana rubricaulis   80. Menyanthes trifoliata   81. Vinca minor

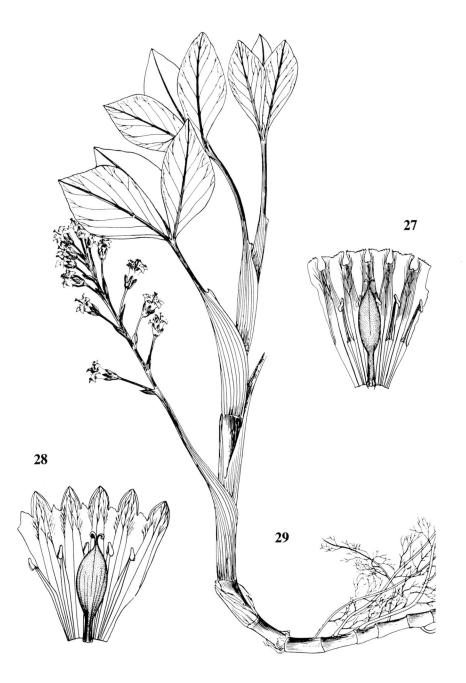

27. *Gentiana andrewsii*, opened corolla ×1
28. *G. rubricaulis*, opened corolla ×1
29. *Menyanthes trifoliata* ×½

APOCYNACEAE   Dogbane Family

This family is poorly represented in temperate regions but has about 200 genera in warmer climates. A number of species are sources of chemicals useful in medicine and as poisons. An increasing number of authors include the Asclepiadaceae in the Apocynaceae. In all our *native* species of both families, the fruit (a follicle) splits open to release seeds bearing a tuft of long silky hairs; the sap of most species is a milky latex.

REFERENCES

Boivin, Bernard. 1966. Les Apocynacées du Canada. Nat. Canad. 93: 107–128 (also in Ludoviciana 1).
Rosatti, Thomas J. 1989. The Genera of Suborder Apocynineae (Apocynaceae and Asclepiadaceae) in the Southeastern United States. Jour. Arnold Arb. 70: 443–514.
Woodson, Robert Everard, Jr. 1938. Apocynaceae. N. Am. Fl. 29: 103–192.

KEY TO THE GENERA

1. Plant a trailing subwoody evergreen creeper; flowers solitary in leaf axils; corolla blue, with limb (1.5) 2–3 cm broad; seeds glabrous . . . . .1. **Vinca**
1. Plant ± erect, herbaceous and deciduous (but fibrous); flowers in small terminal (and sometimes axillary) cymes; corolla white to rosy, less than 1 cm broad; seeds with tuft of silky hairs . . . . .2. **Apocynum**

1. Vinca

1. **V. minor** L.   Periwinkle; "Myrtle"
Map 81. A native of Europe, much cultivated as a ground-cover, often spreading along roadsides and into woods and thickets, sometimes abundant and long-persisting, out-competing native vegetation.

2. Apocynum   Dogbane

The flowers are fragrant and attractive to many insects. A beautiful metallic-iridescent leaf-beetle, *Chrysochus auratus* Fabr., is known as the "dogbane beetle" because of its regular association with *Apocynum* (and its larvae feed on the roots). American Indians in the Great Lakes region (as elsewhere) used the tough stems as a source of fiber (technological, not dietary).

REFERENCE

Anderson, Edgar. 1936. An Experimental Study of Hybridization in the Genus Apocynum. Ann. Missouri Bot. Gard. 23: 159–168.

30. *Apocynum cannabinum* ×½; flower ×5

KEY TO THE SPECIES

1. Corolla (5) 5.5–8 (9) mm long when mature, with spreading to recurved lobes, pink to rosy (at least in lines within) when fresh; leaves ovate to broadly elliptic, widely spreading or drooping ....1. **A. androsaemifolium**
1. Corolla 2.5–4 (4.5) mm long, with erect lobes when mature, white to greenish when fresh; leaves oblong to broadly lanceolate, ± strongly ascending ......2. **A. cannabinum**

1. **A. androsaemifolium** L.   Spreading Dogbane
    Map 82. Dry sandy or rocky woodlands with such associates as oak, hickory, pine, aspen, sassafras, bracken, sweet-fern; does well after disturbance such as fire or logging, and characteristic of clearings, old fields, shores, roadsides, railroads, fencerows and borders of woods, occasionally sand prairies.
    Variable in several characters, including pubescence and shape of sepals. The mature corolla usually runs about 3 times as long as the calyx, although it is relatively shorter in plants with unusually long narrow calyx lobes. The flowers at maturity are ± pendent or spreading, and in well developed plants occur in both terminal and axillary cymes.
    Plants intermediate with the next species in corolla length and color, leaf shape and aspect, flower position, and field appearance are presumably hybrids, as concluded on experimental evidence by Anderson. Other intermediate plants combine small flowers (as in *A. cannabinum*) with spreading habit and axillary cymes (as in *A. androsaemifolium*). The hybrid has long been called *A.* ×*medium* Greene, but the name *A.* ×*floribundum* Greene is four years older and applies to the same interspecific hybrid, as noted by Woodson (who apparently overlooked the nomenclatural significance of the prior name). The types (ND) support that synonymy. Some introgression with the parent species doubtless occurs, making herbarium specimens of the hybrid, especially, difficult to recognize.

2. **A. cannabinum** L.   Fig. 30   Indian-hemp
    Map 83. Often in similar habitats to the preceding species, although more frequent on wet soils: shores, thickets, river banks, at least seasonally wet marshes and meadows; but also in disturbed ground, along railroads and highways, in fields.
    Like the preceding, variable especially in pubescence. Plants with leaf blades on the main stem ± tapered to distinct petioles are typical var. *cannabinum*. Plants with the main blades ± subcordate and sessile or nearly so are var. *hypericifolium* A. Gray [also known as *A. sibiricum* Jacq.]. Plants with subcordate petiolate blades or otherwise intermediate between the alleged varieties support the skepticism with which they are now often viewed. Farwell found plants with whorled leaves (3 at a node) in Wayne and Oakland counties in 1914 (see Rep. Michigan Acad. 17: 167–170. 1917).

The mature corolla is usually about twice as long as the calyx, or shorter, and the flowers are ± erect in terminal cymes (at ends of main stem and principal branches), rarely axillary. The distinctive slender follicles may average a bit longer in this species, but the range is nearly the same in both, about (5) 7.5–16 (22) cm. However, the seeds (excluding hairs) in *A. androsaemifolium* do not exceed 3 mm in length, while in *A. cannabinum* they are distinctly longer, ca. (3.2) 3.5–5.5 mm.

ASCLEPIADACEAE   Milkweed Family

This family is sometimes included in the Apocynaceae, from which (sens. str.) it differs most conspicuously in having the pollen coherent in pollinia, the two carpels united only by their stigmas (to which the anthers are also attached), and the filaments united into a tube with a well developed corona. (In the Apocynaceae, the styles are united at least part of their length.) In our species, the flower has a prominent crown or *corona*, developed from appendages of the filaments. A structure called the *translator*, derived from dried secretions of the reproductive parts, connects pollinia from adjacent pollen sacs of two different anthers (fig. 32) and aids the removal of the pollinia by insects whose legs tangle with it. Like the Apocynaceae, the milkweeds are much better represented in tropical and subtropical regions.

REFERENCES

Noamesi, Gottlieb K., & Hugh H. Iltis. 1958 ["1957"]. Preliminary Reports on the Flora of Wisconsin. No. 40. Asclepiadaceae—Milkweed Family. Trans. Wisconsin Acad. 46: 107–114.

Rosatti, Thomas J. 1989. The Genera of Suborder Apocynineae (Apocynaceae and Asclepiadaceae) in the Southeastern United States. Jour. Arnold Arb. 70: 443–514.

82. Apocynum androsaemifolium

83. Apocynum cannabinum

84. Vincetoxicum nigrum

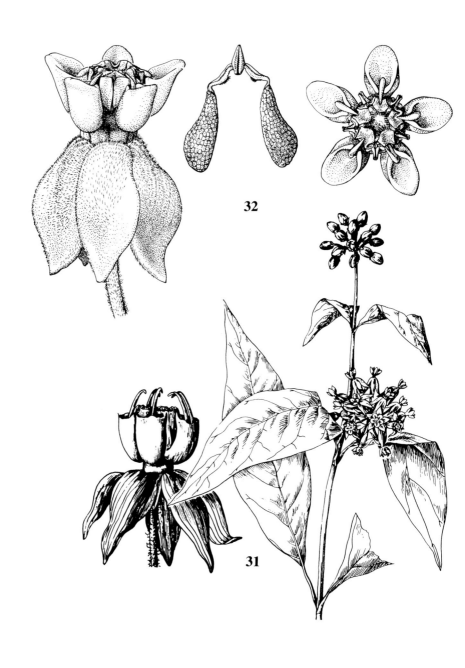

31. *Asclepias exaltata* ×½; flower ×4
32. *A. syriaca*, flower ×5; pollinia & translator ×16; flower from above ×5

KEY TO THE GENERA

1. Plant a climbing vine; inflorescence a cyme; corolla lobes spreading or ascending . . . . .1. **Vincetoxicum**
1. Plant erect or ascending, not twining; inflorescence an umbel; corolla lobes strongly reflexed at maturity . . . . .2. **Asclepias**

1. Vincetoxicum

REFERENCES

Kirk, Malcolm. 1985. Vincetoxicum spp. (Dog-Strangling Vines): Alien Invaders of Natural Ecosystems in Southern Ontario. Plant Press (Ontario) 3: 130–131.
Moore, Raymond J. 1959. The Dog-Strangling Vine Cynanchum medium, Its Chromosome Number and Its Occurrence in Canada. Canad. Field-Nat. 73: 144–147.

KEY TO THE SPECIES

1. Corolla lobes pubescent above; flowering peduncles less than 2 cm long (to 2.5 cm in fruit) . . . . .1. **V. nigrum**
1. Corolla lobes glabrous above; flowering peduncles mostly 2–2.5 cm long . . . . .2. **V. rossicum**

1. **V. nigrum** (L.) Moench
    Map 84. Native in southwestern Europe, locally established as an escape from cultivation, with potential for becoming a serious weed. Cultivated since at least 1880 in the East Lansing area and now well established in woods, thickets, and fencerows there as well as elsewhere.
    Also known as *Cynanchum nigrum* (L.) Pers., which is an illegitimate name (see McNeill in Nat. Canad. 108: 238–239. 1981). If this plant is included in the broad genus *Cynanchum*, the correct name is *C. louiseae* Kartesz & Gandhi.

2. **V. rossicum** (Kleopow) Barbar.   Dog-strangling Vine
    Map 85. Another European species, first collected in Michigan in 1968 in Berrien County and in 1970 in Oakland County.
    This species is included by Gleason and Cronquist (1991) in the white- or yellow-flowered *V. hirundinaria* Medikus. The corollas in both our species are dark purplish, but they run toward black in the preceding species and tend to be somewhat paler in *V. rossicum*, which also has longer corolla lobes (at least twice as long as broad)—and only half as many chromosomes.

2. **Asclepias**   Milkweed
    Our milkweeds have distinctive and easily recognized flowers (fig. 32), borne in hemispherical to spherical umbels. The corona consists of prominent erect cup- or hood-like structures, in most species surrounding a little horn or prolonged tooth. Measurements of flower length always include mature reflexed corolla lobes. More technical differences involving the

hood and horns and the fused reproductive parts (*gynostegium*) may be found in Woodson (1954).

So far as known, *A. meadii* A. Gray, *A. quadrifolia* Jacq., and *A. variegata* L. do not range into Michigan despite old published reports, which presumably represent misidentifications.

REFERENCES

Kephart, Susan R., Robert Wyatt, & Deborah Parrella. 1988. Hybridization in North American Asclepias. I. Morphological Evidence. Syst. Bot. 13: 456–473.

Macior, Lazarus Walter. 1965. Insect Adaptation and Behavior in Asclepias Pollination. Bull. Torrey Bot. Club 92: 114–126.

Nicolson, Dan, & Norman H. Russell. 1955. The Genus Asclepias in Iowa. Proc. Iowa Acad. 62: 211–215.

Pearson, Norma L. 1947. Variations in Floss Characteristics Among Plants of Asclepias syriaca L. Having Different Types of Pods. Am. Midl. Nat. 38: 615–637.

Schwartz, David M. 1987. Underachiever of the Plant World. Audubon 89(5): 46–61.

Sparrow, F. K. 1946. Types of Pods of Asclepias syriaca Found in Michigan. Jour. Agr. Res. 73: 65–80.

Timmons, F. L. 1946. Studies of the Distribution and Floss Yield of Common Milkweed (Asclepias syriaca L.) in Northern Michigan. Ecology 27: 212–225. (Cf. also Natl. Geogr. Mag. 86: 695. 1944.)

Wisse, David F. 1988. Life in a Milkweed Community. Michigan Nat. Res. 57(5): 40–47.

Woodson, Robert E., Jr. 1954. The North American Species of Asclepias L. Ann. Missouri Bot. Gard. 41: 1–211.

Woodson, Robert E., Jr. 1964. The Geography of Flower Color in Butterflyweed. Evolution 18: 143–163 + 1 pl.

KEY TO THE SPECIES

1. Leaves very narrowly linear (less than 2.5 (4) mm broad), mostly whorled or nearly so ....1. **A. verticillata**
1. Leaves all or mostly over 4 mm broad (usually much over), alternate or opposite
  2. Leaves all or mostly alternate
    3. Flowers (yellow-) orange, ca. 11–15.5 mm long, the corona hoods enclosing an elongate awn-like horn; umbels mostly terminal; sap clear ....2. **A. tuberosa**
    3. Flowers greenish, purple-tinged, ca. 7–9 mm long, the corona hoods without a horn; umbels axillary; sap milky ....3. **A. hirtella**
  2. Leaves all or mostly opposite
    4. Blades of leaves ± cordate or subcordate, sessile (or on petioles shorter than the basal lobes of blade)
      5. Horns shorter than the hoods (thus largely concealed); umbels usually 2 or more, with glabrous pedicels ....4. **A. sullivantii**
      5. Horns clearly exserted beyond the hoods; umbels solitary, terminal, with pedicels all or mostly puberulent at least toward base. ....5. **A. amplexicaulis**
    4. Blades of leaves ± rounded or tapered to a short or long petiole
      6. Horns conspicuously surpassing hoods; leaves glabrous or with thin puberulence beneath
        7. Flowers white or greenish, (13) 14–17 (18) mm long, on pedicels 2.5–4.5 cm long at maturity; follicles ca. 9.5–13 (15) cm long ....6. **A. exaltata**
        7. Flowers rich rose-purple (rarely albino), 7–10.5 mm long, on pedicels 0.8–1.6 (2.4) cm long; follicles ca. (5.5) 6.5–9 cm long ....7. **A. incarnata**

6. Horns shorter than hoods (or none); leaves with ± dense and uniformly distributed short pubescence beneath (sometimes glabrate in *A. viridiflora*)
  8. Flowers 8–11 (12) mm long; corolla lobes whitish, cream, or greenish (at most the lobes tinged with purplish beneath)
    9. Hoods distinctly exceeding anthers, with a horn; pedicels much longer than flowers; plants small and slender, the stem less than 3 (4) mm thick, with 4–6 (7) pairs of leaves . . . . .8. **A. ovalifolia**
    9. Hoods slightly shorter than winged tips of anthers, without a horn; pedicels mostly about equalling or shorter than the flowers; plants stouter, the stems usually 3–5 mm thick toward base, with 5–12 pairs of leaves . . . . . .9. **A. viridiflora**
  8. Flowers 10.5–25 mm long; corolla lobes rose to purple on both surfaces (plants with small and/or pale flowers may be separated from the previous couplet by their stature too large for *A. ovalifolia* and the pedicels and hoods too long for *A. viridiflora*)
    10. Hoods ca. 10–13 mm long, including a prolonged tongue-like apex; pedicels densely white-tomentose . . . . .10. **A. speciosa**
    10. Hoods ca. (3) 3.5–7 mm long, without a prolongation; pedicels glabrate to tomentose
      11. Corolla lobes pubescent outside; plant usually with umbels in 3 or more axils; hoods (3) 3.5–5 mm long; flowers 10.5–15 mm long . . . . .11. **A. syriaca**
      11. Corolla lobes glabrous; plant with a terminal umbel and sometimes 1 or 2 additional umbels in upper axils; hoods ca. 5–7 mm long; flowers (9) 13–17 (18) mm long . . . . .12. **A. purpurascens**

1. **A. verticillata** L.   Whorled Milkweed
    Map 86. Dry sandy or gravelly places, including prairies, oak woodland, roadsides, and fields.
    A very distinctive slender plant, the leaves numerous and nearly filiform (often revolute) and the flowers white to greenish (sometimes purple-tinged) in umbels near the top of the plant.

2. **A. tuberosa** L.   Plate 3-A   Butterfly-weed
    Map 87. Dry barrens and woodland openings with oak, sassafras, and/or pines, including jack pine plains; sandy roadsides and fields.

85. Vincetoxicum rossicum   86. Asclepias verticillata   87. Asclepias tuberosa

The flowers are usually some shade of orange, but may be clear yellow or bicolored. Leaf shape is quite variable and was used as a principal basis for the subspecies recognized by Woodson, who would place our material mostly in ssp. *interior* Woodson except for outlying plants in the northern and northwestern Lower Peninsula, for which he established the weakly distinguished ssp. *terminalis* (TL: Burt Lake, Cheboygan Co.).

As our only milkless milkweed—with alternate leaves besides — this is a readily recognized plant. It develops a massive, almost woody tuberous root, from which numerous ± lax stems may arise. The flowers are indeed very attractive to butterflies and other insects.

3. **A. hirtella** (Pennell) Woodson   Prairie Milkweed
    Map 88. Dry to moist sandy open ground such as fields, prairies, and borders of oak woods.

    The pedicels (and generally other vegetative parts) are rather densely hispidulous. The leaves are usually very numerous (even crowded), ca. 4–12 (22) mm wide.

4. **A. sullivantii** A. Gray   Smooth Milkweed
    Map 89. Moist prairies and relics of such habitat along roadsides and railroads.

5. **A. amplexicaulis** Sm.
    Map 90. Dry, sandy, open ground: fields and prairies, savanas, clearings in oak woods, banks and bluffs, roadsides and railroads.

    The margins of the leaves are ± crisped.

6. **A. exaltata** L.   Fig. 31.   Poke Milkweed
    Map 91. This is our only truly woodland species, usually in dry or rocky woods and savanas with oaks and other trees such as hickory, pine, and aspen; banks and borders, even roadsides; much less often in rich beech-maple woods or swamp forest.

    The leaf blades are glaucous beneath, rather thin in texture, broadly elliptic, and somewhat acuminate at both ends. They resemble the leaves of *Phytolacca* (pokeweed) and the species has been known as *A. phytolaccoides* Pursh.

7. **A. incarnata** L.   Swamp Milkweed
    Map 92. This is our only milkweed of truly wet ground, often in several inches of water: edges of rivers and streams; shores, wet prairies; openings in conifer swamps, fens (less often bogs); depressions in woods, swales and ditches, meadows.

    One of our showiest species in its deep color. The leaves are usually smaller and narrower than in the preceding species, green (not glaucous) beneath.

8. **A. ovalifolia** Decne.

Map 93. Very local in savana and thin woods of oak and jack pine and/or aspen, becoming more common after fire.

As in so many milkweeds, the leaves are rather variable in shape, but the slender stature and cream flowers make this rare and delicate prairie species easy to recognize. Besides a terminal umbel, there are usually one or two additional umbels in the upper leaf axils. Coarse leafy plants with pale flowers keying here are probably forms of the common milkweed, *A. syriaca* (see below).

9. **A. viridiflora** Raf.   Green Milkweed

Map 94. Dunes and other dry sandy sites including prairies, borders of oak woods, road cuts and railroads.

The leaves range from linear or lanceolate to nearly oblong. The umbels are axillary, very compact, and usually more numerous than in the preceding species.

This species and *A. hirtella* have often been placed in a separate genus, *Acerates*, characterized by the absence of a horn in the corona hoods.

10. **A. speciosa** Torrey   Showy Milkweed

Map 95. This species is native from western Minnesota westward and is occasionally adventive eastward, as along a roadside west of Trout Lake,

88. Asclepias hirtella

89. Asclepias sullivantii

90. Asclepias amplexicaulis

91. Asclepias exaltata

92. Asclepias incarnata

93. Asclepias ovalifolia

where it was found in 1984 by Penskar, Henson, Brodowicz, and Albert. Here, it was hybridizing with *A. syriaca*.

The hoods of the hybrid are ca. 5.5–7 mm long, quite intermediate between those of the parents, and the appearance of the plant is intermediate in other aspects as well.

11. **A. syriaca** L.   Fig. 32   Common Milkweed
    Map 96. Dry to somewhat moist, usually sandy, often ± disturbed areas: roadsides and railroads, shores and dunes, fields, waste ground; openings in aspen and pine woodlands.

The flowers are occasionally whitish or cream, and such plants, especially if they have unusually small flowers, might key to *A. ovalifolia*; but that species is much more delicate in stature. Flowers in the large umbels sometimes shrivel in drying so that they appear to be less than 10.5 mm long, but well pressed flowers can usually be found at least that long.

The follicles ("pods") are sometimes smooth (as in *A. purpurascens*), but usually are ± prominently warty or tuberculate. Much of the variation in shape and surface is well shown by Sparrow (1946). There is some correlation between pod shape and the character of the *coma* or tuft of hairs ("floss") on the seed (Pearson 1947). During World War II, milkweed floss was discovered to possess buoyant properties similar to those of kapok, previously imported for use in life-preserver jackets. When the supply of kapok (from the Dutch East Indies) was cut off, the U. S. Department of Agriculture encouraged the harvesting and use of milkweed. The Milkweed Floss Division of War Hemp Industries of the Commodity Credit Corporation was headquartered at Petoskey, Michigan, where a processing plant "ginned" the seeds, removing the floss. The acreage of milkweed in Emmet County (of which Petoskey is the county seat) made it "probably the most densely infested county in the United States" (Timmons 1946). In 1943, the total production in that county was estimated to be almost 200 tons of pods, which would yield over 36 tons of dry floss. However, the amount actually harvested in Emmet County in 1943, by school children and others, was a little over half of the estimated production. Over 12,000 tons of milkweed pods were gathered during the last two years of the war, from Michigan and elsewhere, but then it again became cheaper to use kapok and, subsequently, synthetic materials.

Other possible commercial uses of milkweed, e.g., for latex (rubber), fiber, or even fuel, have never amounted to even as much as the short-lived wartime floss operation. Wild foods devotees, however, find the young shoots, flower buds, and young pods tasty, as a vegetable or (the pods) a pickle. The parts require careful processing to remove bitterness (including the cardiac glycosides that render insects feeding on milkweed unpalatable—or worse—to some predators, such as birds).

Hybridizes with *A. speciosa* (see above) and rarely with *A. exaltata*, as at the E. S. George Reserve in Livingston County (Kephart et al. 1988).

12. **A. purpurascens** L.   Purple Milkweed

Map 97. Dry woodland (especially oak) and thickets; shores, prairies.

A handsome species, with rich-colored flowers. The leaves tend to be more acute than in *A. syriaca* and usually somewhat less prominently pinnate-veined (the net-venation more evident). The leaves of *A. syriaca* are very prominently pinnate-veined and usually ± obtuse. But the two species are sometimes difficult to distinguish. An old collection from St. Clair County looks suspiciously like a hybrid with *A. syriaca*, and others from Jackson County also may show some influence of that species.

CONVOLVULACEAE   Morning-glory Family

The dodders (*Cuscuta*) have often been included in this family.

REFERENCES

Lewis, Walter H., & Royce L. Oliver. 1965. Realignment of Calystegia and Convolvulus (Convolvulaceae). Ann. Missouri Bot. Gard. 52: 217–222.
Wilson, Kenneth A. 1960. The Genera of Convolvulaceae in the Southeastern United States. Jour. Arnold Arb. 41: 298–317.

KEY TO THE GENERA

1. Stigma at most with 2–3 small lobes; calyx ca. 6–23 mm long, fully exposed . . . . .1. **Ipomoea**
1. Stigma deeply cleft into 2 lobes ca. 2.5–4 mm long; calyx less than 6 mm long or covered by two large bracts
    2. Calyx fully visible; corolla ca. 2.2 (2.5) cm long or shorter; stigma lobes linear-filiform . . . . .2. **Convolvulus**
    2. Calyx all or mostly covered by 2 large closely subtending bracts; corolla ca. 2.5–6.5 cm long; stigma lobes oblong-elliptic . . . . .3. **Calystegia**

1. **Ipomoea**   Morning-glory

I. batatas (L.) Lam. is the cultivated sweet-potato (often called "yam," a name more strictly applying to the monocot genus *Dioscorea*). This is a

94. Asclepias viridiflora   95. Asclepias speciosa   96. Asclepias syriaca

trailing species with edible tuberous roots and pink to purple flowers, known in Michigan chiefly if not entirely from grocery stores.

KEY TO THE SPECIES

1. Corolla salverform, red, with stamens and pistil conspicuously exserted; sepals with subulate subterminal appendages; leaves deeply and ± palmately divided into narrow lanceolate segments ....1. **I. ×multifida**
1. Corolla funnelform, blue to purple or white, with stamens and pistil shorter; sepals various; leaves not divided or only broadly 3–5-lobed
  2. Stems glabrous or nearly so; sepals obtuse to rounded, at most hispidulous toward margin beneath (more uniformly hispidulous above) ....2. **I. pandurata**
  2. Stems pubescent; sepals acute or acuminate, with long stiff hairs beneath, especially toward base
    3. Calyx ca. 17–25 mm long, the sepal lobes narrowed to prolonged linear-acuminate tips; peduncles rarely as long as calyx (flowers often nearly sessile) .. ..3. **I. hederacea**
    3. Calyx ca. 11–15 (16) mm long, the sepal lobes ± oblong and acute; peduncles longer than the calyx ....4. **I. purpurea**

1. **I. ×multifida** (Raf.) Shinners   Cardinal Climber
    Map 98. An annual widely grown as an ornamental vine. It was developed in cultivation as a hybrid between *I. coccinea* L. (red morning-glory) and *I. quamoclit* L. (cardinal climber or cypress vine). The latter has narrowly pinnatisect leaves (suggesting those of a *Myriophyllum*), while *I. coccinea* has entire to dentate leaves; both are natives of tropical America and, like the hybrid, sometimes escape cultivation as short-lived adventives in colder climates. *I. ×multifida* was collected on newly bulldozed gravel in Jackson County in 1957 (*Horne 4*, MICH); it or its parents may be expected to appear briefly elsewhere in the state. These taxa have sometimes been placed in a separate genus, *Quamoclit*.

2. **I. pandurata** (L.) G. Meyer   Wild Sweet-potato;
                                       Man-of-the-earth
    Map 99. Woods and thickets, open fields, roadsides, and sandy ground. Native barely as far north as Michigan.

97. Asclepias purpurascens   98. Ipomoea ×multifida   99. Ipomoea pandurata

Perennial from a very large deep tuber, the stems clambering over shrubs and other vegetation. The large flowers (to 8 cm long) are white with purplish centers, very showy in July and August.

3. **I. hederacea** Jacq.   Ivyleaf Morning-glory
Map 100. Native of tropical America, becoming weedy northward in fencerows and fields.

An annual twining vine with blue (to purple) flowers less than 5 cm long and leaves usually deeply 3-lobed. Some plants of *I. purpurea* with slightly shorter calyx, or deeply 3-lobed leaves, or shorter peduncles (but not displaying all these features) have been reported as *I. hederacea*, but all of them lack the prolonged sepal tips of this species.

4. **I. purpurea** (L.) Roth   Common Morning-glory
Map 101. Another tropical American annual vine, popular in cultivation and occasionally escaped to fields, waste places, roadsides, and weedy thickets.

The leaves are usually not lobed; cultivars vary in flower color.

2. Convolvulus

1. **C. arvensis** L.   Fig. 33   Field Bindweed
Map 102. A deep-rooted pesky Eurasian weed of gardens and lawns, along roadsides and railroads; in fields, fencerows, parking lots, and waste ground generally. Apparently not found in Michigan until 1890, but by the end of that decade the species had been collected in at least six counties throughout the Lower Peninsula.

Quite variable in pubescence and in leaf shape; the flowers are usually white, occasionally pink.

3. Calystegia   Bindweed
The species placed in this genus have long been included by most authors in *Convolvulus*.

100. Ipomoea hederacea   101. Ipomoea purpurea   102. Convolvulus arvensis

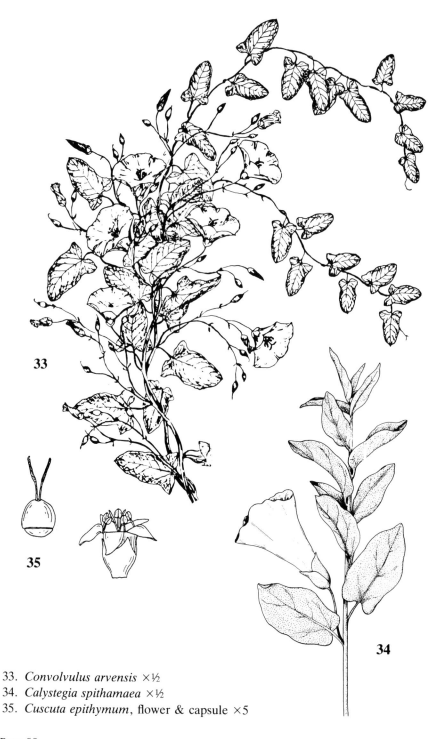

33. *Convolvulus arvensis* ×½
34. *Calystegia spithamaea* ×½
35. *Cuscuta epithymum*, flower & capsule ×5

REFERENCE

Brummitt, R. K. 1980. Further New Names in the Genus Calystegia (Convolvulaceae). Kew Bull. 35: 327–334.

KEY TO THE SPECIES

1. Flowers ca. 3 (4) cm long or shorter, double (stamens and pistil absent); bracts ca. 11–15 (22) mm long . . . . .1. **C. hederacea**
1. Flowers ca. (4) 4.5–6.5 (7.2) cm long, single; bracts (12) 15–26 (32) mm long
    2. Stem erect (or somewhat twisted toward tip, well above flowers), short (less than 0.5 m), usually ± densely pubescent; flowers white, produced only on peduncles arising within 10 (15) cm of the ground . . . . .2. **C. spithamaea**
    2. Stem twining, elongate; flowers pink or white, produced all along the usually glabrous or glabrate stem . . . . .3. **C. sepium**

1. **C. hederacea** Wall.   California-rose; Japanese Bindweed
   Map 103. A species of eastern Asia, occasionally escaping from cultivation to roadsides and railroads.
   The flowers are pink and our specimens are all the small, sterile, double-flowered cultivar 'Flore Pleno' (California-rose), although forms with fertile, single flowers (Japanese bindweed) are sometimes cultivated. This plant has at times been called *Convolvulus japonicus*, *Calystegia pubescens*, *Convolvulus pellitus*, and other names.

2. **C. spithamaea** (L.) Pursh   Fig. 34   Low Bindweed
   Map 104. Dry sandy or rocky ground, especially in aspen, oak, and/or jack pine woodland, often with bracken; does well after fire, and is often found on banks, in clearings, or on bluffs where there is some disturbance.
   This attractive but often overlooked native species does not really twine, although the erect stem may twist a little toward the tip. The flowers are (4) 4.5–6.5 (7.2) cm long, borne only on the non-twisted lower part of the stem, and the petioles are very short compared to those of the next species.

3. **C. sepium** (L.) R. Br.   Plate 3-B   Hedge Bindweed
   Map 105. Thickets and banks along streams, rivers, and shores; marshes and meadows; ditches; borders of swamps; often in weedy places and usually in little if any shade; the twining stems often forming great tangled masses on other plants.
   Authors have differed in sorting out the variation in this species. Three of the native subspecies recognized by Brummitt are attributed to Michigan, and in addition the typical European ssp. *sepium* is rarely adventive in North America. Subspecies (or varieties) seem not clearcut. The introduced ssp. *sepium* has white flowers, averaging slightly smaller than in ssp. *angulata* Brummitt, which has white flowers as large as in *C. spithamaea*; the common ssp. *americana* (Sims) Brummitt differs in having pink flowers. The other native subspecies, *erratica* Brummitt, has the bracts and

sepals apparently in a single series and the leaf blades with a nearly closed sinus; it is also pink-flowered. The species shows much variation in leaf and bract shape, and in pubescence, which is usually sparse or none but may be rather dense.

CUSCUTACEAE   Dodder Family

This family, consisting of a single genus, has often been placed in the Convolvulaceae, from which it is probably derived and from which it differs in anatomical and chemical characters as well as the completely parasitic habit. The leafless seedlings of dodder die, soon after emerging from the ground, unless they happen to touch and then attach to a suitable host plant. These rapidly growing parasites may spread from one species to another above ground, twining their white or yellowish to orange stems over their hosts, to which they become attached by numerous small haustoria ("suckers"). Once a dodder is united by its haustoria, the first portion of the stem withers and the plant is no longer connected with the ground; thereafter, the parasite obtains all its food and water from its host.

1. Cuscuta   Dodder
Our species of *Cuscuta* are all annuals, and mostly native. Only a few species of the genus (including some introduced in North America as seed contaminants) cause serious damage to crop plants, but the whole genus is tarred with the appellation of "noxious weeds."

T. G. Yuncker, long the recognized authority on the taxonomy of this difficult genus, kindly examined available Michigan specimens for this project in 1959; the key is based largely on his work and the line drawings are taken from it. More recent annotations by T. Béliz have also been helpful.

Measurements of flower length are taken from the base of the perianth to the tips of the corolla lobes. Numbers of perianth parts are not always constant, and several flowers should be examined to learn the usual number for a given specimen.

103. Calystegia hederacea

104. Calystegia spithamaea

105. Calystegia sepium

REFERENCES

Lee, W. O., & F. L. Timmons. 1958. Dodder and Its Control. U. S. Dep. Agr. Farmers' Bull. 2117. 20 pp. [Not for identification, but discusses the nature and life history of these weeds in more detail than later bulletins said to supersede this one.]

Crins, William J., & Bruce A. Ford. 1988. The Parasitic Dodders (Cuscuta: Cuscutaceae) in Ontario. Canad. Field-Nat. 102: 209–215.

Yuncker, Truman George. 1921. Revision of the North American and West Indian Species of Cuscuta. Illinois Biol. Monogr. 6: 91–231.

Yuncker, T. G. 1923. The Genus Cuscuta in Michigan. Pap. Michigan Acad. 1: 185–189.

Yuncker, Truman G. 1965. Cuscuta. N. Am. Fl. II(4). 51 pp.

KEY TO THE SPECIES

1. Stigmas slender, linear, not enlarged; styles (including stigmas) much longer than ovary and fruit; fruit circumscissile near its base; plants introduced, on legumes . . . . .1. **C. epithymum**
1. Stigmas capitate; styles mostly no longer than the ovary; fruit breaking open irregularly, not circumscissile; plants native, on a wide variety of hosts
  2. Sepals separate, subtended by bracts of similar shape and texture (but with outcurved tips); flowers sessile, in very dense, usually rope-like masses wound around the host . . . . .2. **C. glomerata**
  2. Sepals united basally, not subtended by bracts; flowers sessile or pediceled, but the inflorescence not rope-like in habit
    3. Flowers all or mostly with calyx and corolla 4-lobed; fruit usually becoming ± depressed-globose at maturity
      4. Lobes of corolla obtuse or rounded at tip, much shorter than the tube; calyx mostly shorter than corolla tube; mature styles ca. (0.7) 1–1.5 mm long, slightly shorter than the fruit; old dry corolla often persistent on summit of fruit . . . . .3. **C. cephalanthi**
      4. Lobes of corolla acute, triangular (or appearing rounded because of incurving in *C. coryli*), about as long as or longer than the tube; calyx about equalling or surpassing corolla tube at flowering time; styles and old corollas various
        5. Mature styles ca. 1 mm or slightly longer; calyx lobes acute; old dry corolla persistent on summit of fruit, or falling off; opening at summit of fruit ringed with a thickened ridge; lobes of calyx and corolla papillate [use 20× lens] . . . . .4. **C. coryli**
        5. Mature styles less than 1 mm long; calyx lobes obtuse to rounded; old dry corolla persistent around base of fruit; opening at summit of fruit not ringed with a thickened ridge; lobes of calyx and corolla smooth (not papillate) . . . . .5. **C. polygonorum**
    3. Flowers all or mostly with calyx and corolla 5-lobed; fruit various (usually globose to slightly elongate in *C. gronovii* & *C. indecora* and depressed-globose in *C. pentagona* & *C. campestris*)
      6. Lobes of corolla broadly rounded, spreading to reflexed at maturity, the tips not incurved (or occasionally so in drying) . . . . .6. **C. gronovii**
      6. Lobes of corolla triangular, acute, their tips usually ± incurved (though lobes themselves may spread)
        7. Corolla lobes papillate [use 20× lens]; calyx usually shorter than corolla tube; inflorescence rather open, the pedicels generally longer than the flowers; perianth ca. 2.5–4 mm long; old withered corolla persistent around fruit . . . . .7. **C. indecora**
        7. Corolla lobes not papillate; calyx as long as corolla tube; inflorescence more compact, the pedicels generally shorter than the flowers; perianth ca. 1–2.5 mm long; old withered corolla persistent at base of fruit

8. Calyx of most flowers appearing angled or knobbed below each sinus; perianth ca. 1.5–2 mm long .... 8. **C. pentagona**
8. Calyx smooth, not angled or knobbed; perianth ca. 2–2.5 mm long ....
.... 9. **C. campestris**

1. **C. epithymum** (L.) L.   Fig. 35   Clover Dodder
   Map 106. Introduced from Europe. Occurs principally on leguminous hosts; the few Michigan collections are all from clover and alfalfa. Besides the specimens on which the map is based, several additional collections on clover may well be this species but are too immature for identification.
   Yuncker cites an 1899 Ingham County collection of another European introduction, parasitic on flax (*Linum*), but efforts to locate any supporting specimen have been unsuccessful. *C. epilinum* Weihe also has a linear stigma and circumscissile fruit, but the style is about equal to the ovary and the calyx lobes are broadly ovate, with distinctly overlapping margins. (In *C. epithymum*, the calyx lobes are triangular, scarcely if at all overlapping.)

2. **C. glomerata** Choisy   Fig. 36
   Map 107. Generally parasitic on Compositae; the Michigan collections are both on *Helianthus* and were made by H. S. Pepoon in 1903 (Cass Co.) and 1906 (Berrien Co.).
   This is a large-flowered species (flowers ca. 4–5 [6] mm long).

3. **C. cephalanthi** Engelm.   Fig. 37
   Map 108. Parasitic on a great diversity of hosts, in Michigan including species of *Cephalanthus, Sambucus, Amphicarpaea, Spiraea, Salix, Equisetum, Boehmeria, Populus, Lycopus, Lythrum, Stachys,* and a number of Compositae (*Eupatorium, Solidago, Aster, Achillea, Cirsium*).
   The fruit and styles somewhat resemble those of the next species, but the perianth is not papillate. The epithet is sometimes carelessly spelled "*cephalanthii,*" but it is a simple Latin genitive (possessive) from the generic name of one of its hosts, not from the Latinized name of some person (Mr. Cephalanth)!

106. Cuscuta epithymum   107. Cuscuta glomerata   108. Cuscuta cephalanthi

4. **C. coryli** Engelm.

Map 109. Parasitic on numerous hosts; those documented in Michigan include species of *Mentha, Euthamia, Aster, Stachys, Ceanothus, Amphicarpaea, Solidago, Bidens, Monarda, Symphoricarpos,* and *Corylus.*

In its distinctive ± fleshy papillate perianth and thickened summit of the fruit, this species resembles *C. indecora,* from which it differs in its 4-parted (though occasionally 5-parted) smaller flowers (ca. 2 mm long before enlargement of fruit). Béliz has in fact included this in *C. indecora.* Frequently at least some of the pedicels equal or exceed the flowers (not fruit) in length, giving the inflorescence a more open aspect than in some other species.

5. **C. polygonorum** Engelm. Fig. 38

Map 110. Named for *Polygonum,* but also on *Cephalanthus* and doubtless other genera.

Distinctive among our species in its very short style, but included by Béliz in *C. pentagona.*

6. **C. gronovii** Schultes Plate 3-C; fig. 39 Common or Swamp Dodder

Map 111. Usually in low marshy or swampy ground. The most frequent hosts include *Impatiens, Salix, Cephalanthus, Decodon,* and *Eupatorium,* but documented Michigan hosts involve species in over 40 genera in at least 25 diverse families: Balsaminaceae, Bignoniaceae, Boraginaceae, Caprifoliaceae, Compositae, Cornaceae, Geraniaceae, Gramineae, Grossulariaceae, Labiatae, Lauraceae, Leguminosae, Lythraceae, Onagraceae, Polygonaceae, Rosaceae, Rubiaceae, Salicaceae, Saururaceae, Scrophulariaceae, Solanaceae, Umbelliferae, Urticaceae, Verbenaceae, and Violaceae. This American species has become naturalized as a weed in Europe.

This is our most common species, and a highly variable one. The flowers are often somewhat glandular-warty, especially toward the summit of the fruit, and this aspect in combination with the rounded corolla lobes may be helpful in identification. The flowers are (2) 2.2–4.2 mm long; the old

109. Cuscuta coryli

110. Cuscuta polygonorum

111. Cuscuta gronovii

withered corolla generally surrounds the fruit, or is sometimes persistent at the base of it. The styles are ca. 1–2.2 mm long. The calyx is shorter than the corolla tube, except in var. *latiflora* Engelm., to which Yuncker assigned two collections from St. Joseph County. One collection from Monroe County (*Denton 1290* in 1966, MICH) has recently been identified by Béliz as *C. rostrata* Engelm., a species of the mountains in the southeastern states. It is flowering, with a few young fruit, and I suspect it to be an unusual *C. gronovii*, for there is no evidence of a conspicuous beak on the fruit although the perianth appears somewhat angled as in *C. rostrata*.

7. **C. indecora** Choisy
Map 112. The only Michigan collection (see Yuncker 1923) was on *Ambrosia artemisiifolia* at Shelby in 1919 (*Wagner 3149*, NY).
See remarks under *C. coryli*. The specimen shows well all the key characters of *C. indecora* sens. str.

8. **C. pentagona** Engelm.
Map 113. Collected in Michigan only by the Haneses, from sandy fields and woods, parasitizing *Euphorbia corollata*, grasses, *Rubus*, and *Ceanothus*.

9. **C. campestris** Yuncker   Field Dodder
Map 114. Considered an important pest on field crops, such as clover, alfalfa, and sugar beets in Michigan, but also (according to Hanes) on *Impatiens, Chamaedaphne,* and *Bidens*.
Some authors have included this species in *C. pentagona*, but I follow Crins and Ford, who do not.

POLEMONIACEAE   Phlox Family

Members of this family are sometimes superficially confused with the Caryophyllaceae, but differ in having the petals united, the style solitary,

112. Cuscuta indecora   113. Cuscuta pentagona   114. Cuscuta campestris

and the ovary 3-locular. The gross aspect of the flowers in a cymose inflorescence, together with the simple, opposite, and entire leaves (in *Phlox*), does indeed suggest a "pink."

REFERENCES

Grant, Verne, & Karen A. Grant. 1965. Flower Pollination in the Phlox Family. Columbia Univ. Press, New York. 180 pp.
Smith, Dale M., & Donald A. Levin. 1967 ["1966"]. Preliminary Reports on the Flora of Wisconsin No. 57 Polemoniaceae—Phlox Family. Trans. Wisconsin Acad. 55: 243–253.
Wilson, Kenneth A. 1960. The Genera of Hydrophyllaceae and Polemoniaceae in the Southeastern United States. Jour. Arnold Arb. 41: 197–212.

KEY TO THE GENERA

1. Leaves simple, entire
   2. Leaves alternate, the uppermost (large bracts) overtopping the inflorescence; corolla ca. 3–6 mm broad; plant an annual waif of disturbed ground . . . . 1. **Collomia**
   2. Leaves (at least principal ones) opposite, the uppermost (tiny bracts) much shorter than the inflorescence; corolla (8) 11–30 (42) mm broad; plant perennial (1 species annual), native or escaped from cultivation . . . . .2. **Phlox**
1. Leaves pinnately compound or dissected
   3. Corolla blue (except in albinos); leaves compound, with lance-elliptic leaflets . . .
   . . . . .3. **Polemonium**
   3. Corolla red; leaves dissected, with filiform segments . . . . .4. **Ipomopsis**

1. Collomia

REFERENCE

Wherry, Edgar T. 1936. Miscellaneous Eastern Polemoniaceae. Bartonia 18: 52–59.

1. C. linearis Nutt.

Map 115. This species is probably not native in Michigan, although its normal habitat (west and northwest of the Great Lakes) is sandy or gravelly grasslands, shores, clearings, and other ± disturbed ground, so that the native range is not always clear. *C. linearis* is known as an adventive adjacent to Michigan in northern Ontario and in Indiana, so it can be expected in the southwestern Lower Peninsula as well as in the Upper Peninsula, where it was collected in 1933 along a railroad in Menominee (*Grassl 2561*, MICH).

The limb of the pink to bluish corolla is only 3–6 mm broad and the compact inflorescence is well overtopped by the upper leaves (or bracts), so this is not a conspicuous plant even when in bloom.

2. Phlox   Phlox

A familiar and attractive genus, all of our native species known also in cultivation.

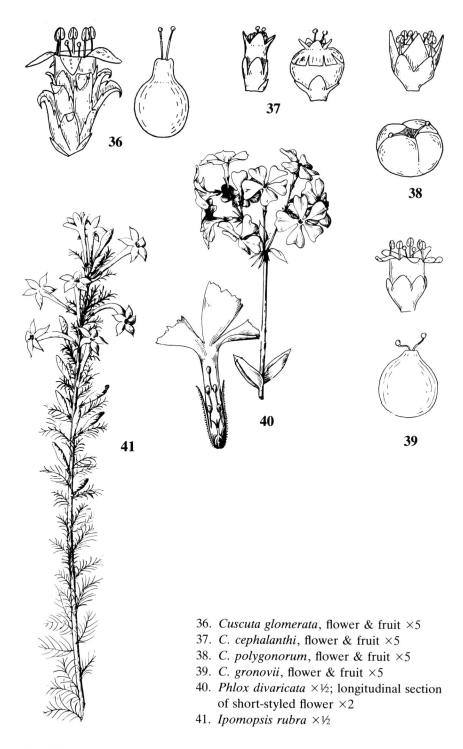

36. *Cuscuta glomerata*, flower & fruit ×5
37. *C. cephalanthi*, flower & fruit ×5
38. *C. polygonorum*, flower & fruit ×5
39. *C. gronovii*, flower & fruit ×5
40. *Phlox divaricata* ×½; longitudinal section of short-styled flower ×2
41. *Ipomopsis rubra* ×½

REFERENCES

Cooperrider, Tom S. 1986. The Genus Phlox (Polemoniaceae) in Ohio. Castanea 51: 145–148.

Levin, Donald A. 1967. Variation in Phlox divaricata. Evolution 21: 92–108.

Levin, Donald A., & Morris Levy. 1971. Secondary Intergradation and Genome Incompatibility in Phlox pilosa (Polemoniaceae). Brittonia 23: 246–265.

Wherry, Edgar T. 1955. The Genus Phlox. Morris Arb. Monogr. 3. 174 pp. [Use of this and other writings of the same author is hindered by his declining to follow the International Code of Botanical Nomenclature regarding designation of infraspecific taxa, types, and synonyms.]

KEY TO THE SPECIES

1. Calyx (except rarely in *P. paniculata*) glabrous and stem (at least below the uppermost internodes) glabrous or nearly so
    2. Leaves with evident anastomosing lateral veins and with hispidulous or ciliate margins; stem green throughout; plant an escape from cultivation . . . . . .1. **P. paniculata**
    2. Leaves with lateral veins obscure or not at all visible and with smooth glabrous margins; stem purple-spotted; plant a rare native of damp places . . . . . . .2. **P. maculata**
1. Calyx and stem puberulent to pilose
    3. Plants ± densely matted or tufted, with branches from nearly woody depressed or trailing stems; leaves rather needle-like (especially in *P. subulata*)
        4. Corolla lobes notched ca. (3) 3.5–7 mm; largest leaves ca. 2.5–4.5 cm long, glabrous or ciliate only at the base . . . . .3. **P. bifida**
        4. Corolla lobes notched less than 2 (2.5) mm; largest leaves (0.8) 1.1–1.6 (2.2) cm long, often ciliate at least to the middle
            5. Style very short (less than 4 mm); stamens all included in the corolla tube; limb of corolla ca. 2–2.4 mm broad . . . . .4. **P. nivalis**
            5. Style longer (sometimes long-exserted); stamens at least partly exserted at maturity; limb of corolla ca. 1.4–1.9 (2.1) cm broad . . . . .5. **P. subulata**
    3. Plants not densely tufted, with erect flowering stems (only *P. divaricata* with depressed subligneous stems and sterile shoots); leaves with flat blades, without subulate tips; styles less than half as long as the calyx
        6. Plant a slender-rooted annual, rarely escaped from cultivation; corolla normally rose-red; leaves on upper part of stem alternate . . . . .6. **P. drummondii**
        6. Plant perennial, native; corolla normally blue to pink or violet; leaves all opposite
            7. Corolla tube pubescent; calyx lobes with awn-like tip at least 0.5–1.5 mm long; larger leaves (5.6) 6–10 (18) times as long as broad; leafy shoots none or very inconspicuous; fresh corollas pink to purple . . . . .7. **P. pilosa**
            7. Corolla entirely glabrous; calyx lobes at most with a subulate tip 0.5 mm long; larger leaves 2.3–5.6 times as long as broad; leafy shoots conspicuous arising from depressed stem; fresh corollas normally blue (if not white) . . . . .8. **P. divaricata**

1. **P. paniculata** L.   Perennial or Garden Phlox

Map 116. Native south of our area, but frequently spreading from cultivation to roadsides, fencerows, and even woods; often near sites of present or former human activity.

Flower color varies from purple to pink or white. The leaves are usually broader than in any of our other naturalized or native species, and certainly are more conspicuously veiny, the anastomosing of lateral veins quite visible. Very rarely the calyx and stem may be pubescent, but such plants can be easily placed by the distinctive leaves and the large dense broad inflorescence.

2. **P. maculata** L. Wild-Sweet-William

Map 117. Ranges mostly to the south and barely enters Michigan, where it is very rare and local in fens and other wet places.

The stems are often sparsely puberulent even below the middle. The flowers are usually purple in an elongate ± cylindrical inflorescence.

Two other species with glabrous calyx and stem and obscure lateral veins (at best) occur adjacent to Michigan in northwest Ohio or northern Indiana and might be discovered as far north as our southernmost counties. These are *P. latifolia* Michaux (see Cooperrider 1986) and *P. glaberrima* L. (see Swink & Wilhelm 1994). *P. latifolia* [long known as *P. ovata* L.] has 3–5 (usually 4) pairs of cauline leaves below the inflorescence and the calyx is ca. 7–10 mm long, while *P. glaberrima* has 8 or more pairs as well as somewhat narrower leaves and the calyx is ca. 4.5–7 mm long. Both differ from *P. maculata* in having an unspotted stem and a broad corymbose inflorescence.

3. **P. bifida** L. C. Beck Sand Phlox

Map 118. Native in dry sandy or rocky ground south from Michigan. Perhaps of questionable status here: collected on railroad ballast near Schoolcraft in 1936 and later years and on the shore of Klinger Lake, St. Joseph County, in 1890.

4. **P. nivalis** Sweet Trailing Phlox

Map 119. Native in the Southeast and Texas. Undoubtedly escaped where found well established in a sandy field near a cemetery (*Swink et al. 8296* in 1989, MOR).

5. **P. subulata** L. Moss-pink; Moss Phlox

Map 120. Native southeast of Michigan, apparently into eastern Ohio, and usually said to range into southern Michigan. However, most if not all of our records surely represent escapes from gardens and cemeteries, where it is frequently planted and whence it often escapes to roadsides, clearings, shores, and banks.

This is much bushier and more compact in appearance than *P. bifida*, with more nodes and densely crowded subulate leaves, especially on sterile shoots. The hairs of the calyx and inflorescence may be gland-tipped or glandless.

6. **P. drummondii** Hooker   Annual Phlox
Map 121. A native of Texas but commonly cultivated. Collected as a waif in a vacant lot in 1984 (*Henson 1725*, MICH).

7. **P. pilosa** L.   Prairie Phlox
Map 122. Dry sandy woods with oak and hickory, even in barren and disturbed sites; also prairies, fens, and borders of swamps.

Our plants are presumably all typical ssp. or var. *pilosa*. Immediately to the west of Michigan occurs var. *fulgida* Wherry [ssp. *fulgida* (Wherry) Wherry], distinguished by longer and eglandular hairs, especially in the inflorescence, and hence with a more lustrous appearance. Detailed studies by Levin and coworkers (e.g., Levin & Levy 1971) indicate a very narrow zone of intergradation roughly following the Indiana/Illinois border.

The type is represented by a white-flowered plant from the east coast of the United States (i.e., from the range of typical var. *pilosa*). Hence, the typical form, though rare, would be f. *pilosa* (white-flowered) and the normal red-purple (but variable) form would require a name (which has apparently never been published). [The name *P. pilosa* f. *albiflora* Mac-Millan applies to the white-flowered form of var. *fulgida*.]

8. **P. divaricata** L.   Fig. 40   Wild Blue Phlox
Map 123. Rich deciduous woods, usually beech-maple-hemlock but sometimes oak-hickory, especially in low areas such as ravines and floodplains. The northernmost record in the state is from St. Martin's Island, Delta County.

Our plants have been referred to typical ssp. or var. *divaricata*. As with the preceding species, however, there is a change immediately to the west, where there is var. *laphamii* A. W. Wood [ssp. *laphamii* (A. W. Wood) Wherry], originally described from Wisconsin. The zone of intergradation is again more or less along the Indiana/Illinois border, but it is a much wider zone than in *P. pilosa* (Levin 1967). In the western variant, the corolla lobes are essentially entire and there are subtle differences in shapes of corolla and leaves; in var. *divaricata*, there is normally a distinct

115. Collomia linearis

116. Phlox paniculata

117. Phlox maculata

notch or sinus in each corolla lobe (though in some individuals the lobes are entire).

In Michigan, both this species and *P. pilosa* have gland-tipped hairs on the calyx. The flowers are usually blue in *P. divaricata*—quite different from those of *P. pilosa* — but white-flowered plants occur: f. *albiflora* Farwell (TL: Ypsilanti [Washtenaw Co.]); plants with reddish-purple flowers were named f. *purpurea* Farwell (TL: Redford [Wayne Co.] and Ypsilanti [Washtenaw Co.]).

3. Polemonium

REFERENCE

Davidson, John F. 1950. The Genus Polemonium [Tournefort] L. Univ. California Publ. Bot. 23: 209–282.

KEY TO THE SPECIES

1. Leaflets ovate to lance-elliptic (often less than twice as long as broad, but sometimes narrower), 17 or fewer; fruit short-stipitate; plant native in southern Michigan (sometimes cultivated) ....1. **P. reptans**
1. Leaflets lance-elliptic (often 2–4 times as long as broad), 19 or more on lower leaves; fruit sessile; plant a rare escape from cultivation ....2. **P. caeruleum**

118. Phlox bifida

119. Phlox nivalis

120. Phlox subulata

121. Phlox drummondii

122. Phlox pilosa

123. Phlox divaricata

1. **P. reptans** L.   Jacob's-ladder
   Map 124. Barely reaches into southern Michigan from the south, but goes farther north in Wisconsin; not native in Ontario. Undoubtedly native on river bluffs and in wet prairies, fens, and tamarack swamps in the southwestern Lower Peninsula. Washtenaw County collections seem likely to be escapes from cultivation.

2. **P. caeruleum** L.   Greek-valerian
   Map 125. A variable Eurasian species, often cultivated. Farwell (*836*, BLH) collected the white-flowered form in 1894 as "Sparingly escaped around gardens" (according to his notes).

4. **Ipomopsis**

This genus has often been included in *Gilia*, but is now generally recognized as distinct on the basis of chromosome number as well as morphological characters. Despite the generic name, which suggests similarity to *Ipomoea*, these plants do not resemble morning-glories!

REFERENCES

Grant, Verne. 1956. A Synopsis of Ipomopsis. Aliso 3: 351–362.
Wherry, Edgar T. 1936. Miscellaneous Eastern Polemoniaceae. Bartonia 18: 52–59.

1. **I. rubra** (L.) Wherry   Fig. 41   Standing-cypress
   Map 126. Native across the southern United States from Texas eastward; adventive northward usually as an escape from cultivation. Collected sporadically in Michigan since the 1880s along roadsides and railroads, in fields, on sandy banks, and once on the Lake Huron shore.
   This is an erect unbranched perennial with bright red salverform corollas, pollinated by hummingbirds. The leaves resemble those of *Myriophyllum*, though of course they are stiffer as well as never truly whorled nor aquatic.

124. Polemonium reptans

125. Polemonium caeruleum

126. Ipomopsis rubra

HYDROPHYLLACEAE   Waterleaf Family

REFERENCES

Shields, Jack W. 1967 ["1966"]. Preliminary Reports on the Flora of Wisconsin No. 58 Hydrophyllaceae—Waterleaf Family. Trans. Wisconsin Acad. 55: 255–259.
Wilson, Kenneth A. 1960. The Genera of Hydrophyllaceae and Polemoniaceae in the Southeastern United States. Jour. Arnold Arb. 41: 197–212.

KEY TO THE GENERA

1. Flowers on solitary pedicels opposite alternate leaves; cauline leaves at lowest nodes opposite . . . . .1. **Ellisia**
1. Flowers in terminal inflorescences; cauline leaves all alternate
  2. Basal and lower cauline leaves long-petioled (petioles over 5 cm long); leaf blades shallowly palmately lobed or deeply pinnately lobed with lateral segments over 12 cm broad . . . . .2. **Hydrophyllum**
  2. Basal and lower cauline leaves sessile or on petioles less than 2.5 [5] cm long; leaf blades pinnatifid, with lateral segments (excluding teeth) less than 0.8 cm broad
   . . . . .3. **Phacelia**

1. Ellisia

REFERENCE

Constance, Lincoln. 1940. The Genus Ellisia. Rhodora 42: 33–39.

1. **E. nyctelea** (L.) L.   Fig. 45
Map 127. A much-branched weak annual native in moist woods both east and west of Michigan, but here apparently adventive in disturbed ground, as around parking areas and gardens.

2. Hydrophyllum   Waterleaf
An occasional specimen may appear ambiguous in leaf shape, but can be placed by checking characters of calyx, pubescence, and underground parts. The earliest (basal) leaves of *H. appendiculatum* and *H. canadense*

127. Ellisia nyctelea   448. Hydrophyllum   129. Hydrophyllum
                                    virginianum   appendiculatum

tend to be pinnately divided although the later (cauline) leaves mostly have a basic palmate pattern; in *H. virginianum* all the leaves are pinnate. Pinnate leaves are usually quite mottled with pale green or grayish, giving the leaf a water-splotched appearance.

REFERENCES

Beckmann, Robert L., Jr. 1979. Biosystematics of the Genus Hydrophyllum L. (Hydrophyllaceae). Am. Jour. Bot. 66: 1053–1061.
Constance, Lincoln. 1942. The Genus Hydrophyllum. Am. Midl. Nat. 27: 710–731.

KEY TO THE SPECIES

1. Cauline leaves all deeply pinnate, with 4–6 lateral lobes; calyx with no appendages or teeth in the sinuses . . . . 1. **H. virginianum**
1. Cauline leaves all or mostly palmately lobed (some at times broadly pinnately 4-lobed); calyx with a tiny appendage or tooth in each sinus (or this often lacking in *H. canadense*)
  2. Sinus of calyx with a distinct reflexed appendage 0.5–1.5 mm long (enlarging in fruit); peduncles and pedicels densely pubescent with very short fine hairs plus stiff hairs 5–10 times as long . . . . 2. **H. appendiculatum**
  2. Sinus of calyx with no appendage or with an erect tooth not over 0.5 mm long; peduncles and pedicels glabrous or with sparse hairs not distinctly of two contrasting lengths . . . . 3. **H. canadense**

1.   **H. virginianum** L.   Virginia Waterleaf
    Map 128. Rich deciduous woods, thickets, and even marshy places; often in the same woods as the next two species but generally blooming before either of them.
    The pubescence of the inflorescence and stem of this species tends to be much more appressed than it is in the next two. The flowers are either white or purplish, in part a function of age.

2.   **H. appendiculatum** Michaux   Fig. 42   Great Waterleaf
    Map 129. Rich deciduous woods (beech, maple, etc.), especially in damp areas and ravines, blooming in late spring.
    The distinct reflexed appendages between the calyx lobes in this species, along with the pubescence, which is consistently spreading and of two very different lengths, make recognition relatively easy. This species is a biennial, with slender taproot, and lacks the stout, ± knotty, elongate rhizomes of the other two, perennial species. The flowers are some shade of blue, purple, or even pink, rarely white.

3.   **H. canadense** L.   Fig. 43   Broadleaved or Canada Waterleaf
    Map 130. Rich deciduous woods, often growing with the preceding species.
    All recorded observations (including specimen labels) indicate that the flowers are white on this species in Michigan (although said to be some-

times pink-purple elsewhere). If there are hairs on the pedicels and peduncles, they are relatively stiff as well as sparse. The stem is usually also glabrate.

3. **Phacelia**   Phacelia; Scorpion-weed

This is a large genus of over 150 species in North America, most of which are western, although a few are centered in the southeastern United States.

REFERENCES

Constance, Lincoln. 1949. A Revision of Phacelia subgenus Cosmanthus (Hydrophyllaceae). Contr. Gray Herb. 168. 48 pp. [Includes *P. purshii.*]

Gillett, George W. 1960. A Systematic Treatment of the Phacelia franklinii Group. Rhodora 62: 205–222.

KEY TO THE SPECIES

1. Stamens long-exserted (2–6 mm) beyond corolla, glabrous ....1. **P. tanacetifolia**
1. Stamens scarcely if at all longer than the corolla, the filaments hairy
  2. Corolla nearly rotate (flat), the lobes fringed; range southern ....2. **P. purshii**
  2. Corolla campanulate, with distinct tube, the lobes entire; range northern (Isle Royale) ....3. **P. franklinii**

1. **P. tanacetifolia** Bentham   Fiddleneck

Map 131. This species, native from central California to northern Mexico, has occasionally been found as a waif beyond that range, presumably as an escape from cultivation, a contaminant in seed, or a hitchhiker on vehicles. The only Michigan collection was made in 1904 at the then recently established Upper Peninsula Experiment Station at Chatham (*Geisman*, MSC; det. Constance); the railroad may have been associated with its introduction, or it may even have been cultivated. (It was collected in 1929 from a commercial garden in Charlevoix County, presumably cultivated.)

The leaves tend to be more finely pinnatifid than in the next two species, with the primary segments bearing at least small lobes or teeth.

130. Hydrophyllum canadense   131. Phacelia tanacetifolia   132. Phacelia purshii

2. **P. purshii** Buckley   Miami-mist

Map 132. Native mostly south of Michigan, perhaps only adventive
northward. Collected in a "moist woods" at Lansing in 1917 (*Yuncker 178*,
US). Frequent as close to Michigan as the Lake Erie islands of Ontario and
Ohio and adjacent Sandusky area of Ohio.

3. **P. franklinii** (R. Br.) A. Gray

Map 133. Ranges mostly northwest of the Great Lakes; found here only
along the north shore of Lake Superior, including Isle Royale, where it is
apparently not common in open rocky and gravelly places.

BORAGINACEAE   Borage Family

Members of this family in our area are generally rough-pubescent herbs
with radially symmetrical flowers, deeply 4-lobed ovary, and alternate
leaves. Any exception to these features is noted in the keys or text. The
inflorescence is technically cymose, but usually develops in a one-sided way
that results in the axis being coiled at the tip; as this straightens when fruit
develops, the inflorescence can resemble a spike or raceme.

Features of the nutlets (one normally developing from each lobe of the
ovary if all ovules have been fertilized) are traditionally important in classification, but in our limited flora features easier to see can usually be emphasized for identification. The species with bristly nutlets, forming readily
animal-dispersed burs, understandably bear these well exposed to fur and
hence easily visible, whereas the species with smooth or merely rugose
nutlets tend to keep them concealed in the calyx, making it difficult for one
to see them when attempting an identification. Hence, the keys here avoid
nutlet characters, especially those of the attachment to the base of the
style. Many species have long flowering periods, so fortunately specimens
with flowers at one end of the inflorescence and at least immature fruit at
the other end are commonly collected. If one has a specimen with only
mature fruit and no flowers, a key stressing nutlets is useful and can be
found in Johnston (1924), Kruschke (1946), and other works.

Brunnera macrophylla (Adams) I. M. Johnston, Siberian bugloss, may
seed in where it is cultivated. It has very small blue flowers but differs from
Myosotis in its rugose rather than smooth nutlets; and it has large basal
leaves with cordate to reniform blades and long petioles.

REFERENCES

Al-Shehbaz, Ihsan A. 1991. The Genera of Boraginaceae in the Southeastern United States.
    Jour. Arnold Arb. Suppl. 1: 1–169.
Ingram, John. 1961. Studies in the Cultivated Boraginaceae 4. A Key to the Genera. Baileya
    9: 1–12. [Includes all of our native and introduced genera except *Onosmodium* and,
    surprisingly, *Symphytum*.]

Johnston, I. M. 1924. Studies in the Boraginaceae,—II. 1. A Synopsis of the American Native and Immigrant Borages of the Subfamily Boraginoideae. Contr. Gray Herb. 70: 3–55.

Kruschke, Emil P. 1946. Preliminary Reports on the Flora of Wisconsin. XXXI [sic, for XXXII]. Boraginaceae. Trans. Wisconsin Acad. 36: 273–290.

KEY TO THE GENERA

1. Corolla bilaterally symmetrical (lobes unequal), with some or all filaments much longer than the lobes . . . . .1. **Echium**
1. Corolla radially symmetrical, with all filaments (and usually anthers) shorter than corolla lobes
   2. Corolla rotate, ca. 16–25 mm broad, the inconspicuous tube saucer-shaped; anthers conspicuous, ca. 5–7 mm long, on very short, appendaged filaments . . . .2. **Borago**
   2. Corolla funnel-shaped to campanulate or salverform (or less than 16 mm broad), the limb ca. 1–22 mm broad; anthers shorter (less than 4 mm), the filaments various
      3. Style conspicuously exserted, longer than calyx and at least the tube of the corolla; anthers ca. 1.5–3.5 mm long; corolla (9) 10–22 (26) mm long, ± cylindrical, even if expanded above the tube the lobes barely if at all spreading; nutlets never bristly; corolla blue (or pink) (dull white–greenish in *Onosmodium*)
         4. Corolla dull white or greenish, the lobes acute and longer than broad; inflorescence with a bract at the base of each pedicel . . . . .3. **Onosmodium**
         4. Corolla blue (or pink), the lobes obtuse or rounded and broader than long; inflorescence with bracts none or only at the base
            5. Corolla glabrous; calyx lobes glabrous or with mostly appressed hairs . . . .
               . . . . .4. **Mertensia**
            5. Corolla puberulent outside; calyx lobes hispid with spreading bristles. . . . .
               . . . . .5. **Symphytum**
      3. Style ± inconspicuous, usually shorter than the calyx and not exserted beyond tube of corolla (exceeding calyx and sometimes corolla tube in long-styled *Lithospermum*, with yellow salverform corollas, and in *Anchusa*); anthers 0.3–2 mm long; corolla funnelform or salverform, the lobes (especially if corolla over 10 mm long) usually spreading; nutlets in some species bristly; corolla variously colored (blue or not)
         6. Stigma distinctly 2-lobed; corolla yellow to orange-yellow or whitish, never blue or reddish; flowers sessile or on very short pedicels (even in fruit the pedicels shorter than the calyx), each subtended by a bract or leaf; nutlets smooth and shiny (or with tiny pits or in *L. arvense* rugose-pitted) . . . . .6. **Lithospermum**
         6. Stigma unlobed (2-lobed in the blue-flowered *Anchusa*); corolla blue to pink or maroon or white (not yellow, except at most for a yellow "eye" or for young corollas of *Myosotis discolor*); flowers usually on pedicels exceeding the calyx when mature, or nutlets conspicuously rough or bristly, or pedicels mostly (or all) bractless
            7. Pedicels each with a bract (though this often supra-axillary)
               8. Corolla 8–16 mm broad; nutlets only ridged or wrinkled, each set into a collar-like ring at the base . . . . .7. **Anchusa**
               8. Corolla less than 5 mm broad; nutlets bearing bristles (or slender prickles) minutely barbed at the tip, without collar-like ring at the base . . . .
                  . . . .[go to couplet 12]
            7. Pedicels all or mostly without bracts
               9. Nutlets smooth or rugose-tuberculate, without elongate bristles, nearly or quite concealed in the calyx; corolla blue or white, often with yellow "eye" at the center (rarely all yellow)

10. Leaves linear, opposite (at least at middle and lower nodes); corolla white (except for yellow eye), ca. 9–10 mm broad; nutlets rugose-tuberculate; plant a rare waif . . . . .8. **Plagiobothrys**

10. Leaves ± elliptic, all alternate; corolla usually blue (except for eye) or sometimes white (rarely pink or yellow), 1–9.5 mm broad; nutlets smooth and shiny; plants (at least some species) widespread . . . . .9. **Myosotis**

9. Nutlets bearing prominent bristles with barbed tips, early surpassing the calyx as they ripen; corolla mostly blue or dull red, sometimes white, without yellow (or other) "eye"

11. Corolla red or upper leaves ± clasping; nutlets oriented horizontally, with bristles over all faces including the conspicuously exposed ventral face . . . .
. . . . .10. **Cynoglossum**

11. Corolla blue (or white) *and* upper leaves not cordate-clasping; nutlets ± erect, with bristles only along the margins of the exposed (dorsal) face (and in one species, on that face)

12. Pedicels promptly becoming deflexed after flowering; leaf blades elliptical, with evident lateral veins; bracts toward base of inflorescence much exceeding the flowers and fruits . . . . .11. **Hackelia**

12. Pedicels ascending even in fruit; leaf blades linear to lanceolate, without evident lateral veins; bracts toward base of inflorescence usually about equalling the flowers and fruits, or shorter . . . . .12. **Lappula**

1. Echium

1. **E. vulgare** L.   Fig. 44   Viper's Bugloss; Blueweed

Map 134. A Eurasian native, now locally an established weed, perhaps once escaped from cultivation. Dry, disturbed, usually sandy or gravelly places: roadsides, railroads, fields, vacant lots. First collected in Michigan in 1897 at Grand Rapids and soon afterwards found in St. Clair County. By 1920 it was scattered throughout the state (even Ontonagon County in 1926); and by the 1940s, a serious weed at least in the Lower Peninsula. Still quite local in the Upper Peninsula.

The bristly hairs, often pustulate-based, on stem, calyx, and usually leaves make some people believe this plant is a kind of thistle. Pink- and white-flowered forms are known although only blue and white ones have been collected in Michigan. The long style is distinctly bifid (0.5–1.5 mm at apex), the corolla is ± pubescent, and the pedicels are all subtended by bracts.

2. Borago

1. **B. officinalis** L.   Borage

Map 135. A native of southern Europe, grown in gardens for its showy flowers or as a potherb, rarely escaped to cultivated ground and waste places.

Like *Echium*, this is a bristly-hairy but showy plant. The flowers of *Borago* are unusually rotate for this family, while in *Echium* they are unusual in being bilateral.

42. *Hydrophyllum appendi-*
   culatum, flower ×2
43. *H. canadense* ×½;
   flower ×2
44. *Echium vulgare* ×½

3. Onosmodium

REFERENCES

Cochrane, Theodore S. 1976. Taxonomic Status of the Onosmodium molle Complex (Boraginaceae) in Wisconsin. Michigan Bot. 15: 103–110.
Turner, Billie L. 1995. Synopsis of the Genus Onosmodium (Boraginaceae). Phytologia 78: 39–60.

1. **O. molle** Michaux   Marbleweed
   Map 136. Collected at Niles in 1838 by the First Survey (GH); not found in Michigan again, but to be sought in dry places or on shores.
   The foliage resembles that of a coarsely pubescent *Lithospermum*, but the long-styled flowers with straight acute corolla lobes are distinctive. Our plant is referred to var. *hispidissimum* (Mack.) Cronquist. Turner has recently associated this variety with the Texan *O. bejariense* A. DC. rather than with *O. molle*.

4. **Mertensia**   Lungwort
   The flowers are pink in bud but become blue as they mature, as in some other Boraginaceae, presumably as a result of changes in acidity (pH).

KEY TO THE SPECIES

1. Plant glabrous throughout, glaucous; corolla (14) 16–22 (26) mm long; calyx lobes obtuse or nearly so; range in southern Lower Peninsula (except as escaped from cultivation) . . . . .1. **M. virginica**
1. Plant with stiff hairs on calyx, pedicels, leaves, and stems; corolla 10–14 mm long; calyx lobes acute; range in northwestern Upper Peninsula . . . . .2. **M. paniculata**

1. **M. virginica** (L.) Link   Virginia Cowslip; "Bluebells"
   Map 137. Rich woods and floodplain forests. The plants from Ingham, Oakland, and Schoolcraft counties are undoubtedly escapes from cultivation; the species can spread readily in woodland gardens.

133. Phacelia franklinii

134. Echium vulgare

135. Borago officinalis

2. **M. paniculata** (Aiton) G. Don   Fig. 48   Tall Lungwort
    Map 138. Coniferous swamps and seepy woods, shaded edges of streams, rocky openings, and rarely in deciduous woods. From the western Upper Peninsula, north shore of Lake Superior, and James Bay, this species ranges west to Oregon and Alaska.

5. **Symphytum**   Comfrey
    The corolla changes from pink to blue as in *Mertensia*.

REFERENCES

Gadella, T. W. J. 1985 ["1984"]. Notes on Symphytum (Boraginaceae) in North America. Ann. Missouri Bot. Gard. 71: 1061–1067.
Ingram, John. 1961. Studies in the Cultivated Boraginaceae 5. Symphytum. Baileya 9: 92–99.

KEY TO THE SPECIES

1. Upper leaves decurrent, with elongate narrow wings extending down the stem; calyx at anthesis ca. (5.5) 6–9 mm long; pubescence spreading but not thorn-like; nutlets smooth and shiny between the ridges on the angles; connective between the anther sacs extending slightly beyond them ....1. **S. officinalis**
1. Upper leaves not or scarcely (less than 1 cm) decurrent; calyx at anthesis ca. 3–4 mm long (soon elongating); pubescence of stem and inflorescence partly of strong recurved prickles or "thornlets" (± flattened and broad at base, like miniature rose thorns); nutlets papillose and coarsely reticulate between the angles; connective shorter than the anther sacs (the anthers appearing bifid at the apex) .....2. **S. asperum**

1. **S. officinale** L.   Common Comfrey
    Map 139. A native of Eurasia, cultivated and sometimes escaping to roadsides, ditches, fields, dooryards, and waste ground.
    A form with yellowish corollas is known but most of our escapes seem to be the usual blue-violet. A white-flowered form was collected in 1992 (*Dritz 1051*, MOR) with the normal one along a ditch west of Coloma in Berrien County.

2. **S. asperum** Lepechin   Prickly Comfrey
    Map 140. A native of the Caucasus region, also cultivated and rarely escaped to roadsides and dooryards.
    Intermediate plants with ± prickly stems (though perhaps the prickles fewer, straighter, and/or less flattened), with calyx intermediate (ca. 4.5–7 mm long at anthesis), and sometimes with ± decurrent leaves are apparently the hybrid *S. ×uplandicum* Nyman, which is more common than pure *S. asperum*. Gadella identified cytotypes with 2n = 36 and 2n = 40 in Michigan, the latter having less harsh pubescence. Specimens referable to the hybrid are known from Cheboygan, Gratiot, Ingham, Iron, and Washtenaw counties.

6. Lithospermum Puccoon; Gromwell

REFERENCES

Baker, H. G. 1961. Heterostyly and Homostyly in Lithospermum canescens (Boraginaceae). Rhodora 63: 229–235.

Cusick, Allison W. 1985. Lithospermum (Boraginaceae) in Ohio, with a New Taxonomic Rank for Lithospermum croceum Fernald. Michigan Bot. 24: 63–69.

Johnston, Ivan M. 1952. Studies in the Boraginaceae, XXIII a Survey of the Genus Lithospermum. Jour. Arnold Arb. 33: 299–366.

Weller, Stephen G. 1985. The Life History of Lithospermum caroliniense, a Long-lived Herbaceous Sand Dune Species. Ecol. Monogr. 55: 49–67.

KEY TO THE SPECIES

1. Plants with a terminal inflorescence of showy light to deep yellow corollas much exceeding the calyx and with limb 9–22 (25) mm broad (inflorescence elongate later in season)
    2. Corolla with tube 22–35 mm long and lobes distinctly erose or fringed; leaves narrowly linear, mostly less than 4 mm wide (a few as wide as 6.5 mm); pubescence on stem entirely of closely appressed hairs ....1. **L. incisum**
    2. Corolla with tube (unexpanded portion) 6–14 mm long and lobes entire (though often crisped); leaves mostly broader; pubescence, at least on stem, all or partly spreading
        3. Calyx lobes 2.2–4.5 (5) mm long (to 6.5 mm in fruit); pubescence of leaves, stems, and bracts soft, usually not pustulate-based ....2. **L. canescens**
        3. Calyx lobes (5) 6–10 (13) mm long; pubescence largely of very stiff slightly pustulate-based hairs ....3. **L. caroliniense**
1. Plants with flowers mostly axillary, the corolla barely if at all (at most 1.5–2 (3) mm) exceeding the calyx, white to pale yellow or greenish, the limb less than 8 mm broad
    4. Leaves without visible lateral veins; nutlets gray-brown, rough-rugose; corolla lobes (not tube) glabrous outside for most or all of their length; plant annual ....
    ....4. **L. arvense**
    4. Leaves with strong to weak (but visible) lateral veins; nutlets smooth, shiny, and white, at most with a few tiny pits; corolla lobes pubescent outside nearly or quite to their tips; plant perennial
        5. Largest leaves or bracts 2–4 cm broad; cauline leaves usually 15–20 or even fewer below the inflorescence; pubescence of leaves and stems all or mostly closely appressed ....5. **L. latifolium**
        5. Largest leaves or bracts (0.5) 0.7–2 cm broad; cauline leaves at least 25–30 below the inflorescence; pubescence, especially on undersides of leaves, mostly somewhat ascending (raised, not appressed) ....6. **L. officinale**

1. L. incisum Lehm. Fringed Puccoon

Map 141. A species of plains, dunes, and dry disturbed ground, mostly to the west of us although rare in southern Ontario. Collected in dry sandy ground on and near Lake Huron in St. Clair County by Dodge 1899–1915, but not recorded in Michigan since.

In *L. incisum*, unlike the next two species, cleistogamous flowers are produced in the leaf axils later in the season, producing abundant, usually finely pitted, shiny white nutlets. The showy flowers are not heterostylous, the stamens in all flowers being inserted near the summit of the corolla tube.

2. **L. canescens** (Michaux) Lehm. Fig. 46 Hoary Puccoon
   Map 142. Sandy prairie remnants; openings in oak and jack pine woodland; edges of woods, roads, and railroads.

The calyx lobes in our material are distinctly shorter than reported in Ohio by Cusick. The texture of the pubescence is sufficiently different from that of the next species to enable an experienced person to separate the two by merely "petting" the foliage. *L. caroliniense* is decidedly rough. The flower color tends to be deeper (more orange) in *L. canescens*. Both this species and the next (as well as some others) have a pigment in their sap, especially in the sturdy taproot, which yields a strong purple-red dye.

3. **L. caroliniense** (J. F. Gmelin) MacMillan Hairy or Yellow Puccoon
   Map 143. Characteristic of sand dunes and shores; also in oak and pine (especially jack pine) woodland, often with *Juniperus*; sandy barrens and ridges, prairie remnants.

Both this species and the preceding are heterostylous, some plants bearing flowers with long styles and short stamens and other plants, short styles with long stamens (i.e., inserted near summit of corolla tube).

136. Onosmodium molle

137. Mertensia virginica

138. Mertensia paniculata

139. Symphytum officinale

140. Symphytum asperum

141. Lithospermum incisum

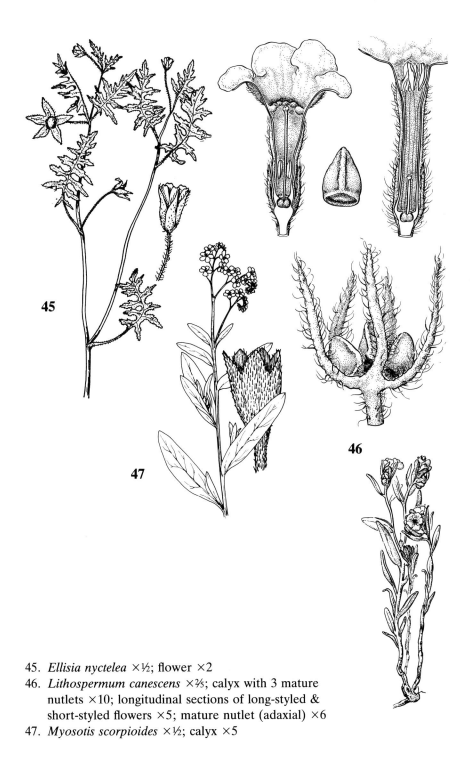

45. *Ellisia nyctelea* ×½; flower ×2
46. *Lithospermum canescens* ×⅖; calyx with 3 mature
    nutlets ×10; longitudinal sections of long-styled &
    short-styled flowers ×5; mature nutlet (adaxial) ×6
47. *Myosotis scorpioides* ×½; calyx ×5

In the narrowest sense this is a species of Coastal Plain areas from Virginia to Florida, Texas, and Mexico, and north to Arkansas. *L. croceum* Fernald (TL: east of Manistique [Schoolcraft Co.]) was described in 1935 for the similar plants ranging from the Great Lakes region to Montana and Kansas. However, despite the separate ranges, most authors have not considered the two species to be separable. Johnston claimed that the differences are simply those which distinguish short- and long-styled flowers of this heterostylous species. Cusick, on the other hand, made *L. croceum* a subspecies of *L. caroliniense*.

4. **L. arvense** L. Corn Gromwell
    Map 144. Introduced from Eurasia and collected in Michigan as early as 1844 (Macomb Co.). Usually in dry, sandy, disturbed ground, especially fields (cultivated or weedy), roadsides, and railroads.
    Now often segregated into a separate genus, as *Buglossoides arvensis* (L.) I. M. Johnston.

5. **L. latifolium** Michaux American Gromwell
    Map 145. Shaded river banks and wooded floodplains; borders of woods; rarely collected in the past half-century.
    This native species has softer, finer, more closely appressed pubescence than either the preceding or the next. The similarly wide-leaved *Hackelia virginiana*, which might be confused when sterile, also differs in having some strongly spreading pubescence.

6. **L. officinale** L. Gromwell
    Map 146. Another Eurasian introduction, an uncommon weed except in calcareous rocks and gravels around the northern ends of Lakes Michigan and Huron. Edges and openings in woods and thickets, often in rocky and gravelly ground including railroads, roadsides, and shores. Collected in Washtenaw County by the First Survey in 1838; next gathered in 1861 on Mackinac Island by Thoreau and Mann—and still especially frequent in the Straits of Mackinac area.

142. Lithospermum
     canescens

143. Lithospermum
     caroliniense

144. Lithospermum
     arvense

The nutlets of this species and the preceding are generally finely but sparsely pitted.

7. **Anchusa** Bugloss; Alkanet
Both of our species are natives of Eurasia, rarely escaped from cultivation.

KEY TO THE SPECIES

1. Calyx cleft about half to two-thirds to the base, the lobes ca. 2–5 mm long and narrowly lanceolate to triangular . . . . 1. **A. officinalis**
1. Calyx cleft nearly to the base, the lobes at least 7 mm long and narrowly linear-attenuate . . . . .2. **A. azurea**

1. **A. officinalis** L.
   Map 147. Roadsides, vacant lots, and other disturbed ground.

2. **A. azurea** Miller
   Map 148. Roadsides, fields, and waste places.
   Quite similar to the preceding, but with distinctively slender, long-attenuate calyx lobes. The bract subtending the flower is similarly shaped, while in *A. officinalis* it is more ovate-lanceolate.

145. Lithospermum latifolium

146. Lithospermum officinale

147. Anchusa officinalis

148. Anchusa azurea

149. Plagiobothrys hirtus

150. Myosotis laxa

8. **Plagiobothrys**

1. **P. hirtus** (Greene) I. M. Johnston   Popcorn-flower
    Map 149. An annual, native to the Pacific Northwest, collected in a
moist field with *Eleocharis obtusa* near Schoolcraft in 1938 by the Haneses
(*118*, GH, WMU; Pap. Michigan Acad. 25: 42. 1940). It was probably
introduced by the railroad running beside the field. Collected again in 1939
at a nearby site. Also known in the eastern United States from fields in
North Carolina.
    Johnston identified the Michigan material as var. *figuratus* (Piper) I. M.
Johnston, although the corolla is a little larger than usually described for
that taxon. The showy white flowers in a densely curled cyme tip make the
"popcorn" name an apt one.

9. **Myosotis**   Forget-me-not
    This is our only genus of Boraginaceae with small blue flowers and
smooth shiny nutlets (although the latter are ± hidden in the calyx). While
the genus is a familiar and easily recognized one, the species are often
difficult. The characters used in most keys do not always hold up, and the
key below may prove unsatisfactory as well. In cases of doubt, as always,
try *both* choices and hope that one of them will lead to an unreasonable
result while the other will make sense. There are pubescence differences
that are difficult to describe but help one with experience or adequate
material to compare for texture, angle, and other aspects of the hairs.
    All of our species except two are Old World introductions, the showier
ones as escapes from cultivation and the tiny-flowered ones as presumably
uninvited weeds.

KEY TO THE SPECIES

1. Hairs on calyx all straight at the tip, ± appressed
  2. Style distinctly shorter than nutlets and calyx; calyx lobes (at least at anthesis)
      about as long as the tube or slightly longer; corolla limb less than 5 mm broad . . .
      . . . . .1. **M. laxa**
  2. Style exceeding the nutlets and usually the calyx; calyx lobes shorter than tube;
      corolla limb 5–9.5 (10) mm broad . . . . .2. **M. scorpioides**
1. Hairs at least toward base of calyx hooked at the tip, ± spreading
  3. Pedicels (at anthesis, or certainly as fruit matures) at least as long as the calyx;
      corolla ca. 2.5–9 mm wide; calyx lobes ± spreading around the fruit; plant usually
      biennial, when well developed with several stems shortly decumbent at their bases
    4. Limb of corolla ± cupped, ca. 2.5–3.5 mm broad . . . . .3. **M. arvensis**
    4. Limb of corolla flat, 5–10 mm broad . . . . .4. **M. sylvatica**
  3. Pedicels shorter than calyx (even in fruit); corolla ca. 1–2 mm wide; calyx lobes
      erect (± parallel beyond the fruit); plant annual, with ± straight slender stems
    5. Calyx appearing 2-lipped at least when mature, the lobes unequal (2 lower lobes
        longer than 3 upper); corolla white; pedicels with mostly slightly spreading (not
        closely appressed) ± curved hairs lacking hooked tips; nutlets 1.1–1.2 mm
        broad . . . . .5. **M. verna**

5. Calyx lobes equal or subequal; corolla blue or yellowish turning blue; pedicels with some hooked hairs or with all hairs closely appressed and straight; nutlets ca. 0.7–0.8 mm broad

   6. Stems bearing flowers nearly to the base, the lowest ones subtended by bracts (or leaves); pedicels and stems mostly with some spreading hook-tipped hairs; corolla blue . . . . .6. **M. stricta**

   6. Stems bearing flowers only on upper part, above any leaves or bracts; pedicels with only very closely appressed straight hairs; stems with only straight-tipped hairs; corolla yellow when young, turning blue . . . . .7. **M. discolor**

1. **M. laxa** Lehm.

Map 150. Wet, seepy shores and banks; cedar swamps.

2. **M. scorpioides** L.   Fig. 47

Map 151. A common garden species from Eurasia, thoroughly naturalized and often abundant in ditches, along streams and rivers, on shores, in swamp forests (deciduous and coniferous) and wet hollows in woods, and in all sorts of wet muck and mud.

3. **M. arvensis** (L.) Hill

Map 152. Another Eurasian species, not as showy as the preceding or the next, but locally common along wooded trails and roadsides, clearings, shores, lawns, old gardens; more often in weedy places than undisturbed woods.

Herbarium specimens of *M. sylvatica* with shriveled or immature corollas are easily misidentified as *M. arvensis*. One should look for corollas displaying a well pressed limb to be sure of having this small-flowered species. I have tried in vain to substantiate correlated differences alleged by some authors to exist in calyx lobes, abundance of hooked hairs on the calyx, attitude of the calyx after flowering, and nutlet size. *M. sylvatica* may have slightly deeper sinuses between the calyx lobes, more open fruiting calyx, calyx with fewer hooked bristles, and smaller nutlets, but these are at best average features that help little with ambiguous individuals. White-flowered plants may be locally common with blue-flowered ones.

4. **M. sylvatica** Hoffm.   Garden Forget-me-not

Map 153. This is the commonest cultivated species, another Eurasian native escaped extensively in dry or damp ground, especially along trails and roads through woods (even spreading into the woods); neglected lawns, shores, dumps, and other ± disturbed sites in sandy, rocky, or rich soils.

A variable species, this one is frequently white-flowered and occasionally pink-flowered. Rarely (especially with white- or pink-flowered plants) the corolla may be a little smaller than usual (± 4 mm broad).

5. **M. verna** Nutt.

Map 154. One of our two native species. Dry sandy or rocky banks (both grassy and wooded), meadows and pastures, oak woodland.

Long known incorrectly as *M. virginica* (L.) BSP. *M. macrosperma* Engelm., sometimes treated as a variety of *M. verna*, is known (rare) from the Lake Erie region of Ontario and Ohio, although primarily a species of farther south. It differs in larger nutlets and large leaf-like bracts at the base of the inflorescence and may one day be discovered in Michigan.

6. **M. stricta** Roemer & Schultes

Map 155. Native of Eurasia. Sandy or gravelly, usually disturbed places including roadsides, fields, parking lots, lawns and grassy banks, pine-oak woodland.

Often known as *M. micrantha* Lehm. Leaves of this species have some hooked hairs on the midrib beneath, while these are lacking in the preceding species and the next.

7. **M. discolor** Pers.

Map 156. An inconspicuous little weed from Europe and adjacent territory. The only verified Michigan collection is from partly shaded ground on a bluff above the Paw Paw River (*Swink & Wetstein 8418* in 1989, MOR).

Often known as *M. versicolor*.

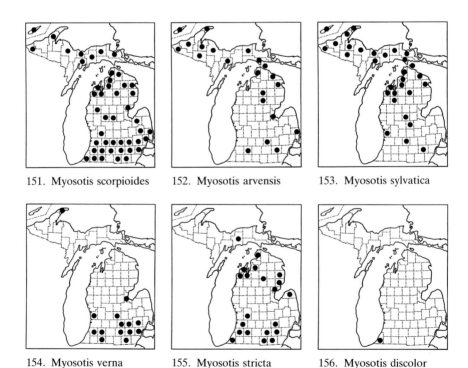

151. Myosotis scorpioides   152. Myosotis arvensis   153. Myosotis sylvatica

154. Myosotis verna   155. Myosotis stricta   156. Myosotis discolor

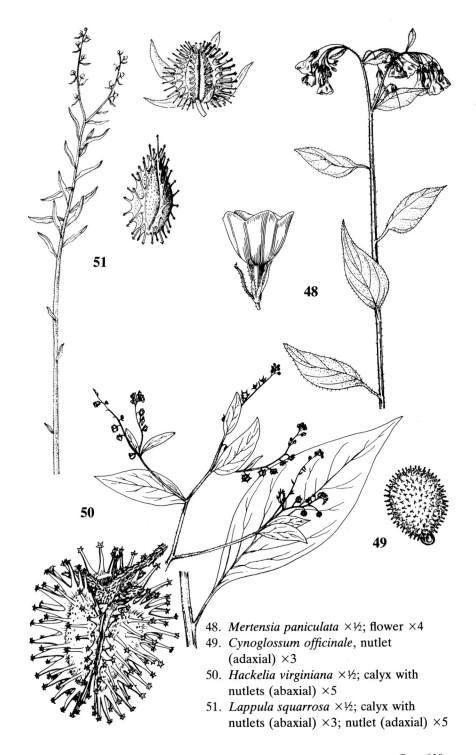

51

48

50

49

48. *Mertensia paniculata* ×½; flower ×4
49. *Cynoglossum officinale*, nutlet
    (adaxial) ×3
50. *Hackelia virginiana* ×½; calyx with
    nutlets (abaxial) ×5
51. *Lappula squarrosa* ×½; calyx with
    nutlets (abaxial) ×3; nutlet (adaxial) ×5

10. Cynoglossum

KEY TO THE SPECIES

1. Corolla dull red or maroon (to purplish); calyx 4–6.5 mm long; stem leafy its entire
   length, with branches of the inflorescence arising from axils of leafy bracts; style
   conspicuously surpassing mature nutlets; nutlets flat on exposed surface, 5.5–8 (9)
   mm long (including prickles), with raised rim on border . . . . .1. **C. officinale**
1. Corolla blue; calyx 1.8–2.7 mm long (to 3 mm in fruit); stem leafy only on lower
   portion, with inflorescence(s) entirely above uppermost leaves and bracts; style
   shorter than mature nutlets, inconspicuous; nutlets convex on exposed surface,
   4–5.5 mm long, without rim . . . . .2. **C. boreale**

1. **C. officinale** L.   Fig. 49   Hound's-tongue
   Map 157. A weedy Eurasian species, not often cultivated, well established in waste ground such as roadsides, fields, dumps, farmyards, and
gravel pits; clearings and disturbed woods; often on shores.
   The corolla is usually ± reddish, becoming more purple in age (or upon
drying). White-flowered plants were collected at the E. S. George Reserve
in Livingston County in 1932. This species is not only much more leafy than
the next but also has more branches to the inflorescence. The upper leaves
are often merely sessile. The pubescence is quite soft, compared with the
very rough *Anchusa officinalis*, which furthermore has bracted pedicels but
is sometimes confused with hound's-tongue.

2. **C. boreale** Fernald   Plate 3-D   Northern Wild Comfrey
   Map 158. Borders, openings, and clearings or under dense shade in
coniferous or mixed woods (fir, cedar, spruce, pine, birch, aspen, occasionally beech and maple), especially in sandy or rocky soil.
   The upper leaves are strongly cordate-clasping, although the lower and
basal ones are petiolate. While the species is easily told at a distance from *C.
officinale*, it is much closer to the southern wild comfrey, *C. virginianum* L.,
from which Fernald distinguished it in 1905 as a new species. Sometimes it is
considered only a variety of *C. virginianum* (Cooperrider, Michigan Bot. 23:
166. 1984), which has slightly larger nutlets, calyx, and flowers.
   In his travels through the Great Lakes region in the spring and early
summer of 1861, Henry David Thoreau noted at least five times (in Minnesota, Michigan, Ontario, and New York) an unknown prickly-fruited
boraginaceous plant with small blue flowers. Although no specimens are
extant, I strongly suspect that Thoreau's unknown borage was *C. boreale*,
not formally described until over 40 years later (see Voss 1978, pp. 81–82).

11. Hackelia   Stickseed; Beggar's-lice
   This genus is sometimes merged into *Lappula*. The pedicels in our species of both genera tend to be supra-axillary and the exposed faces of the
nutlets are rough-papillose besides whatever bristles are present. The tiny
inconspicuous flowers and bur-like fruits do not encourage collectors and
both species are doubtless less scarce than the maps suggest.

REFERENCE

Gentry, Johnnie L., Jr., & Robert L. Carr. 1976. A Revision of the Genus Hackelia (Boraginaceae) in North America, North of Mexico. Mem. New York Bot. Gard. 26(1): 121–227.

KEY TO THE SPECIES

1. Exposed face of nutlets with barb-tipped bristles only around the margins (at most a few tiny ones in the middle); widest cauline leaves (0.8) 1–2.5 (5) cm broad; pubescence of uppermost internodes and branches of the inflorescence closely appressed; corolla blue . . . . .1. **H. deflexa**
1. Exposed face of nutlets covered with several barb-tipped bristles nearly or quite as large as the marginal ones; widest cauline leaves (2.5) 3–5 (9.5) cm broad; pubescence of uppermost internodes and branches of the inflorescence partly spreading-ascending; corolla white . . . . .2. **H. virginiana**

1. **H. deflexa** (Wahlenb.) Opiz

Map 159. Deciduous woods, especially along borders, trails, and clearings; on shores and bluffs; dune thickets and rocky sites.

Eurasian plants differ slightly from our native American ones, which are var. *americana* (A. Gray) Fernald & I. M. Johnston, also known as *H. americana* (A. Gray) Fernald.

Plants with leaves as wide (very rarely) as 5 cm are robust ones (a meter or more tall), while plants of the next species with leaves as narrow as 2.5 cm are small, delicate ones. The difference in flower color is consistent on Michigan material, although as in any blue-flowered species white-flowered individuals of *H. deflexa* might occur. Published descriptions (perhaps copied from one another) declare that *H. virginiana* may have blue flowers, but I have noted only the slightest hint of blue on a few dried flowers that had been white when fresh.

2. **H. virginiana** (L.) I. M. Johnston   Fig. 50

Map 160. Woods and thickets: oak-hickory, beech-maple, floodplain; like the preceding species, favors some disturbance, as along trails and other weedy openings.

157. Cynoglossum officinale

158. Cynoglossum boreale

159. Hackelia deflexa

12. **Lappula**

Apparently we have a single species in the state (unless one includes *Hackelia*; see above). A western species, *L. redowskii* (Hornem.) Greene, has frequently been reported to range into Michigan as a waif, but all specimens so labeled are better referred to *L. squarrosa* (with even immature fruit showing 2 rows of marginal prickles). Each nutlet of *L. redowskii* has only a single row of marginal prickles and hence appears distinctly less prickly than in our species, which also tends to have larger nutlets and stems with more appressed pubescence. Material bearing only flowers would be risky for certain identification if more than one species were suspected in the area.

REFERENCE

Frick, Brenda. 1984. The Biology of Canadian Weeds. 62. Lappula squarrosa (Retz.) Dumort. Canad. Jour. Pl. Sci. 64: 375–386.

1. **L. squarrosa** (Retz.) Dumort.   Fig. 51   Stickseed
   Map 161. Said by Hultén (1971) to be "originally a steppe plant of southern Russia and adjacent Siberia, which has spread as a weed across most of Europe and north America. It is still spreading rapidly. . .". Evidently spread early to Michigan, for the first collections in the state were made in 1838 (Jackson Co.) and 1840 (Macomb Co.). When railroads were thriving, their ballast beds were a characteristic habitat, but the species also occurs in other usually dry (sometimes moist) waste places such as roadsides (especially newly made ones), farmyards, fields, sandy or gravelly shores.

This is an easily overlooked but readily recognized plant, with stiffly erect habit (stems simple or much branched), tiny blue flowers, and numerous narrow leaves. The margins of each nutlet have 2 rows of barb-tipped bristles.

Use of the name *L. echinata* Gilib. for this plant has been extraordinarily persistent, despite several published notes, starting as early as 1945, pointing out that Gilibert's binomials were not validly published. Some authors consequently called this species *L. myosotis* Moench, but the epithet *squarrosa* is older (the still older Linnaean epithet *lappula* may not be used in this genus as it would create a tautonym).

VERBENACEAE   Vervain Family

The Verbenaceae are mostly tropical and subtropical. In a broad circumscription, the family includes many woody plants, such as *Avicennia* spp. (black and gray mangroves), *Callicarpa* spp. (beautyberry), *Lantana* spp., *Tectona grandis* L. f. (teak), and *Vitex* spp. (chaste-tree). All of our species are herbaceous. Following recent classifications (but not Moldenke), the family Phrymaceae (sometimes misspelled Phrymataceae) is included here.

REFERENCES

Moldenke, Harold N. 1980. A Sixth Summary of the Verbenaceae, Avicenniaceae, Stilbaceae, Chloanthaceae, Symphoremaceae, Nyctanthaceae, and Eriocaulaceae of the World as to Valid Taxa, Geographic Distribution and Synonymy. Phytologia Mem. 2. 629 pp. [Cites the author's copious previous work on these families.]

Tans, William E., & Hugh H. Iltis. 1980 ["1979"]. Preliminary Reports on the Flora of Wisconsin No. 67. Verbenaceae—the Vervain Family. Trans. Wisconsin Acad. 67: 78–94.

KEY TO THE GENERA

1. Flowers and fruit distinctly paired (on opposite sides of spike, with lower internodes at least as long as calyx tube); upper calyx teeth bristle-like, lower teeth ± deltoid; fruit a single elongate achene, in strongly reflexed calyx .1. **Phryma**
1. Flowers and fruits alternate (± crowded in spikes, the lower internodes usually [except in fruiting *V. urticifolia*] shorter than calyx tubes); upper calyx teeth similar to the lower, all deltoid to lanceolate but not bristle-like; fruit of 2 or 4 nutlets, in ascending calyx
    2. Inflorescences all axillary, subglobose to short-cylindric at anthesis, much shorter than peduncles (which usually equal or exceed the leaves); corolla 4-lobed; nutlets 2 . . . . .2. **Phyla**
    2. Inflorescences all or mostly terminal, elongate at anthesis, exceeding the peduncles; corolla 5-lobed; nutlets 4 . . . . .3. **Verbena**

1. Phryma

Often recognized in a separate, monotypic family, the only species occurring in eastern Asia as well as eastern North America. I have often wondered about the student who was keying this plant as an "unknown" and reported the family to be Phrymaceae but declared that he couldn't get it any further!

REFERENCE

Thieret, John W. 1972. The Phrymaceae in the Southeastern United States. Jour. Arnold Arb. 53: 226–233.

1. **P. leptostachya** L.   Figs. 52, 53   Lopseed
Map 162. Usually in rich deciduous woods, especially low moist areas in

160. Hackelia virginiana   161. Lappula squarrosa   162. Phryma leptostachya

beech-maple forest, but also in drier woods with oak and sometimes with conifers.

This species is easily recognized by the distant paired flowers, reflexed fruit, and broad, opposite, petioled leaves. The corolla appears pinkish, being basically white inside with pink tinge at least outside. Hooked tips on the bristle-like mature upper sepal lobes suggest animal dispersal. The reflexed fruit, unlike that of *Polygonum virginianum*, is not under tension.

2. Phyla
Often included in the large genus *Lippia*.

1. **P. lanceolata** (Michaux) Greene   Fig. 54   Fog-fruit
   Map 163. River banks, shores, mud flats, marsh borders; often on seasonally flooded sites.

3. Verbena   Vervain
All of our species except *V. canadensis* are known to hybridize with one or more other species, but with us only *V. hastata* × *V. urticifolia* is at all frequent.

The corollas are normally some shade of blue to purple or white, with rose-colored forms occurring rarely. The fruit consists of four nutlets or mericarps, but the style is terminal unlike the basal style of the Boraginaceae and Labiatae.

REFERENCES

Barber, Susan C. 1982. Taxonomic Studies in the Verbena stricta Complex (Verbenaceae). Syst. Bot. 7: 433–456. [Includes all of our species except *V. canadensis*.]
Umber, Ray E. 1979. The Genus Glandularia (Verbenaceae) in North America. Syst. Bot. 4: 72–102.

KEY TO THE SPECIES

1. Spikes (even excluding corollas) at least (8) 12 mm thick; plant ± spreading-decumbent, with incised-pinnatifid leaves
   2. Bracts shorter than calyx; corolla showy, with long exserted tube and limb ca. 1 cm or more broad; style greatly exceeding calyx, conspicuous after shedding of corolla; calyx hairs (or some of them) gland-tipped ....1. **V. canadensis**
   2. Bracts much longer than calyx; corolla inconspicuous, shorter than the bracts, with limb less than 0.5 cm broad; style shorter than calyx, very inconspicuous; calyx hairs without gland-tips ....2. **V. bracteata**
1. Spikes (at least excluding corolla) less than 8 mm thick; plant stiffly erect, with leaves unlobed or lobed only at the base (pinnate only in *V. officinalis*)
   3. Leaves linear-lanceolate, 5.5–12 times as long as broad, the blade less than 1.5 cm wide and tapering into ± indistinct petiole; plant glabrous or with scattered, mostly appressed stiff hairs ....3. **V. simplex**
   3. Leaves ovate-lanceolate or broader, 1.3–6 (6.3) times as long as broad, (1.3) 2–6.5 cm wide, sessile to subsessile or mostly with distinct petiole; plant with hairs on stems and leaves numerous and/or ± spreading

4. Leaves (except uppermost) deeply pinnately lobed or incised; flowers at maturity remote in spike; stem glabrous or sparsely scabrid . . . . .4. **V. officinalis**
4. Leaves at most (*V. hastata*) with 1–2 lobes at base of the blade; flowers remote or crowded; stems glabrous or pubescent
  5. Plant ± densely hoary-pubescent; leaves sessile (or the lower subsessile on winged peioles), 1.3–3 times as long as broad; limb of corolla ca. 7–11 mm broad; calyx over 3.5 mm long . . . . .5. **V. stricta**
  5. Plant with more sparse pubescence, not hoary; leaves distinctly petioled, 2–6.3 times as long (including petiole) as broad; limb of corolla ca. 2.5–4 mm broad; calyx less than 3.5 mm long
    6. Spikes with flowers (except in bud) and fruit distinctly separated on the axis; corolla white, the limb ca. 2.5 mm or less broad; style ca. 0.4–0.6 mm long (or less), scarcely if at all exceeding calyx; calyx (1.5) 1.8–2.3 (2.5) mm long; leaves all unlobed . . . . .6. **V. urticifolia**
    6. Spikes with flowers and fruit (except sometimes lowest ones) densely overlapping; corolla blue to purple (rarely rose), the limb ca. 2.8–4 mm broad; style 1–1.5 mm long, noticeably exceeding calyx after shedding of corolla; calyx 2.3–3 mm long; leaves often with 1–2 lobes at base of blade . . . . .7. **V. hastata**

1. **V. canadensis** (L.) Britton   Rose Verbena; Creeping Vervain
    Map 164. Apparently native (despite its name) to the south and, especially, southwest of Michigan. Cultivated in gardens and presumably escaped where found in this region in sandy ± disturbed ground.
    Sometimes placed in a separate genus, as *Glandularia canadensis* (L.) Nutt.

2. **V. bracteata** Lag. & Rodr.   Fig. 55   Prostrate or
                                                       Creeping Vervain
    Map 165. Essentially a weed in our area, perhaps in part escaped from cultivation; probably native only to the south and west of us. Apparently first established in Michigan on the west side of the state (collected in Manistee Co. in 1882, in several others including Delta Co. in the 1890s), now widespread in sandy or gravelly soil along roadsides and railroads, in parking lots and fields, and other relatively raw recently disturbed ground.
    A hybrid with *V. stricta*, known as *V.* ×*deamii* Moldenke, was collected in Ottawa County in 1896 (but the specimen was not seen by Moldenke

163. Phyla lanceolata   164. Verbena canadensis   165. Verbena bracteata

until 1982). It has the foliage and habit of *V. bracteata* but much more slender and elongate spikes, with small bracts, as in *V. stricta*.

3. **V. simplex** Lehm.

Map 166. Plants on limestone pavements (alvars) at Drummond Island and on thin soil over limestone in Lenawee and Monroe counties are presumably native; elsewhere in the state, collections from sandy fields, waste places, and railroads may at least in part be adventive.

This species has spikes almost as slender and remotely flowered as in *V. urticifolia*, which, however, has broad leaf blades and distinct petioles.

4. **V. officinalis** L.   European Vervain

Map 167. A native of Europe, collected in 1893 by Farwell (*1004a*, BLH) and said by him to be "frequently seen in waste places . . . at Detroit" (Asa Gray Bull. 4: 63. 1896). Apparently not collected since in Michigan.

5. **V. stricta** Vent.   Fig. 56   Hoary Vervain

Map 168. Although Michigan is within the native range of this species as usually described, its status here is doubtful. The only collections I have seen from before 1900 are from waste ground at Grand Rapids in 1896. There are very few collections until the 1940s, after which there has been apparently rapid spread in sandy fields and roadsides, along railroads, and in barren disturbed ground generally. Only a very few collections are said to have come from wet ground near marshes and ponds.

The relatively large flowers and hoary foliage make this one of our most attractive species.

6. **V. urticifolia** L.   White Vervain

Map 169. Deciduous woods of various kinds, especially along borders and trails; swamp forests and thickets; roadsides and fencerows.

Hybrids with the next species have been named *V. ×engelmannii* Moldenke and are intermediate in size and color of corolla, length of

166. Verbena simplex   167. Verbena officinalis   168. Verbena stricta

52. *Phryma leptostachya* ×½
53. *P. leptostachya*, fruit ×5
54. *Phyla lanceolata* ×½
55. *Verbena bracteata* ×½
56. *V. stricta* ×½

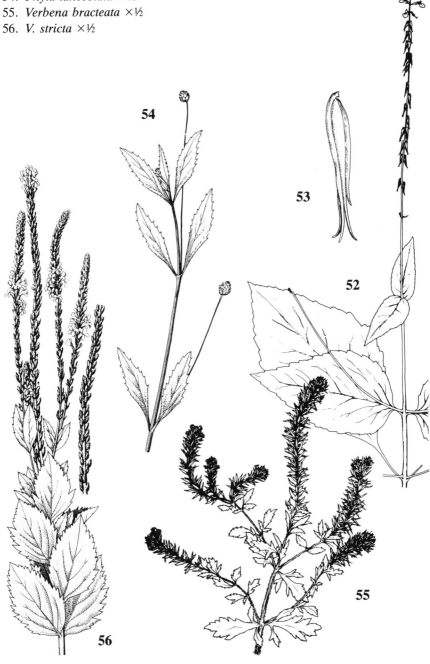

style, and denseness of spikes. They have been identified from Barry, Calhoun, Cass, Ingham, Kalamazoo, Monroe, Oakland, St. Clair, and Wayne counties.

7. **V. hastata** L.   Blue Vervain
   Map 170. Marshes, ditches, wet shores and stream banks; thickets and openings in swamps.
   Usually a tall, conspicuous plant with somewhat candelabrum-like inflorescence, but may be as short as 15 cm with a single spike. Plants of this species were once used to make a beverage with reputed medicinal value.
   White-flowered plants have been named f. *albiflora* Moldenke, but have not yet been reported from Michigan. Pink- or rose-flowered plants are known from the state and were twice named f. *rosea*: by C. I. Cheney in 1902 and by O. A. Farwell in 1923 (TL: Trenton [Wayne Co.]); however both names are illegitimate because of var. *rosea* N. Coleman (1874, TL: presumably near Grand Rapids [Kent Co.]). There appears to be no legitimate name at rank of form for the plants with striking rose corollas. An apparent hybrid with *V. stricta* [*V. ×rydbergii* Moldenke] was collected in 1959 at the George Reserve near Pinckney, Livingston County. It has the aspect of *V. hastata*, but slightly larger calyces and corollas and hoary pubescence on the undersides of the leaves.

LABIATAE (LAMIACEAE)   Mint Family

This is a rather large, well known, and easily recognized family. Most (not all) of our species have a 4-angled ("square") stem (but square stems occur in several other families as well), and all have simple opposite leaves. The ovary is ± strongly 4-lobed, producing (if all ovules are fertilized) 4 nutlets. Most species contain aromatic oils, associated with resinous dots on the leaves and other organs. In most species the flowers are in small to

169. Verbena urticifolia

170. Verbena hastata

171. Pycnanthemum
     tenuifolium

dense axillary inflorescences, but these are crowded toward the summit of the stem in some species, forming what appears to be a terminal head or spike (usually not a true head or spike). In some cases, the little cymules are somewhat corymbose.

The corolla is typically bilaterally symmetrical and 2-lipped, but the upper lip may be much reduced. In some genera the flowers are essentially radially symmetrical (regular). The calyx ranges from 2-lipped to regular and is often easier to use in keys as it is more persistent and easily seen than other flower parts. A dense zone of hairs forming a beard in the throat of the calyx (at summit of tube) is usually conspicuous when present, visible between the bases of the calyx lobes. The keys here do not stress the relative length and position of the stamens or the shape of the corolla lips, as these characters are rarely well displayed on pressed specimens and most collectors are careless about recording such helpful information on their labels. The fertile stamens are 4 in most genera, but 2 in a few (sometimes with staminodia).

A few cautions may help to interpret the keys, which as always are artificial: Before concluding that flowers have only 2 fertile stamens, examine enough fresh ones to be sure that anthers have not simply fallen off. Before measuring flower (or corolla) length, or extent of exsertion of corolla beyond the calyx teeth, be sure that the corolla has not separated from the receptacle with its base clinging higher in the calyx tube.

Several genera in which certain characters used in the keys may be ambiguous or easily misinterpreted are entered at more than one place. Enough characters have generally been included in the keys so that after trying both choices in a puzzling couplet, the user should find that only one of them leads to a reasonable conclusion.

Many mints are used for flavoring, as herbs and spices, or for their reputed medicinal properties. None are documented as seriously toxic to humans, but several, especially *Glechoma hederacea*, are poisonous to livestock and other animals. Some mints have variants with a more citrus-like aroma than the usual "minty" sort of fragrance.

REFERENCES

Gill, L. S. 1981. Taxonomy, Distribution and Ecology of the Canadian Labiatae. Feddes Repert. 92: 33–93.
Koeppen, Robert C. 1958. Preliminary Reports on the Flora of Wisconsin. No. 41. Labiatae—Mint Family. Trans. Wisconsin Acad. 46: 115–140.
Waterman, Ann H. 1960. The Mints (Family Labiatae) of Michigan. Publ. Mus. Michigan St. Univ. 1: 269–302.

KEY TO THE GENERA

1. Corolla (but not necessarily calyx) radially symmetrical or nearly so (slightly oblique), with 4–5 lobes of about equal length (if not always width)
  2. Leaves entire (or with an occasional irregular tooth)

3. Flowers white (usually dotted with color), in small involucrate heads borne in a corymb-like inflorescence; ovary deeply 4-lobed, with style inserted all the way to the base; stems and pedicels glabrous or with non-glandular pubescence . . . . 1. **Pycnanthemum**

3. Flowers blue, mostly pediceled in axils of leaves or bracts (or on 1–3-flowered peduncles); ovary shallowly lobed, with terminal style; stems and pedicels with glandular pubescence . . . . 2. **Trichostema**

2. Leaves with blades regularly toothed or lobed for at least half their length

4. Stamens 2; flowers all sessile in axillary clusters; bruised foliage not (or very slightly) aromatic; nutlets with corky rims or teeth at summit; corolla white . . . . 3. **Lycopus**

4. Stamens 4; flowers short-pediceled, whether axillary or on terminal spikes; bruised foliage strongly aromatic; nutlets smooth or (*Perilla*) reticulate; corolla white to purple

5. Flowers clearly in terminal and axillary racemes, the solitary pedicels quite visible though shorter than the calyx and subtending bract; plants annual; leaf blades coarsely toothed, ca. 1.5 times as long as broad, or shorter, and also less than 1.5 times as long as petiole . . . . 4. **Perilla**

5. Flowers densely whorled in terminal spike-like and/or axillary clusters, appearing not at all racemose; plants perennial; leaf blades various

6. Corolla over 5 mm long, 5-lobed; flowers all in terminal spikes at least 1–5 cm thick; plants without rhizomes . . . . 5. **Agastache**

6. Corolla 5 (5.7) mm long or less, 4-lobed; flowers in axillary clusters or if in terminal spikes, these less than 1.5 cm thick; plants rhizomatous . . . . 6. **Mentha**

1. Corolla clearly bilaterally symmetrical, 1- or 2-lipped

7. Upper lip of corolla much reduced or apparently absent (deeply split, each half connate with lower lip); style not fully gynobasic (though ovary 4-lobed)

8. Corolla deep blue (rarely pink or white); bracts of inflorescence ovate or obovate; lower lip of corolla 3-lobed, the middle lobe sometimes notched but scarcely (less than twice) as large as lateral lobes . . . . 7. **Ajuga**

8. Corolla pink-magenta (rarely white); bracts of inflorescence narrowly linear-lanceolate; lower lip of corolla 5-lobed, the middle lobe much larger than the 2 tiny lobes on each side of the lip . . . . 8. **Teucrium**

7. Upper lip of corolla well developed and evident; style gynobasic (inserted deep into the base of the 4 lobes)

9. Calyx of essentially 2 obtuse lips, the upper with a ± erect, flattened to conical, transverse protuberance; corolla blue . . . . 9. **Scutellaria**

9. Calyx various but without such a protuberance; corolla various

10. Flowers in dense heads or whorls subtended by 4 or more leafy bracts of similar and conspicuous size . . . . 10. **Monarda**

10. Flowers in heads or not, but subtending bracts inconspicuous or only 2

11. Most if not all flowers in distinct *terminal* inflorescences (heads, spikes, or racemes) at the ends of stems and leafy branches, the uppermost whorls of flowers exceeding any bracts

12. Inflorescence ± corymbose: flat-topped or dome-like and *branched*, with flowers or few-flowered heads on evident stalks

13. Calyx densely bearded in the throat; leaf blades less than 2.8 (3) cm long, ± truncate or rounded to a distinct petiole; calyx lobes and bracts in the heads ± deep red-purple . . . . 11. **Origanum**

13. Calyx naked in the throat; leaf blades over 2.8 cm long, or linear, or nearly or quite sessile (or all of these); calyx lobes and bracts green to white-pubescent (at most pink-tipped) . . . . 1. **Pycnanthemum**

12. Inflorescence a raceme (or panicle) or elongate spike, or flowers sessile to short-pediceled in a ± dense head

14. Inflorescence racemose (paniculate), each mature flower on a pedicel greatly exceeding its minute bract and distinctly longer than the calyx; lower lip of corolla yellow, fringed; fertile stamens 2, long-exserted . . . . .12. **Collinsonia**

14. Inflorescence spicate or of crowded heads, the pedicels none or shorter than both bracts and calyx; lower lip rarely yellow, rarely fringed (never both); fertile stamens 4 (except in *Blephilia* & *Salvia*)

15. Fertile (anther-bearing) stamens 2 [see also notes in text on these species]

16. Flowers sessile or nearly so, numerous in dense whorls at each node of inflorescence, at least the uppermost 3–4 whorls contiguous (forming a many-flowered—sometimes tiered—head); plants native in moist woods, thickets, and drier openings. . . . . .13. **Blephilia**

16. Flowers on readily visible pedicels ca. 2–4 (6) mm long, only 2–12 developed at each node of the inflorescence, not crowded into heads (the internodes of inflorescence apparent); plants in our area all of waste ground and fields . . . . .14. **Salvia**

15. Fertile (anther-bearing) stamens 4

17. Leaves entire or nearly so (teeth, if any, obscure and broadly rounded)

18. Calyx densely and conspicuously bearded in the throat, strongly 2-lipped (upper lip with 3 short teeth, lower lip deeply bifid); leaves less than 1.5 cm long . . . . .15. **Thymus**

18. Calyx scarcely or not at all bearded in the throat and (except in *Prunella*) with 5 elongate (over 12.5 mm) teeth or lobes (i.e., regular or only weakly 2-lipped); leaves all or almost all over 1.5 cm long

19. Leaf blades ± linear, over 4 times as long as broad

20. Stem and calyx glabrous or nearly so; mature stamens not exceeding upper lip of corolla; lower lip of corolla about equalling the upper lip . . . . .20. **Stachys (hyssopifolia)**

20. Stem and calyx finely pubescent on all sides; mature stamens well exserted beyond corolla; lower lip of corolla distinctly longer than the upper lip . . . . .16. **Hyssopus**

19. Leaf blades less than 4 (very rarely 4.5) times as long as broad

21. Floral bracts broadly ovate, overlapping, obtuse or rounded (except for short-acuminate or mucronate tip); upper lip of calyx with 3 very short (less than 1 mm) teeth or excurrent bristles . . . . .17. **Prunella**

21. Floral bracts linear-setaceous; upper lip of calyx with 3 well developed narrow lobes similar to those of lower lip . . . . .18. **Clinopodium**

17. Leaves clearly and regularly toothed

22. Flowers only 2 per node of the inflorescence (a simple or branched elongate slender raceme or spike); calyx at flowering time with obscure nerves (at fruiting, reticulate-veiny and inflated), puberulent; leaves and internodes glabrous. . . . . .19. **Physostegia**

22. Flowers 3 or more at all or most nodes of the inflorescence; calyx strongly longitudinally nerved, variously glabrous or pubescent; leaves (at least beneath) and usually stem pubescent

23. Calyx regular, with essentially equal lobes, the tube 5–15-nerved

24. Calyx lobes with terete, non-green, spiny tip about half or more the total length of the lobe. . . . . .26. **Galeopsis**

24. Calyx lobes with non-green tip, if any, much less than half the length of the lobe

25. Nerves of calyx 15; leaves (except sometimes the uppermost) with petioles ca. 1–6 (9) cm long............................5. **Agastache**

25. Nerves of calyx 5 or 10; leaves sessile or with petioles up to 2.2 (2.8) cm long ....20. **Stachys**

23. Calyx ± 2-lipped, with upper lobe distinctly broader or with some sinuses distinctly deeper than the others, the tube 15-nerved

26. Blades (at least of middle and upper leaves) ± tapered to base; calyx nearly glabrous (except for hairs on nerves and margins), with upper lobe distinctly broader than the other 4 and appearing at least slightly longer, with (in common species) a small hairy callus at each sinus; corolla blue or pink (rarely white) ....21. **Dracocephalum**

26. Blades subcordate or cordate; calyx softly pubescent, with all lobes of ± equal width but upper 2 lobes longer (at least in part because of deeper sinuses, which lack a callus); corolla white, dotted with purplish ....22. **Nepeta**

11. Most or usually all flowers in distinct *axillary* inflorescences subtended by leaves or leafy bracts that exceed them, the intact stem usually ending in small leaves (not flowers)

27. Calyx lobes each tipped with a rigid, terete, non-green spine ca. half as long as the rest of the lobe, or longer

28. Lobes of calyx 10, the spine with recurved tip; stems and leaves (especially beneath) densely white-tomentose............................23. **Marrubium**

28. Lobes of calyx 5, the spine straight; stems and leaves glabrate or with usually straight hairs

29. Leaves, at least on middle and lower part of stem, 3-lobed; calyx tube 5-nerved; anthers opening longitudinally (sacs parallel) ....24. **Leonurus**

29. Leaves toothed but not lobed; calyx tube 10 (–15)-nerved; anthers opening transversely (sacs divergent)

30. Calyx ca. 5–7 mm long, about equalling the corolla, the lobes ± erect; corolla only short-pubescent above; nutlets sharply 3-angled, pubescent on apex; internodes finely pubescent, the longest hairs scarcely 0.5 mm...25. **Chaiturus**

30. Calyx mostly 8–14 (17) mm long at maturity, shorter than the corolla, the lobes becoming widely spreading; corolla with upper lip usually long-pilose above; nutlets rounded (very obscurely angled), glabrous throughout; internodes in common species with stiff spreading-retrorse hairs ...
....26. **Galeopsis**

27. Calyx lobes without spiny tips

31. Tube of calyx with 5 (–10) strong nerves (weaker nerves may also be present); throat of calyx naked

32. Leaf blades entire, linear to narrowly elliptic and tapered to the base

33. Pubescence of stem (especially toward summit) and calyx including many short gland-tipped hairs; leaves with scarcely visible resinous dots beneath; calyx strongly 2-lipped; stamens long-exserted ....
....2. **Trichostema (dichotomum)**

33. Pubescence all eglandular; leaves with prominent dark resinous dots on both surfaces; calyx nearly regular; stamens included in the corolla ....
....27. **Satureja**

32. Leaf blades regularly toothed or crenate, reniform to ovate, mostly truncate to subcordate at base

34. Flowers and fruit on definite pedicels ca. 2 mm or more long ......28. **Melissa**

34. Flowers and fruit sessile or nearly so ....29. **Lamium**

31. Tube of calyx with 13–15 strong nerves; throat of calyx bearded or naked

35. Calyx with beard sparse or none in the throat; leaf blades ± ovate to reniform, broadly rounded to cordate at base
   36. Floral bracts elliptic, mostly with distinct lateral veins . . . . .28. **Melissa**
   36. Floral bracts linear-setaceous, at most the midnerve evident
      37. Leaves entire (or very obscurely toothed); stem pubescent with long, usually dense spreading hairs . . . . .18. **Clinopodium**
      37. Leaves prominently crenate-toothed; stem (except sometimes at nodes) glabrous or very minutely pubescent . . . . .30. **Glechoma**
35. Calyx strongly bearded in the throat; leaf blades linear or ± elliptic and tapered to the base
   38. Corolla scarcely if at all (at most 1 (2) mm) longer than the calyx lobes; fertile stamens 2 . . . . .31. **Hedeoma**
   38. Corolla 2–8 mm longer than the calyx lobes; fertile stamens 4
      39. Blades of middle and upper leaves linear, sessile, entire; calyx lobes without cilia; stem, pedicels, and calyx (except for beard) glabrous . . . .
      . . . . .32. **Calamintha**
      39. Blades of all leaves ± elliptic, short-petiolate, often obscurely or irregularly toothed; calyx lobes all ciliate; stem, pedicels, and calyx finely pubescent . . . . .33. **Acinos**

1. **Pycnanthemum** Mountain Mint

These are very aromatic plants, a pleasure to crush in the field and scenting even a herbarium case. The corollas are white, generally dotted or tinged with a pink or purple shade.

REFERENCE

Grant, Elizabeth, & Carl Epling. 1943. A Study of Pycnanthemum (Labiatae). Univ. California Publ. Bot. 20: 195–240.

KEY TO THE SPECIES

1. Main stem (except sometimes at nodes) pubescent only on the angles or glabrous; leaves all or mostly not over 7 mm wide (rarely to 13 mm); outermost bracts of heads pubescent (if at all) only on lower surface and margins
  2. Calyx lobes with distinct subulate tip glabrate at least on upper side; stem glabrous . . . . .1. **P. tenuifolium**
  2. Calyx lobes with tip not subulate but densely white-pubescent; stem pubescent on angles with spreading and/or incurved hairs . . . . .2. **P. virginianum**
1. Main stem pubescent on all sides; leaves often over 7 mm wide; outermost bracts of heads finely pubescent on both surfaces (only on lower surface in *P. muticum*)
  3. Leaf blades on distinct petioles at least 2.5 mm long; inflorescence somewhat open, the branches evident . . . . .3. **P. incanum**
  3. Leaf blades sessile or essentially so; inflorescence dense and head-like
    4. Principal leaves less than 3 times as long as broad; outermost bracts of heads glabrous on upper surface . . . . .4. **P. muticum**
    4. Principal leaves over 3 times as long as broad; outermost bracts of heads pubescent on both surfaces
      5. Leaves pubescent beneath on surface as well as veins; involucral bracts merely acute . . . . .5. **P. pilosum**
      5. Leaves pubescent beneath only on main veins; involucral bracts shortly acuminate . . . . .6. **P. verticillatum**

1. **P. tenuifolium** Schrader

Map 171. Quite local, in sandy fields, damp meadows, grassy areas, and wet prairies.

The leaves are narrowly linear; but some specimens of *P. virginianum*, which normally has slightly broader leaves, have them narrow enough to be within the range of *P. tenuifolium*.

Long known as *P. flexuosum* (Walter) BSP., a name actually applying to a species of the Southeast formerly called *P. hyssopifolium* Bentham.

2. **P. virginianum** (L.) B. L. Rob. & Fernald

Map 172. Fens, prairies, marshes, sedge meadows, tamarack swamps, swales, depressions such as old lakebeds; fields, sandy banks; less often in wooded areas.

This is our only common species in the genus, and a fairly distinctive one, with the angles of the main stem usually copiously pubescent and the sides rarely with a few small hairs. The leaves are variable in size, shape, and pubescence. In one collection, the outer involucral bracts are very minutely pubescent and in another the stem is unusually glabrate on the angles.

3. **P. incanum** (L.) Michaux

Map 173. Ranges mostly south and east of Michigan. Collected in 1920 "on an unimproved lot" in Detroit—presumably adventive (*Farwell 5627*, BLH; Pap. Michigan Acad. 1: 97–98. 1923).

The distinctive ovate-elliptic leaf blades, densely and finely white-pubescent beneath and on short petioles, together with a more open inflorescence, characterize this species.

4. **P. muticum** (Michaux) Pers.

Map 174. A southern and eastern species, collected north of Dayton in Berrien County (*Schulenberg* in 1972, MOR); also listed from Ottawa County by Grant and Epling.

172. Pycnanthemum
    virginianum

173. Pycnanthemum
    incanum

174. Pycnanthemum
    muticum

One of three sheets of a *Pycnanthemum* from Huron County was annotated by Epling as ?*P. muticum* × *pilosum*. The other two sheets, however, were annotated as *P. verticillatum* by Grant. Two sheets from Berrien County were also cited by Grant and Epling as suspected *P. muticum* × *pilosum*; such a hybrid would look rather like *P. verticillatum* [as do the sheets of the same collections in MICH (*Dodge 548 & 549*, in 1917)]. *P. muticum* differs from *P. verticillatum* in larger principal leaves, ± ovate, a third or more as broad as long.

5. **P. pilosum** Nutt.
Map 175. Our few specimens, insofar as any habitat is stated, are from a shaded river bank (Berrien Co.) and upland roadside and pasture (Monroe Co.).

Sometimes considered a variety of the next, as *P. verticillatum* var. *pilosum* (Nutt.) Cooperr. Our few specimens seem distinct enough, and in a genus where hybridization is suspected, the significance of intermediates is not always clear. Surely *P. pilosum, P. verticillatum,* and *P. muticum* are very similar, quite likely intergrading and/or hybridizing (see Cooperrider in Michigan Bot. 23: 166–167. 1984).

6. **P. verticillatum** (Michaux) Pers.
Map 176. Moist sandy shores, fields, roadsides, and borrow pits.
See comments under the two preceding species.

2. **Trichostema**   Blue-curls
The first species is sometimes segregated into a separate genus, as *Isanthus brachiatus* (L.) BSP., but in accord with Lewis and recent checklists, it is left here in *Trichostema*.

REFERENCE

Lewis, Harlan. 1945. A Revision of the Genus Trichostema. Brittonia 5: 276–303.

175. Pycnanthemum   176. Pycnanthemum   177. Trichostema
     pilosum   verticillatum   brachiatum

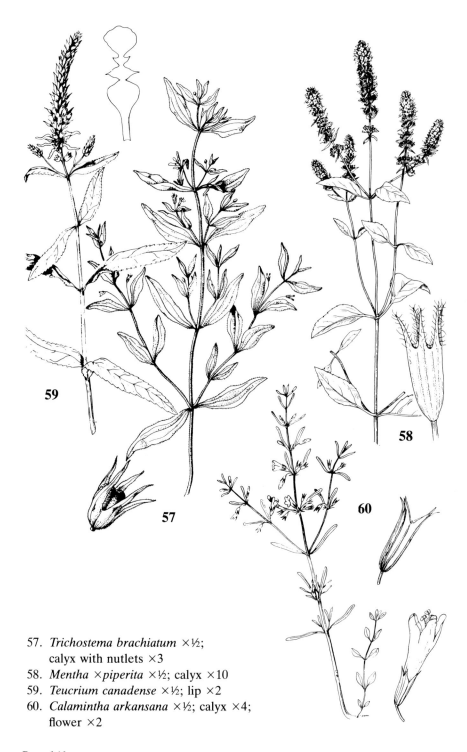

57. *Trichostema brachiatum* ×½;
    calyx with nutlets ×3
58. *Mentha* ×*piperita* ×½; calyx ×10
59. *Teucrium canadense* ×½; lip ×2
60. *Calamintha arkansana* ×½; calyx ×4;
    flower ×2

KEY TO THE SPECIES

1. Calyx lobes all of ± the same length.............................1. **T. brachiatum**
1. Calyx lobes or lips of very different length.........................2. **T. dichotomum**

1. **T. brachiatum** L.   Fig. 57   False Pennyroyal
    Map 177. A calciphile of limestone areas, as at Drummond Island and Thunder Bay Island, and also weedy but very local in sandy fields or other sterile ground.
    The foliage (at least at Drummond Island) has an aromatic, almost lemon-like fragrance. I do not know the basis for Steyermark's statement (1963) that this species has a 5-lobed ovary.

2. **T. dichotomum** L.   Bastard-pennyroyal
    Map 178. Dry sandy fields, prairie relics, oak barrens.
    The inflorescence may have a turpentine-like fragrance.

3. **Lycopus**   Bugleweed; Water-horehound
    The species of this genus tend to be so variable that overlapping measurements in the keys may appear discouraging. In fact, the species are rather easily distinguished, and specimens in the range of overlap for one character will usually be clearcut on the basis of others (or of the same structures elsewhere on the plant; e.g., calyx lobes may differ at different nodes). The corky borders of the nutlets doubtless contribute to their buoyancy, and dense rows of seedlings are a common sight on shores and banks where the fruits have been washed. Measurements of leaf length used in length/width ratios include the petiole (if any).

REFERENCES

Andrus, Mark E., & Ronald L. Stuckey. 1981. Introgressive Hybridization and Habitat Separation in Lycopus americanus and L. europaeus at the Southwestern Shores of Lake Erie. Michigan Bot. 20: 127–135.

178. Trichostema
    dichotomum

179. Lycopus uniflorus

180. Lycopus virginicus

Henderson, Norlan C. 1962. A Taxonomic Revision of the Genus Lycopus (Labiatae). Am. Midl. Nat. 68: 95–138.

Hermann, Frederick J. 1936. Diagnostic Characters in Lycopus. Rhodora 38: 373–375 + pl. 439.

Stuckey, Ronald L. 1969. The Introduction and Spread of Lycopus asper (Western Water Horehound) in the Western Lake Erie and Lake St. Clair Region. Michigan Bot. 8: 111–120.

Stuckey, Ronald L., & W. Louis Phillips. 1970. Distributional History of Lycopus europaeus (European Water-horehound) in North America. Rhodora 72: 351–369.

Webber, J. M., & P. W. Ball. 1980. Introgression in Canadian Populations of Lycopus americanus Muhl. and L. europaeus L. (Labiatae). Rhodora 82: 281–304.

KEY TO THE SPECIES

1. Calyx lobes less than 1 mm long, ± deltoid, at most barely as long as the mature nutlets
  2. Plant arising from a soft corm-like tuber barely under ground; stamens conspicuously exserted beyond corolla tube; nutlets not over 1.5 mm long, with tubercles or teeth only on or near outer rim of apex, the group of 4 nutlets thus concave or bowl-like across the tops; plant common throughout the state . . . . .1. **L. uniflorus**
  2. Plant without a tuberous base; stamens included; nutlets ca. 1.3–2.1 mm long, with teeth all across the apex, the group of 4 nutlets thus square or even convex across the tops of the teeth; plant very rare, in southern Michigan . . . . .2. **L. virginicus**
1. Calyx lobes (1) 1.2–2.8 mm long, acuminate to subulate at tip, distinctly exceeding the nutlets
  3. Nutlets ± tuberculate, toothed, or knobby on the apex; leaves (at least the middle ones) either clearly sessile with sharply and regularly serrate margin or tapered to petiolate base with shallowly and remotely serrate margin
    4. Leaves strongly tapering to base, ± petiolate; bracts of inflorescence minute, linear, much shorter than calyces; nutlets not over 1.5 mm long . . . . .3. **L. rubellus**
    4. Leaves with ± obtuse, sessile base; bracts of inflorescence lanceolate, the larger ones equalling or exceeding the adjacent calyces; nutlets 1.7–2.1 mm long . . . . .
    . . . . .4. **L. asper**
  3. Nutlets with smooth corky rim along outer margin of apex, but not tuberculate or knobby; leaves tapered to petiolate or subpetiolate base, prominently dentate (or often narrowly lobed toward base)
    5. Principal cauline leaves (2.1) 2.5–3.1 times as long as broad, coarsely dentate with forward margin of middle teeth at ± a right angle to midrib; calyx lobes (1.5) 1.8–2.8 mm long; leaves at least sparsely strigose above . . . . .5. **L. europaeus**
    5. Principal cauline leaves (1.5) 2.3–5.5 (8) times as long as broad, the lower ones often deeply pinnatifid into linear lobes, the teeth (or lobes) mostly with forward margin at an acute angle to the midrib; calyx lobes (1) 1.3–1.8 (2) mm long; leaves usually glabrous or merely scabrous above . . . . .6. **L. americanus**

1.   **L. uniflorus** Michaux

Map 179. Usually in habitats similar to those of our other common species, *L. americanus*, and often growing with or near it; shores, swamps (cedar, tamarack, hardwoods); wet prairies, fens, bogs; ditches, borrow pits, ponds; wet thickets, meadows, stream banks.

Some calyx lobes, especially on fruiting material, may be as long as 1.4 mm but still without exceeding the nutlets. Henderson has referred specimens from Gratiot and Huron counties to *L.* ×*sherardii* E. S. Steele, a hybrid with the next species; these look very close to *L. uniflorus* although the ends of the nutlets are a bit more toothed than usual. A few other specimens may be the same. A collection from Lenawee County (*Smith 1900* in 1986, MICH) has the short calyx lobes of *L. uniflorus* but smooth nutlets (and pinnatifid lower leaves) of *L. americanus* and is presumably a hybrid. Some collections from Monroe County (*Voss 14902* in 1976, MICH; *Reznicek 7839* in 1986, MICH, MSC) have the vegetative aspect of *L. asper* but the calyx and nutlets of *L. uniflorus*.

For those who do not keep this species distinct from the next, the correct name is *L. virginicus* var. *pauciflorus* Bentham.

2. **L. virginicus** L.
Map 180. Our several collections, all fruiting, are from wooded floodplains.

This is a much more robust species than the preceding, with densely subglobose infructescences nearly or quite concealing the bracts and calyces. The nutlets are larger than in *L. uniflorus*. Alleged differences in pubescence do not seem reliable. Most Michigan specimens labeled in the past as *L. virginicus* are referable to other species, chiefly *L. uniflorus*. Henderson mapped *L. virginicus* only well to the south and east of Michigan.

3. **L. rubellus** Moench
Map 181. Swampy woods and floodplains, less often in open moist ground.

The tiny bracts of the inflorescence are a good character for telling this from all of our other species except the two preceding ones.

4. **L. asper** Greene
Map 182. A species native west of the Great Lakes, first noted in Michigan in the 1890s, around grain elevators at Port Huron. It is now frequent in the western Lake Erie area and local elsewhere in wet ground, especially ± disturbed shores and ditches.

181. Lycopus rubellus

182. Lycopus asper

183. Lycopus europaeus

Another species with sessile leaves, *L. amplectens* Raf., a Coastal Plain disjunct, is known from northern Indiana and may turn up in southwestern-most Michigan. It has tiny linear bracts in the inflorescence and relatively few-toothed leaves ca. 3 times as long as broad, or shorter.

5. **L. europaeus** L.

Map 183. A European species discovered in Michigan in 1968, but doubt-less in the Lake Erie area somewhat before then. It can be expected to spread in damp somewhat disturbed places.

Rather easily distinguished from *L. americanus* by the leaf shape, al-though this is hard to express in words. Hybrids between the two species are known elsewhere around Lake Erie and will surely be found in Michigan.

6. **L. americanus** W. P. C. Barton   Plate 3-E

Map 184. Very common throughout the state, nearly always in moist to wet places: shores, edges of marshes and ponds, fens and springy areas, ditches and swales, river and stream margins, swamps, wet gravel pits and other excavations (or filled ground).

Variable in leaf form, pubescence, and other characters. The stems are ± densely hairy at the nodes that lack flowers, but the internodes vary from glabrous to hairy. The upper surface of the leaves is usually glabrous, often scabrous, but may have a few hairs especially toward the base—a feature normally associated with *L. europaeus*. Both species have a very sharply 4-angled stem and nutlets usually 1.2–1.7 mm long. Both lack surficial run-ners, though they are rhizomatous, while our other species (except *L. asper*) frequently produce slender elongate runners.

4. Perilla

1. **P. frutescens** (L.) Britton

Map 185. A native of southeast Asia, sometimes cultivated for its oily seeds or bronzy foliage, and locally escaped as a weed in eastern North America. Collected in 1994 (*Reznicek 9976*, MICH) along a sidewalk in Ann Arbor.

See also comments under *Collinsonia* (12).

5. Agastache   Giant-hyssop

These are tall mints with stout sharply square stems. Our species are all known in cultivation. The flowers, especially in *A. foeniculum*, may appear nearly regular.

REFERENCE

Lint, Harold, & Carl Epling. 1945. A Revision of Agastache. Am. Midl. Nat. 33: 207–230.

KEY TO THE SPECIES

1. Leaf blades whitened beneath with fine dense felt-like pubescence; calyx including
   lobes pubescent, the lobes deeply colored (blue) like the corolla . . . . . . .1. **A. foeniculum**
1. Leaf blades green beneath, not hidden by pubescence (individual hairs visible);
   calyx usually glabrous, at least on lobes, the lobes green or whitish (to rosy)
   2. Stems glabrous or minutely pubescent on middle internodes; leaf blades pubescent
      across the lower surface; calyx lobes at anthesis all less than 2 mm long; corolla
      yellowish . . . . .2. **A. nepetoides**
   2. Stems with at least a few long hairs on middle internodes; leaf blades pubescent
      mainly on the veins beneath; calyx lobes at anthesis averaging at least 2 mm long;
      corolla rosy . . . . .3. **A. scrophulariifolia**

1.   **A. foeniculum** (Pursh) Kuntze
   Map 186. Native west of Michigan, but presumably recently spread (perhaps from cultivation) into the Upper Peninsula, where first collected in
1934 in a "prairie" near Lake Roland, Keweenaw County (*Farwell 10231*,
BLH, MSC; Pap. Michigan Acad. 23: 131–132. 1938). Later collected in
dry fields and openings in other counties.
   Bruised foliage has a strong anise-like aroma.

2.   **A. nepetoides** (L.) Kuntze
   Map 187. Upland ± open deciduous woods (oak, beech-maple); also
meadows, fencerows, thickets, lowland woods.

3.   **A. scrophulariifolia** (Willd.) Kuntze   Fig. 61
   Map 188. Woodland borders, clearings, moist woods and floodplains.
   The calyx lobes, besides running a little longer than in the preceding
species, are also more narrowly lance-acute, less veiny, and less firm in
texture; but these are rather subjective characters.

6.   **Mentha**   Mint
   There has been some shifting in application of names in this genus (see
Tucker et al. 1980), complicated by a strong tendency to hybridize and a
long history of cultivation of various taxa, including hybrids.

184. Lycopus americanus   185. Perilla frutescens   186. Agastache foeniculum

REFERENCES

Barnett, LeRoy. 1984. Mint in Michigan. Michigan Hist. 68(2): 16–20. [History of mint-growing in the state.]

DeWolf, Gordon P. 1954. Notes on Cultivated Labiates. 2. Mentha. Baileya 2: 3–11.

Gill, L. S., B. M. Lawrence, & J. K. Morton. 1973. Variation in Mentha arvensis L. (Labiatae). I. The North American Populations. Jour. Linn. Soc. Bot. 67: 213–232.

Harley, R. M., & C. A. Brighton. 1977. Chromosome Numbers in the Genus Mentha L. Jour. Linn. Soc. Bot. 74: 71–96.

Nelson, Ray. 1950. Verticillium Wilt of Peppermint. Michigan Agr. Exp. Sta. Tech. Bull. 221. 259 pp.

Tucker, Arthur O., Raymond M. Harley, & David Fairbrothers. 1980. The Linnaean Types of Mentha (Lamiaceae). Taxon 29: 233–255.

Tucker, Arthur O., & David E. Fairbrothers. 1990. The Origin of Mentha ×gracilis (Lamiaceae). I. Chromosome Numbers, Fertility, and Three Morphological Characters. Econ. Bot. 44: 183–213.

KEY TO THE SPECIES

1. Flowers in axils of ordinary leaves (these, like the internodes, only gradually reduced upwards on the stem) . . . . .1. **M. arvensis**
1. Flowers all or mostly crowded into terminal inflorescences, with most of their internodes obscured
  2. Leaves ± densely pubescent (even tomentose) beneath
    3. Leaf blades less than twice as long as broad, the tip and teeth obtuse or rounded, with many branched hairs beneath . . . . .2. **M. suaveolens**
    3. Leaf blades about twice or more as long as broad, acute, sharply toothed, with branched hairs few or none beneath . . . . .3. **M. ×villosa**
  2. Leaves glabrous or nearly so
    4. Principal leaves sessile or with petioles less than 3 mm long; inflorescences (including corollas) ca. 7–9 (10) mm thick, mostly 3.5–8 times as long . . . . .4. **M. spicata**
    4. Principal leaves with petioles (2.5) 3–12 mm long; inflorescences (at least principal central one) ca. 12–15 mm thick and 2–3 times as long as broad or shorter
      5. Calyx strictly glabrous; leaves all or mostly obtuse to rounded at apex; foliage with lemon-like aroma . . . . .5. **M. aquatica**
      5. Calyx at least slightly pubescent (on the lobes); leaves all clearly acute at apex; foliage with peppermint aroma . . . . .6. **M. ×piperita**

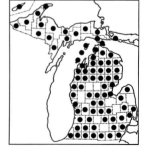

187. Agastache nepetoides   188. Agastache   189. Mentha arvensis
                                 scrophulariifolia

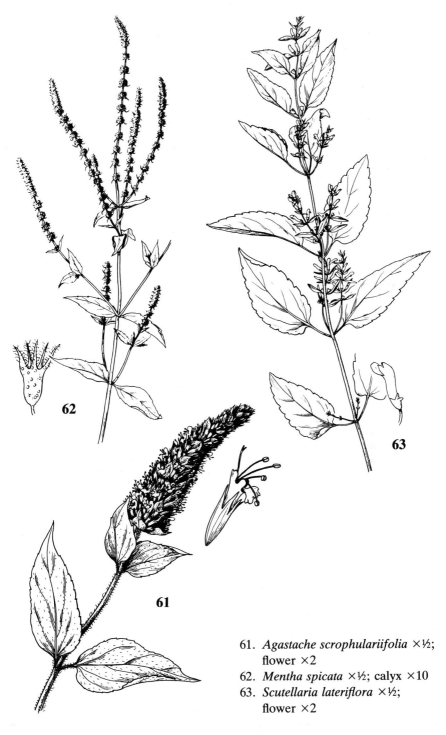

61. *Agastache scrophulariifolia* ×½;
    flower ×2
62. *Mentha spicata* ×½; calyx ×10
63. *Scutellaria lateriflora* ×½;
    flower ×2

1. **M. arvensis** L.   Wild Mint
    Map 189. Our only native *Mentha*. Low ground and wet places gener-
ally, whether disturbed or undisturbed: marshes and wet shores, borders of
rivers and streams, fens and wet prairies, swamps (hardwood, cedar, tamarack), thickets (often with willows and alder), beach pools, ditches and
swales, meadows and pastures.
    North American plants vary in pubescence, calyx and leaf shape, essential
oils, reproductive behavior (some clones male-sterile), and other characters,
but all apparently have a chromosome number of 2n = 96. European plants
are also variable and there is no absolute morphological distinction from
American ones, although the chromosome number is 2n = 72. Some authors
maintain the North American plants as a distinct species, *M. canadensis* L.,
and some have reported European plants as adventive in North America
although most if not all of these reports presumably are based on morphological variants of the American cytotype (see Gill et al.). Those who desire to
treat American plants as a single infraspecific taxon may call it *M. arvensis*
var. *canadensis* (L.) Kuntze or *M. arvensis* ssp. *canadensis* (L.) H. Hara.
    Plants resembling *M. arvensis*, but differing in somewhat smaller leaves
(large bracts) toward the summit of the stem and in supposedly more
glabrate stem and calyces, may be called *M.* ×*gracilis* Sole, apparently a
hybrid with *M. spicata*. *M.* ×*gracilis* is notoriously variable and often
closely resembles *M. arvensis*. *M. gentilis* L. is a name long applied to this
hybrid, but it is placed by Tucker et al. in the synonymy of *M. arvensis*
instead. *M.* ×*cardiaca* (S. F. Gray) Baker is a later name that has sometimes been applied to the hybrid. It has been collected several times in
Michigan; records are from Benzie, Crawford, Emmet, Gogebic, Kalamazoo, Macomb, Marquette, Monroe, Oakland, St. Clair, Washtenaw, and
Wayne counties.

2. **M. suaveolens** Ehrh.   Apple or Pineapple Mint
    Map 190. A European species, sometimes cultivated and rarely escaped
to disturbed ground and along roadsides.

190. Mentha suaveolens

191. Mentha ×villosa

192. Mentha spicata

Page 154

The leaves are quite rugose. Long known as *M. rotundifolia* (L.) L., a name now thought to apply to a diploid hybrid, *M. longifolia* (L.) L. × *M. suaveolens*.

3. **M. ×villosa** Hudson
Map 191. This is one of several mints cultivated for its oils and fragrant foliage. It is occasionally established as an escape on shores, along stream banks and roadsides, in fields and waste places.
This sweet-scented intermediate plant is a triploid hybrid of *M. spicata* × *M. suaveolens*.

4. **M. spicata** L.   Fig. 62   Spearmint
Map 192. This familiar plant is presumably Eurasian in origin and may have arisen long ago in cultivation as an allotetraploid derived from *M. ×rotundifolia* and selected for glabrousness. Escaped to roadsides, banks, shores, thickets, fields, and waste ground; in damp or dry sites, but only occasionally along streams.
Michigan ranks sixth in the nation (as of 1993) in commercial production of spearmint oil, which is so extensively used for flavoring and aroma.

5. **M. aquatica** L.   Lemon Mint
Map 193. Lemon mint is now considered a variety or cultivar of the Old World *M. aquatica*, and rarely escapes from cultivation. Collected in open ground in Alpena and Chippewa counties in 1913 and 1914 respectively by Dodge (both determinations confirmed by Morton and by Tucker). More recently, collected in Alcona County (*D. Garlitz 263* in 1986, MICH) along a creek in an aspen-birch woodlot.
Lemon mint, with a citrus-like aroma, has often been called *M. citrata* Ehrh. or treated as a nothovariety of peppermint.

6. **M. ×piperita** L.   Fig. 58   Peppermint
Map 194. Escaped and often very well established along rivers and streams, in marshes and ditches, on shores, in open areas of swamps; and in moist pastures, clearings, thickets, and waste ground.
Peppermint originated as a naturally occurring sterile hybrid between *M. aquatica* and *M. spicata*, and has been selected in cultivation—like spearmint—for glabrousness. The volatile oil distilled from peppermint plants is used to flavor gum, toothpaste, medicine, candy, and other articles. Commercial growing of peppermint in Michigan began in the mid 19th century. By the end of that century, Michigan produced 90% of the world's supply of mint oil—all from within a 90-mile radius of Kalamazoo. The town of Mentha was established in Van Buren County in 1870, but its post office finally closed in 1954. A fungal disease, "Verticillium wilt," drastically reduced mint production in Michigan, but at present nearly as much peppermint as spearmint is grown in the state.

7. **Ajuga**   Bugleweed

KEY TO THE SPECIES

1. Plant strongly stoloniferous; stem glabrous or if pubescent, the hairs mostly on opposite sides of the stem (alternating on each internode); leaves glabrous or nearly so, especially beneath .... 1. **A. reptans**
1. Plant not stoloniferous (but with rhizome); stem and leaves ± equally pubescent on all surfaces .... 2. **A. genevensis**

1. **A. reptans** L.

Map 195. Native of Europe and adjacent areas, widely cultivated as a sometimes aggressive ground-cover. Locally established along roads and in gravel pits, meadows, thickets, lawns, and yards.

2. **A. genevensis** L.

Map 196. A species of similar origin to the preceding; a very local weed of roadsides and yards.

Hybrids between these two species are known [*A.* ×*hybrida* Kerber].

8. **Teucrium**

REFERENCES

McClintock, Elizabeth, & Carl Epling. 1946. A Revision of Teucrium in the New World, with Observations on Its Variation, Geographical Distribution and History. Brittonia 5: 491–510.
Shinners, Lloyd H. 1963. The Varieties of Teucrium canadense (Labiatae). Sida 1: 182–183.

1. **T. canadense** L.   Fig. 59   Wood-sage

Map 197. Damp shores, marshes, meadows, prairies, river and stream margins; shaded woodland and thickets, floodplains (wooded or grassy).

Our specimens, almost without exception, have gland-tipped hairs on the calyx, ranging from sparse to abundant. They are also variable in other aspects of pubescence and in shape of calyx lobes; recognition of varieties—

193. Mentha aquatica   194. Mentha ×piperita   195. Ajuga reptans

much less distinct segregate species—does not seem warranted on these characters. Plants with glandular calyces have been called var. *occidentale* (A. Gray) E. M. McClint. & Epling [= *T. occidentale* A. Gray]. Among the several trivial segregates named, *T. menthifolium* E. P. Bicknell was originally described from Michigan (TL: Alma [Gratiot Co.], *C. A. Davis* in 1890 and 1892).

The corolla is normally pale pink, but white-flowered plants do occur in the state.

9. **Scutellaria** Skullcap

This is a very distinctive genus in several morphological characters, perhaps the most obvious of which is the little crest (scutellum—literally a little shield) across the top of the calyx, the origin of the generic name.

REFERENCES

Epling, Carl. 1942 The American Species of Scutellaria. Univ. California Publ. Bot. 20: 1–145.
Fritsch, Peter. 1992. Scutellaria nervosa (Lamiaceae), a Species of Skullcap New to Michigan. Michigan Bot. 31: 37–38.
Gill, L. S., & J. K. Morton. 1979. Scutellaria churchilliana—Hybrid or Species? Syst. Bot. 3: 342–348.

KEY TO THE SPECIES

1. Flowers all in axillary or terminal racemes (i.e., in axils of small bracts); principal leaves with petiole over 4 mm long (3–4 mm only on depauperate individuals)
   2. Racemes chiefly axillary; flowers (4.5) 5.5–7.5 (8.5) mm long . . . . 1. **S. lateriflora**
   2. Racemes chiefly terminal; flowers ca. 11–16 [20] mm long
      3. Principal cauline leaves with blade lacking glandular dots beneath, strongly cordate, the lobes extending well below any tissue decurrent on the petiole . . . . .2. **S. ovata**
      3. Principal cauline leaves with blade glandular-dotted beneath, truncate or rounded at the base, usually with tissue decurrent on the petiole
         4. Calyx and bracts of inflorescence with some long, spreading, gland-tipped hairs . . . . .3. **S. elliptica**
         4. Calyx and bracts with only very short, ± appressed or incurved, eglandular hairs . . . . .4. **S. incana**
1. Flowers in the axils of cauline leaves ± gradually (if at all) reduced in size; principal leaves with petioles none or less than 4 mm long
   5. Corolla (13) 16–22 mm long; mid-cauline leaves (1.8) 2.5–5.5 (8) cm long, including distinct petiole 0.5–3.4 (4) mm long. . . . . .5. **S. galericulata**
   5. Corolla 6–11 mm long; mid-cauline leaves ca. 1–4 cm long, sessile or on petioles less than 0.7 mm long
      6. Stem (middle internodes) glabrous or with a few scattered hairs; principal leaf blades ca. (1.5) 2.2–4 cm long, with a few definite teeth on each side; corolla white to pale blue, ca. 9–11 mm long . . . . .6. **S. nervosa**
      6. Stem puberulent or ± pilose; leaf blades 1–1.8 cm long, essentially entire; corolla deep blue to purple, ca. 6–9 mm long. . . . . .7. **S. parvula**

1. **S. lateriflora** L.   Fig. 63   Mad-dog Skullcap

Map 198. Hardwood and conifer swamps, thickets, wet shores, meadows, river banks, ditches and swales, marshes, and bogs. In much the same diversity of wet habitats as the equally common *S. galericulata*, but the latter seems more often to be in open places and *S. lateriflora* more often in shaded ones.

An occasional plant has gland-tipped hairs on the calyx, which is usually eglandular. Also rare are plants with white corollas: f. *albiflora* (Farw.) Fernald (TL: Grosse Isle [Wayne Co.]). Depauperate plants very rarely have the racemes reduced to a single flower, but are identifiable by the very small size of the flowers together with petiolate leaves.

Plants clearly intermediate with *S. galericulata*, with intermediate-shaped petiolate leaves and larger flowers (ca. 10–12 mm) than in *S. lateriflora*, borne singly or in few-flowered racemes, are presumably hybrids, *S.* ×*churchilliana* Fernald. The only Michigan collections seen are from Gogebic County (*Bessey & Darlington 3027* in 1919, MICH) and Marquette County (*Bingham* in 1942; *Wells & Thompson* in 1972, both BLH)—some considerable distance from previously reported stations in Maine and adjacent Canada.

2. **S. ovata** Hill

Map 199. The only Michigan collection I have seen is from the Bankson Lake area (*Nieuwland* in 1918, ND).

Plants from the northern part of the range of this variable species are referred to var. *versicolor* (Nutt.) Fernald. Besides the Michigan locality, it reaches northernmost Ohio, just across the Michigan line, and northeastern Indiana. See also comments under *S. incana* below.

3. **S. elliptica** Sprengel

Map 200. Deciduous woods (oak, hickory, sassafras, basswood, beech, maple).

Plants from southwestern Michigan and northern Indiana are disjunct from the rest of the more southern range. Our plants have straight spreading hairs on the stem and are thus var. *hirsuta* (Short) Fernald. The protuberance on the calyx is particularly large in this species.

4. **S. incana** Biehler

Map 201. A species of dry woods and thickets. It can presumably now be considered extinct in this state.

Three specimens attributed to Michigan, without further locality or date, have been confirmed as this species by Epling (in 1938) and by Collins (in 1976). One sheet (GH) is labeled in the hand of Asa Gray. Two sheets (MICH) are labeled in later hands, but all three may well represent collections of the First Survey, whose report for 1838 did list *S. cordifolia* (= *S. ovata*)—and indeed one of the two MICH specimens was labeled as

S. versicolor (now included in *S. ovata*), while the GH sheet was labeled as *S. canescens* (now included in *S. incana*). *S. incana* is known from northern Ohio and Indiana, so is not wholly unexpected in Michigan. Farwell's report of *S. cordifolia* from Michigan (Rep. Michigan Acad. 20: 186. 1918) was based on a specimen of *S. elliptica*.

5. **S. galericulata** L.   Marsh Skullcap

Map 202. Wet or marshy shores and banks; borders of streams, rivers, lakes, and ponds; swampy woods, thickets, and clearings; bogs, sedge meadows, cedar swamps; ditches and swales.

The calyx and corolla are puberulent. White- and pink-flowered forms are known, but the white one seems not yet to be reported from Michigan. The flowers are solitary in the axils of normal to reduced leaves and both flowers at a node turn to the same side, giving a distinct lateral appearance. Depauperate plants with leaves as short as in the next two species have them over twice as long as broad and thus can be distinguished. (In our plants of *S. nervosa* and *S. parvula*, the mid-cauline leaves are no more than twice as long as broad.)

American plants of this circumpolar species have sometimes been recognized as a separate species, *S. epilobiifolia* A. Ham.; or they may be called *S. galericulata* var. *pubescens* Bentham.

196. Ajuga genevensis

197. Teucrium canadense

198. Scutellaria lateriflora

199. Scutellaria ovata

200. Scutellaria elliptica

201. Scutellaria incana

6. S. nervosa Pursh

Map 203. A southern species, known from nearby Ohio and adjacent Ontario, but not collected in Michigan until 1989 (*Fritsch 1141*, MICH), in a floodplain forest.

Northern plants have been referred to var. *calvifolia* Fernald. Attempts to replace the species name with *S. integrifolia* L. were rendered irrelevant by changes in the Code in 1987 regarding lectotypification.

7. S. parvula Michaux

Map 204. Usually associated with calcareous areas, such as limestone pavements and gravels.

Most Michigan collections of this rather dwarf species may be called var. *parvula*, but ones from a gravelly bank along a highway in Menominee County (in 1984) and from a railroad embankment at East Lansing (in 1922 and 1923) may be distinguished (by some definitions) as var. *leonardii* (Epling) Fernald, sometimes treated at specific rank as *S. leonardii* Epling. This differs from typical *S. parvula* in having only tiny incurved eglandular hairs on the stem and eglandular hairs on the calyx; in var. *parvula* there are straight, erect, gland-tipped hairs on stem and calyx (overtopping any shorter hairs). Authors have not agreed on how to treat the variation in this group, nor on what characters to stress in defining taxa.

10. Monarda

REFERENCES

Scora, Rainer W. 1967. Interspecific Relationships in the Genus Monarda (Labiatae). Univ. California Publ. Bot. 41. 71 pp.

Sherff, Earl Edward. 1945. Monarda fistulosa L. and Its Two White-flowered Forms. Torreya 45: 68.

KEY TO THE SPECIES

1. Flowers at least partly (except on depauperate plants) in dense axillary whorls; corolla yellowish, dotted with purple, the upper lip strongly arched . . . . . . 1. **M. punctata**
1. Flowers all in terminal heads at ends of main stem and branches; corolla bluish, purplish, or red, the upper lip ± straight on top at maturity
  2. Corolla 31–45 mm long, bright red, the upper lip glabrate (especially toward tip) or sparsely pubescent; calyx glabrate at throat (summit of tube within) . . . 2. **M. didyma**
  2. Corolla less than 30 mm long, lavender (rarely white), the upper lip pubescent with tuft of longer hairs at tip; calyx densely bearded at throat (the hairs conspicuous between the teeth) . . . . .3. **M. fistulosa**

1. **M. punctata** L.　　Fig. 64　　　　　　　　Dotted or Horse Mint

Map 205. Sand dunes, sandy fields and relic prairies, oak and pine woodland; also along roadsides, railroads, and waste places; seems to do well with some disturbance. Presumably native in western Michigan as far north as dunes in Leelanau County; quite probably adventive elsewhere.

Specimens from this region are all referred to the pubescent var. *villicaulis* (Pennell) E. J. Palmer & Steyerm.

2. **M. didyma** L. Oswego-tea; Bee-balm

Map 206. Native in rich woods on banks and floodplains of the Clinton and Black rivers in Macomb and St. Clair counties—and thence eastward. An escape from cultivation elsewhere in the state, along roadsides and in thickets.

M. ×media Willd. is supposedly a hybrid between *M. didyma* and *M. fistulosa*. Such plants have large colorful flowers, as in *M. didyma*, but more purple than red and with the throat of the calyx prominently bearded. The status of Michigan collections of *M. ×media* is not clear; they could be direct escapes from cultivation or they could have resulted from hybridization between garden plants of *M. didyma* and wild *M. fistulosa*. Such plants have been collected in Alpena, Kent, St. Clair, and Wayne counties.

3. **M. fistulosa** L. Wild-bergamot

Map 207. Usually in dry, open, sandy, gravelly, or rocky ground such as oak or jack pine woodland, prairies, fields, and roadsides; occasionally in sedge meadows or other damp places; often at edges of woods and thickets, on open stream and lake banks and stabilized dunes; spreading into waste places.

We have two rather insignificant varieties in the state: var. *mollis* (L.) Bentham, with only minute pubescence on the lower surface of the leaves, and var. *fistulosa*, with some long straight hairs on the same surface. The latter is occasional throughout the state; var. *mollis* is, however, more common except in the western Upper Peninsula. White-flowered forms are uncommon; named first was f. *albescens* Farwell (TL: Metamora [Lapeer Co.]). If var. *mollis* is recognized as distinct, its white-flowered form is f. *albiflora* (Farwell) Sherff (TL: La Salle [Monroe Co.]). In addition, a collection from Warren Woods, Berrien County (*Hebert* in 1932, ND) appears to be var. *clinopodia* (L.) Cooperr. (see Michigan Bot. 23: 406.

202. Scutellaria
     galericulata

203. Scutellaria nervosa

204. Scutellaria parvula

1984). This taxon (often recognized at the rank of species) differs in its long petioles, more moderately hirsute calyx throat, and absence of longer hairs at the tip of the upper lip of the corolla. Its habitat is quite different, for instead of the usual dry upland of the other varieties, it grows in damp floodplain forests.

11. **Origanum**

REFERENCES

DeWolf, Gordon P. 1954. Notes on Cultivated Labiates. 3. Origanum and Relatives. Baileya 2: 57–66.
Tucker, Arthur O., & Elizabeth D. Rollins. 1989. The Species, Hybrids, and Cultivars of Origanum (Lamiaceae) in the United States. Baileya 23: 24–27.

1. **O. vulgare** L.   Oregano; Wild-marjoram
Map 208. Native principally in the Mediterranean region, occasionally escaped from herb gardens to roadside banks, refuse piles, and waste ground.

The chief habitat in Michigan is on pizza. The commercial Greek/ Turkish oregano, however, is not the same subspecies commonly cultivated in America. And "Mexican oregano" is obtained from *Lippia graveolens* Kunth in the Verbenaceae.

O. vulgare has calyx lobes ± equal. Other species, with more irregular calyx, are sometimes grown.

12. **Collinsonia**

REFERENCE

Shinners, Lloyd H. 1962. Synopsis of Collinsonia (Labiatae). Sida 1: 76–83.

1. **C. canadensis** L.   Fig. 65   Stoneroot; Horse-balm
Map 209. Oak-hickory and sassafras or, more often, rich beech-maple or even swampy deciduous woods.

205. Monarda punctata   206. Monarda didyma   207. Monarda fistulosa

A stout plant (to 1.5 m tall), large-leaved, from a large hard rhizome. The flowers and crushed foliage have a strong lemon-like fragrance. *Perilla frutescens* is sometimes mistaken as this species, but is annual and has white to pink corollas (not yellow), with 4 stamens (not exserted), large bracts, and generally shorter pedicels.

13. Blephilia

These attractive and very aromatic mints are not often noticed and furthermore tend to be superficially confused with more common species such as *Clinopodium vulgare* and *Prunella vulgaris*—both of which generally have straighter and often sparser hairs on the stem and inflorescence than does *Blephilia* (especially *B. hirsuta*).

A few specimens from Ionia, Kent, and Ottawa counties resemble in some aspects *B. subnuda* Simmers & Kral, recently described as endemic to Alabama. However, Kral has examined these specimens and considers them to fall within the range of variation displayed by the two common species. The characters of all three taxa are thoroughly presented by Simmers and Kral.

REFERENCE

Simmers, Richard W., & Robert Kral. 1992. A New Species of Blephilia (Lamiaceae) from Northern Alabama. Rhodora 94: 1–14.

KEY TO THE SPECIES

1. Bracts of inflorescence ovate, with short-acuminate tip, appressed, the lower ones of each whorl nearly or fully concealing the calyces beneath; stem unbranched; upper leaves sessile or with a short (to 4 or rarely 7 mm) petiole . . . . .1. **B. ciliata**
1. Bracts of inflorescence mostly narrowly linear-subulate to lanceolate, becoming ± reflexed, not at all concealing the calyces; stems usually branched; upper leaves with petiole 9–18 (27) mm long . . . . .2. **B. hirsuta**

208. Origanum vulgare

209. Collinsonia
        canadensis

210. Blephilia ciliata

64. *Monarda punctata* ×½;
   flower & calyx ×2
65. *Collinsonia canadensis* ×½;
   flower ×2; calyx ×5
66. *Blephilia ciliata* ×½; calyx ×2

1. **B. ciliata** (L.) Bentham   Fig. 66
Map 210. Oak woodlands and borders; thickets, banks, and clearings; meadows, borders of fens, thin soil over limestone.

The broad appressed bracts in the inflorescence are suggestive of *Prunella vulgaris*, which (besides having 4 stamens) tends to have longer petioles and the upper lip of the calyx only shallowly 3-toothed instead of cleft into long copiously ciliate teeth.

2. **B. hirsuta** (Pursh) Bentham
Map 211. Rich woods, swamp forests, floodplains.

If the number of stamens is not evident, one using the keys might confuse this species with *Clinopodium vulgare*, which has shorter petioles.

14. Salvia   Sage

KEY TO THE SPECIES

1. Calyx 3-lobed (i.e., the upper lip entire, the lower 2-lobed); leaf blades linear-lanceolate, tapered at base, glabrous or minutely pubescent
   2. Flowers mostly 2 (rarely 4) at each node of the inflorescence, at most 1 cm long; calyx minutely pubescent on the nerves, about equalling or exceeding the corolla tube . . . . 1. **S. reflexa**
   2. Flowers mostly 6 or more at each node, ca. 1.2–2 cm long; calyx pubescent throughout, distinctly shorter than the corolla tube .2. **S. azurea**
1. Calyx 5-lobed (upper lip 3-toothed); leaf blades various
   3. Leaf blades all tapering to petioles, ± narrowly elliptic; upper lip of corolla straight; plant shrubby . . . . .3. **S. officinalis**
   3. Leaf blades, at least the lower ones, truncate to cordate at base, ± ovate; upper lip of corolla strongly arched; plant herbaceous . . . . .4. **S. pratensis**

1. **S. reflexa** Hornem.   Rocky Mountain Sage
Map 212. Native west of the Great Lakes and with us rarely adventive in farmyards and waste ground.

The light blue corollas are ephemeral.

211. Blephilia hirsuta

212. Salvia reflexa

213. Salvia azurea

2. **S. azurea** Lam.   Blue Sage
Map 213. Native in western North America, rarely escaped from cultivation to waste ground eastward.

Plants such as ours with short recurved hairs on the stem have been referred to var. *grandiflora* Bentham [*S. pitcheri* Bentham—named for Dr. Zina Pitcher (cf. *Cirsium pitcheri*)].

3. **S. officinalis** L.   Garden Sage
Map 214. Native of the Mediterranean region, long persistent at garden sites but rarely escaped. Well known as a condiment and aromatic herb.

The leaves in this species are very regularly and finely crenulate-toothed, while in the next they tend to be more coarsely, irregularly toothed.

4. **S. pratensis** L.   Meadow Clary
Map 215. A European species with attractive flowers, rarely escaped from cultivation to roadsides and fields.

15. **Thymus**   Thyme
A large and complex Eurasian genus, the species often misidentified and with tendencies to hybridize. *T. vulgaris* L. is the common herb; it has ± tomentose leaves with revolute margins. *T. serpyllum* L. with stems pubescent on all sides is rarely cultivated in North America; most garden plants (as well as escapes) are other species, including the two below.

REFERENCES

Flannery, Harriet B. 1984. Thymus serpyllum Misapplied Name, Misunderstood Plant. Herbarist 50: 26–35.
Roussine, N. 1962 ["1961"]. Note sur les Espèces du Genre Thymus aux États-Unis d'Amérique. Naturalia Monspel. Sér. Bot. 13: 59–61.

KEY TO THE SPECIES

1. Stem 4-angled, pubescent entirely or primarily on the angles; leaves elliptic to somewhat spatulate, with strong lateral veins beneath . . . . .1. **T. pulegioides**
1. Stem ± terete, pubescent on the entire surface; leaves ± linear, with weak lateral veins . . . . .2. **T. pannonicus**

1. **T. pulegioides** L.
Map 216. A Eurasian species, escaped from cultivation to roadsides, clearings, lawns, and waste places. Probably planted on a roadside bank in Roscommon County.

The inflorescence is more compact (and clearly terminal) than in the next species; the stem pubescence is short (less than 0.5 mm) and ± recurved. In the somewhat similar *T. praecox* Opiz, the strong lateral veins anastomose and form an equally strong marginal vein around the apex of the leaf; the stem is hairy on all or at least two opposite sides. It may be expected as an escape.

2. **T. pannonicus** All.
Map 217. A native of Eurasia, rarely escaped from cultivation.

Compared with the preceding species, the inflorescence is more elongate and interrupted, with axillary clusters as well as a terminal one. The stems are usually pubescent with long spreading hairs, especially near the summit. The Macomb County material (escaped near Richmond, *Dodge* in 1904) is somewhat ambiguous, with stem pubescence short and leaves less linear than usual, but the stems are pubescent on all sides, the venation weak, and the inflorescence elongate.

Sometimes known as *T. marschallianus* Willd., a species now generally included in *T. pannonicus*.

16. **Hyssopus**

1. **H. officinalis** L. Hyssop
Map 218. Native to southern Europe, occasionally escaped from cultivation to roadsides, banks, and other disturbed ground. Collected as early as the 1830s in Michigan, but I have seen no specimens gathered since the 1890s except for a 1918 one from Emmet County.

214. Salvia officinalis

215. Salvia pratensis

216. Thymus pulegioides

217. Thymus pannonicus

218. Hyssopus officinalis

219. Prunella vulgaris

Long grown for reputed medicinal value or as a condiment, but common names are unreliable for information. This species is not the plant called "hyssop" in the Bible.

17. Prunella

REFERENCE

Nelson, Andrew P. 1965. Taxonomic and Evolutionary Implications of Lawn Races in Prunella vulgaris (Labiatae). Brittonia 17: 160–174.

1. **P. vulgaris** L.   Fig. 67   Self-heal; Heal-all
Map 219. Grows in almost every sort of damp or dry, disturbed or undisturbed area, including upland woodlands as well as swamp forests (hardwoods, cedar, tamarack), especially along borders and trails (and old logging roads), in clearings, and after logging; roadsides, parking lots, campgrounds, recreation areas; fields and meadows, along rivers and streams, in fens; on rocky outcrops including limestone pavements. Native in Eurasia and apparently also in North America, although we also have weedy variants thoroughly established.

Any effort to distinguish significant varieties or subspecies in this polymorphic taxon is abandoned by recent authors (e.g., in *Flora Europaea*, and *Hortus Third*). Some strains have a creeping habit that appears adapted to a lawn environment. While stems of such plants may be only a few cm tall when flowering, on favorable sites plants may be 0.5 m tall. Unlike most mints, *Prunella* is not aromatic, but long ago it had reputed medicinal properties. It is too aggressive to be popular in most gardens.

The corolla is normally blue-violet (or even pinkish), but white-flowered plants [f. *albiflora* Britton] are occasional throughout the state. The middle lobe of the lower lip is fringed. *Blephilia ciliata* might be superficially confused with *Prunella* because of the broad, conspicuous floral bracts, but that species has entire corolla lobes, 2 stamens, all 5 calyx lobes well developed and more densely ciliate, and paler flowers.

18. Clinopodium

This genus was long included in *Satureja*, in which case our species is called *S. vulgaris* (L.) Fritsch.

REFERENCES

DeWolf, Gordon P. 1955. Notes on Cultivated Labiates 4. Satureja and Some Related Genera. Baileya 2: 142–150.
Shinners, Lloyd H. 1962. Calamintha (Labiatae) in the Southern United States. Sida 1: 69–75.

1. **C. vulgare** L.   Fig. 68   Wild-basil; Dog-mint
Map 220. Deciduous or pine woodland, sometimes in swamps but especially in dry sandy or rocky clearings, trails, or otherwise disturbed areas;

gravel ridges, dunes, and shores; old fields, roadsides, gravel pits, and other disturbed ground.

Usually but not always considered to be a native circumpolar species, North American plants being the weakly distinguished var. *neogaea* (Fernald) C. F. Reed. Often weedy in habit. The leaves have short petioles (all or mostly less than 7 mm). These, and the more evident pedicels in the inflorescence, will readily distinguish this species from *Blephilia hirsuta*, with which it might be confused if the number of stamens is not obvious. Rarely the corolla is white rather than pink-purple.

19. **Physostegia**

REFERENCE

Cantino, Philip D. 1982. A Monograph of the Genus Physostegia (Labiatae). Contr. Gray Herb. 211: 1–105.

1. **P. virginiana** (L.) Bentham   Fig. 69   False Dragonhead; Obedient Plant

Map 221. Swamps, shores, river floodplain forests, wet thickets, low open ground; also roadsides, ditches, near habitations, and other sites where probably escaped from cultivation. Specimens from Alpena, Grand Traverse, and Luce counties were surely escaped from cultivation, and some of the others, especially northern ones, are apparently of similar origin. The species appears indigenous north at least to the Grand, Pine, and Au Sable rivers in the Lower Peninsula and the Menominee River in the Upper Peninsula.

All native Michigan specimens seen by Cantino he referred to ssp. *virginiana* [including those sometimes identified as *P. formosior* Lunell or *P. speciosa* (Sweet) Sweet]. An odd collection from waste ground at Detroit (*Farwell 5093½*, GH, BLH) with only obscure teeth on the leaves, reported (Pap. Michigan Acad. 1: 97. 1923) as *Dracocephalum denticulatum* [= *P. aboriginorum* Fernald = *P. leptophylla* Small sensu Cantino] would be

220. Clinopodium vulgare   221. Physostegia virginiana   222. Stachys hyssopifolia

67. *Prunella vulgaris* ×½; flower ×2
68. *Clinopodium vulgare* ×½; calyx ×3
69. *Physostegia virginiana* ×½; leaf ×½

quite geographically misplaced for that southern Atlantic Coastal Plain species. Farwell thought it might be escaped, and furthermore its odd collection number is suspicious; Cantino determined it as aberrant material of *P. virginiana* or perhaps a hybrid escaped from cultivation. Cultivated variants of this species may include some results of hybridization.

20. **Stachys** Hedge-nettle

Petiole length and internode pubescence should be checked at the middle of the stem.

REFERENCES

Cooperrider, T. S., & R. F. Sabo. 1969. Stachys hispida and S. tenuifolia in Ohio. Castanea 34: 432–435.
Epling, Carl. 1934. Preliminary Revision of American Stachys. Feddes Repert. Beih. 80. 75 pp.
Mulligan, Gerald A., & Derek B. Munro. 1990 ["1989"]. Taxonomy of Species of North American Stachys (Labiatae) Found North of Mexico. Nat. Canad. 116: 35–51.
Nelson, John B. 1981. Stachys (Labiatae) in Southeastern United States. Sida 9: 104–123.

KEY TO THE SPECIES

1. Leaves glabrous, sessile or subsessile, ± linear, entire to lightly toothed . . . .
. . . . .1. **S. hyssopifolia**
1. Leaves pubescent (at least beneath), sessile or petioled, lanceolate to ovate, regularly serrate
 2. Stem internodes glabrous or pubescent only on the angles; calyx glabrous or pubescent (eglandular) primarily on the nerves and ciliate lobes; leaves with petioles 3–22 (28) mm long . . . . .2. **S. tenuifolia**
 2. Stem internodes pubescent on sides (at least 2) and angles; calyx pubescent throughout, usually including short gland-tipped hairs as well as longer ones; leaves sessile or with petioles to 4 (very rarely 8) mm long . . . . .3. **S. palustris**

1. **S. hyssopifolia** Michaux Fig. 70

Map 222. Sandy shores (especially recently exposed ones) and fields, wet depressions, meadows and prairies, even marly-peaty boggy places.

Our plants are all the narrow-leaved var. *hyssopifolia*, a taxon largely of the Atlantic Coastal Plain from Massachusetts to Georgia, disjunct in Michigan and adjacent northwest Indiana.

2. **S. tenuifolia** Willd.

Map 223. Marshes, wet prairies, shores, floodplain swamps, thickets, river banks.

Including *S. hispida* Pursh, sometimes regarded as a distinct species. Mulligan and Munro maintain *S. hispida* as a tetraploid with leaves sessile or nearly so (while *S. tenuifolia* is diploid or tetraploid with definitely petiolate leaves). Even in a broad sense, separation of this species from the next is often difficult and some authors (e.g., Boivin) have given it only varietal status. Some specimens from the western Upper Peninsula and

Page 171

west-central Lower Peninsula (particularly Ottawa County) approach *S. palustris*; and the St. Clair county material mapped as *S. palustris* does approach *S. tenuifolia* (and was determined as *S. hispida* by Epling).

3. **S. palustris** L.   Fig. 71

Map 224. Low ground, including shores, meadows, thickets, openings in swamps, river and creek borders, ditches.

Quite variable in pubescence, but there is insufficiently clear distinction between the varieties that have been named on the basis of relative lengths of glandular and non-glandular hairs and shape of leaves. Plants with hairs on the sides of the stem much shorter than those on the angles may come close to *S. tenuifolia* although the latter usually has the angle hairs stouter, more strongly pustulate-based, and more often curved. The more abundant straight softer hairs, including short gland-tipped ones, on the calyces and stems of *S. palustris* are about the only distinction from *S. tenuifolia*, together with the tendency to shorter petioles.

Mulligan and Munro restrict the name *S. palustris* to hexaploid plants locally naturalized from Eurasia (and not reported by them from Michigan). They refer our native tetraploid plants to *S. pilosa* Nutt., while placing the diploid *S. nuttallii* Bentham south of Michigan.

21. Dracocephalum

REFERENCE

DeWolf, Gordon P., Jr. 1955. Notes on Cultivated Labiates 7. Dracocephalum. Baileya 3: 115–129.

KEY TO THE SPECIES

1. Terminal inflorescence densely crowded, compact, the long bracts with prolonged spine-tipped teeth; calyx mostly 10 mm or more long ....1. **D. parviflorum**
1. Terminal inflorescence elongate after flowering, the small bracts scarcely if at all exceeding the pedicels, entire, spineless; calyx ca. 6–8 mm long.......2. **D. thymiflorum**

223. Stachys tenuifolia   224. Stachys palustris   225. Dracocephalum parviflorum

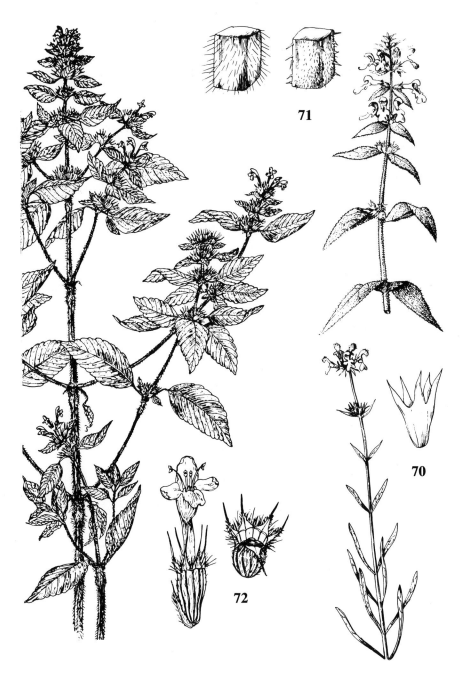

70. *Stachys hyssopifolia* ×½; calyx ×3
71. *S. palustris* ×½; portions of stems ×2
72. *Galeopsis tetrahit* ×½; flower & fruit with nutlets ×2½

1. **D. parviflorum** Nutt.   Fig. 73   Dragonhead
    Map 225. A northern species, presumably native in rocky openings on
Isle Royale and some other sites in the western Upper Peninsula; probably
native also in the jack pine plains but apparently adventive where found in
fields, along roadsides, and in waste ground elsewhere in the state.
    The spiny floral bracts and calyx lobes, with nearly concealed flowers,
give the heads a forbidding appearance. The calyx lobes have prominent
cross-veins between the nerves in this genus, unlike the next genus, *Nepeta*,
which has no evident cross-veins on its small calyx.

2. **D. thymiflorum** L.
    Map 226. A Eurasian species, found only locally as a weed in North
America. Collected by Billington in 1916 on the shore of Cass Lake (and
reported by Farwell as *Ocimum basilicum* L., basil; Rep. Michigan Acad.
19: 260. 1918). More recently found as an adventive in the Great Lakes
region on Manitoulin Island, Ontario (Morton & Venn 1984).

22. **Nepeta**

REFERENCES

DeWolf, Gordon P., Jr. 1955. Notes on Cultivated Labiates 6. Nepeta. Baileya 3: 99–107.
Tucker, Arthur O., & Sharon S. Tucker. 1988. Catnip and the Catnip Response. Econ. Bot.
    42: 214–231.

1. **N. cataria** L.   Fig. 74   Catnip; Catmint
    Map 227. A Eurasian species, thoroughly naturalized and known from
Michigan as early as the First Survey (1838). Common in all kinds of waste
places and weedy areas including dumps, parking lots, around buildings
and old foundations, fields and thickets, roadsides and railroads, shores,
and flower beds.
    This familiar strongly aromatic plant is rather gray-green in aspect, from
the pubescence on the stems, undersides of the leaves, and inflorescences.
In mild seasons, some individuals may continue flowering into December.

226. Dracocephalum   227. Nepeta cataria   228. Marrubium vulgare
     thymiflorum

It was formerly used for medicinal purposes, but is now appreciated primarily by cats, which have long been observed to respond to it by a unique pattern of behavior, leading from sniffing and chewing to rolling and suggestion of intoxication or hallucination. Not all species of felines respond, nor are all individuals of a species responsive. Cats with a friendly personality are considered to react with the most animation, but young kittens do not respond. (See Tucker & Tucker 1988.) I never weed all the catnip from my yard, so that the premises will appear hospitable to visitors that purr.

23. **Marrubium**

1. **M. vulgare** L.   Horehound
    Map 228. This Old World species was long cultivated and used to flavor candies, medicines, and beverages, with alleged medicinal effects. Apparently much less cultivated (and hence escaped) than formerly; collected in Michigan as early as the First Survey (1838), but I have seen only one collection (Barry County) made since 1935. Typical habitats include pastures, fields, roadsides, dooryards, old homesites, and waste places.

24. **Leonurus**

1. **L. cardiaca** L.   Fig. 76   Motherwort
    Map 229. A native of Europe and adjacent Asia, cultivated and considered to have medicinal properties. In dry or swampy, often ± shaded waste ground: roadsides, railroads, dumps and refuse heaps, around buildings, farmyards, parking lots; dry woods (e.g., oak), weedy floodplains, river banks, thickets, sandy banks and fields.
    This is a tall, hard-stemmed plant, with the spiny calyx lobes spreading to reflexed at maturity. The nutlets are strongly 3-angled, flat and pubescent at the apex. The upper lip of the corolla is long-pilose, the buds thus appearing like hairy plugs in the calyx. A few specimens collected in the fall bear leaves (late-formed?) with rather rounded lobes and short sinuses, resembling those of a *Ribes*, rather than the usual shape with 3 long sharp-pointed lobes.

25. **Chaiturus**

1. **C. marrubiastrum** (L.) Rchb.
    Map 230. A Eurasian species, locally established as a weed in North America. Our few collections are from river banks and nearby waste ground.
    Except in China and Russia, this species has usually been included in *Leonurus*—where Linnaeus originally placed it and with which it shares triangular nutlets with pubescent ends. However, it differs in enough ways that the increasing recognition of *Chaiturus* seems justified.

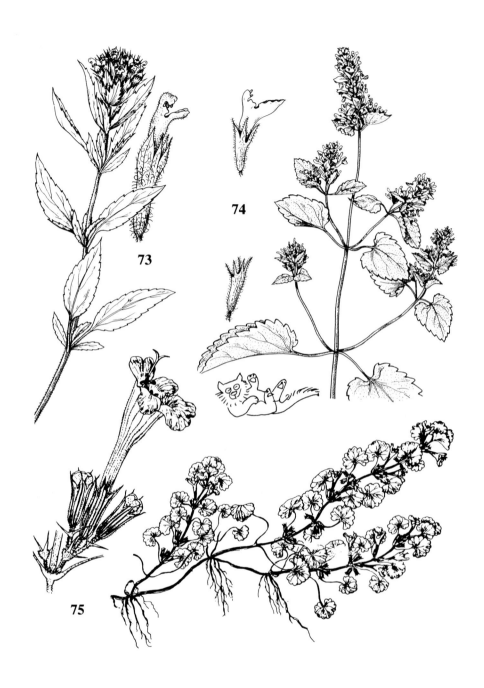

73. *Dracocephalum parviflorum* ×½; flower ×2
74. *Nepeta cataria* ×½; flower & mature calyx ×2
75. *Glechoma hederacea* ×½; inflorescence ×2½

26. Galeopsis

REFERENCE

O'Donovan, J. T., & M. P. Sharma. 1987. The Biology of Canadian Weeds. 78 [sic, for 80].
Galeopsis tetrahit L. Canad. Jour. Pl. Sci. 67: 787–796.

KEY TO THE SPECIES

1. Stem with stiff, straight, ± reflexed hairs.............................1. **G. tetrahit**
1. Stem with only curved or curled hairs (and tiny glandular ones)..........2. **G. ladanum**

1. **G. tetrahit** L.   Fig. 72   Hemp-nettle
   Map 231. A Eurasian species (or complex), reported from Michigan as
early as the First Survey (1838) although the oldest collection seen is from
1850 (Macomb County). Especially common in the northern part of the
state, in moist woods and thickets, particularly in logging trails, borders,
and clearings; also along rivers, streams, and gravelly shores; dooryards,
refuse heaps, and other waste ground.
   Much of our material is var. *bifida* (Boenn.) Lej. & Courtois, sometimes
recognized as a closely related species, *G. bifida* Boenn., with the middle
lobe of the lower lip ± notched rather than entire and squarish. However,
lip shape seems not to be correlated with other differences sometimes cited
(cuneate rather than rounded base of leaf blade and shorter calyx lobes in
bifida). Some botanists (e.g., Eric Hultén) considered that *bifida* was the
only taxon in the state.
   Corolla color is pink or white, both forms being frequent and at times
growing together. The density of the hairs on the retrorse-hispid stem is
also variable.

2. **G. ladanum** L.
   Map 232. Another Eurasian species, cited for Michigan in the 19th
century. However, the only specimens encountered were collected along a
disturbed cobble trail near the shore of Lake Huron (*D. Garlitz 856* in
1989, MICH; *1174* in 1990, GH, UMBS).

229. Leonurus cardiaca   230. Chaiturus   231. Galeopsis tetrahit
                       marrubiastrum

27. Satureja

In many works, this genus is treated more broadly than here, including *Clinopodium, Calamintha,* and *Acinos*, which an increasing number of authors are now recognizing as distinct.

REFERENCE

DeWolf, Gordon P. 1955. Notes on Cultivated Labiates 4. Satureja and Some Related Genera. Baileya 2: 142–150.

1. **S. hortensis** L.   Summer Savory

Map 233. A native of the Mediterranean region, an annual cultivated as a savory herb and rarely escaped from gardens to roadsides and other waste places.

28. Melissa

1. **M. officinalis** L.   Lemon-balm

Map 234. A native of southern Europe and perhaps Asia, long cultivated for its aromatic oil and rarely escaped to waste places and thickets.

The bruised foliage has a very pleasant citrus odor.

29. Lamium   Dead-nettle

All four species are natives of Europe, two of them known in cultivation and the other two uninvited weeds.

KEY TO THE SPECIES

1. Corolla ca. 19–26 mm long, the tube curved upward; plants perennial; leaves acute and all petioled
  2. Corolla white, the upper lip ciliate with long hairs (at least 0.5 mm) and usually pilose above as well; terminal tooth of largest cauline leaves prolonged or ± acuminate, often 9 mm or more long .... 1. **L. album**
  2. Corolla pink to purplish [albinos rare], the upper lip with hairs less than 0.5 mm; terminal tooth of largest cauline leaves less than 9 mm long .... 2. **L. maculatum**
1. Corolla ca. 9–19 mm long, the tube straight; plants annual; leaves acute to rounded in outline, all or only the lower ones petioled
  3. Leaves subtending flowers sessile, ascending or at most horizontal, slightly broader than long (± reniform in overall outline) .... 3. **L. amplexicaule**
  3. Leaves subtending flowers short-petioled, reflexed, the blade about as long as broad or a little longer .... 4. **L. purpureum**

1. **L. album** L.

Map 235. Although the species is said to be naturalized in eastern North America, I have seen only one Michigan collection, from sandy, weedy ground at the edge of a pine plantation in the Nichols Arboretum, Ann Arbor (*Reznicek 6699 & Catling* in 1982, MICH). An older report from Cheboygan County is based on a specimen of *Galeopsis tetrahit*.

2. **L. maculatum** L.

Map 236. Another large-flowered cultivated species, occasionally escaped to roadsides, railroads, and waste places.

The leaves are blotched with a pale streak along the midrib.

3. **L. amplexicaule** L.   Fig. 77

Map 237. A weed of lawns, flower beds, barnyards; around buildings, sidewalks, and railroads; fields and sandy or gravelly disturbed ground. First collected in Michigan in 1869 (Detroit) and in the 1880s (East Lansing).

4. **L. purpureum** L.

Map 238. Like the preceding species, a frequent weed of lawns, gardens, road shoulders, railroads, farmyards, dumps, vacant lots, and other waste places, usually sandy or gravelly; also at woods borders and clearings, along trailsides, on river banks.

30. **Glechoma**

The generic name was not consistently spelled by Linnaeus, and two spellings have long been in use. The one adopted here was officially conserved in 1987 to settle the matter (in a way more satisfactory than merging the genus into *Nepeta*, as has sometimes been done).

232. Galeopsis ladanum

233. Satureja hortensis

234. Melissa officinalis

235. Lamium album

236. Lamium maculatum

237. Lamium amplexicaule

1. **G. hederacea** L.   Fig. 75   Ground-ivy; Gill-over-the-ground;
Creeping Charlie

Map 239. A Eurasian species, which can be cultivated as a ground cover but is too aggressive for most uses, escaping to cover more ground than desired. Now a pernicious weed of lawns, roadsides, river and stream banks, shores, borders of woods and thickets, barnyards, and all sorts of waste ground. The earliest Michigan collections seem to be from Ann Arbor in 1870 and Port Huron in 1888, where Dodge noted that the species was "Becoming very common everywhere. A bad lawn weed."

Variable in flower size (9–20 mm long) and also in stature, with low creeping plants rooting at the nodes and hence almost impossible to exterminate in lawns, or with tall luxuriant stems in dense taller vegetation along ditches.

Sometimes confused, in haste, with *Lamium*, but easily distinguished by the distinctly pedicellate flowers (as well as the strongly 15-nerved calyx).

31. **Hedeoma**   False or Mock Pennyroyal

The generic name is feminine, in accord with Art. 62.2(b) of the Code (and previous Recommendations; see also Taxon 43: 105. 1994).

Our two species are both annuals, of weedy habit. They parallel in leaf shape the distinction between *Calamintha* and *Acinos*, but these (even if corollas and stamen number are not obvious) can be easily distinguished from *Hedeoma* by the nature of the pubescence on the calyx lobes.

REFERENCE

Irving, Robert S. 1980. The Systematics of Hedeoma (Labiatae). Sida 8: 218–295.

KEY TO THE SPECIES

1. Leaf blades linear, sessile, entire; cilia present on all calyx lobes . . . . .1. **H. hispida**
1. Leaf blades ± elliptic, short-petioled, often weakly and irregularly toothed; cilia present on only the lower 2 calyx lobes . . . . .2. **H. pulegioides**

1. **H. hispida** Pursh

Map 240. Sandy fields (often in quite bare patches), clearings, roadsides, gravel pits, and banks; railroad beds; prairie remnants; usually associated with some disturbance.

In linear leaves and a distended (gibbous) calyx, this species might be confused with *Calamintha arkansana*, which, however, has 4 stamens, much larger flowers, and glabrous foliage.

2. **H. pulegioides** (L.) Pers.

Map 241. Damp or dry fields and pastures; woodlands (oak or beech-maple), especially in openings and sometimes abundant after logging or other disturbance.

The 3 calyx lobes forming the upper lip are not only essentially without cilia but also more triangular-acute than the ciliate strongly narrowed lobes in the superficially similar *Acinos arvensis*.

32. Calamintha

This genus has frequently been included in the genus *Satureja*.

REFERENCE

Shinners, Lloyd H. 1962. Calamintha (Labiatae) in the Southern United States. Sida 1: 69–75.

1. **C. arkansana** (Nutt.) Shinners   Fig. 60   Calamint
     Map 242. Locally abundant in damp or springy flats and hollows or hummocks (depending on high or low water levels) among dunes and on rocky shores or edges of thickets, seldom far from the shores of Lakes Michigan and Huron, although rarely at inland lakes. From calcareous places in northern Michigan, ranges mostly south and southwest, to Texas.
     This species is also known as *Satureja arkansana* (Nutt.) Briquet. It is sometimes included in *C. glabella* (Michaux) Bentham, a species with pubescent nodes and no stolons. *C. arkansana* produces leafy stolons with rounded leaves quite unlike the linear leaves on the mid and upper parts of erect shoots. This is a very aromatic little plant, the new seedlings early in the season easily detected when one crushes them under foot while looking for, say, *Primula mistassinica* or *Viola nephrophylla* in the same habitat. It ordinarily reaches the peak of flowering late in July, when it is especially attractive with flowers larger than the leaves.
     Plants with pure white corollas [for which no name has been validly published] occur with normal purple-flowered ones, and corollas of intermediate shade may be found in the same populations.

33. Acinos

This genus, like the preceding, and along with *Clinopodium*, has frequently been included in *Satureja*.

238. Lamium purpureum   239. Glechoma hederacea   240. Hedeoma hispida

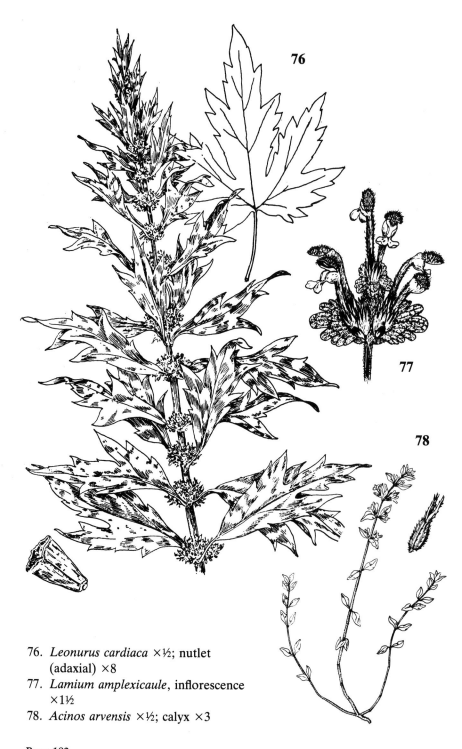

76. *Leonurus cardiaca* ×½; nutlet
    (adaxial) ×8
77. *Lamium amplexicaule*, inflorescence
    ×1½
78. *Acinos arvensis* ×½; calyx ×3

REFERENCES

DeWolf, Gordon P. 1955. Notes on Cultivated Labiates 4. Satureja and Some Related Genera. Baileya 2: 142–150.
Lawrence, G. H. M. 1961. The Name of the Basil-thyme, Acinos arvensis. Baileya 9: 125.

1. **A. arvensis** (Lam.) Dandy   Fig. 78   Mother-of-thyme;
Basil-thyme

Map 243. A native of Europe and especially well established in counties near the Straits of Mackinac, where its preferred calcareous habitat abounds. Collected by the First Survey in Kalamazoo County in 1839 but not again gathered in Michigan until 1909–1917, at sites in Emmet and Cheboygan counties. By the 1950s it was showing up more commonly, in sandy, gravelly, and rocky clearings, roadsides, fields, and shores; also along trails and railroads.

This little blue-flowered annual closely resembles *Hedeoma pulegioides*, but differs in stamen number (4), larger flowers, and having all 5 calyx lobes ciliate. In *Satureja*, this species is called *S. acinos* (L.) Scheele.

SOLANACEAE   Nightshade Family

Many species in this family are unpleasant-smelling and/or clammy-pubescent. Many are poisonous or thought to be so, although persons vary in their responses and plants may vary with age or geography. Strong drugs are obtained from some species. It is not a family to trust when "living off the land," for "living" may turn into the alternative. On the other hand, notable edible species include the white potato [*Solanum tuberosum*], eggplant [*S. melongena*], tomato [*Lycopersicon esculentum*], and species of *Capsicum* (the hot peppers, sweet peppers, paprika, Tabasco and chili peppers, and their kin—all natives of the New World, but none established outside of cultivation in our region).

241. Hedeoma pulegioides   242. Calamintha   243. Acinos arvensis
arkansana

REFERENCES

Fassett, Norman C. 1944 ["1943"]. Preliminary Reports on the Flora of Wisconsin. XXXI. Solanaceae. Trans. Wisconsin Acad. 35: 105–112.

Hawkes, J. G., R. N. Lester, & A. D. Skelding (eds.). 1979. The Biology and Taxonomy of the Solanaceae. Linn. Soc. Symp. 7. 738 pp.

Heiser, Charles B., Jr. 1969. Nightshades The Paradoxical Plants. Freeman, San Francisco. 200 pp. [Reprinted 1987 by Dover with new title: The Fascinating World of the Nightshades.]

Jacobs, Brian F., & W. Hardy Eshbaugh. 1983. The Solanaceae of Ohio: A Taxonomic and Distributional Study. Castanea 48: 239–249.

KEY TO THE GENERA

1. Flowers all or mostly more than 1 at a node, in peduncled inflorescences or on axillary pedicels (at least on older woody stems in *Lycium*)
    2. Plant woody; leaves all simple, unlobed, entire, glabrous, less than 1.8 (4) cm wide; flowers on slender glabrous pedicels in axils of foliage leaves (opposite or fascicled at least on older stems) . . . . .1. **Lycium**
    2. Plant herbaceous (or somewhat woody toward base); leaves entire, sinuate to deeply lobed, or compound, at least slightly pubescent, the largest (2) 3–22 cm wide; flowers in peduncled inflorescences or on hairy pedicels
        3. Lobes of corolla much shorter than the tube; leaves entire
            4. Corolla greenish yellow with elongate tube; flowers in branched inflorescence; fruit a capsule . . . . .2. **Nicotiana**
            4. Corolla white (with yellow star-like center), ± flat (rotate); flowers all or mostly fascicled in axils of leaves; fruit a berry closely enclosed in the enlarged calyx . . . . .3. **Leucophysalis**
        3. Lobes of corolla nearly as long as the tube or longer; leaves (except youngest) usually irregularly sinuate to deeply lobed or compound
            5. Corolla yellow; plant not spiny; anthers opening by longitudinal slits . . . . .4. **Lycopersicon**
            5. Corolla white, purple, or if yellowish the plant spiny; anthers opening by terminal pores . . . . .5. **Solanum**
1. Flowers all one per node (solitary in the axils of leaves or bracts) or between nodes (developing a terminal, bracted spike in *Hyoscyamus*)
    6. Plant with either auriculate-clasping leaves or with calyx ca. (3) 3.5–9 (11.5) cm long; fruit a capsule
        7. Cauline leaves auriculate-clasping at the base; corolla less than 3.5 cm long, reticulate-veiny (purple on yellow-green petals); calyx not over 2.5 cm long but completely enclosing the smooth circumscissile fruit . . . . .6. **Hyoscyamus**
        7. Cauline leaves petiolate, not clasping; corolla ca. 7.5–20 cm long with no reticulate pattern; calyx ca. (3) 3.5–9 (11.5) cm long but shed at maturity to expose the conspicuously spiny dehiscent fruit . . . . .7. **Datura**
    6. Plant with neither auriculate-clasping leaves nor calyx over 3.5 cm long (except in fruit in some species of *Physalis*); fruit a capsule (not often seen) or a berry
        8. Corolla yellow (whitish in one rarely escaped species); lobes of calyx shorter than the tube . . . . .8. **Physalis**
        8. Corolla blue to red or purple or sometimes white; lobes of calyx distinctly longer than the tube
            9. Flowers over 4 cm long, erect or ascending; fruit a capsule, not covered by the calyx; stems clammy-pubescent with multicellular hairs . . . . .9. **Petunia**
            9. Flowers less than 4 cm long, nodding; fruit a berry, covered by or at least with conspicuous enlarged calyx; stem glabrous to pubescent

10. Leaves coarsely toothed; calyx lobes broadly ovate, overlapping; berry enclosed in enlarged calyx; plant annual . . . . .10. **Nicandra**
10. Leaves entire; calyx lobes ovate-lanceolate, scarcely overlapping at base; berry merely subtended by enlarged calyx lobes; plant perennial . . . . .11. **Atropa**

1. **Lycium**   Matrimony Vine

KEY TO THE SPECIES

1. Narrowed part (before flaring) of mature corolla tube hidden in the calyx tube; calyx ca. 3–3.5 mm long . . . . .1. **L. chinense**
1. Narrowed part of mature corolla tube about equalling or slightly exceeding the calyx tube (visible between the lobes, at least); calyx ca. 3.5–4.5 mm long . . . . .2. **L. barbarum**

1. **L. chinense** Miller
Map 244. A native of eastern Asia, rarely escaped from cultivation to dry banks, roadsides, railroads, fields.

Not easily distinguished from the next species; the distinctions used here are derived from those of Stearn in *Flora Europaea*. The leaves tend to be nearly obtuse or even rounded, compared to the very acute leaves of *L. barbarum*.

2. **L. barbarum** L.   Fig. 79
Map 245. Apparently a native of Asia, widely cultivated (e.g., for hedges, because of the spines) and occasionally escaped to waste ground, including roadsides, railroads, sites of old buildings, and dry open fields.

As now recognized, this species includes *L. halimifolium* Miller. The corolla lobes in our specimens run ca. 4–7 mm long—a bit longer than often stated.

2. **Nicotiana**   Tobacco
Besides the one species treated here and probably no longer to be found in the state, *N. longiflora* Cav. (with a very long corolla tube) was collected at East Lansing 1894–1901; as the species was not listed in Beal's *Michigan*

244. Lycium chinense

245. Lycium barbarum

246. Nicotiana rustica

Flora (1905), we can assume that it was only in cultivation there. Likewise, *N. tabacum* L., now the common commercial tobacco species, is not known out of cultivation—nor much *in* cultivation—in Michigan, although it is grown in southern Ontario. This species is a cultigen originating in pre-Columbian times and now widespread. As Heiser has aptly written (1969): "For a plant that provides neither food nor drink and is considered harmful by a great many people, tobacco has achieved a remarkable success. While some people have considered it, along with corn and potatoes, as one of the New World's greatest gifts to the Old, others have suggested that it was the Indian's revenge."

1.   **N. rustica** L.   Wild Tobacco
Map 246. Probably originating in South America, but early spread by native Americans (who cultivated it) into North America. Our few collections are all from waste places, those for Kent and Washtenaw counties before 1900 and one for Ingham County in 1927.

This species has wingless petioles, while the tobacco that has replaced it in cultivation, *N. tabacum*, has short winged petioles or sessile decurrent leaves. *N. rustica* has a higher nicotine content and was cultivated in Europe even before *N. tabacum*.

3.   **Leucophysalis**

REFERENCE

D'Arcy, William G., Kathy Pickett, & Richard C. Keating. 1990. Investigation into Leucophysalis grandiflora. Wildflower 3(2): 20–26.

1.   **L. grandiflora** (Hooker) Rydb.   Plate 3-F
Map 247. Grows in a fairly narrow belt from Quebec and northern New England across the northern Great Lakes region and into Saskatchewan. To find it, one needs to understand its restricted habitat: sandy, gravelly, or rocky disturbed areas (including floodplains). It thrives a year or two after bulldozing, clearing, or fire, and survives for only 2–3 years if the area is not kept disturbed. Dumps, gravel pits, and roadsides may have extensive stands, along with construction sites. I have been unable to find any specimens to document old reports from Iosco and Clare counties.

Although usually said to be an annual, may live for an additional year or two. A large bushy plant covered with sometimes as many as 300 white flowers (3) 3.5–5 (8) cm wide is a handsome sight, but the foliage is malodorous and clammy-pubescent, so the plant has its unpleasant features. Small plants may bear only one flower or a few. The blooming season is long, and large plants, or a good stand, may be relied upon for flowers from late May until early September.

Another species of this genus grows in western North America and there are others in eastern Asia. Taxonomists now seem agreed that these are distinct from *Chamaesaracha*. Both these genera have sometimes been included in a broad concept of *Physalis*.

4. Lycopersicon

1. **L. esculentum** Miller   Fig. 82   Tomato
  Map 248. Waste places and disturbed ground, including dumps, railroad embankments, shores, river banks. This very well known garden plant originated in the Western Hemisphere (Mexico or South America). It grows readily from seed, and appears frequently as a waif. The seeds may survive sewage treatment and be dispersed in sludge.
  The tomato is an excellent example of a true berry: a fleshy, many-seeded, indehiscent fruit. Of that there can be no doubt. Whether one also wants to call it a vegetable, for which there is no precise botanical definition (unless one includes *all* plant material), is a matter of taste. Michigan ranks third among the states in its crop of tomatoes for processing (and 12th in fresh market tomatoes).
  Sometimes included in the next genus, as *Solanum lycopersicum* L. For over a century, until the name *L. esculentum* was conserved in 1987, the correct (though not always used) name in *Lycopersicon* was *L. lycopersicum* Karsten—not technically a tautonym (in which generic name and specific epithet are identical), but so close as to seem undesirable in view of the prohibition of tautonyms. (See Taxon 36: 74. 1987; and 37: 434 & 439; 1988.)

5. Solanum
  Most species are known or reputed to be poisonous, and even potatoes which have turned green when exposed to light should be thoroughly boiled (discarding the water) before eating. Toxicity varies, and may be less in fully ripe fruits, but experimenting with *Solanum* fruits is not recommended. Eggplant, *S. melongena* L., does not escape from cultivation.

247. Leucophysalis
     grandiflora

248. Lycopersicon
     esculentum

249. Solanum rostratum

REFERENCES

Bassett, I. J., & D. B. Munro. 1985. The Biology of Canadian Weeds. 67. Solanum pty-canthum [sic] Dun., S. nigrum L. and S. sarrachoides Sendt. Canad. Jour. Pl. Sci. 65: 401–414. [Includes key to all weedy species (omitting only *S. tuberosum* of those below).]

Bassett, I. J., & D. B. Munro. 1986. The Biology of Canadian Weeds. 78. Solanum caro-linense L. and Solanum rostratum Dunal. Canad. Jour. Pl. Sci. 66: 977–991.

Edmonds, Jennifer M. 1986. Biosystematics of Solanum sarrachoides Sendtner and S. physalifolium Rusby (S. nitidibaccatum Bitter). Jour. Linn. Soc. Bot. 92: 1–38.

Hughes, Meredith Sayles. 1991. Potayto, Potahto—Either Way You Say It, They A'peel. Smithsonian 22(7): 138–140; 142; 144–146; 148–149.

Lawrence, George H. M. 1960. The Cultivated Species of Solanum. Baileya 8: 20–35.

Schilling, Edward E. 1981. Systematics of Solanum sect. Solanum (Solanaceae) in North America. Syst. Bot. 6: 172–185.

KEY TO THE SPECIES

1. Plants spiny or prickly; pubescence of stems and leaves otherwise all or mostly stellate
    2. Calyx spiny, fully enclosing the berry when mature; corolla yellow; leaves very deeply lobed or even compound; plant a tap-rooted annual . . . . .1. **S. rostratum**
    2. Calyx without spines (may be coarsely stellate-hairy), not enclosing the berry; corolla white to purple; leaves simple, with only an irregularly sinuate or coarsely toothed margin; plant perennial with deep, often rhizome-like roots . . . .
    . . . . .2. **S. carolinense**
1. Plants without spines or prickles; pubescence of simple hairs
    3. Plant a subshrubby perennial vine, with elongate sprawling or climbing stems; corolla purple (occasionally albino); fruit red at maturity . . . . .3. **S. dulcamara**
    3. Plant an erect (straight or bushy) herbaceous annual; corolla white; fruit green or black at maturity
        4. Leaves pinnately compound or lobed at least halfway to the midrib
            5. Corolla more than 1 cm broad; leaves pinnately compound (tiny leaflets alternat-ing with much larger ones); plant with large underground tubers (potatoes), only rarely fruiting . . . . .4. **S. tuberosum**
            5. Corolla less than 1 cm broad; leaves pinnately lobed; plant without tubers, usually fruiting copiously . . . . .5. **S. triflorum**
        4. Leaves ± irregularly toothed or sinuate (or occasionally entire), neither com-pound nor lobed
            6. Plant (especially stems and inflorescence) covered with ± dense, spreading, multicellular, glandular hairs; mature fruit dark green to green-brown, ca. half covered by the calyx . . . . .6. **S. physalifolium**
            6. Plant glabrate, at most sparsely pubescent with very short ± appressed eglandular hairs; mature fruit black, scarcely if at all covered by the calyx . . . . .
            . . . . .7. **S. ptychanthum**

1. **S. rostratum** Dunal   Buffalo-bur

Map 249. Native west of the Great Lakes, but an unpleasant weed in our area. Farmyards, pastures (uneaten), flower beds, roadsides, gravel pits, fields, dump heaps, railroads, and other disturbed ground, though appar-ently not lasting long in a given site. The earliest Michigan collection I have seen is from Lansing in 1871, but there are no others until several counties in the 1890s.

This species seems to be less often a cause of poisoning than the next, perhaps because its spine-covered fruits are so unattractive to ingest. It was the original native host of the Colorado potato beetle. Some recent authors have included this species in *S. cornutum* Lam., but that position appears to be both taxonomically and nomenclaturally incorrect (see Whalen in Gentes Herb. 11: 397–406. 1979).

2. **S. carolinense** L.   Fig. 83   Horse-nettle

Map 250. A native of the southeastern United States, spread northward as an obnoxious (also noxious) weed. Sandy fields, pastures, a weed in waste ground and gardens; roadsides and railroads; floodplains and river banks. First collected in Michigan in the 1890s in Kent, Muskegon, St. Joseph, and Wayne counties.

The fruit is poisonous when eaten by people and farm animals, and the foliage serves as an alternate host for various insects and diseases of crop plants. Occasionally there are a very few spines on a calyx.

3. **S. dulcamara** L.   Fig. 80   Nightshade; Bittersweet

Map 251. Introduced from Eurasia and so thoroughly naturalized in North America that to many people it appears as if native. The species was apparently widespread in the southern Lower Peninsula by the 1890s although only three counties are represented by collections before that decade, all in the 1860s and 1870s: Genesee, Ionia, and Washtenaw. Not listed by the First Survey for 1838. Now all too abundant in swamp forests (conifers and hardwoods), depressions and clearings in deciduous woods, edges of ponds and marshes, river banks, thickets and shores, waste ground, dumps, roadsides, railroads, fields, cultivated ground, and gardens.

All our plants have at least a little pubescence on the leaves and stems, especially young ones. The larger leaves are almost always deeply lobed, but smaller ones may not be. The attractive flowers with purple corollas and cone of yellow anthers are familiar to many. The corolla is occasionally white: f. *albiflorum* Farwell (TL: Trenton [Wayne Co.]). Walpole described a plant from Ypsilanti as red-flowered. The ripe fruit is an attractive red.

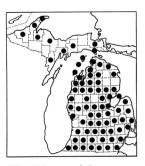

250. Solanum carolinense   251. Solanum dulcamara   252. Solanum tuberosum

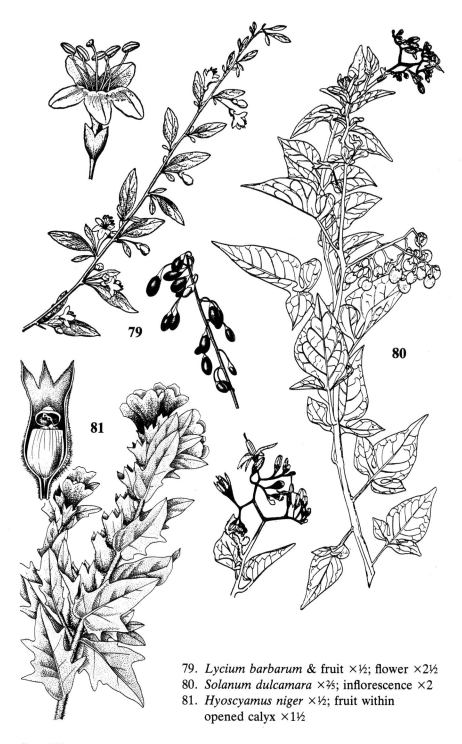

79. *Lycium barbarum* & fruit ×½; flower ×2½
80. *Solanum dulcamara* ×⅖; inflorescence ×2
81. *Hyoscyamus niger* ×½; fruit within opened calyx ×1½

Plants emit an unpleasant odor when bruised and all parts are more or less poisonous if ingested.

This is the original Old World "bittersweet," but in North America that common name has more often been used for *Celastrus*, a genus not native in Europe and in quite a different family.

4. **S. tuberosum** L.   Potato

Map 252. The potato's ancestors were originally native in highlands of South America, but our well known vegetable was developed in cultivation over thousands of years. Not really established anywhere out of cultivation, but surviving, where dumped, long enough to flower or for as long as 5 years in abandoned fields; occasionally found in ditches and waste ground. Included here despite dubious credentials as established.

Whether termed "Idaho," "Michigan," "Maine," "Irish," or other sort, the white potato is *Solanum tuberosum*. In both volume and sales dollars, potatoes are Michigan's leading produce commodity. Michigan is the nation's largest producer of summer or "new" potatoes and potatoes for chip manufacturing. For readable essays on the history of the potato, see Chapter 2 in Heiser (1969) and Hughes (1991).

5. **S. triflorum** Nutt.   Cut-leaved Nightshade

Map 253. A malodorous native west of the Great Lakes, adventive eastward and not found in Michigan until 1984 (*Reznicek 7454*, MICH), as a weed at the edge of a tilled garden in Ann Arbor.

6. **S. physalifolium** Rusby   Hairy or Argentinian Nightshade

Map 254. A native of South America, weedy into North America, Europe, Australia, and New Zealand. First reported from Michigan (Michigan Bot. 16: 136–137. 1977—as *S. sarrachoides*) from Cheboygan County, where it has now been found in additional places, but earlier collections have subsequently turned up under other names in herbaria: Livingston County (1965) and Berrien County (1973). In any event, this is evidently a recent introduction in newly bulldozed or cultivated ground and waste places.

Edmonds (1986) distinguished this species, with calyx lobes at anthesis ca. 1–2 mm long and flowers 4–8 in the inflorescence, from *S. sarrachoides* Sendtner. The latter species has larger calyx lobes but fewer flowers in the inflorescence.

7. **S. ptychanthum** Dunal   Black Nightshade

Map 255. Native in eastern North America, but weedy in habit. A pioneer on ground recently bulldozed, cleared, burned, or exposed by lowering of a pond or river; roadsides, railroads, old fields, shores, gravel pits, dump heaps, dooryards, and other waste ground; a weed in gardens.

This is the species long called *S. americanum* Miller, a name that has been typified in such a way that it now applies to a more southern species. Both belong to the complex of *S. nigrum* L., a Eurasian species not so common in North America. Our specimens all appear to be *S. ptychanthum* (the epithet usually misspelled *ptycanthum* by recent authors—but not by Dunal in Lamarck), which has umbelliform inflorescences, seeds 1.3–1.6 mm long, and anthers 1.4–1.9 mm long. In *S. nigrum* the inflorescence is racemiform, the seeds and anthers are larger; the berries are dull; and plants are hexaploid, while in *S. ptychanthum* the berries are shiny and plants diploid. The southern *S. americanum*, as now defined, is similar to *S. ptychanthum* but has numerous white flecks on the immature fruit and fewer (not over 5) hard granules mixed with the numerous seeds in the mature berries.

6. Hyoscyamus

1. **H. niger** L. Fig. 81 Henbane
  Map 256. A Eurasian species, locally established as an escape from cultivation but not recommended as an ornamental! Known from Michigan since the First Survey (probably 1837) and found in the early 1860s at Mackinac Island (where well established), Detroit, and Ann Arbor. Waste places, including newly made roadsides.
  This is a fetid, sticky-pubescent plant all parts of which are poisonous to ingest; for many centuries used medicinally (or worse), the source of the potent drug alkaloid hyoscyamine. The stem conspicuously elongates in fruit, so that it appears to terminate in a large spike or raceme.

7. Datura Jimson-weed
  Species of *Datura* are very poisonous plants, all their parts containing strong alkaloids, including atropine, hyoscyamine, and scopolamine. These are useful in medicine, under carefully controlled conditions, but cultivation for the attractive large flowers is not recommended, even if one concedes the problem of overpopulation. Children have been reported as poisoned even by sucking nectar from the corolla tube.

REFERENCE

DeWolf, Gordon P., Jr. 1956. Notes on Cultivated Solanaceae 2. Datura. Baileya 4: 12–23.

KEY TO THE SPECIES

1. Stem and petioles densely puberulent; fruit dehiscing irregularly, on recurved pedicel; flowers ca. 12–20 cm long; leaves entire or with an occasional broad tooth . . . . .1. **D. inoxia**
1. Stem and petioles glabrous or nearly so; fruit dehiscing regularly, on erect pedicel; flowers ca. (7.5) 8–10 cm long; leaves coarsely but clearly toothed throughout . . . .
  . . . . .2. **D. stramonium**

1. **D. inoxia** Miller

Map 257. Native to southwestern United States and adjacent Mexico, but occasionally established as a weed or escape from cultivation across the continent (and abroad). Cultivated ground, farmyards, alleys, roadsides.

Including *D. meteloides* Dunal, long misidentified as *D. metel* L., a Chinese species not known in our region outside of cultivation and differing in being as glabrous as *D. stramonium*. *D. inoxia* (the epithet often misspelled *innoxia*) has long been used in ceremonial rites in its native range. The plants cultivated and escaped in our region have sometimes been segregated as *D. wrightii* Regel.

2. **D. stramonium** L.   Fig. 84

Map 258. Probably native originally in the Western Hemisphere and early spread from eastern North America to Europe and elsewhere. Waste places including dumps, vacant lots, farmyards, roadsides, and disturbed mudflats and shores (especially recently exposed ones).

A tall annual plant, with very long-lived seeds and appearing quite dramatically when appropriate disturbance or exposure causes long-dormant seeds to grow. The flowers may be white or purple; the name *D. tatula* L. was applied to the latter form, but as the difference has been shown to result from a single gene it is hardly worthy of recognition except as f. *tatula* (L.) B. Boivin.

253. Solanum triflorum

254. Solanum
    physalifolium

255. Solanum
    ptychanthum

256. Hyoscyamus niger

257. Datura inoxia

258. Datura stramonium

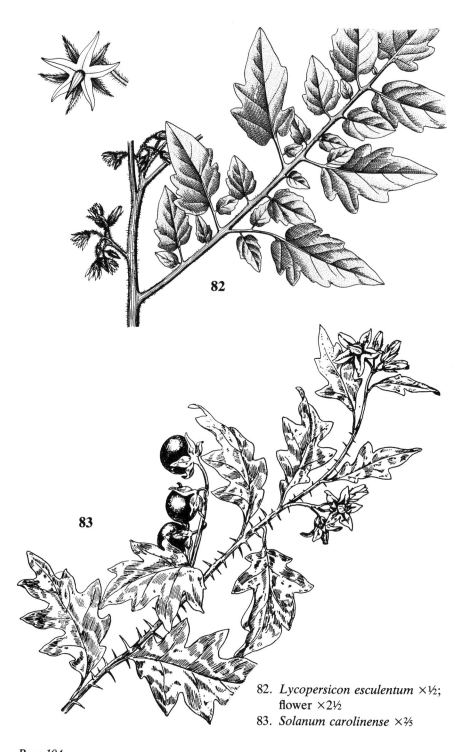

82. *Lycopersicon esculentum* ×½;
    flower ×2½
83. *Solanum carolinense* ×⅖

8. Physalis   Ground-cherry; Husk-tomato

The berry in some species resembles a miniature tomato (and, unlike many Solanaceae, is similarly non-poisonous); it is, however, completely surrounded by the enlarged calyx. The species are not always easy to tell apart, and there is not full agreement on what the species are. Ordinarily, determining whether a plant is annual or perennial is a difficult key character, especially when only herbarium material is available, but in *Physalis* that character can be particularly helpful and often relatively easy to use. The annual species have a small root, easily pulled from the ground and hence usually present on collected specimens; the perennial species have a large root or rhizome often *not* gathered (except by truly conscientious collectors).

REFERENCE

Waterfall, U. T. 1958. A Taxonomic Study of the Genus Physalis in North America North of Mexico. Rhodora 60: 107–114; 128–142; 152–173.

KEY TO THE SPECIES

1. Corolla whitish, with distinct lobes and sinuses; calyx at maturity bright red or orange, over 3.5 cm long . . . . 1. **P. alkekengi**
1. Corolla yellow (brown-centered), without distinct sinuses, only minute lobes; calyx at maturity green (or pale orange), less than 3.5 (3.8) cm long
  2. Stems essentially glabrous or with a few antrorse hairs (especially toward summit); anthers blue
    3. Pedicels, even in fruit, less than 1 cm long; plant annual . . . . .2. **P. philadelphica**
    3. Pedicels 1–3 cm long; plant perennial . . . . .3. **P. longifolia**
  2. Stems rather evenly spreading-hairy or with at least short retrorse hairs; anthers blue or yellow
    4. Hairs (or little bristles) of stem at least in large part retrorse or recurved, none of them glandular or sticky; anthers yellow, 1.6–2.5 (2.8) mm long . . . . . . .4. **P. virginiana**
    4. Hairs of stem ± dense and multicellular, all spreading, at least some sticky or glandular; anthers various
      5. Plants annual; flowers 6–9 [11] mm long; anthers blue, ca. 1.5–2 mm long, mostly on slender filaments; mature fruiting calyces not over 2.5 (3) cm long, on pedicels less than 1 cm long . . . . .5. **P. pubescens**
      5. Plants perennial; flowers (10) 12–18 (20) mm long; anthers yellow or tinged with blue, (2.5) 2.8–4 mm long, on broadly dilated filaments; mature fruiting calyces ca. 2.5–3.7 cm long, on pedicels ca. 1.5–2.7 (4.5) cm long . . . . .6. **P. heterophylla**

1. P. alkekengi L.   Chinese-lantern-plant

Map 259. A Eurasian species, grown as an ornamental (for its calyx) and rarely escaping or spreading by rhizomes if not seeds. Near old gardens, railroads, and waste ground.

Dry plants keep well for winter bouquets.

2. P. philadelphica Lam.   Tomatillo

Map 260. A Mexican native, cultivated for the edible fruit and rarely escaped to disturbed ground. The Emmet County material cited by Waterfall is labeled as cultivated.

This is the plant long misidentified as *P. ixocarpa* Hornem., which is not known in America. (See Ann. Missouri Bot. Gard. 60: 668–669. 1973; & Bol. Soc. Brot. 44: 343–367. 1970.)

3. **P. longifolia** Nutt.

Map 261. Waste places (usually dry and sandy), including roadsides, railroads, gravel pits; fields, shores, and meadows; a weed in gardens and cultivated ground; borders of woods and disturbed areas; rarely in swampy ground.

Unlike our other species, the pedicels are ± densely covered with antrorse, sometimes curved, pubescence. Included here is *P. subglabrata* Mack. & B. F. Bush, now considered not a distinct species; some authors, following Waterfall, have included all these in *P. virginiana*.

4. **P. virginiana** Miller   Fig. 85

Map 262. Less weedy in habit than some of our species, in prairies, oak and jack pine savanas; also sandy open fields, roadsides, and railroads.

Usually the pubescence on stems and pedicels is little more than tiny retrorse bristles, giving the plant a strongly scabrous (hardly hairy) aspect. Michigan specimens labeled in the past as *P. lanceolata* Michaux (not now recognized as a good species) are mostly *P. virginiana*.

5. **P. pubescens** L.

Map 263. Common farther south and perhaps only adventive in Michigan, in weedy places.

Included here are the plants referred by some authors (apparently in error) to *P. pruinosa* L. Generally recognizable by its smaller flowers and calyces, together with shorter anthers and pedicels, although an exception may occur to any feature mentioned in the key. A glabrate variety has (perhaps fortunately!) not been collected in Michigan.

259. Physalis alkekengi   260. Physalis philadelphica   261. Physalis longifolia

84. *Datura stramonium* ×½
85. *Physalis virginiana* ×½;
portion of stem ×6; calyx ×1½

6. **P. heterophylla** Nees   Fig. 86

Map 264. Dry sandy fields, hillsides, and banks; along trails in woods, grassy clearings, meadows; waste places, dumps, gravel pits, roadsides, railroads, parking lots; weedy sites near buildings, old gardens, cultivated ground, and on shores; invades wooded areas after logging and fire.

This is our commonest *Physalis*, variable in pubescence but usually with stems densely covered with long, spreading, sticky, multicellular hairs, and with long anthers and pedicels, large fruiting calyces, and deep underground parts.

9. Petunia

REFERENCE

Ferguson, Margaret C., & Alice M. Ottley. 1932. Studies on Petunia III. A Redescription and Additional Discussion of Certain Species of Petunia. Am. Jour. Bot. 19: 385–405.

1. **P. ×atkinsiana** Loudon   Petunia

Map 265. Our petunias, so familiar as ornamentals, are natives of South America, rarely escaped temporarily to roadsides, fields, shores, and other such sites.

The common garden petunia is an annual of hybrid origin, involving perhaps three species, and has been so extensively manipulated and selected that drawing lines between species and hybrids is difficult. The collective name *P. ×atkinsiana* [= *P. ×hybrida* Vilm.] is used here for the few recorded escapes from cultivation. Another position would be to consider this and its parents—if one were sure what they are—as a single species (cf. the Kartesz *Checklist*).

10. Nicandra

1. **N. physalodes** (L.) Gaertner   Apple-of-Peru; Shoo-fly Plant

Map 266. A native of Peru, cultivated for its blue flowers and rarely a waif in waste ground (our few collections all from before 1940).

262. Physalis virginiana   263. Physalis pubescens   264. Physalis heterophylla

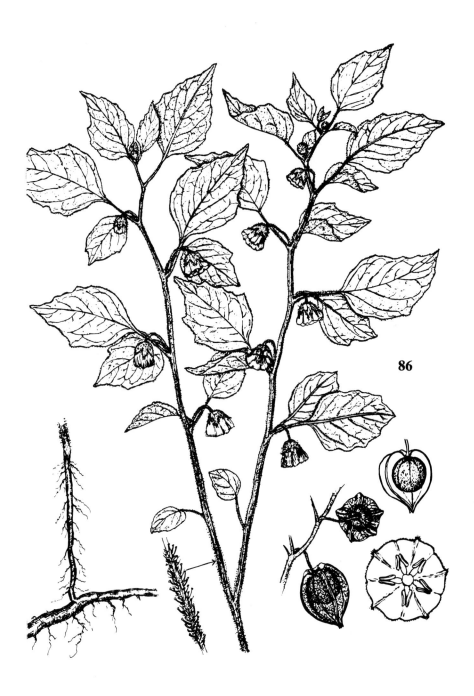

86

86. *Physalis heterophylla*, underground parts, fruit, and calyx opened with fruit ×½; opened flower ×1½

11. Atropa

1. **A. belladonna** L.   Deadly Nightshade; Belladonna
    Map 267. Native of the Old World, cultivated as a drug plant (though no longer much in the United States), and very rarely escaped. Farwell (*1481* in 1894, BLH; Asa Gray Bull 4: 46 & 62. 1896) found waifs in waste places at Detroit. The status of a "1931" collection from Ann Arbor is not certain, but the fact that the collector's original identification was "Ruellia (?)"—a plant in the Acanthaceae—suggests that he was innocent of knowingly gathering a garden plant (but the date may be an error for 1913, when the collector was a graduate student in botany at the University of Michigan).
    The sap is red and the berry a glossy black in this perennial and very poisonous species. It is a source of atropine, scopolamine, and other alkaloids used as sedatives, antispasmodics, and, especially, mydriatics (to dilate the pupil of the eye). Everyone who has had an eye examination knows the effect of atropine (or its corresponding synthetics).

SCROPHULARIACEAE   Snapdragon Family

The corolla in this family is typically 2-lipped and strongly bilaterally symmetrical. However, in several genera the flowers are nearly regular (radially symmetrical), with the corolla lobes similar in size and shape. Among our genera, the fertile stamens are 5 only in *Verbascum*. The fertile (anther-bearing) stamens are 2 or 4 in the other genera, sometimes with 1 or more sterile stamens (staminodia), which may be different from the filaments of the fertile ones (e.g., fig. 97). These structures are often not easy to see, especially on small or pressed specimens, as they tend to be inside the corolla tube, and they are not stressed in the keys here—though important in the classification of the family.

A number of the genera are hemiparasitic, the plants containing chlorophyll and bearing leaves but also depending (or at least existing) in part on roots of neighboring plants. A few of the many publications pertaining to this habit are cited, especially if the work was done in the Great Lakes

265. Petunia ×atkinsiana   266. Nicandra physalodes   267. Atropa belladonna

region. Good color photos of several of these parasites are provided by Musselman and Mann (1978). The hemiparasitic genera are all in the subfamily Rhinanthoideae, although they do not appear consecutively here, where genera are in the same sequence as they are most conveniently keyed. They tend to blacken when dried and to wilt very quickly when picked (although readily freshening when cut stems are placed in water). The type genus in this subfamily is represented by the circumboreal *Rhinanthus crista-galli* L. (sens. lat.), yellow-rattle, which is somewhat aggressive toward the southern edge of its range and has become established in thickets and openings at some places along the north shore of Lake Superior in Ontario. It has opposite, toothed leaves on an erect stem, yellow 2-lipped corollas, and the calyx laterally compressed even when inflated in fruit. Some day it may well show up in northern Michigan.

I am grateful to Tom S. Cooperrider and Frank S. Crosswhite for allowing me to study their manuscripts (at the time, unpublished) on this family in Ohio and Wisconsin, respectively.

REFERENCES

Musselman, Lytton J., & William F. Mann, Jr. 1977. Host Plants of Some Rhinanthoideae (Scrophulariaceae) of Eastern North America. Pl. Syst. Evol. 127: 45–53.

Musselman, Lytton J., & William F. Mann, Jr. 1978. Root Parasites of Southern Forests. U. S. Dep. Agr. For. Serv. Genl. Tech. Rep. SO-20. 76 pp.

Pennell, Francis W. 1935. The Scrophulariaceae of Eastern Temperate North America. Acad. Nat. Sci. Philadelphia Monogr. 1. 650 pp. + 2 maps.

Salamun, Peter J. 1951. Preliminary Reports on the Flora of Wisconsin. XXXVI. Scrophulariaceae. Trans. Wisconsin Acad. 40(2): 111–138.

Sutton, David A. 1988. A Revision of the Tribe Antirrhineae. Oxford Univ. Press, Oxford. 575 pp. + 4 microfiche (344 pp).

KEY TO THE GENERA

1. Cauline leaves all or mostly alternate on fertile stems (lowermost leaves sometimes opposite and rosette of larger basal leaves sometimes present)
  2. Corolla nearly regular, the lobes equalling or exceeding the tube
    3. Fertile stamens 5 (but not of equal length); corolla ca. 1–3 cm broad, yellow (or white); plant tall, stiffly erect . . . . .1. **Verbascum**
    3. Fertile stamens 2; corolla less than 1 cm broad, blue (or white); plant low, often creeping . . . . .19. **Veronica**
  2. Corolla bilateral, ± 2-lipped, the lobes distinctly shorter than the tube (including spur, if any)
    4. Stem trailing or sprawling; leaf blades not over 1.5 times as long as broad; corolla with basal spur; capsules ± globose, ca. 3–4.5 mm in diameter
      5. Stem, leaves, and calyx glabrous; leaf blades palmately veined and lobed or scalloped; seeds with thin raised reticulate ornamentation . . . . .2. **Cymbalaria**
      5. Stems, leaves, and calyx pilose; leaf blades pinnately veined, usually hastate (1–2 pairs of small basal lobes); seeds covered with dense convolute-rounded ornamentation . . . . .3. **Kickxia**
    4. Stem erect; leaf blades (or their principal lobe) over 1.5 times as long as broad; corolla spurred (*Chaenorrhinum* & *Linaria*) or not; capsules various, mostly longer than broad

6. Corolla with a slender basal spur projecting back between the lower calyx lobes (figs. 90 & 91)
  7. Flowers all solitary in axils of leaves (nearly to base of plant); corolla pale purple and white; leaves, calyx, and stem with ± dense gland-tipped hairs ... . .4. **Chaenorrhinum**
  7. Flowers in compact or elongate terminal inflorescences (ca. half or less the height of the plant); corolla yellow, red, or blue; leaves, calyx, and usually stem glabrous and eglandular or nearly so . . . . .5. **Linaria**
6. Corolla without spur (at most swollen or saccate at base)
  8. Corolla ca. half or more covered by the calyx
    9. Bracts of inflorescence green; stamens 2, strongly exserted beyond the corolla; basal leaves with ± cordate blades and distinct petioles; mid-cauline leaves bract-like, 1–2 (very rarely 3.5) cm long, obscurely toothed . . . . . .6. **Besseya**
    9. Bracts of inflorescence cream, yellow, or red at least apically; stamens 4, barely if at all exceeding the upper lip of corolla; basal leaves, if any, neither cordate nor petioled; mid-cauline leaves usually 2–6 (9) cm long, deeply lobed or completely entire . . . . .7. **Castilleja**
  8. Corolla much less than half covered by the calyx
    10. Leaves deeply pinnately lobed . . . . .8. **Pedicularis**
    10. Leaves unlobed (at most shallowly toothed)
      11. Leaves at least remotely crenate or toothed; corolla symmetrical (not swollen) at base, fully open at the throat . . . . .9. **Digitalis**
      11. Leaves entire; corolla swollen at base, closed at the throat by the palate
        12. Corolla at least 2.5 cm long, much exceeding the broadly ovate calyx lobes . . . . .10. **Antirrhinum**
        12. Corolla not over 1.5 cm long, barely if at all exceeding the linear calyx lobes. . . . . .11. **Misopates**
1. Cauline leaves all or mostly opposite (rarely whorled) on fertile stems (may be alternate beneath flowers)
  13. Inflorescence terminal and branched (± paniculate); stamens 4 fertile plus 1 staminodium
    14. Leaves below the inflorescence distinctly petioled; corolla brownish, less than 12 mm long; staminodium broad (ca. 1–2 mm) and flat at the free apex (mostly adnate to the upper lip), glabrous . . . . .12. **Scrophularia**
    14. Leaves below the inflorescence sessile; corolla white to purple-violet, ca. 15–30 (45) mm long; staminodium slender, elongate (of similar diameter and length as the style), close to lower lip of corolla, bearded at the apex (fig. 97) ...13. **Penstemon**
  13. Inflorescence a spike or raceme (no branched stalks), or flowers all axillary; stamens 2 or 4 fertile, in most genera with no staminodium (or only a very rudimentary one)
    15. Leaves (especially middle and lower ones) deeply pinnately toothed or lobed ca. one-third or more the distance to the midrib; corolla yellow or cream, ca. 1.5–5 cm long
      16. Flowers in dense spike-like racemes at the ends of stems and branches; stamens ± enclosed in the hooded upper lip of corolla; calyx and corolla both strongly 2-lipped . . . . .8. **Pedicularis (lanceolata)**
      16. Flowers axillary or in racemes; stamens not enclosed in upper corolla lip; calyx and corolla 5-lobed but only weakly 2-lipped
        17. Flowers (and fruit) less than 6 mm long . . . . .14. **Leucospora**
        17. Flowers (and usually fruit) over 10 mm long
          18. Corolla ca. 1.5 cm long; flowers nearly sessile (pedicels ca. 2 mm or less); calyx with eglandular pubescence . . . . .15. **Dasistoma**
          18. Corolla ca. 2.5–4.5 cm long; flowers on pedicels 2–13 mm long; calyx glabrous or with glandular or eglandular pubescence . . . . .16. **Aureolaria**

15. Leaves of main stem toothed or entire but not so deeply pinnately toothed or lobed (uppermost leaves or bracts may have small basal lobes); corolla yellow or not, in most species less than 1.5 cm long

  19. Sepals separate nearly or quite to the base; fertile stamens 2 or 4

    20. Corolla ca. 2.3–3.5 cm long; sepals broadly ovate-orbicular, overlapping; stamens 4 fertile plus a filamentous elongate staminodium . . . . 17. **Chelone**

    20. Corolla less than 1.5 cm long; sepals linear-lanceolate to somewhat ovate, not conspicuously overlapping; stamens 2 or 4 (including any staminodia)

      21. Corolla with a spur projecting back at the base; plant with ± dense gland-tipped hairs; leaves linear; fertile stamens 4 . . . . 4. **Chaenorrhinum**

      21. Corolla not spurred; plant glabrous or with eglandular hairs (or if with gland-tipped hairs, the leaves not linear); fertile stamens 2 (staminodia filamentous, reduced, or none)

        22. Leaves in whorls of 3–6 (7), sharply toothed; inflorescence of 1–several dense elongate slenderly tapering spikes or spike-like racemes; corolla tube much longer than the lobes . . . . 18. **Veronicastrum**

        22. Leaves opposite, entire or toothed; inflorescence racemose or flowers solitary in axils of alternate or opposite bracts or leaves; corolla tube various

          23. Corolla often nearly regular, the tube shorter than the lobes (usually a flat limb); flowers in axillary racemes or solitary in axils of bracts or leaves; sepals 4 . . . . 19. **Veronica**

          23. Corolla 2-lipped, the tube much longer than the lobes; flowers solitary in axils of opposite leaves; sepals 5

            24. Pedicels minutely glandular-pubescent, with a pair of sepal-like bractlets at their summit, subtending the calyx; leaves often gland-dotted beneath; corolla bright yellow to white; capsules ovoid or globose . . . . 20. **Gratiola**

            24. Pedicels smooth and glabrous, without bractlets; leaves not gland-dotted; corolla whitish to purple; capsule distinctly ellipsoid . . . . 21. **Lindernia**

  19. Sepals (at least at anthesis) fused ca. one-third or more the length of the calyx; fertile stamens 4

  25. Flowers (all or many of them, especially lower ones) in the axils of alternate bracts in a distinct terminal or racemose inflorescence

    26. Pedicels about equalling or longer than the calyx and much longer than the minute bracts; calyx glabrous; plants strongly stoloniferous . . . . 22. **Mazus**

    26. Pedicels shorter than the calyx and bracts (or none); calyx usually pubescent (except in *Euphrasia stricta*); plants ± erect

      27. Corolla 15–22 mm long; calyx lobes ca. one-third as long as tube, or shorter; leaves (at least the larger ones) (3) 4–8 cm long . . . . 23. **Buchnera**

      27. Corolla ca. 10 mm long or shorter; calyx lobes ca. half as long as tube; leaves not over 3 cm long

        28. Leaves pinnately veined (with a prominent midrib), 3–6 times as long as wide; mucros at ends of anther sacs equal in length; corolla uniformly and densely puberulent on the outside . . . . 24. **Odontites**

        28. Leaves ± palmately veined (with 3–5 prominent veins from the base), less than twice as long as wide; mucros at end of anther sacs very unequal (one much longer than the others); corolla glabrous or ± pubescent on the outside . . . . 25. **Euphrasia**

  25. Flowers all solitary in the axils of opposite (or whorled) leaves or bracts

  29. Corolla nearly regular, the 5 lobes all about the same size and shape

    30. Corolla bright yellow; calyx tube, pedicels, and/or stems with hairs of uniform or mixed lengths (not of 2 distinct lengths and only rarely completely glabrous)

31. Capsule (like the calyx tube, pedicels, and stem) ± densely pubescent with short uniform eglandular hairs; corolla ca. 3–4 cm long . . . .
. . . . .16. **Aureolaria (virginica)**
31. Capsule glabrous; calyx tube, pedicels, and/or stems with viscid or minute gland-tipped hairs; corolla less than 2.5 cm long (except in *M. guttatus* if sought here despite 2-lipped corolla) . . . . .28. **Mimulus** (couplet 3)
30. Corolla pink to purple (white in albinos); calyx and other parts glabrous (at most scabrous) or with hairs of distinctly different lengths
32. Leaves lanceolate or broader, the uppermost often with a pair of basal lobes; stem terete, with stiff retrorse hairs longer than the very short pubescence; calyx pubescent . . . . .26. **Tomanthera**
32. Leaves narrowly linear (up to 3.5 mm wide), without basal lobes; stem usually angled (not in *A. gattingeri*), glabrous or antrorse-scabrous; calyx glabrous . . . . .27. **Agalinis**
29. Corolla strongly 2-lipped and bilaterally symmetrical, the upper lip (2-lobed) and lower lip (3-lobed) differentiated in size and shape
33. Lobes less than a third the total length of the calyx, or corolla bright yellow (or both conditions) . . . . .28. **Mimulus**
33. Lobes ca. half or more the total length of the calyx; corolla blue, purple, white, or cream (never pure yellow)
34. Flowers and fruit on pedicels about equalling or exceeding the calyx; corolla with at least the lower lip deep blue . . . . .29. **Collinsia**
34. Flowers and fruit sessile or on pedicels distinctly shorter than the calyx; corolla with whitish to pink or magenta ground color (plus dark spots and/ or yellow markings)
35. Leaves ca. 3–16 (26) times as long as broad, including a short petiole . .
. . . . .30. **Melampyrum**
35. Leaves less than twice as long as broad, mostly sessile . . . . .25. **Euphrasia**

1. **Verbascum**   Mullein

This is a large Eurasian genus (ca. 250 species); all our species are adventive or escapes from cultivation, and identification of the uncommon ones is to some extent tentative. Hybrids are frequent in *Verbascum*, generally intermediate between their parents as distinguished in the key below.

268 Verbascum blattaria   269. Verbascum lychnitis   270. Verbascum phlomoides

REFERENCES

Gross, Katherine L., & Patricia A. Werner. 1978. The Biology of Canadian Weeds. 28. Verbascum thapsus L. and V. blattaria L. Canad. Jour. Pl. Sci. 58: 401–413.

Murbeck, Sv. 1933–1934. Monographie der Gattung Verbascum. Lunds Univ. Årsskr. II 29(2): 1–630 + 31 pl.

Reinartz, James A. 1984. Verbascum densiflorum in Southeast Wisconsin. Rhodora 86: 96–99.

Voss, Edward G. 1967. The Status of Some Reports of Vascular Plants from Michigan. Michigan Bot. 6: 13–24. [Verbascum, pp. 21–23.]

Wagner, W. H., Jr., Thomas F. Daniel, & Michael K. Hansen. 1980. A Hybridizing Verbascum Population in Michigan. Michigan Bot. 19: 37–45.

KEY TO THE SPECIES

1. Flowers one in axil of each bract, on a pedicel about equalling or exceeding the bract; leaves and stems at middle of plant glabrous or nearly so (upper portion often with sparse stalked or sessile glands) . . . . .1. **V. blattaria**
1. Flowers mostly 2 or more in axil of each bract, on pedicels shorter than the bracts; leaves and stem ± densely stellate-pubescent, not glandular
  2. Upper cauline leaves only slightly if at all decurrent
    3. Calyx 2.5–4 mm long, shorter than the longest pedicels; corolla less than 2 cm broad; pubescence of stems and upper surface of leaves ± thin, readily deciduous at maturity, leaving green surface; stem angled or ridged, especially above . . . . .2. **V. lychnitis**
    3. Calyx ca. 5.5–8 mm long, exceeding the pedicels; corolla over 3 cm broad; pubescence of stems and both surfaces of leaves very dense, persistent; stems not angled. . . . . .3. **V. phlomoides**
  2. Upper cauline leaves decurrent, with a narrow wing running halfway to all the way to the next lower node
    4. Stigma shortly decurrent down the style; lower bracts and upper leaves ± caudate (abruptly narrowed to prolonged tip); corolla ca. 3–5 cm broad, very flat . . . . .4. **V. densiflorum**
    4. Stigma strictly terminal, across the apex of the style; lower bracts and leaves not caudate (at most acuminate); corolla ca. 1.2–2.5 (2.8) cm broad, ± concave . . . . .5. **V. thapsus**

1. **V. blattaria** L.   Moth Mullein
    Map 268. Reported by the First Survey, but the earliest extant collections date from 1865 (Ingham Co.), 1871 (Muskegon Co.), and 1888 (Kent Co.); mostly spreading in the 20th century to roadsides and railroads, fields and pastures, waste ground (including newly disturbed sites and vacant lots), borders of woods (and clearings).
    The corolla is usually yellow, but frequently white [f. *erubescens* Brügger], very rarely pink.

2. **V. lychnitis** L.   Fig. 87   White Mullein
    Map 269. Sandy, gravelly, or rocky roadsides, fields, and waste ground.
    The fresh pubescence, besides being sparser than in our other species (except *V. blattaria*), is rather white. The species is easily recognized by its

smaller flowers and much-branched inflorescence of slender stalks. Some specimens collected in 1962 and 1963 (*Clover*, MICH, UMBS) near Lake Michigamme are a little more pubescent than usual, perhaps showing some introgression from another species, presumably *V. thapsus*.

3. V. phlomoides L.

Map 270. First collected in Michigan in 1941 by the Haneses along a roadside west of Schoolcraft (Pap. Mich. Acad. 45: 39. 1943). Found in 1963 along an expressway interchange on the west side of Ann Arbor (Voss 1967); another population a few miles farther west was studied by Wagner et al. (1980), who found sterile hybrids with *V. thapsus* [*V.* ×*kerneri* Fritsch] to be rather common. Other reports of *V. phlomoides* from the state refer to the next species.

4. V. densiflorum Bertol.

Map 271. Old fields and roadsides.

Very similar to the preceding in its large flowers and tendency to a much-branched candelabrum-like inflorescence on robust plants, but with strongly decurrent leaf bases, prolonged acuminate or caudate tips on the lower bracts of the inflorescence and the upper cauline leaves, and distinctly crenate leaf margins. Plants in cultivation derived from the Barry County population grew to a height of 3 m, with corollas as broad as 5 cm. Both this species and the preceding are handsome tall plants in the garden, with a long blooming season.

Long known as *V. thapsiforme* Schrader, a later name.

5. V. thapsus L.   Fig. 88   Mullein; Flannel Plant

Map 272. Everywhere in waste ground, including filled land, vacant lots, clearings; roadsides and railroads; shores and open banks. Quickly forms dense stands in forest clearings (after logging or other activity). Reported by the First Survey and already in the 1860s known from Detroit to Isle Royale.

 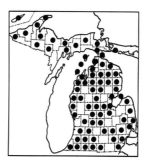 

271. Verbascum
     densiflorum

272. Verbascum thapsus

273. Cymbalaria muralis

A familiar tall weed; the densely flowered inflorescence produces a thick rod-like spike of fruits, which often lasts through the winter. The first-year rosettes of densely felted leaves are also well known to all field botanists.

2. Cymbalaria

1. **C. muralis** P. Gaertner, Meyer, & Scherb.   Fig. 89   Kenilworth-ivy
    Map 273. Introduced from Europe, locally established as an escape from cultivation to shores and waste ground around buildings and along sidewalks.

3. Kickxia

1. **K. elatine** (L.) Dumort.   Canker-wort
    Map 274. An Old World native, locally naturalized in North America but not collected in Michigan until 1975, in railroad yards south of New Buffalo (*Schulenberg & Wilhelm 75–997*, MOR), where it was still thriving in 1993.

4. Chaenorrhinum

REFERENCES

Arnold, Robert M. 1981a. Population Dynamics and Seed Dispersal of Chaenorrhinum minus on Railroad Cinder Ballast. Am. Midl. Nat. 106: 80–91.
Arnold, Robert M. 1981b. Weeds That Ride the Rails. Nat. History 90(8): 59–65.
Widrlechner, Mark P. 1983. Historical and Phenological Observations on the Spread of Chaenorrhinum minus Across North America. Canad. Jour. Bot. 61: 179–187.

1. **C. minus** (L.) Lange   Plate 3-G   Dwarf-snapdragon
    Map 275. Originally native to southern Europe, this species was first noted in North America in 1874 at Camden, New Jersey, whence it spread rapidly along the network of railroads across the continent. It seems partial to cinders of railroad ballast, where it is evidently tolerant of very rapid drainage as well as the sulfurous or other toxic aspects of the substrate and the disturbance of traffic and maintenance operations. Dispersal of seeds is accomplished more effectively by a passing train than by strong winds (Arnold 1981a, 1981b). With the decline of railroading, *Chaenorrhinum* lost out to competitors on stabilized railroad beds as they reverted to weedbeds. However, the removal of rails and ties recreated the necessary disturbed conditions for seed germination and survival, so the little dwarf-snapdragon made a comeback at some sites as they were converted from rails to trails. Better stands of *Chaenorrhinum* are now often found in gravel pits, newly graded roadsides and clearings, dumps (a species of habitat almost as endangered as railroads), gravelly shores, and such scenes of disturbance.

The first Michigan collections are from railroad yards at Port Huron in 1901; next, the species was collected in Oakland (1914), Wayne (1917), and Washtenaw (1918) counties. By the early 1920s it was found in Monroe, Emmet, and Luce counties (where it may, of course, have been long overlooked).

The lower leaves are frequently opposite. This little linear-leaved annual may be only a few cm tall when it flowers, or it may be much taller (close to 0.5 m) and bushy.

5. **Linaria** Toadflax

REFERENCES

Alex, J. F. 1962. The Taxonomy, History, and Distribution of Linaria dalmatica. Canad. Jour. Bot. 40: 295–307.

DeWolf, Gordon P., Jr. 1956. Notes on Cultivated Scrophulariaceae 3. Linaria. Baileya 4: 102–114.

Saner, Marc A., David R. Clements, Michael R. Hall, Douglas J. Doohan, & Clifford W. Crompton. 1995. The Biology of Canadian Weeds. 105 [sic, for 104]. Linaria vulgaris Mill. Canad. Jour. Pl. Sci. 75: 525–537.

Sutton, David A. 1988. A Revision of the Tribe Antirrhineae. Oxford Univ. Press, Oxford. 575 pp. + 4 microfiche (344 pp.).

KEY TO THE SPECIES

1. Plant a slender annual with cauline leaves less than 1.7 mm broad; corolla blue (or pink-purple)
    2. Corolla ca. 1.3–2.2 cm long (including spur); seeds with strongly wrinkled surface . . . . .1. **L. spartea**
    2. Corolla 0.6–1.1 cm long; seeds smooth or weakly pebbled . . . . .2. **L. canadensis**
1. Plant a sturdy perennial with cauline leaves all or mostly at least (1.5) 1.7 mm broad; corolla yellow
    3. Leaves ovate, ca. 3 times as long as broad or shorter; corolla (including spur) (23) 27–40 (43) mm long . . . . .3. **L. dalmatica**
    3. Leaves linear, tapered to the base, ca. 10 times as long as broad or longer; corolla (18) 23–28 (35) mm long . . . . .4. **L. vulgaris**

274. Kickxia elatine

275. Chaenorrhinum minus

276. Linaria spartea

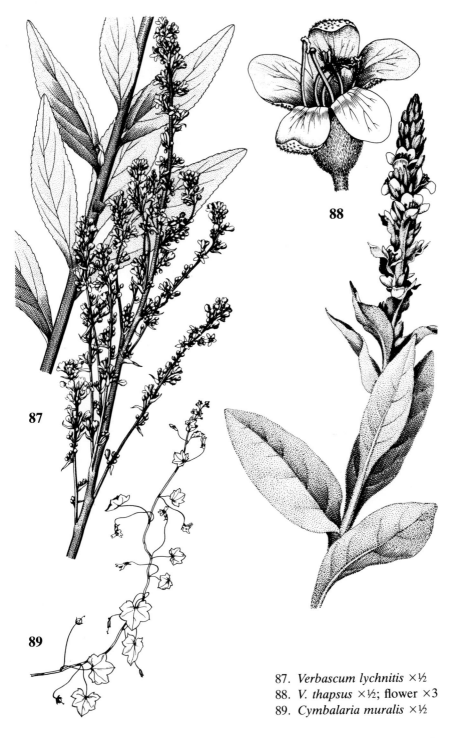

87. *Verbascum lychnitis* ×½
88. *V. thapsus* ×½; flower ×3
89. *Cymbalaria muralis* ×½

1. **L. spartea** (L.) Chaz.

Map 276. A native of southwestern Europe, collected in 1984 in a vacant lot along Lake Michigan north of Menominee (*Henson 1726*, MICH), presumably as a waif or escape from cultivation.

This species usually has a yellow corolla, but the rare red-pink form is what was found in Michigan. The lowermost leaves are opposite, the cauline leaves less than 1 mm wide, the raceme glandular-puberulent, and the flowers few with corolla lips apparently not divergent. The sometimes cultivated *L. incarnata* (Vent.) Sprengel [long misidentified as *L. bipartita* (Vent.) Willd.] normally has red-purple flowers, but they are more numerous, with spreading lips, and the leaves are all alternate.

2. **L. canadensis** (L.) Dum. Courset   Fig. 90   Blue Toadflax

Map 277. Dry, open, sandy or rocky ± barren ground; oak and sassafras woodland and jack pine plains; beds of dried lakes.

Placed by Sutton in a new genus, *Nuttallanthus*. Surprisingly frequently misidentified in herbaria as *Lobelia kalmii*, with which it shares narrow leaves and blue bilaterally symmetrical flowers—but not habitat, superior ovary, spurred corolla, absence of milky juice, or other characters!

3. **L. dalmatica** (L.) Miller   Fig. 91   Dalmatian Toadflax

Map 278. A native of southern Europe, first collected in Michigan in the 1940s and now locally established along roadsides and railroads, on sand dunes, at borders of woods, and in fields and waste places such as vacant lots.

The later name *L. macedonica* Griseb. applies to this same species, which some recent authors further combine as a subspecies of *L. genistifolia* (L.) Miller, a taxon tending to smaller flowers and relatively narrow leaves. This is a very attractive tall plant for cultivation, with flowers much as in the next species but averaging larger and with glaucous foliage, the clasping leaves much broader than in *L. vulgaris*.

4. **L. vulgaris** Miller   Fig. 92   Butter-and-eggs

Map 279. A Eurasian species, widely naturalized and sometimes cultivated. The oldest Michigan collection I have seen is by D. Cooley in 1844 (presumably from Macomb or Oakland county); listed for Mackinac Island by Thoreau in 1861. Before 1890 it was known from Kent to Keweenaw counties. Long a familiar and attractive weed along roadsides and railroads; on shores; invading woods; and thriving in fields, vacant lots, gravel pits, and dry disturbed or waste ground generally.

The common name aptly refers to the yellow flowers with bright orange area on the lower lip. The linear leaves, mostly 1.5–4 (5) mm wide, are often rather densely crowded.

6. Besseya

1. **B. bullii** (Eaton) Rydb.   Plate 4-A   Kitten-tail
    Map 280. Sandy grasslands and hillsides with sparse oaks. Ranges
rather locally from Michigan and southwest Ohio to Iowa and southeastern Minnesota.
    This species was named in 1840 for George Bull, a botanical assistant in
1838 and 1839 on the First Geological Survey of Michigan under Douglass
Houghton. The type locality is "prairies" of Michigan, with no more precise data. No type specimen has ever been located, and in fact I have seen
no First Survey collections of the species at all. Six years later (1846), the
same species was independently described as *Synthyris houghtoniana*
Bentham, based on material collected by Houghton in Minnesota on the
Schoolcraft expedition of 1832. The species has sometimes been assigned to
the genus *Wulfenia*.
    The inconspicuous yellow flowers are in a long spike-like inflorescence
at the end of an unbranched stem bearing many small leaves.

7. Castilleja   Indian Paintbrush
    This genus is much more diverse and the species more difficult to iden-
tify in western North America. We have it easy with only two readily
distinguished species. The common name derives from the conspicuously
colored tips of the bracts, as if dipped in paint. The corollas themselves are
much less conspicuous. The plants are hemiparasitic, i.e., roots must attach
to those of a host plant (not specific) for full growth and development.

REFERENCES

Malcolm, William M. 1962. Culture of Castilleja coccinea (Indian Paint-brush), a Root-
    parasitic Flowering Plant. Michigan Bot. 1: 77–79.
Malcolm, William M. 1966. Root Parasitism of Castilleja coccinea. Ecology 47: 179–186.

KEY TO THE SPECIES

1. Bracts of inflorescence cream-colored or yellowish (lower ones often purplish);
   mid-cauline leaves entire, unlobed, glabrous....................1. **C. septentrionalis**
1. Bracts and calyx tipped with bright red, orange, or yellow; mid-cauline leaves
   deeply lobed (rarely simple), hairy ....2. **C. coccinea**

1. **C. septentrionalis** Lindley   Northern Paintbrush
    Map 281. Essentially an eastern North American subarctic species, rang-
ing from Newfoundland to Great Bear Lake, south into New England and
the Lake Superior area, where it grows locally in Ontario, Michigan, and
the northeasternmost county (Cook) of Minnesota. Rock crevices, ledges,
openings, thin woods, and sandy banks near Lake Superior.

2. **C. coccinea** (L.) Sprengel   Fig. 93

Map 282. Calcareous sandy or gravelly shores, including interdunal flats and conifer thickets; river and stream banks, swamps (especially cedar), occasionally fens; marshy ground, meadows, springy and marly places, crevices and shallow soil at limestone outcrops; damp jack pine and oak scrub.

In f. *lutescens* Farw. (TL: Goodison [Oakland Co.]), the calyx and bracts are bright yellow rather than red; it is found occasionally, usually with the common form but sometimes in a stand by itself. Farwell also described a f. *alba* for a white form with his yellow plant.

8. **Pedicularis**   Lousewort

Like the preceding genus, these plants are hemiparasitic on a wide range of hosts. Species diversity is much greater in western and northern areas than in our region.

REFERENCES

Lackney, V. K. 1982 ["1981"]. The Parasitism of Pedicularis lanceolata Michx., a Root Hemiparasite. Bull. Torrey Bot. Club 108: 422–429.

277. Linaria canadensis

278. Linaria dalmatica

279. Linaria vulgaris

280. Besseya bullii

281. Castilleja septentrionalis

282. Castilleja coccinea

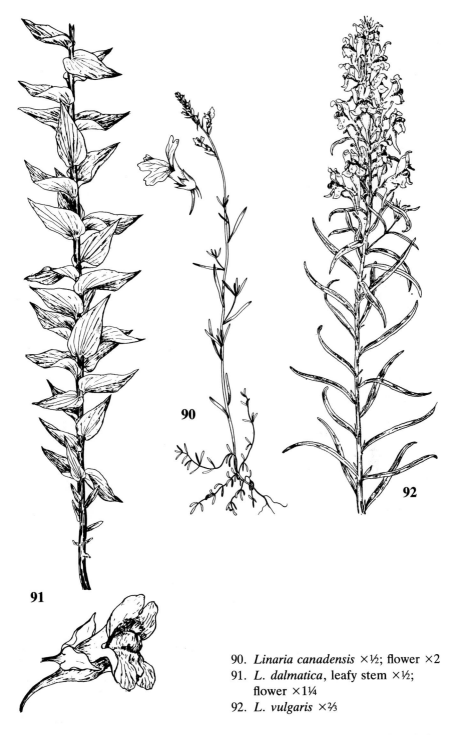

90. *Linaria canadensis* ×½; flower ×2
91. *L. dalmatica*, leafy stem ×½;
 flower ×1¼
92. *L. vulgaris* ×⅔

Piehl, Martin A. 1963. Mode of Attachment, Haustorium Structure, and Hosts of Pedicularis canadensis. Am. Jour. Bot. 50: 978–985.

Piehl, Martin A. 1965. Studies of Root Parasitism in Pedicularis lanceolata. Michigan Bot. 4: 75–81.

KEY TO THE SPECIES

1. Cauline leaves all or mostly alternate and lobed more than halfway to midrib; plant with stem hairy especially upwards, blooming in spring or early summer (May–June depending on season and latitude); upper lip of corolla with a prolonged tooth on each side below the entire or shallowly toothed apex; capsule ca. 2–3 times as long as calyx . . . . .1. **P. canadensis**
1. Cauline leaves opposite (rarely all alternate), lobed (or coarsely toothed) less than halfway to midrib; plant with stem (at least above) glabrous or nearly so, blooming in late summer and fall (Aug.–Sept.); upper lip of corolla with at most a shallow tooth or angle on each side of the entire or notched apex; capsule scarcely if at all longer than calyx . . . . .2. **P. lanceolata**

1. **P. canadensis** L.   Fig. 94

Map 283. Dry woods and woodland (oak, pine, aspen, red maple); also rich hardwoods including beech-maple, especially in openings; less often in conifer swamps, meadows, and grasslands.

The corollas are yellow [f. *canadensis*] or maroon [f. *praeclara* A. H. Moore] or yellow on the lower lip and maroon on the upper [f. *bicolor* Farw. (TL: near Farmington Junction [Oakland Co.])]. All color states may occur together.

2. **P. lanceolata** Michaux

Map 284. Much more of a wetland or low ground species than the preceding: borders of marshes, swamps, ponds, and lakes; river banks, thickets, and springy slopes; fens (even in sphagnum), meadows, and wet prairies.

The corolla is pale yellow or cream—not as variable as in the preceding species. Even more than in the preceding, the margins of the leaf lobes are edged with hard white ceramic-like deposits.

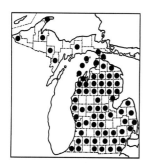  

283. Pedicularis canadensis   284. Pedicularis lanceolata   285. Digitalis lutea

93. *Castilleja coccinea* ×½; flower with bract ×2
94. *Pedicularis canadensis* ×½; flower ×1; fruit ×2
95. *Scrophularia lanceolata* ×½; flower & fruit ×3

9. **Digitalis** Foxglove

The foxgloves are familiar, usually tall and large-flowered garden ornamentals. *D. purpurea* is especially well known as the source of a cardiac medicine, first reported in 1785 by the British botanist and physician William Withering, after a decade of research on an effective folk medicine. *D. lanata* is also used as a drug source. These are poisonous plants, the line between therapeutic and toxic doses rather narrow.

Corolla color in the key refers to the general ground color, not to spots or other markings.

REFERENCE

Werner, Klaus. 1962. Die Kultivierten Digitalis-Arten. Kulturpfl. Beih. 3: 167–182 + 2 pl.

KEY TO THE SPECIES

1. Pedicels and axis of the inflorescence glabrous; corolla ca. 1.3–2.1 cm long, white to
   yellowish . . . . .1. **D. lutea**
1. Pedicels and axis of inflorescence densely glandular-pubescent; corolla ca. (1.9)
   2.3–5 cm long, white, yellow, or purple
   2. Corolla ca. (1.9) 2.3–3 cm long, the lower lip with very greatly prolonged middle
      lobe; inflorescence densely covered with long (over 1 mm) hairs . . . . .2. **D. lanata**
   2. Corolla ca. (2.8) 3.5–5 cm long, the lower lip with middle lobe scarcely longer than
      the lateral lobes; inflorescence (pedicels and axis) densely covered with short hairs
      (less than 0.5 mm)
      3. Calyx lobes linear-lanceolate; corolla yellow; leaves glabrous above . .3. **D. grandiflora**
      3. Calyx lobes broadly ovate; corolla purple (rarely white); leaves pubescent (rarely
         glabrate) above . . . . .4. **D. purpurea**

1. **D. lutea** L. Straw Foxglove

Map 285. A native of central and southern Europe, established in a pine plantation in Ann Arbor Township (*Nimke 718* in 1985, MICH).

2. **D. lanata** Ehrh. Grecian Foxglove

Map 286. A native of the Balkans and adjacent area, escaped along railroads and roadsides and in open meadows and woodland.

286. Digitalis lanata

287. Digitalis grandiflora

288. Digitalis purpurea

3. **D. grandiflora** Miller   Yellow Foxglove
    Map 287. A native of Europe and western Asia, escaped from cultiva-
tion along roadsides and thickets. There has long (for at least four decades)
been an especially striking colony along U. S. Highway 41 in northern
Baraga County.
    Including *D. ambigua* Murray. The inflorescence has the densest cover
of short gland-tipped hairs of all our species.

4. **D. purpurea** L.   Common Foxglove
    Map 288. A native of western Europe, escaped into woods, especially
borders and clearings. Introduced in the 1880s at Whitefish Lake in western Alger County, thence spreading and becoming well established in forested areas.

10.   **Antirrhinum**

REFERENCE

DeWolf, Gordon P., Jr. 1956. Notes on Cultivated Scrophulariaceae 2. Antirrhinum and
    Asarina. Baileya 4: 55–68.

1. **A. majus** L.   Snapdragon
    Map 289. A native of southern Europe, very familiar in cultivation but
only rarely escaped to dumps and other waste places.
    Extensive breeding and selection have produced countless strains of
snapdragon of diverse color and stature. Other species, of more prostrate
habit or with smaller flowers, are also cultivated.

11.   **Misopates**

1. **M. orontium** (L.) Raf.   Lesser Snapdragon; Weasel's-snout
    Map 290. Weedy in habit in its native Eurasian range, occasionally
escaped from cultivation in North America. Collected in 1886 as escaped
along a roadside in Grand Rapids (*Fallass*, ALBC).

289. Antirrhinum majus   290. Misopates orontium   291. Scrophularia
                                                               lanceolata

The leaves are linear or very narrowly elliptical and the stem is ± densely glandular-pubescent in and near the inflorescence. Often included in the preceding genus as *A. orontium* L.

12. Scrophularia   Figwort

KEY TO THE SPECIES

1. Staminodium greenish or yellowish, the free apex usually distinctly broader than long; mature capsules (5) 6–9 mm long; larger leaf blades truncate to rounded at base, never cordate ....1. **S. lanceolata**
1. Staminodium dark brown or purple (very rarely green), the free apex scarcely if at all broader than long; mature capsules 4–6 (7) mm long; larger leaf blades often cordate ....2. **S. marilandica**

1. S. lanceolata Pursh   Fig. 95

Map 291. Often forms dense stands, with stems up to 2 m tall; very conspicuous along roadsides and railroads; often in woods, especially clearings, openings, edges, and old roads; fields, fencerows, and thickets; shores and swamp borders; often thrives after some disturbance.

The width of the free part of the staminodium is variable, but usually broader than in the next species (in which it seldom exceeds 1 mm). Both species bear a large number of flowers over a long blooming season (some flowers open while others have formed mature fruit). Hence, it may be difficult to determine that *S. lanceolata* is an earlier-blooming plant, the first flowers appearing in May or June, while in *S. marilandica* they usually appear in July or August (though at least one collection, from Lenawee County, is blooming the last week of June).

2. S. marilandica L.

Map 292. Less common than the preceding species but in similar habitats: riverbank thickets and floodplains; woods (especially at borders and in clearings); roadsides.

Not always easily distinguished from *S. lanceolata*. The sides of the stem in *S. marilandica* are usually ± grooved or channeled, while in the other species the sides are usually flat or slightly convex—and the leaves of the latter are often more irregularly toothed (even somewhat incised toward the base). A variant with staminodium (and corolla) green was discovered by Farwell at Ypsilanti in 1919.

13. Penstemon   Beard-tongue; Penstemon

P. gracilis is apparently indigenous in Dickinson County—the easternmost point of its range. Northern records of our other species may represent plants adventive from farther south in the state.

KEY TO THE SPECIES

1. Corolla ca. 3.3–4.5 cm long; upper leaves (if not all) broadly ovate, ± obtuse to rounded at apex and subcordate-clasping at base, strongly glaucous .... 1. **P. grandiflorus**
1. Corolla not over 3 cm long; leaves narrower, very acute, often not glaucous
  2. Throat of corolla inflated, distinctly differentiated from the narrower tube (fig. 96); lower (3) lobes of corolla slightly longer than upper lobes; anthers usually at least sparsely bearded; stem glabrous below the inflorescence or rarely locally puberulent
    3. Anthers glabrous; calyx lobes linear-lanceolate .... 2. **P. calycosus**
    3. Anthers with at least a few bristly hairs (rarely glabrous) on the back (near attachment of the filament); calyx lobes lanceolate-ovate .... 3. **P. digitalis**
  2. Throat of corolla not sharply differentiated, only gradually expanded from the tube (fig. 97); lower lobes of corolla distinctly longer than the upper; anthers glabrous; stem ± uniformly puberulent to glandular-hairy from base to near the inflorescence
    4. Stem with numerous hairs 0.5–1 mm long (or even longer); lower lip humped, nearly closing throat of corolla; corolla (20) 22–28 (30) mm long, purple-violet but not streaked with purple lines (nectar guides) .... 4. **P. hirsutus**
    4. Stem with few or no hairs as long as 0.5 mm; lower lip not humped, the throat open; corolla (14) 16–20 [22] mm long, streaked with nectar guides [usually not visible on dry specimens]
      5. Longest (lower) branches of inflorescence usually at least 2 cm long before first pedicels; cauline leaves puberulent on both surfaces (at least on main veins and margins); corolla white .... 5. **P. pallidus**
      5. Longest branches of inflorescence less than 2 cm before first pedicels; cauline leaves glabrous; corolla pale violet .... 6. **P. gracilis**

1. **P. grandiflorus** Nutt.   Large Beard-tongue
    Map 293. Native to prairies west of the Great Lakes. Collected once along a roadside in Saginaw County (*Smith 2926* in 1990, MICH), presumably as an escape from cultivation.
    A very stout, unmistakable species.

2. **P. calycosus** Small
    Map 294. Fields and openings, apparently very rare in Michigan. The

292. Scrophularia marilandica

293. Penstemon grandiflorus

294. Penstemon calycosus

collection cited by Pennell from Macomb County is in fact labeled as from Illinois.

The corolla tends more often to pink-violet in this species. A collection by Farwell from Belle Isle in 1905 has hairy anthers but strongly linear-lanceolate sepals and was thought by Pennell to be a hybrid with the next species (Pap. Michigan Acad. 26: 18–19. 1941).

3. **P. digitalis** Sims   Fig. 96   Foxglove Beard-tongue
Map 295. Fields, meadows, and prairie remnants; openings in dry wood-lands (oak, aspen); along roadsides and railroad beds.

The corolla is white or creamy, usually ± lined or tinged with light pink-violet. The inflorescence is usually ± densely glandular-pubescent.

Both this species and the preceding are sometimes merged into the more southeastern *P. laevigatus* Aiton.

4. **P. hirsutus** (L.) Willd.   Fig. 97   Hairy Beard-tongue
Map 296. Sandy, barren, open, usually dry ground, including prairies, oak woodland and borders, fields, roadsides; stream and river banks, rocky ground.

The rare white-flowered form has been called f. *albiflorus* Farw. (TL: Flat Rock [actually Rockwood, Wayne Co.]). Steps have been taken (Taxon 44: 637–638. 1995) to avoid having to place the name *P. hirsutus* in the synonymy of *P. laevigatus*, causing the present species to take a different name.

5. **P. pallidus** Small
Map 297. Like most of our species in this genus, found in dry fields and along roadsides.

6. **P. gracilis** Nutt.
Map 298. Rocky ledges and oak woodland along the Menominee River, discovered by Henson in 1989 (*2979*, MICH) and the next year at an additional site by Henson and Reznicek.

295. Penstemon digitalis   296. Penstemon hirsutus   297. Penstemon pallidus

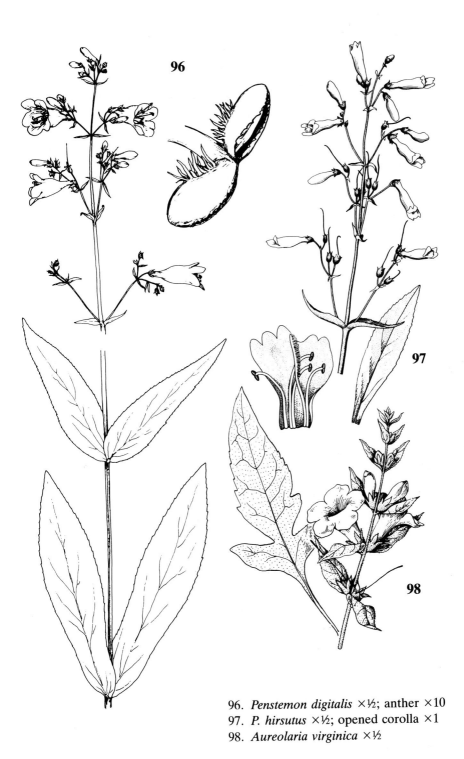

96. *Penstemon digitalis* ×½; anther ×10
97. *P. hirsutus* ×½; opened corolla ×1
98. *Aureolaria virginica* ×½

14. Leucospora

This genus is sometimes included in *Conobea*.

1. L. multifida (Michaux) Nutt.   Fig. 114

Map 299. Not known from Michigan until 1991, when discovered by Reznicek and Penskar (*8870*, MICH) in moist or wet prairie/savana in Wayne County. The next two years, discovered by them and by Brodowicz to be locally common in intermittent wet borrow pits, disturbed sandy soil, and prairie/oak savana in both Monroe and Wayne counties.

This is a very distinctive, much-branched, pubescent little annual, with bluish-pink flowers on slender axillary pedicels.

15. Dasistoma

REFERENCE

Piehl, Martin A. 1962. The Parasitic Behavior of Dasistoma macrophylla. Rhodora 64: 331–336.

1. D. macrophylla (Nutt.) Raf.   Fig. 99   Mullein-foxglove

Map 300. Barely ranges into Michigan, there being only two or three known sites in deciduous forest in Berrien County.

This species is hemiparasitic, its roots attaching to those of diverse forest trees. In the absence of flowers (which are nearly regular), one might confuse the upper part of specimens with the similarly pubescent *Aureolaria virginica*, although the latter, besides its much larger flowers, usually has slightly longer pedicels and a larger capsule (over 1 cm long, compared to 1 cm or less in *D. macrophylla*). Our other two species of *Aureolaria* are easily distinguished from *Dasistoma* by having the pubescence glandular (*A. pedicularia*) or absent (*A. flava*).

This species is sometimes retained in the genus *Seymeria*, where Nuttall first placed it.

298. Penstemon gracilis   299. Leucospora multifida   300. Dasistoma macrophylla

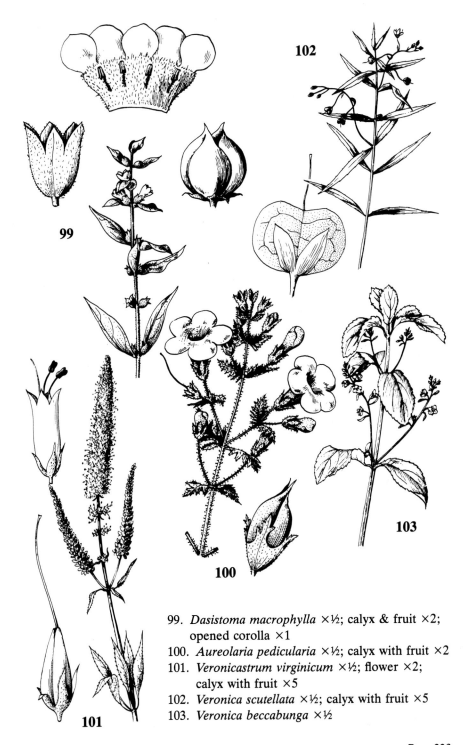

99. *Dasistoma macrophylla* ×½; calyx & fruit ×2;
    opened corolla ×1
100. *Aureolaria pedicularia* ×½; calyx with fruit ×2
101. *Veronicastrum virginicum* ×½; flower ×2;
    calyx with fruit ×5
102. *Veronica scutellata* ×½; calyx with fruit ×5
103. *Veronica beccabunga* ×½

16. **Aureolaria**   False Foxglove

This genus has sometimes been included under *Gerardia*, a name officially rejected in 1959, to be replaced by *Agalinis* (though many have been
slow to accept the change). The plants have long been known as
hemiparasitic on oaks; the first two species below are perennials and said to
parasitize white oaks while the third, an annual or winter annual, occurs on
black oaks. However, *A. pedicularia* has been found to parasitize Ericaceae
in the Southeast, and these species may be more widespread on woody
plants than once thought.

REFERENCES

Ballard, Harvey E., Jr., & Richard W. Pippen. 1991. An Intersubgeneric Hybrid of
    Aureolaria flava and A. pedicularia. Michigan Bot. 30: 59–63.
Pennell, Francis W. 1928. Agalinis and Allies in North America,—I. Proc. Acad. Nat. Sci.
    Philadelphia 80: 339–449.
Werth, Charles R., & James L. Riopel. 1979. A Study of the Host Range of Aureolaria
    pedicularia (L.) Raf. (Scrophulariaceae). Am. Midl. Nat. 102: 300–306.

KEY TO THE SPECIES

1. Stem, outside of calyx, and fruit all glabrous . . . . .1. **A. flava**
1. Stem, calyx, and fruit pubescent
  2. Pubescence eglandular; pedicels 2–5 mm long, straight; calyx lobes entire, merely
    obtuse or rounded at tip . . . . .2. **A. virginica**
  2. Pubescence, especially of calyx and pedicels, conspicuously gland-tipped; pedicels
    8–25 (30) mm long, ± curved; calyx lobes with prolonged toothed or pinnatifid
    tip . . . . .3. **A. pedicularia**

1. **A. flava** (L.) Farw.

Map 301. Oak openings, sandy oak and oak-hickory woodland, with
jack pine and aspen often present, woods borders and clearings. The oaks
in its neighborhood are sometimes in the black oak group rather than white
oak.

An apparent hybrid with *A. pedicularia* was collected in Bay County in
1931 (*Dreisbach 7414*, MICH). The outside of the calyx is nearly or quite
glabrous but the lobes are ± dentate. The foliage is intermediate in lobing,
and the stem has eglandular pubescence. Similar hybrids were found in
1981 and 1982 in Kalamazoo County (Ballard & Pippen).

2. **A. virginica** (L.) Pennell   Fig. 98

Map 302. In the same habitats as the preceding, with oak of both groups,
pine, and/or hickory.

See comments under *Dasistoma macrophylla*, above.

3. **A. pedicularia** (L.) Raf.   Fig. 100

Map 303. Oak, oak-pine, and oak-hickory woodland, including old
wooded dunes; transition zones between savana and marsh; appears to be

more faithful to presence of black oaks than *A. flava* is to presence of white oaks.

Most specimens are referable to the densely glandular var. *ambigens* (Fernald) Farw. Typical var. *pedicularia* has only sparse glandular hairs in the upper stem pubescence. In var. *intercedens* Pennell, such hairs are more numerous but still sparse. While Pennell has confirmed the presence of all three varieties in Michigan, our specimens seem scarcely worthy of distinction.

17. **Chelone**   Turtlehead

The common name is apt, for the flower does indeed resemble the protruding head of a turtle.

REFERENCE

Crosswhite, Frank S. 1965. Variation in Chelone glabra in Wisconsin (Scrophulariaceae). Michigan Bot. 4: 62–66.

KEY TO THE SPECIES

1. Corolla pink-purple throughout; staminodium white; sepals finely and closely ciliolate; leaf blades ca. 3–5 times as long as broad .... 1. **C. obliqua**
1. Corolla white or greenish yellow; staminodium greenish; sepals obscurely if at all ciliolate; leaf blades (3.7) 4.5–15 times as long as broad .... 2. **C. glabra**

1. **C. obliqua** L.   Plate 4-B

Map 304. First discovered in Michigan by S. Alexander (in 1904, MSC; Rep. Michigan Acad. 10: 86. 1908; 12: 97. 1910) along the Huron River east of Ann Arbor, where it has not been rediscovered. Also grows in thickets along the same river northwest of Ann Arbor. Apparently very rare, and listed as endangered in the state. Not common, and widely scattered throughout its range, which is mostly well to the south.

Our plants are tall, very leafy, in dense clumps, with handsome flowers open toward the end of September and early October—later than usual for *C. glabra*.

301. Aureolaria flava

302. Aureolaria virginica

303. Aureolaria pedicularia

2. **C. glabra** L.

Map 305. Low ground along streams, rivers, ponds, and lakes; swamps (coniferous or hardwood), especially in openings and borders; fens, damp fields and meadows, wet shores, marshes, thickets.

Our plants are mostly the relatively narrow-leaved form named var. *linifolia* N. Coleman (TL: [presumably near Grand Rapids, Kent Co.]), a taxon of dubious significance. Plants with leaves relatively broader (length/width as in *C. obliqua*), while scarce, occur throughout the state, even to Isle Royale (as does var. *linifolia*).

18. Veronicastrum

1. **V. virginicum** (L.) Farw.   Fig. 101   Culver's-root
    Map 306. Prairie remnants, fens, and meadows; river banks; deciduous woodlands (especially with oaks); and adjacent roadsides.

This species has sometimes been included in the genus *Veronica*, but the corolla tube is much longer than the lobes and the sepals are usually 5. The stamens and style are strongly exserted and the corolla is usually white, though sometimes pink. The leaves are glabrous or pubescent beneath. The underground parts were long used medicinally.

19. Veronica   Speedwell; Brooklime

Except for the first four species below, which are rather conspicuous perennial escapes from cultivation, these are mostly unprepossessing plants, often scorned for their taxonomic complexity and nomenclatural confusion as well as for their tiny flowers. However, en masse even the smallest plants can present a striking scene of blue when in full blooom. Some species have usually white rather than blue corollas, and in some, especially *V. scutellata*, the corolla is often a shade of pink.

The fruit is flattened and ± heart-shaped (or at least notched at the apex). The style is usually readily visible on nearly mature fruit, when it often provides useful key characters, although there are sometimes exceptions to the customary lengths—in both actual measurements and relation

304. Chelone obliqua

305. Chelone glabra

306. Veronicastrum
      virginicum

104. *Veronica persica* ×1; calyx with fruit ×10
105. *V. peregrina* ×½; calyx with fruit ×4
106. *V. arvensis* ×½; foliage ×2½; flower ×7½

to the two capsule lobes between which the style arises. Some species have what definitely appear to be racemes, with distinctly pediceled flowers; others have more spike-like inflorescences, with very short pedicels. Again, measurements may be less discriminating than one would like. Pedicels near the base of an inflorescence tend to be longer (mature earlier) than those toward the apex of the stem, and the mature pedicels should be examined when their length must be known. In the species with a single terminal inflorescence, over half the plant may be occupied by flowers—whether one interprets them as being in a spike or raceme or as solitary in the axils of leaves. The real cauline leaves are opposite, but flowers usually arise from the axils of alternate leaves (or bracts, depending on their gradation in size and one's definitions).

REFERENCES

Crins, William J., Donald A. Sutherland, & Michael J. Oldham. 1987. Veronica verna (Scrophulariaceae), an Overlooked Element of the Naturalized Flora of Ontario. Michigan Bot. 26: 161–165.

DeWolf, Gordon P., Jr. 1956. Notes on Cultivated Scrophulariaceae 4. Veronica. Baileya 4: 143–159.

Les, Donald H., & Ronald L. Stuckey. 1985. The Introduction and Spread of Veronica beccabunga (Scrophulariaceae) in Eastern North America. Rhodora 87: 503–515.

Muenscher, W. C. 1949. Veronica filiformis a Weed of Lawns and Gardens. Rhodora 51: 365.

KEY TO THE SPECIES

1. Flowers (and fruits) in definite racemes on peduncles arising mostly from axils of ordinary opposite or whorled foliage leaves (sometimes terminal in *V. longifolia*, with sharply toothed leaves), with bracts at base of pedicels extending distinctly less far than the flowers and much smaller than the foliage leaves; plants perennial
  2. Stem below the inflorescence at least minutely pubescent with eglandular hairs
    3. Blades of principal cauline leaves 2.5–8 times as long as broad, with numerous sharp teeth; tube of corolla nearly or quite as long as the lobes ....1. **V. longifolia**
    3. Blades of principal cauline leaves ca. twice or less as long as broad, entire to crenulate or shallowly toothed; tube of corolla very much shorter than the lobes.
      4. Leaf blades entire or shallowly crenulate, glabrous ....11. **V. serpyllifolia**
      4. Leaf blades definitely toothed, hairy beneath (and often above)
        5. Pedicels up to 2.5 mm long (in fruit); corollas (when the limb is spread flat) ca. 5–8 mm broad or less; calyx 1.7–3.5 mm long; hairs on capsule ± dense, mostly gland-tipped ....2. **V. officinalis**
        5. Pedicels (2) 2.5–7.5 mm long (in anthesis and fruit); corollas ca. (9) 10–14 mm broad; calyx 3.5–5.5 mm long; hairs (if any) on capsule without glands
          6. Capsule ± elliptical, as long as broad or slightly longer; calyx lobes broadest well below the middle; pubescence ± evenly distributed around the stem below the nodes; stems stiff and upright, over 3 dm tall ....3. **V. austriaca**
          6. Capsule broadly obcordate, distinctly broader than long; calyx lobes broadest at or above the middle; pubescence much denser in 2 lines below the nodes; stems slender, prostrate, with ascending shoots up to 3 dm tall ...4. **V. chamaedrys**
  2. Stems glabrous or with some gland-tipped hairs (rarely with eglandular pubescence in *V. scutellata*, with very narrow leaves under 1 cm wide and 6–20 times as long as wide)

7. Leaves sessile, linear-lanceolate, not over 1 cm broad and ca. (6) 8–20 times as long as broad, entire or remotely denticulate; mature pedicels reflexed, mostly (2) 2.8–6 times as long as the bracts at their base; calyx 2–3.5 mm long, shorter than the mature fruit; racemes all or mostly alternate (arising in the axil of only one of any pair of opposite leaves) . . . . .5. **V. scutellata**

7. Leaves petioled or sessile, mostly (2) 3–5 times as long as broad, entire or toothed; mature pedicels usually spreading or ascending, mostly about twice as long as the bracts or shorter; calyx 2.5–4 (5.5) mm long, about equalling or slightly exceeding the mature fruit; racemes all or mostly opposite

    8. Leaves (except rarely the lower) sessile, often ± clasping; stems and axis of inflorescence often glandular . . . . .6. **V. anagallis-aquatica**

    8. Leaves all distinctly petioled; stems and axis eglandular . . . . .7. **V. beccabunga**

1. Flowers (and fruits) solitary in axils of ordinary leaves or in terminal spikes or racemes with bracts at base of pedicels (or flowers) usually equalling or surpassing the flowers and gradually reduced from the foliage leaves (leaves or bracts sub-tending flowers usually alternate); plants annual (except *V. serpyllifolia* and *V. filiformis*)

  9. Flowers all in the axils of ordinary (but often alternate) foliage leaves, on pedicels (5) 7–22 (25) mm long and equalling or exceeding the subtending leaves; stems prostrate

    10. Leaf blades reniform, broader than long, well-developed ones cordate at base; plant perennial, mat-forming with prostrate stems rooting at the nodes, not [or very rarely?] setting fruit; calyx lobes broadest beyond the middle, rounded at apex, with 1 vein . . . . .8. **V. filiformis**

    10. Leaf blades mostly longer than broad, all truncate to tapered at the base; plant annual, ± prostrate but with taproot and not forming mats, setting fruit at maturity; calyx lobes broadest below the middle, ± acute, at maturity with 1–5 strong veins

      11. Capsule lobes divergent and usually angled at apex; corolla ca. 7.5–8.5 (9.5) mm broad . . . . .9. **V. persica**

      11. Capsule lobes not divergent (i.e., their axes nearly parallel) and broadly rounded at apex; corolla at most ca. 5–7 (7.5) mm broad . . . . .10. **V. polita**

  9. Flowers in the axils of bracts much smaller than the lower foliage leaves, on pedicels ca. 0.5–4.5 (8) mm long and almost always shorter than the subtending bracts; stems (at least the flowering portion) erect or ascending from decumbent bases

    12. Blades of leaves and bracts glabrous or essentially so, mostly entire or obscurely crenulate (with minutely notched margins)

      13. Flowers (at least the lower ones) on pedicels ca. 2–4.5 (8) mm long at maturity; stem with minute ± upcurved pubescence (rarely glabrate or with spreading hairs—especially toward the inflorescence) .11. **V. serpyllifolia**

      13. Flowers (and fruit) nearly sessile (pedicels ca. 0.5–1 mm long); stems glabrous or with gland-tipped hairs . . . . .12. **V. peregrina**

    12. Blades of leaves and bracts clearly pubescent and/or ciliate, entire to toothed or lobed

      14. Leaves and bracts entire or merely crenate or toothed; style slightly shorter than the lobes of the capsule to slightly exceeding them . . . . .13. **V. arvensis**

      14. Leaves (at least the upper) and often lower bracts ± deeply pinnately lobed toward the base; style various

        15. Style ca. 1–1.3 mm long, distinctly surpassing the capsule lobes; corolla ca. 4–5 mm broad; fruiting pedicels ca. 2–3.5 mm long . . . . .14. **V. dillenii**

        15. Style ca. 0.5 mm long or less, shorter than the capsule lobes; corolla ca. 3 mm broad; fruiting pedicels ca. 1–2 (very rarely 2.5) mm long . . . . .15. **V. verna**

1. **V. longifolia** L.   Garden Veronica

Map 307. A tall, handsome Eurasian species occasionally escaped from cultivation to roadsides, fields, waste ground, and other open moist to dry places.

2. **V. officinalis** L.   Common Speedwell

Map 308. A Eurasian species rather frequently escaped from cultivation and well established, forming large prostrate colonies often in natural-appearing habitats. Woods (deciduous, coniferous, mixed, dry to swampy), especially along trails and in clearings; pine plantations; occasionally in disturbed open ground and on rocky outcrops; less often a weed in yards and waste ground.

The low habit and obcordate fruit are similar to those of *V. chamaedrys*, but besides the smaller flowers and shorter pedicels, *V. officinalis* has leaves more shallowly toothed, more broadly obtuse or even rounded at the apex, and with lateral veins obscure instead of strong all the way to the teeth (or sinuses). The inflorescence is usually rather densely glandular-pubescent.

3. **V. austriaca** L.

Map 309. Another Old World species but only rarely escaped from cultivation, as along roadsides and near old homesites.

Our material is ssp. *teucrium* (L.) D. A. Webb [or var. *teucrium* (L.) Bolòs & Vigo], often recognized as a distinct species and sometimes called *V. latifolia* L. The calyx lobes in this group are often 5, but as noted in *Flora Europaea* may be 4—as in the Michigan specimens, which have the 5th sepal rudimentary at best.

4. **V. chamaedrys** L.   Fig. 107   Germander Speedwell

Map 310. A Eurasian native, sometimes cultivated and apparently more often escaping than the preceding two species. Lawns, roadsides, hillside banks, waste ground, trails, invading coniferous and deciduous woods.

This species has a more open raceme than the preceding (the flowers more widely spaced).

307. Veronica longifolia   308. Veronica officinalis   309. Veronica austriaca

5.   **V. scutellata** L.   Fig. 102   Marsh Speedwell

Map 311. Marshes, hardwood swamps and wet thickets (e.g., alders and willows); meadows, wet depressions, swales, and ditches; borders of streams, rivers, and ponds.

The narrow remotely denticulate leaves (barely more than a gland for a "tooth") are similar in shape to those of *Salix exigua*. Occasionally the stems (and leaves) bear fine spreading pubescence [f. *villosa* (Schum.) Pennell] but the leaf shape will readily distinguish this species from those above. Mixed collections of pubescent and glabrous plants indicate that the two forms may grow together. A zigzag axis of the raceme, with its long filiform pedicels soon reflexing, and the general sprawling delicate habit are characteristic of the species.

6.   **V. anagallis-aquatica** L.   Water Speedwell

Map 312. Wet sandy or muddy shores and ditches, especially in springy places; in flowing water of streams and rivers as well as on their banks; rarely on floating mats in peatlands.

Following a number of recent authors who have given up trying to distinguish American from Old World plants of this species, I include here *V. catenata* Pennell, often applied to the American element (to which also the names *V. comosa* and *V. salina* have sometimes been misapplied). The typical Old World element was collected in 1936 in Keweenaw County (*Hermann 7998*, MICH, GH). It differs in more erect-ascending pedicels, relatively narrower sepals, and capsules a bit narrower than long.

This is quite a variable species. Most of our specimens are ± glandular-pubescent in the raceme and on the upper part of the stem: var. *glandulosa* Farw. (TL: near Royal Oak [Zoo Park, Oakland Co.]). The lower leaves and leaves on basal offshoots sometimes are short-petioled. Rarely plants have alternate racemes, as in *V. scutellata*. Submersed plants (especially in flowing water) frequently have very limp and rather elongate aquatic leaves; these, of course, are opposite so such plants ought not be confused with a *Potamogeton*, which they might otherwise resemble.

310. Veronica chamaedrys

311. Veronica scutellata

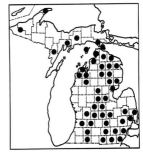
312. Veronica anagallis-aquatica

7. **V. beccabunga** L.   Fig. 103   Brooklime
    Map 313. Like the preceding, a wetland species but not developing an
underwater aquatic form as *V. anagallis-aquatica* does. Edges of creeks,
rivers, and ponds; marshy shores; swales and ditches; swamps of cedar,
other conifers, and hardwoods (often on hummocks or mossy logs); clearings in moist woods.
    Our plants are almost all the native var. *americana* Raf. [= *V. americana*
Bentham; there is no evidence that Bentham's binomial was based on
Rafinesque's varietal name]. These have the leaf blades broadest below the
middle and the general outline thus more acute toward the apex; the style
is usually only slightly if at all shorter than the capsule. In the Eurasian var.
beccabunga, the blades are broadest beyond the middle and are thus more
broadly rounded at the apex; the styles are distinctly shorter than the
capsule. These have traditionally been recognized as distinct but closely
related species, but I am following 1983 annotations by Sellers in not treat-
ing the distinction at so high a rank. The only clear specimens of var.
beccabunga I have seen from the state are from Mackinac Island (*Gleason,
Jr.* in 1935, MICH; *Potzger 7193* in 1936, BUT); other Michigan specimens
cited by Les and Stuckey have not been so annotated by Sellers and some
other Michigan specimens referred in the past to the exotic *V. beccabunga*
he has determined as the native taxon.

8. **V. filiformis** Sm.   Creeping Speedwell
    Map 314. A native of Asia Minor, rather aggressively escaping from
cultivation to lawns and unmowed grass. First collected in Michigan in 1978
in Kalamazoo and Washtenaw counties, but doubtless established somewhat earlier.
    The attractive blue flowers are 6.5–12.5 mm wide and on their elongate
pedicels stand out above the creeping stems with their small leaves (blades
less than 1 cm long). The capsule is said (in Europe) to be "subglabrous
except for long hairs on keels"; fruit seems rarely if ever to be found in
North America. The next two species have glandular hairs on the capsules.

9. **V. persica** Poiret   Fig. 104   Bird's-eye Speedwell
    Map 315. Native to Europe and western Asia, locally established along
roadsides; in vacant lots; and in farms, orchards, gardens, lawns, and vine-
yards. Collected in the 1890s in Shiawassee and St. Clair counties, and then
in 1913 and 1914 in Van Buren and Oakland counties; much later becoming
a more widespread weed.
    This species has often been called *V. tournefortii* Vill., a misapplication
of that name.

10. **V. polita** Fries
    Map 316. Another little prostrate Eurasian species, its smaller flowers
on shorter pedicels making it less attractive than the previous two. Col-

lected in Ann Arbor (where it is still a local weed) in 1870, but not again in the state until the Haneses found it in 1937 as a lawn weed in Schoolcraft. Since then, becoming recognized as more widespread in lawns, cemeteries, fields, and disturbed ground.

The mature pedicels are ca. (5) 7–10 [18] mm long, running shorter than in the preceding two species, in which they are ca. 11–22 (25) mm. Sometimes included in *V. agrestis* L., but the capsules of our plants have very short eglandular hairs in addition to the slightly longer gland-tipped hairs, and the style distinctly exceeds the lobes. The capsules of *V. agrestis* have only the longer glandular hairs and the style equals or is shorter than the lobes. This species has also sometimes been called *V. didyma* Ten., a name of uncertain application.

11. **V. serpyllifolia** L.   Fig. 108   Thyme-leaved Speedwell
Map 317. A widespread Eurasian weed, well established in woods (especially along borders, trails, old logging roads, openings, and clearings); meadows, pastures, farmyards, roadsides, and other disturbed ground; fields, lawns, gardens; banks of rivers and streams. Collected as early as 1842 (presumably Macomb or Oakland county), 1857 (Detroit), 1860 (Ann Arbor), and 1861 (Mackinac Island); by 1890, locally throughout the state.

313. Veronica beccabunga

314. Veronica filiformis

315. Veronica persica

316. Veronica polita

317. Veronica serpyllifolia

318. Veronica peregrina

The flowers are usually whitish. A very few old 19th century collections from the Lake Superior area are apparently the native boreal and western var. *humifusa* (Dickson) Vahl, sometimes recognized as a distinct species. This variety has many straight hairs slightly longer than the antrorsely incurved ones, especially in the inflorescencee (on axis and pedicels), giving that part of the plant a more hairy look than the lower portion of the stem; the flowers (deeper blue) and fruits run a little larger than in var. *serpyllifolia*.

12. **V. peregrina** L.　　Fig. 105　　　　　　　　Purslane Speedwell
Map 318. One of our few native veronicas, but usually ± weedy in habit. The oldest Michigan collections I have seen are from Oakland County (1852) and Washtenaw County (1861); by the 1890s it had also been gathered in Ionia, Shiawassee, Gratiot, Wayne, St. Clair, and Kent counties. So there is some reason to be suspicious that the species has spread considerably over the past century. It grows in cultivated fields, flower beds, lawns; damp disturbed ground such as logging trails, gravel pits, shores; on limestone pavements, outcrops, and gravels; open moist swales and streamsides; waste ground.
Frequently confused with the preceding species by those not familiar with both. Besides the characters in the key, *V. peregrina* usually has nearly linear leaves, somewhat succulent in aspect, in contrast to the broadly rounded, sometimes even orbicular, blades that are usual in *V. serpyllifolia*. Furthermore, the latter is perennial. Scattered throughout the state are populations with short gland-tipped hairs on the stem (especially above) and in the inflorescence; these are var. *xalapensis* (Kunth) Pennell.

13.　**V. arvensis** L.　　Fig. 106　　　　　　　　Field Speedwell
Map 319. A common little Eurasian weed, reported by the First Survey and collected in several counties before 1890 (Macomb, Washtenaw, Kent, Ingham, St. Clair, Grand Traverse, Ionia). Along roads and trails, disturbed areas in woods, on shores and rock ledges; in damp fields and gardens; and waste ground generally, such as parking lots, vacant lots, roadsides.
The style is ca. 0.5–1 mm long: often intermediate between style lengths of the next two species, which resemble it. Pedicels are up to 2 mm long.

14.　**V. dillenii** Crantz
Map 320. A European species, first found in Michigan by Farwell in 1928 (*8188*, MICH, BLH), who reported it "along roadsides, on sandy hillsides, in sandy fields and cultivated grounds everywhere about Ortonville" (Am. Midl. Nat. 12: 131. 1930). It is still at Ortonville and elsewhere in Oakland County, and in 1992 was found in a disturbed sandy area in Crawford County (*Chittenden 450 & Peil*, MICH, MSC).
Some authors have included this species in *V. verna*.

15. **V. verna** L.

Map 321. Yet another little Eurasian weed, first reported from several Michigan counties in 1977 (Michigan Bot. 16: 137). The earliest collection in the state dates from 1959 (Isle Royale) and the species is now doubtless more widespread than the map suggests. It is easily overlooked, but can be found in dry sandy, gravelly, or rocky bare places in lawns, parking lots, driveways, roadsides, campgrounds, and pineland.

Plants as short as 2 cm may bloom. Small individuals with poorly developed leaves (e.g., with only a single pair of basal lobes) are sometimes difficult to recognize, especially when growing with *V. arvensis* (rather than with better developed plants of *V. verna*!). Sometimes the lobes of the capsule are low and rather divergent (the notch unusually shallow); the style, though still very short, may then about equal the lobes.

20. **Gratiola**   Hedge-hyssop

KEY TO THE SPECIES

1. Leaves with conspicuous dark glandular dots on both surfaces; corolla bright yellow; capsule at most scarcely 3 mm long ....1. **G. aurea**
1. Leaves with glandular dots obscure or none; corolla yellowish, cream, or white; capsules 3–7 mm long
   2. Pedicels at most ca. 1.5 times as long as the calyx ....2. **G. virginiana**
   2. Pedicels at maturity at least twice as long as the calyx ....3. **G. neglecta**

1. **G. aurea** Pursh   Plate 4-C

Map 322. First discovered in Michigan in 1968 in Gogebic County (*Voss 12699*, MICH, UMBS, MSC; *12793*, MICH, GH) as the submersed f. *pusilla* Fassett, in which the stems and leaves are of different texture than in terrestrial plants; and the conspicuous glands are lacking (see Key A of General Keys for characters). Later found with the handsome yellow flowers, when water levels were lower (*Henson 2575 & 2577* in 1988, MICH). Thus far known in Gogebic County only from Clark, Long, and Snap Jack lakes, but to be expected at other sandy-bottom, softwater lakes with such

319. Veronica arvensis   320. Veronica dillenii   321. Veronica verna

associates as *Elatine minima, Eriocaulon aquaticum* [*E. septangulare*], and *Lobelia dortmanna*. However, on Drummond Island, it grows on the shores of Potagannissing Bay (*Stephenson* in 1988, MICH, MSC).

2. **G. virginiana** L.

Map 323. Discovered in Michigan in 1981 (*Ng 600 & 645*, AUB, MSC) in a prairie marsh south of Niles; also in a periodically flooded sand flat near New Buffalo (*Dritz 935* in 1991, MOR).

The pedicels are not only shorter than in the next species, but also thicker, giving a quite different aspect to the plant. The capsules are globose and somewhat larger, on average, than the more ovoid fruit of *G. neglecta*.

3. **G. neglecta** Torrey

Map 324. Muddy river banks, depressions in fields, and wet sandy disturbed places.

The elongate (especially in fruit) filiform pedicels tend to be a little more conspicuously glandular-pubescent than those of the preceding species.

21. **Lindernia**

REFERENCE

Cooperrider, Tom S., & George A. McCready. 1975. On Separating Ohio Specimens of Lindernia dubia and L. anagallidea (Scrophulariaceae). Castanea 40: 191–197.

1. **L. dubia** (L.) Pennell   Fig. 109   False Pimpernel

Map 325. An inconspicuous annual of bare mud or wet sand on shores, river banks, marsh and pond margins, interdunal flats, borrow pits, and old woodland trails; thriving on exposure after lowering of water levels and seldom seen in times of high water. (The seeds are evidently dispersed or rejuvenated from seed banks by fluctuations in water level.)

A few specimens from the southern counties of the Lower Peninsula can be referred to var. *anagallidea* (Michaux) Cooperr., with slender pedicels

322. Gratiola aurea

323. Gratiola virginiana

324. Gratiola neglecta

exceeding the subtending rather broad, somewhat clasping leaves. Plants with pedicels shorter than the leaves, which are ± tapered to their bases, represent var. *dubia*. The two have often been regarded as distinct species, but there are too many intermediates in the Midwest even to recognize them as varieties.

22. **Mazus**

1. **M. reptans** N. E. Br.

Map 326. A native of Asia, cultivated as a perennial ground cover for its attractive blue flowers; occasionally becoming a lawn and garden weed.

The plant has abundant, very slender stolons, and roots freely at the nodes. It is thus quite different in habit from the erect annual *M. japonicus* (Thunb.) Kuntze, although often misidentified as the latter [for which Van Steenis (Nova Guinea n.s. 9: 31. 1958) has suggested that the correct name (admittedly uncertain) may be *M. pumilus* (Burman f.) Steenis].

Plants of *Kickxia* and *Cymbalaria* with similar prostrate habit might be keyed here if one fails to note the spur on the corolla of those two; however, the corollas of *Mazus reptans*, besides lacking a spur, are much larger (over 1 cm and as much as 2.6 cm long).

23. **Buchnera**

1. **B. americana** L.   Fig. 111   Blue-hearts

Map 327. A southern species of sandy open woodland and prairies, not collected in Michigan since the First Survey gathered it in Kalamazoo and Calhoun counties in 1838. Two sheets in the Cooley collection (MSC), without locality, may have come from Macomb County in 1843.

The species is hemiparasitic on the roots of a great variety of trees (Musselman & Mann 1978) and presumably other plants but can mature without a parasitic attachment. The corolla is purplish to blue and nearly regular.

325. Lindernia dubia   326. Mazus reptans   327. Buchnera americana

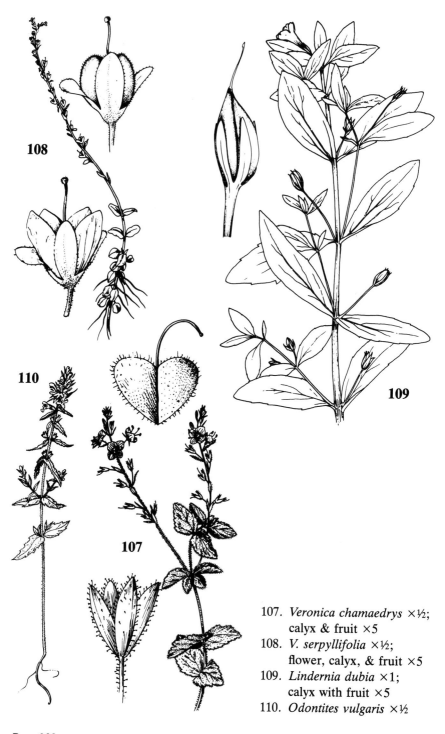

108

110

107

109

107. *Veronica chamaedrys* ×½;
     calyx & fruit ×5
108. *V. serpyllifolia* ×½;
     flower, calyx, & fruit ×5
109. *Lindernia dubia* ×1;
     calyx with fruit ×5
110. *Odontites vulgaris* ×½

24. **Odontites**

1. **O. vulgaris** Moench   Fig. 110   Eyebright
    Map 328. Adventive from Europe, gradually spreading in north-eastern North America as a local weed. First collected in Michigan in 1981 (*Henson 1366*, MICH, MSC) along a highway (M-35) in Delta County; subsequently found at additional sites along the same highway in both Delta and Menominee counties, and it can be expected to spread farther.
    There has been much nomenclatural confusion over this taxon. It has often been called *O. serotina* Dumort., a name dating from 1827 (a supposed basionym by Lamarck is illegitimate). However, there are older names, some of which are illegitimate although used in some floras and lists. Another taxonomic position is to treat this taxon as only a subspecies of another, which has longer floral bracts: *O. vernus* ssp. *serotinus* (Dumort.) Corb. (Furthermore, since 1987 the Code has required that all generic names ending in *-ites* be treated as masculine, a provision previously only a recommendation.)

25. **Euphrasia**   Eyebright
    This is a very difficult genus taxonomically, the species very similar to each other and often not easily recognizable, especially when individuals are small or immature. Collections previously reported from Isle Royale as *E. arctica* Rostrup and *E. disjuncta* Fernald & Wiegand are now both referred to *E. nemorosa*.
    The corollas are white or pale blue-purple, with darker purple lines and a bright yellow spot at the base of the lower lip. *Euphrasia* is one of the many hemiparasitic genera in the Scrophulariaceae, attaching to the roots of a wide range of hosts; plants may, however, develop to maturity (though less vigorously) without a parasitic attachment. There is no information on parasitism specifically in the Great Lakes region.

328. Odontites vulgaris

329. Euphrasia stricta

330. Euphrasia hudsoniana

REFERENCES

Downie, Stephen R., Andrée Quintin, & John McNeill. 1988a. Le Statut des Euphrasia borealis, E. nemorosa et E. stricta dans l'Est de l'Amérique du Nord: Une Analyse Numérique. Canad. Jour. Bot. 66: 2208–2216.

Downie, Stephen R., & John McNeill. 1988b. Description and Distribution of Euphrasia stricta in North America. Rhodora 90: 223–231.

Sell, P. D., & P. F. Yeo. 1970. A Revision of the North American Species of Euphrasia L. (Scrophulariaceae). Jour. Linn. Soc. Bot. 63: 189–234.

Yeo, P. F. 1964. The Growth of Euphrasia in Cultivation. Watsonia 6: 1–24.

Yeo, P. F. 1978. A Taxonomic Revision of Euphrasia in Europe. Jour. Linn. Soc. Bot. 77: 223–334.

KEY TO THE SPECIES

1. Calyx, bracts, and leaves glabrous or at most scabrous on some nerves; calyx lobes and leaf teeth tapered into prolonged bristle tips . . . . .1. **E. stricta**
1. Calyx, bracts, and leaves pubescent; calyx lobes acute to somewhat bristle-tipped
  2. Cauline leaves (not subtending flowers) strongly cuneate at base, the teeth acute to blunt; calyx pubescent principally on the nerves and lobes; plants sometimes flowering as low as the second node . . . . .2. **E. hudsoniana**
  2. Cauline leaves broadly cuneate to rounded at base, the teeth rounded to blunt; calyx rather densely hairy throughout; plants flowering only as low as the fifth node . . . . .3. **E. nemorosa**

1.   **E. stricta** J. F. Lehm.   Fig. 112

Map 329. A European species, spreading westward from New England and first collected in Michigan in 1983 (Delta Co.)—about a decade after it was widely collected as a common weed in the Lake Superior region of Ontario (but misidentified as the next species) and on Manitoulin Island in northern Lake Huron. Likely to continue spreading in lawns, clearings, trails, roadsides, old railroad grades.

This is a fairly easy species to identify, with glabrous leaves and with more definite (almost hair-like) bristle tips on the teeth of the leaves and calyx than the next two species—although the bristle tips are a little less pronounced than on some Ontario material. The plants often present a rather "skinny" aspect, with longer internodes and smaller leaves than other species, and a close look reveals to the naked eye a somewhat bristly aspect from the prolonged teeth. Furthermore, the bracts tend to be more strongly erect, compared to a spreading habit in the other species. Robust specimens may be as tall as 20–40 cm, while on Isle Royale our native species do not exceed 20 cm in height (though they may be larger elsewhere).

2.   **E. hudsoniana** Fernald & Wiegand

Map 330. Rock crevices and ledges at several Isle Royale sites.

Often not easily distinguished from *E. nemorosa*. The characters in the key are those that seem to work best on most collections from Lake Superior determined by Yeo, but even he has expressed doubt about some individual specimens. One must often consider a larger population. This species appears intermediate between our other two in the conspicuousness of sharp tips on bracts and calyx teeth.

3. **E. nemorosa** (Pers.) Wallr.
Map 331. Rock crevices and ledges at Scoville Point on Isle Royale and nearby Edwards Island.
Yeo (1978) includes *E. curta* (Fries) Wettst., to which he previously referred material from Lake Superior, as a dwarf hairy state of this widespread circumpolar species.

26. **Tomanthera**
This genus is sometimes included in the next (see Canne 1980) as the differences apparent in our species do not hold up well when South American ones are considered. However, tradition is here maintained for our very distinctive species, presumed extinct in Michigan.

REFERENCES

Canne, Judith M. 1980. Seed Surface Features in Aureolaria, Brachystigma, Tomanthera, and Certain South American Agalinis (Scrophulariaceae). Syst. Bot. 5: 241–252.
Knoop, Jeffrey D. 1988. Tomanthera auriculata (Michx.) Raf. Extant in Ohio. Ohio Jour. Sci. 88: 120–121.
Pennell, Francis W. 1928. Agalinis and Allies in North America.— I. Proc. Acad. Nat. Sci. Philadelphia 80: 339–449. [*Tomanthera*, pp. 439–446.]

1. **T. auriculata** (Michaux) Raf.   Eared False Foxglove
Map 332. Apparently originally native from Ohio, southern Michigan, and southern Wisconsin southward and westward, but rare and local everywhere that it still grows. Collected in Michigan only by the First Survey in a sandy bur oak opening at White Pigeon, St. Joseph County (in 1837, MICH) and in dry openings near Edwardsburg, Cass County (in 1838, NY). Long thought to be extirpated in Ohio, until rediscovered there in 1985, so perhaps there is hope that it will turn up again in Michigan.
With at least equal justice, this species may be classified as *Agalinis auriculata* (Michaux) S. F. Blake.

27. **Agalinis**   Gerardia
The generic name *Gerardia* was rejected at the 1959 International Botanical Congress. It may not be used for any plant (except familiarly as a common name, to which no rules apply). It is to be replaced in the Scrophulariaceae by *Agalinis*, formerly considered by many authors as a segregate from *Gerardia*. Other segregates retain their generic names if recognized at that rank (see above: 15. *Dasistoma*; 16. *Aureolaria*; 26. *Tomanthera*).

Our species are all rather similar: somewhat delicate-looking plants with attractive pink-purple corollas falling in the afternoon, very slender linear usually scabrous leaves, and tiny tap root. This is another hemiparasitic group, with roots attaching to diverse hosts, especially graminoids.

REFERENCES

Brodowicz, William. 1990. Noteworthy Collections. Michigan Bot. 29: 28–30. [*A. gattingeri* & *A. skinneriana*]

Canne, Judith M. 1979. A Light and Scanning Electron Microscope Study of Seed Morphology in Agalinis (Scrophulariaceae) and Its Taxonomic Significance. Syst. Bot. 4: 281–296.

Pennell, Francis W. 1929. Agalinis and Allies in North America—II. Proc. Acad. Nat. Sci. Philadelphia 81: 111–249.

KEY TO THE SPECIES

1. Mature pedicels 1–5.5 mm long . . . . .1. **A. purpurea**
1. Mature pedicels 7–28 (30) mm long
  2. Calyx tube with at most the longitudinal nerves conspicuous (fig. 113); plants usually blackening when dried; seeds dark brown; widest leaves often 1–3 (3.5) mm broad . . . . .2. **A. tenuifolia**
  2. Calyx tube conspicuously reticulate-veined (fig. 115); plants remaining ± green when dried; seeds pale or light brown; widest leaves at most 1 mm broad
    3. Stem strongly angled (ridged); corolla with all lobes glabrous outside (though ciliate); flowers usually all or mostly on pedicels in axils of opposite leaves or bracts on the main stem (rarely branched) . . . . .3. **A. skinneriana**
    3. Stem nearly or quite terete; corolla with 3 lower lobes pubescent outside; flowers all or mostly on branches, often appearing terminal on these . . . . .4. **A. gattingeri**

1.   **A. purpurea** (L.) Pennell

Map 333. Sandy, gravelly, and rocky shores (of Great Lakes and inland lakes and ponds) and interdunal swales, especially after lowering of water levels; fens, sedge meadows, bogs; sand prairies and wet calcareous banks.

This, our commonest and most widespread species in the genus, is also our most easily recognized one, thanks to its short and relatively stout pedicels. The leaves are less than 1.5 mm (very rarely 2.7 mm) broad. Flowering plants may be as short as 5 cm or as tall as 60 cm; in any event, the relatively large ratio of corolla to foliage makes a stand of *Agalinis* a very colorful sight. In a given longitude, the farther north one goes, the smaller the corollas in this species, as here treated in a broad sense.

In Michigan, individuals of the large-flowered end of this cline are uncommon, and variable in other respects, north to Newaygo and Saginaw counties. Typical *A. purpurea* has corollas at least 2 cm long, styles ca. 1.5 cm or more long, and calyx lobes less than 2 mm long; such plants are known very locally in the southernmost two tiers of counties. Throughout the state are plants often segregated as *A. paupercula* (A. Gray) Britton, with corollas less than 2 cm, styles ca. 1 cm or shorter, and calyx lobes up to 3.5 or even 4 mm long; such plants may be called *A. purpurea* var. *parviflora* (Bentham) B. Boivin [or ssp. *parviflora* (Bentham) Löve &

Löve]—including the northernmost, smallest-flowered plants originally named *A. paupercula* var. *borealis* Pennell.

However, some plants of the small-flowered northern variety may have calyx lobes as short as 0.5 mm; some plants with short styles and long calyx lobes may have large corollas; and other plants with large corollas and long styles may have long calyx lobes. These observations, and similar ones by others, confirm Pennell's statement (1929) that intergrades between *purpurea* and *paupercula* "occur rather plentifully" in their geographic zone of contact. (See also Farwell, Am. Midl. Nat. 9: 276. 1925.) Specimens can be separated on the basis of corolla size, style length, or calyx lobe length—but the results will be different in each case.

2. **A. tenuifolia** (Vahl) Raf.   Fig. 113

Map 334. Damp open or even marshy ground: sandy ditches, borrow pits, shores, prairies, and meadows; fens, river banks.

This is our second commonest species, with very slender elongate pedicels obviously different from the preceding. The length of the calyx lobes is quite variable.

3. **A. skinneriana** (A. W. Wood) Britton   Fig. 115

Map 335. Usually considered a rare species, and certainly so in the Great Lakes region. The only documented collection from Michigan is from an open sandy depression in Algonac State Park (*Brodowicz* in 1988, MICH).

Good stems of this species are very narrowly winged on the 4 corner angles, with a pale rounded ridge on each side between them. The branching and flowering characters that help to distinguish this from the next are rather obscure, and depauperate, unbranched plants of *A. gattingeri* may resemble *A. skinneriana*. Persons who know them in the field note that they have different aspects, evident when they grow together. The observations of Canne (1979) on seed morphology would seem to support recognition of these as distinct species. The corollas of *A. skinneriana* are said to

331. Euphrasia nemorosa

332. Tomanthera auriculata

333. Agalinis purpurea

be very pale, with scarce if any evidence of spots or lines, in contrast to the deeper pink ones of *A. gattingeri*, with red spots and two yellow lines in the throat.

4. **A. gattingeri** (Small) Small

Map 336. Open sandy places, as on higher ground near marshes and in old borrow pits.

While the stems may have low ridges, they are not sharply angled or winged. One of the best known sites for this species and *A. skinneriana* has long (since around 1900) been the delta islands on the Ontario side of the St. Clair River. Discovery of these species in St. Clair County, Michigan, has been recent, by Brodowicz et al. in 1988.

28. **Mimulus** Monkey-flower

REFERENCES

Beadle, Sandy J., & Susan R. Crispin. 1990. The Elusive Monkey-flower. Michigan Nat. Resources 59(3): 42–43.

Bliss, Margaret. 1983. A Comparative Study of the Two Varieties of Mimulus glabratus (Scrophulariaceae) in Michigan. M. S. thesis, Univ. Michigan. 58 pp.

Bliss, Margaret. 1986. The Morphology, Fertility and Chromosomes of Mimulus glabratus var. michiganensis and M. glabratus var. fremontii (Schrophulariaceae [sic]). Am. Midl. Nat. 116: 125–131.

Grant, Adele Lewis. 1924. A Monograph of the Genus Mimulus. Ann. Missouri Bot. Gard. 11: 99–388 + pl. 3–10.

KEY TO THE SPECIES

1. Corolla bluish to purple; stems 4-angled; leaves pinnately veined (with lateral veins arising all along the midrib); stems and pedicels glabrous
   2. Leaves petioled; pedicels (even in fruit) shorter than the calyx . . . . .1. **M. alatus**
   2. Leaves sessile, ± clasping; pedicels longer than the calyx . . . . .2. **M. ringens**
1. Corolla yellow; stems terete or many-ridged but not square in section; leaves nearly or quite palmately veined (with lateral veins all near or at the base of the blade); stems and pedicels glabrous or pubescent

334. Agalinis tenuifolia   335. Agalinis skinneriana   336. Agalinis gattingeri

3. Plant ± densely viscid-pubescent (with many hairs exceeding 0.5 mm); calyx lobes
   slightly unequal, half to fully as long as the tube; corolla nearly regular . . . .
   . . . . .3. **M. moschatus**
3. Plant glabrous or with minute, often gland-tipped hairs up to 0.5 mm long; calyx
   lobes very unequal, only the upper large one ever as much as half as long as the
   tube; corolla 2-lipped
   4. Corolla ca. 3–4.5 cm long, the throat nearly closed by the uparching lower lip;
      style ca. 20–25 mm long; plant pubescent with minute eglandular hairs on calyx,
      pedicels, and stems . . . . .4. **M. guttatus**
   4. Corolla less than 3 cm long, the throat open; style less than 12 mm long; plant
      glabrous or with minute usually glandular hairs (rarely as long as 0.5 mm) . . . . .
      . . . . .5. **M. glabratus**

1. **M. alatus** Aiton

Map 337. A wetland species more southern in range, collected in Michigan three times, all on August 27, 1916, if the labels be trusted: by B. F. Chandler (MICH, MSC, BLH, WUD) along Lake St. Clair at Gauklers Point, in Erin Tp., Macomb County (just north of Wayne County); by Cecil Billington (WMU) in adjacent Grosse Pte. Tp., Wayne County; and by O. A. Farwell (*4412*, BLH) on the Detroit Zoo tract in southern Oakland County (see Rep. Michigan Acad. 19: 249. 1918; Pap. Michigan Acad. 2: 40. 1923). Because they have the most precise locality, I suspect that the Macomb County specimens are correctly labeled and that the other collectors (who often did botanize together) were present but not sure of the source of their specimens. An alternative interpretation, of course, is that the species was indeed collected at these three sites, in as many counties, on the same day and never before or since anywhere in the state.

The corolla was said to be unusually pale or white on the Chandler specimens.

2. **M. ringens** L.   Fig. 116.

Map 338. Wet, usually ± open places: marshes, swales and ditches, swamps, shores, fens; borders of lakes, ponds, rivers, and streams; does well in disturbed wetlands.

337. Mimulus alatus

338. Mimulus ringens

339. Mimulus moschatus

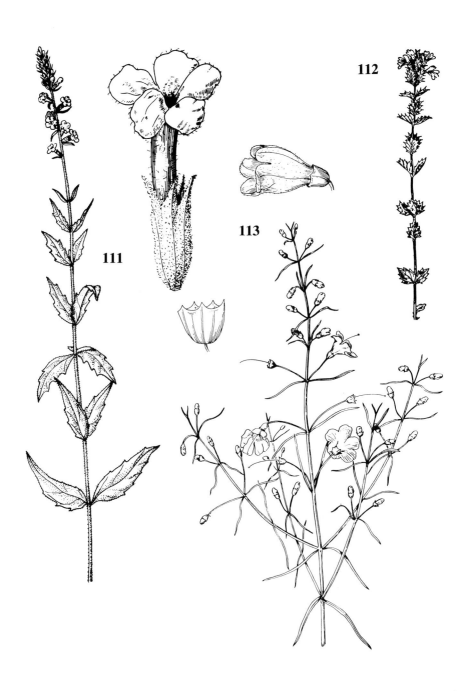

111. *Buchnera americana* ×½; flower ×4
112. *Euphrasia stricta* ×¾
113. *Agalinis tenuifolia* ×½; calyx ×3; flower ×2

3.   **M. moschatus** Lindley   Muskflower

Map 339. Muddy or wet ditches, creeks, springy banks, borders of swamps and ponds; moist openings, trails, and roadsides in woods. This species is found in western North America from British Columbia southward, primarily in the mountains. There is some question whether all the disjunct eastern populations are indigenous (see Michigan Bot. 20: 67–68. 1981).

This is a lax, ± prostrate plant, with musky odor and very clammy pubescence.

4.   **M. guttatus** DC.

Map 340. Discovered in Michigan in 1987 (*Henson 2299*, MICH, UMBS & *Reznicek 7962*, MICH, MSC, BLH, GH) along a springy ditch and in adjacent seepy mixed woods with *Saxifraga pensylvanica* along a Forest Service road about 8 miles northeast of Bergland. As for the preceding species, whether this is a native disjunct from western North America or an introduction is an open question.

Our very showy specimens represent only one of the many variants known of this species, so common and widespread in the West, where populations differ in nature (or absence) of pubescence, size of flowers, and other characters. Plants are usually much more erect and taller than our other two yellow-flowered species.

5.   **M. glabratus** Kunth   Plate 4-D

Map 341. Both of our varieties are calciphiles, growing along marly springs, in cold streams through cedar swamps, on calcareous shores, and in associated ditches. At least var. *michiganensis* may overwinter in flowing water.

This is usually a prostrate, mat-forming species, although shoots (especially in var. *michiganensis*) may be quite erect from ± decumbent bases. The flowering season begins in June and can run into October.

Most of our plants belong to var. *jamesii* (Bentham) A. Gray, with styles ca. 3–5 mm long and flowers averaging smaller, (6) 8–15 mm long, than in var. *michiganensis* (Pennell) Fassett, with styles ca. (7) 8–11 mm long and corollas 14–22 (24) mm long. Depauperate plants of the latter and vigorous plants of the former can overlap in flower size (and more often in leaf size), so that style length is the most convenient and reliable character to distinguish them.

The "Michigan monkey-flower," var. *michiganensis*, has been on the Federal (and hence the state) list of endangered species since 1990. Its total range, so far as known, is in the Straits of Mackinac and Grand Traverse regions of Michigan (about 20 sites altogether in Benzie, Leelanau, Emmet, Cheboygan, and Mackinac counties and Beaver Island). Within this range, both varieties are known. There has been much research on the life history, ecology, and distribution of this rare taxon, which may be derived from a genetic variant of var. *jamesii* or as a result of hybridization with *M. guttatus*. Fruit is very rarely formed and most reproduction must be vegetative—but it can be luxuriant. The type locality for this variety, wholly endemic to Michigan, is near Topinabee, Cheboygan County.

The more common var. *jamesii* is nevertheless seldom seen, being quite local. [It was long known as var. *fremontii* (Bentham) A. L. Grant, but under the Code as amended in 1981 it must be known by the autonym (var. *jamesii*) created when *M. jamesii* var. *fremontii* was originally published.] Other varieties grow in the western United States and Mexico.

29. Collinsia

The leaves are sometimes whorled at some nodes. The flower may appear to have a 2-lobed lower lip, the keel-like middle lobe obscured. The fruit is a globose-ovoid capsule.

KEY TO THE SPECIES

1. Corolla 10–17 mm long, the upper lip white; principal leaves 7–26 mm broad (widest below the middle); style (persisting on immature fruit) 8–11 mm long; range in southern Lower Peninsula . . . . .1. **C. verna**
1. Corolla 5–8 mm long, the upper lip often at least partly blue; principal leaves 2–7 mm broad (widest at or beyond the middle); style 3.5–5 mm long; range in Upper Peninsula . . . . .2. **C. parviflora**

1. **C. verna** Nutt.   Blue-eyed-Mary

Map 342. Rich deciduous woods, especially in ravines and low moist areas.

This is one of our most attractive wildflowers, striking in its bicolored corolla.

2. **C. parviflora** Lindley   Blue-lips

Map 343. A species disjunct in range from farther west (Michigan Bot. 20: 72–73. 1981). In Schoolcraft County it may be adventive, on "weedy sand barrens" (*Gillis 2455* in 1958, MSC). Otherwise, in crevices and gravelly soil on exposed rock outcrops.

340. Mimulus guttatus

341. Mimulus glabratus

342. Collinsia verna

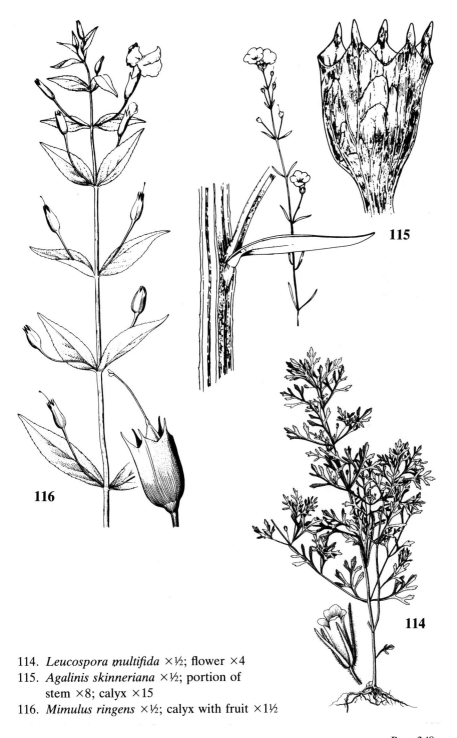

115

116

114

114. *Leucospora multifida* ×½; flower ×4
115. *Agalinis skinneriana* ×½; portion of
 stem ×8; calyx ×15
116. *Mimulus ringens* ×½; calyx with fruit ×1½

30. Melampyrum

REFERENCES

Cantlon, J. E., E. J. C. Curtis, & W. M. Malcolm. 1963. Studies of Melampyrum lineare. Ecology 44: 466–474.
Piehl, Martin A. 1962. The Parasitic Behavior of Melampyrum lineare and a Note on Its Seed Color. Rhodora 64: 15–23.

1. **M. lineare** Desr.   Fig. 118   Cow-wheat
Map 344. Surely one of our most abundant and ubiquitous plants, especially in the northern part of the state, in both wet and dry woods (deciduous or coniferous) and peatlands, favoring drier and only moderately shaded sites but ranging from bare rock cliffs to wet fens.

The fruit is strongly flattened laterally and sharply pointed. Leaves vary from as narrowly linear as those of *Agalinis* to ovate, often with some narrow teeth toward the base of bracteal leaves (subtending flowers). The corolla is whitish, both lips tipped with yellow. This species has been documented in Michigan as parasitizing roots and rhizomes of many herbaceous and woody species (even perhaps on *Sphagnum*) (Piehl 1962; Cantlon et al. 1963). In Europe this is a much larger and very complex genus, with over 20 species.

For a plant so abundant and readily accessible, *M. lineare* has long suffered neglect as to its habits. Parasitism was not documented until the early 1960s. Piehl also pointed out that the relatively large (for Scrophulariaceae) "black" seeds become black (like the foliage) only upon drying from an original tan. Furthermore, the seeds have an elaiosome, which indicates dispersal by ants, as long known for some European species of the genus. (Any need for ant dispersal has apparently not kept this species from populating even small islands in the Great Lakes.) Pennell (1935, p. 507) reported failure to see insects visiting the flowers and believed the species to be "wholly self-pollinated." However, butterfly and bee visitors had been reported in wildflower books long before. In the sandy coniferous woods along Lake Superior in Luce County, I have seen numerous small bumblebees busily visiting *Melampyrum* flowers in late August, and likewise pink-edged sulfur butterflies probing the flowers with their proboscises. While such visits in themselves do not prove pollination, they are surely suggestive.

BIGNONIACEAE   Bignonia or Trumpet-creeper Family

This is a large family, almost entirely of woody plants, with a few species native in the North Temperate zone. A number of Bignoniaceae are cultivated for their large colorful flowers.

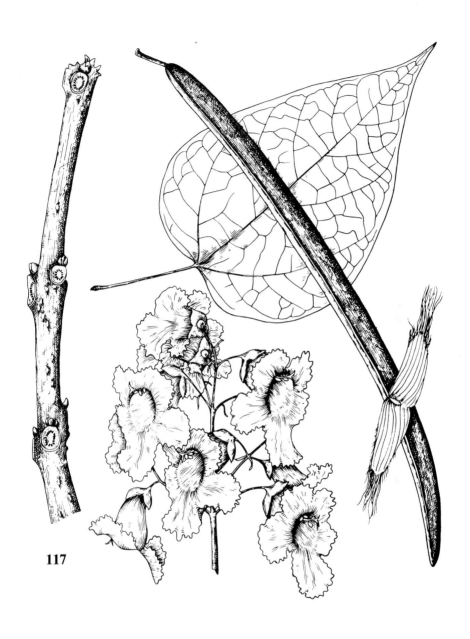

117

117. *Catalpa speciosa*, inflorescence ×½; winter twig & seed ×1¼; capsule
×⅝; leaf ×⅜

KEY TO THE GENERA

1. Plant a shrubby vine with pinnately compound leaves (toothed leaflets) and red-orange flowers ....1. **Campsis**
1. Plant a tree with simple entire leaves and white (but spotted) flowers ....2. **Catalpa**

1. Campsis

1. **C. radicans** (L.) Bureau   Trumpet-creeper; Trumpet-flower
    Map 345. Native south of Michigan, here escaped from cultivation to roadsides, railroads, fencerows, and thickets.
    This shrubby vine may produce aerial roots, but not tendrils. The bright corollas are (5.5) 6–8 (8.5) cm long with long orange tube and short red limb; the capsule is elongate, larger in diameter than in *Catalpa* but somewhat shorter. Mail-order firms sometimes promote this plant as a "hummingbird vine," for the flowers are indeed well adapted to that pollinator.

2. Catalpa
Two species native to the southern United States are planted northward and escape from cultivation, although only one seems to be documented in Michigan. The leaves are usually whorled (or nearly so), sometimes opposite or even alternate on some shoots (Grover).

REFERENCES

Grover, Frederick O. 1942. The Phyllotaxy of Catalpa. Rhodora 44: 414–416.
Stephenson, Andrew G., & Wm. Wayt Thomas. 1977. Diurnal and Nocturnal Pollination of Catalpa speciosa (Bignoniaceae). Syst. Bot. 2: 191–198.

1. **C. speciosa** (Warder) Engelm.   Fig. 117   Catalpa; Cigar-tree
    Map 346. Widely planted as far north as the Straits of Mackinac, and sprouting only occasionally in waste places from the copious winged seeds produced. Fields, dumps, fencerows, railroads, roadsides, parking lots, thickets, and borders of woods.
    Insofar as our non-planted specimens are flowering or fruiting, they

343. Collinsia parviflora

344. Melampyrum lineare

345. Campsis radicans

appear to be this species, although *C. bignonioides* Walter has frequently been reported from the state. *C. speciosa* is usually said to have larger corollas (limb 4–6 cm broad), odorless leaves, thicker capsules (12–15 mm), and bark on old trees fissured and ridged. *C. bignonioides* is said to have smaller flowers (corolla limb 2–3 cm broad), leaves ill-smelling when crushed, slenderer capsules (5–10 mm), and old bark thin and scaly. However, Swink and Wilhelm found distinctions frustrating in the Chicago area and referred all their specimens to *C. speciosa*. Steyermark contrasted 11 characters in Missouri and observed that "all the characters in the key must be considered for final determination" because the species are so difficult to distinguish. Herbarium specimens lacking fruit and/or carefully pressed flowers as well as notes on odor and bark are inadequate for critical study. *C. speciosa* is often said to have the leaf tips more acuminate, and indeed they are so on all our specimens. The hairs at the ends of the seeds are connivent in a roadside collection from Kalamazoo County, and this is supposed to be a character of *C. bignonioides*; in *C. speciosa*, the hairs remain separate, in a broad brush. But whether this specimen was from a planted tree is not clear.

Pollination studies near Ann Arbor determined that this species is an obligate outcrosser, pollinated by bees during the day and by a diversity of moths during the night (Stephenson & Thomas).

PEDALIACEAE   Sesame Family

Cronquist and other recent authors include in this family the plants sometimes treated as a New World segregate, Martyniaceae—to which the only species ever found in our flora belongs. The Pedaliaceae are a small, mostly tropical family and include *Sesamum indicum* L., the source of sesame seeds, probably the most familiar form of the family in Michigan supermarkets and bakeries. The seeds are produced by an annual grown for their edible oil and reputed medicinal properties since at least 1600 B.C.—perhaps the earliest crop used by man for an edible oil.

346. Catalpa speciosa

347. Proboscidea
     louisianica

348. Conopholis
     americana

1. Proboscidea

REFERENCES

Hevly, Richard H. 1969. Nomenclatural History and Typification of Martynia and Proboscidea (Martyniaceae). Taxon 18: 527–534.
Thieret, John W. 1977. The Martyniaceae in the Southeastern United States. Jour. Arnold Arb. 58: 25–39.

1. **P. louisianica** (Miller) Thell.   Fig. 119.   Unicorn Plant
Map 347. The native range of this species is apparently from southern-most United States into Mexico, but it has spread widely as a weed even to the latitude of Michigan, where it has been collected (rarely) as a farm and garden weed and in roadside ditches. The only extant 20th century collections are from Washtenaw, Wayne, and Oakland counties, made in 1918, 1955, and 1960, respectively. Dodge in 1900 reported it as an occasional weed at Port Huron; his label for an 1899 collection declared it "Plentiful" in a cornfield. What was presumably this species was reported from Detroit as early as 1767 (see Voss 1978, p. 1). Farwell reported it as rare in waste places at Detroit at the turn of the century, but his specimens have not survived.

This is a bushy, clammy-pubescent, bad-smelling annual with large whitish flowers, the corolla tinged or spotted with pink and the fruit a peculiar object often used in novelties and winter flower arrangements. The fruit is a drupe with woody endocarp prolonged into an arched beak (which may be twice as long as the body) that splits into two halves, each hooked at the tip, like two large claws. These are obnoxious in the wool of sheep and anywhere else they do not belong. The leaves are palmately veined, entire, ± cordate, all or mostly opposite.

OROBANCHACEAE   Broom-rape Family

This family, apparently derived from the Scrophulariaceae, consists entirely of root parasites, lacking chlorophyll and thus wholly dependent on their hosts (unlike our hemiparasitic Scrophulariaceae, which have some green leaves). Some species are important agricultural pests in the Old World, but our native species seem to be of no economic importance.

Seeds of these plants germinate, so far as known, only when they are close enough to a host plant root to be stimulated by some product secreted by that host.

REFERENCE

Thieret, John W. 1971. The Genera of Orobanchaceae in the Southeastern United States. Jour. Arnold Arb. 52: 404–434.

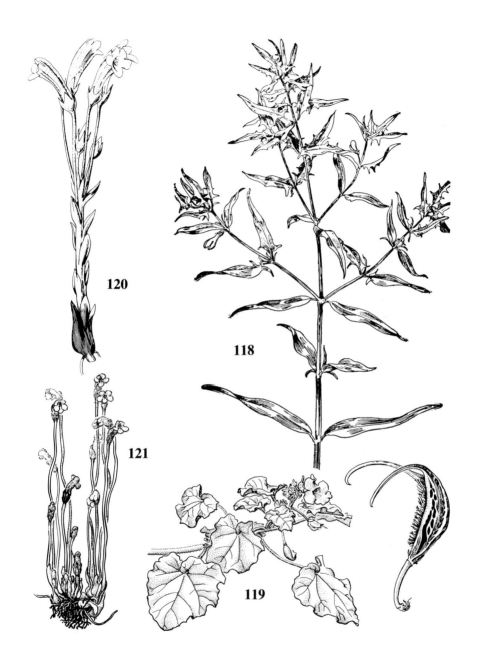

118. *Melampyrum lineare* ×½
119. *Proboscidea louisianica* ×¼; fruit (dry endocarp) ×½
120. *Orobanche fasciculata* ×1
121. *O. uniflora* ×⅖

KEY TO THE GENERA

1. Flowers in a dense, thick, scaly spike, the bracts subtending them 3–8 mm broad and nearly or quite exceeding the calyx; stem below the inflorescence ca. 4–10 mm thick, ± densely covered with bract-like leaves at least as large as the floral bracts; stamens exserted . . . . .1. **Conopholis**
1. Flowers solitary or well separated, their bracts at the bases of long pedicels or up to 2 mm wide and scarcely touching the calyx; aerial stems if any 1.5–4.5 (6) mm thick below the inflorescence, with scattered bracts; stamens sometimes exserted (*Epifagus*) or included in the corolla (*Orobanche*)
    2. Pedicels none or much shorter than the flowers, essentially glabrous; flowers ca. 11 mm long or a little shorter, blooming in late summer or early fall; stem often much-branched; calyx glabrous or nearly so, the lobes ± deltoid and glandular along the margin . . . . .2. **Epifagus**
    2. Pedicels equalling or exceeding the flowers, densely glandular-pubescent on all sides; flowers 1.6–2.5 cm long, blooming in early to mid summer; stems unbranched; calyx glandular-pubescent, the lobes ± acuminate . . . . .3. **Orobanche**

1. Conopholis

REFERENCES

Baird, Wm. Vance, & James L. Riopel. 1986. Life History Studies of Conopholis americana (Orobanchaceae). Am. Midl. Nat. 116: 140–151.
Buddell, George F., II, & John W. Thieret. 1981. Squawroot: It Toils Not. Explorer 23(2): 28–29.
Haynes, Robert R. 1971. A Monograph of the Genus Conopholis (Orobanchaceae). Sida 4: 246–264.

1. **C. americana** (L.) Wallr.   Plate 4-E   Squaw-root
Map 348. Often common locally in deciduous or mixed woods, wherever there is oak, including beech-maple, oak-hickory, or northern hardwoods stands, and even sometimes in rather swampy sites; occasionally no oak is evident in the immediate vicinity.

The dense inflorescence, which some have likened to a pine cone in gross appearance, usually occupies half or more the height of the plant, which overall is somewhat cream-colored to yellowish brown when fresh, except for brown tips on the bracts and leaves. The only known host is oak (*Quercus* spp., especially *Q. rubra* in Michigan). A large woody gall is formed where the parasite's root is attached to the oak root. After about 4 years of underground growth, *Conopholis* sends up thick annual flowering stems for several more years.

2. Epifagus

REFERENCES

Thieret, John W. 1970 ["1969"]. Notes on Epifagus. Castanea 34: 397–402.
Thieret, John W. 1975. Beechdrops . . . Benign Parasite. Explorer 17(4): 12–14.
Williams, Charles E., & Robert K. Zuck. 1986. Germination of Seeds of Epifagus virginiana (Orobanchaceae). Michigan Bot. 25: 103–106.

1. **E. virginiana** (L.) W. P. C. Barton   Plate 4-F   Beech-drops
Map 349. In woods with beech (and usually other trees, such as sugar maple, hemlock, oak). The rich deciduous woods at the Cut River in Mackinac County is among the sites where fine stands are well displayed, blooming in late summer (late August–September). Beech-drops is often said to grow "under" beech (*Fagus*), but that is a half-truth; more importantly, as its generic name declares, it grows *over* (upon) the roots of beech, its only known host plant.

The dead dry stems of the previous season are reasonably conspicuous through the winter and into the next summer. Most of the flowers are usually cleistogamous (setting fruit in the bud), but nearly always there are at least a few open (chasmogamous) flowers toward the ends of the flowering branches. Simple stems are usually recognized in the field as depauperate individuals of a population; most plants are branched, the flowering portion occupying more than half the height of the individual— which may be as tall as 60 cm.

Like *Corallorhiza maculata*, this species produces various color forms besides the most common reddish purple form (including stripes on the chasmogamous flowers). On Beaver Island I have collected a yellow form with white flowers. W. H. Wagner has collected an "ivory white" form in Allegan County; and S. R. Crispin, a "cream colored" one in Muskegon County. Sometimes the purple color is in stripes much narrower than usual on the stem and flowers (the background whitish), and more frequently a shade of brown is the dominant color. The pedicels are minutely glandular-pubescent on the side next to the stem, otherwise glabrous; the upper part of the stem (especially among the flowers) is usually glandular-pubescent also, but the lower part is glabrous or nearly so.

3. **Orobanche**   Broom-rape; Cancer-root

REFERENCE

Reuter, Barbara C. 1986. The Habitat, Reproductive Ecology and Host Relations of Orobanche fasciculata Nutt. (Orobanchaceae) in Wisconsin. Bull. Torrey Bot. Club 113: 110–117.

KEY TO THE SPECIES

1. Plant with an above-ground stem ca. 3–12 (15) cm tall, bearing a cluster of long pedicels at the summit; bracts at base of pedicels glandular-pubescent (like the pedicels and calyx) . . . . .1. **O. fasciculata**
1. Plant with no stem developed above ground, the long pedicels arising from a basal caudex; bracts at the base glabrous . . . . .2. **O. uniflora**

1. **O. fasciculata** Nutt.   Fig. 120
Map 350. Sand dunes along the Lake Michigan shore. There is a strange gap in the distribution between western Michigan and the dunes of Indiana

and Illinois. The range is otherwise to the west, as far as the Pacific. The only documented host in Michigan appears to be *Artemisia campestris* sens. lat. on sand dunes, although other hosts have been reported elsewhere. I have excavated *O. fasciculata* attached to an *Artemisia* root a full meter from the host stem.

The flowers are similar to those of the next species, with which this one is sometimes confused. An erect buried stem in *O. uniflora* may look like an aerial stem of *O. fasciculata* if one has only a collected specimen out of its habitat.

2. **O. uniflora** L.   Fig. 121

Map 351. Local, at edges of conifer thickets along dunes; even more local inland, in both dry sandy open areas and rich woods. Although this species has been reported throughout its range (covering most of the United States and southern Canada) from a diversity of hosts, *Aster* and *Solidago* seem to be the only verified ones in Michigan.

Flowering pedicels are often 1–2 dm tall, although pedicels as short as 2 cm may bear a flower. The corollas are creamy white to bluish or purplish (at least in part) with two bright yellow lines (folds) extending from the sinuses of the lower lip into the throat. While not a "pretty" plant, it is interesting in its ghostly way, and a crowded clump of 40 or 50 pedicels with flowers in full bloom can be impressive. More often, the pedicels are only solitary or a few blooming at any one time, although others may elongate and provide a long blooming season in early summer.

LENTIBULARIACEAE   Bladderwort Family

While the Orobanchaceae are considered a wholly parasitic derivative from the Scrophulariaceae, the Lentibulariaceae are considered a wholly carnivorous one (meaning that all members of the family are carnivorous, not that they are solely dependent on carnivory for nourishment). The strongly 2-lipped corolla with a spur from the base of the lower lip is similar to that of many Scrophulariaceae and is held above the water on aquatic

349. Epifagus virginiana

350. Orobanche fasciculata

351. Orobanche uniflora

species. The considerable literature on carnivorous plants can be consulted for chapters or sections on this family and its habits.

REFERENCE

Tans, William. 1987. Lentibulariaceae: The Bladderwort Family in Wisconsin. Michigan Bot. 26: 52–62.

KEY TO THE GENERA

1. Leaves ovate or broadly elliptic, entire, in a compact basal rosette; corolla blue-violet; calyx 5-lobed . . . . 1. **Pinguicula**
1. Leaves (or leaf-like structures) filiform and mostly buried in the substrate or dissected into filiform or very narrowly linear segments and aquatic (some buried); corolla yellow to pink-purple; calyx deeply 2-lobed . . . . 2. **Utricularia**

1. Pinguicula

Enzymatic secretions from glands on the upper surface of the leaves aid in digestion of insects and other little creatures that try to land or stroll on the sticky surface. The yellowish green color of the leaves (which curl inward along the margin) enhances the slippery, buttery aspect, and the leaves are said to coagulate milk. "Butterwort" is a doubly appropriate common name.

REFERENCE

Catling, Paul M., & Sheila M. McKay. 1971. In Search of the Butterwort. Ontario Nat. 9(2): 10–13.

1. **P. vulgaris** L.   Fig. 122   Butterwort

Map 352. This is a circumpolar boreal species, ranging southward in North America to the northern United States. Other species occur in the southern states and Mexico. *P. vulgaris* is a distinct calciphile, growing on alkaline rocks (Isle Royale and Pictured Rocks) and sands (as on interdunal flats and hollows), marly flats, and occasionally in marly fens and on moist rock outcrops inland from the Great Lakes, although most sites in Michigan are along the cool shores.

The rather bright-looking rosettes of leaves are very distinctive and almost always are found near *Primula mistassinica* (although the primrose also grows many places where butterwort does not). The flowers are solitary on scapes 1.5–12 (15) cm tall and there may be as many as 9 scapes from a single rosette of leaves.

2. Utricularia   Bladderwort

Bladderworts are a particularly interesting and diverse genus of aquatic plants, well represented in Michigan by north-temperate standards (half the American species grow in the state). Over 200 species occur in the world—some epiphytic (with large simple leaves) and a considerable number terrestrial. Distinctions between roots, stems, and leaves are difficult; in fact, none of the species produce what can be called, anatomically, a true root. However, absence of a root does not mean the aquatic species are always drifting free; some species, as in the rootless *Ceratophyllum*, are well anchored in the substrate by modified structures of different origins but which function as roots. The structures here called "leaves," following Taylor, have been treated by some authors as branches.

Some of the aquatic species produce turions (or "winter buds") at the apices of branches. These consist of very short internodes with tightly compacted overlapping dissected leaves and a mucilaginous matrix. Turions are often conspicuous toward the end of the summer and in *U. vulgaris*, at least, have been shown to be induced by environmental change to short day lengths. Brought indoors to an aquarium, late-summer turions quickly open up, the leaves unfolding and the internodes elongating. In nature, the turions normally sprout in the spring from their over-wintering condition (for details, see Winston & Gorham 1979).

The short-stalked "bladders" which give these plants their name are actually little traps. Hairs at the opening of the bladder serve as triggers and when contacted by a little aquatic organism, mechanically cause the trap to spring open, drawing in a rush of water with whatever comes along: insects, crustaceans, tiny fish fry, even green plants (water-meal, *Wolffia* spp.; Roberts 1972). Enzymes inside the traps or resident bacteria are agents of digestion (see Lloyd 1942).

Several species may grow close to each other or even with vegetative parts intertangled—a pain to collectors and to persons who have to deal with dried herbarium specimens that were not properly floated out and separated when fresh. *U. cornuta* and *U. resupinata* are often together at the edges of softwater lakes. For many years I showed my class in the fall vegetative material of six or seven species in Silver Lake, northwestern Washtenaw County—without having to move the boats at all. The most impressive sight for me was in late August of 1989 when a patterned peatland (soon afterward acquired by The Nature Conservancy) in northern Luce County displayed six species *in bloom* (*U. cornuta, U. geminiscapa, U. gibba, U. intermedia, U. purpurea,* and *U. vulgaris*) plus *U. minor* (on which no flowers happened to be noticed). Although the foliage in this genus is well designed for aquatic life, the flowers in all species are held above the water by long peduncles, so they can be readily admired by pollinators and botanists.

REFERENCES

Crow, G. E., & C. B. Hellquist. 1985. Aquatic Vascular Plants of New England: Part 8. Lentibulariaceae. New Hampshire Agr. Exp. Sta. Bull. 528. 19 pp.
Lloyd, Francis Ernest. 1942. The Carnivorous Plants. Chronica Botanica, Waltham, Mass. 352 pp. [*Utricularia*, ch. 13–14, pp. 213–270.]

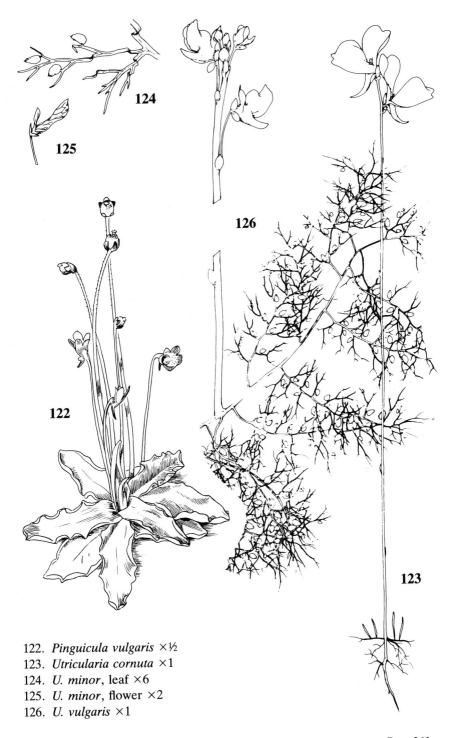

122. *Pinguicula vulgaris* ×½
123. *Utricularia cornuta* ×1
124. *U. minor*, leaf ×6
125. *U. minor*, flower ×2
126. *U. vulgaris* ×1

Haber, Erich. 1979. Utricularia geminiscapa at Mer Bleue and Range Extensions in Eastern Canada. Canad. Field-Nat. 93: 391–398.

Roberts, Marvin L. 1972. Wolffia in the Bladders of Utricularia: An "Herbivorous" Plant? Michigan Bot. 11: 67–69.

Taylor, Peter. 1989. The Genus Utricularia—a Taxonomic Monograph. Kew Bull. Add. Ser. 14. 724 pp. [Note extended review by Catling in Nat. Canad. 116: 288–289. 1991 ["1989"].]

Taylor, Peter. 1991. Utricularia in North America North of Mexico. Carniv. Pl. Newsl. 20(1–2): 9–20.

Winston, Robert D., & Paul R. Gorham. 1979. Turions and Dormancy States in Utricularia vulgaris. Canad. Jour. Bot. 57: 2740–2749.

KEY TO THE SPECIES

1. Leaves (often not seen or collected) simple, filiform or very narrowly linear, entirely or mostly embedded in the substrate ("terrestrial")
    2. Calyx less than 2 mm long; bracts at base of pedicels and on scape peltate (attached at the middle); corolla yellow, ca. 5–6.5 mm long (smaller on cleistogamous flowers); plant very rare, in southwestern Lower Peninsula . . . . .1. **U. subulata**
    2. Calyx at least 2 mm long; bracts attached at their base; corolla purple or yellow (if yellow, then over 12 cm long); plants widespread
        3. Corolla purple, ca. 6.5–9 mm long, the spur 1.5–3.5 mm long and upcurved; flowers solitary on scape; bracts tubular around scape, slightly flared at summit; calyx 2–3 mm long . . . . .2. **U. resupinata**
        3. Corolla bright yellow, 9–17 mm long (excluding the spur), the spur (5) 6–10 (11) mm long and pointed downward; flowers (1) 2–6 on scape; calyx (upper lip) (3) 3.5–5.5 mm long . . . . .3. **U. cornuta**
1. Leaves at least once-forked, in most species more dissected, all or mostly on stems in water
    4. Leaves (branches) whorled; bladders lacking external hairs, borne at the ends of the ultimate leaf segments; corolla pink-purple with yellow marks at base of lower lip . . . . .4. **U. purpurea**
    4. Leaves (except for floating whorl in *U. radiata*) alternate (though divided at their base, so that 2–3 main segments may arise on *one side* of the stem, not truly whorled); bladders with whisker-like hairs at the mouth, borne on sides or base of leaf segments (or on separate branches); corolla yellow (pale and sordid in *U. minor* and often drying ± purplish in all species)
        5. Plant bearing a whorl of leaves with strongly inflated petioles on the peduncle near water surface, very rare, in southwestern Lower Peninsula . . . . .5. **U. radiata**
        5. Plant without whorled or inflated leaves, ± common throughout the state
            6. Bladders borne on leafless branches (normally whitish and buried) separate from the leaves; leaf segments strongly flattened throughout, with marginal (as well as apical) spicules . . . . .6. **U. intermedia**
            6. Bladders borne on normal green leaves; leaf segments (at least ultimate ones) terete or thread-like except in *U. minor* (which lacks marginal spicules) and occasionally *U. vulgaris*
                7. Principal leaves mostly forked 1–4 (6) times (including a division into 2 or 3 main segments at the base), up to 0.9 (1.4) cm long (plants consequently very slender in aspect), the leaf segments without marginal spicules
                    8. Leaf segments flat; corolla pale or dingy yellow, with purplish tinge or stripes toward base of lower lip and only an obscure saccate spur; lower lip held ± horizontal, elongate-oblong (as result of downcurling of margins); bracts of scape and inflorescence auriculate; leaves mostly forked (2) 3–4 (6) times; flowers (2) 3–8 on scape . . . . .7. **U. minor**

8. Leaf segments hair-like; corolla bright yellow; lower lip (like the upper) turned upward, broadly rounded, the corolla saddle-shaped in aspect, with definite spur; bracts not auriculate (though clasping); leaves mostly forked 1 (2) times; flowers 1 (2) on scape . . . . .8. **U. gibba**

7. Principal leaves mostly forked 5–17 (22) times, (1) 1.3–5 (6.5) cm long (plants consequently more bushy in aspect), the leaf segments with minute marginal spicules

9. Flowers (chasmogamous) ca. 5–8.5 mm long, on pedicels subtended by narrowly elliptic-oblong bracts unlobed at the base; cleistogamous flowers often present late in summer, on pedicels 2.5–7.5 mm long on submersed stems; bladders 0.8–1.7 (2) mm long; leaves (1) 1.3–2 cm long, extremely delicate and limp, forked 5–9 (11) times; plant of softwater lakes and acid pools in peatlands . . . . .9. **U. geminiscapa**

9. Flower ca. 9–16 (18) mm long, on pedicels subtended by ovate ± auriculate or cordate bracts; cleistogamous flowers none; bladders 0.8–5 (6) mm long; leaves (1) 1.5–5 (6.5) cm long, coarser [see text], usually forked 6–17 times; plant of diverse ponds, lakes, and streams (in all but swiftly flowing waters) .
. . . . .10. **U. vulgaris**

1.   **U. subulata** L.

Map 353. Although this tiny plant is considered to be the most widespread species of *Utricularia* in the world, in the United States it was thought to be restricted to the Coastal Plain from Massachusetts to Texas until it was discovered around the head of Lake Michigan in the 1980s. The first Michigan collection was made by K. Klick in an interdunal flat at Warren Dunes State Park, Berrien County, in 1987 (*2766*, MICH). Penskar found it in similar habitats in Muskegon and Allegan counties in 1992 (MICH).

2.   **U. resupinata** Bigelow   Plate 5-A

Map 354. Flowering in shallow water and on wet or recently exposed sandy, peaty, or marly shores of lakes and ponds; often forming extensive vegetative mats in deeper water. Usually (but not always) blooms when water levels are low and temperatures are high. Sometimes forms pink carpets or bands at the sandy-peaty edges or bottoms of a small receding lake. The finest flowering may be seen on a bed of knee-deep muck—but the sight is worth it!

352. Pinguicula vulgaris   353. Utricularia subulata   354. Utricularia resupinata

This species may grow with *U. cornuta* around the edges of softwater lakes, and in years of higher water both exist as carpets of dense leaves and stems, buried except for the green tips of leaves that rise above the substrate. Fragments of such vegetative mats often break loose and drift about. Young leaves can be distinguished from those of other small bottom-living aquatics by their inrolled (circinate) tips. Mature leaves may be as long as 7.5 cm. Telling these two "terrestrial" species from each other is more difficult from sterile material, but apparently *U. resupinata* rather consistently bears two rootlet-like branches from the horizontal stem at each leaf-bearing node, while in *U. cornuta* the leaf-bearing nodes produce 0–4 branches (a character indicated by Lloyd and requiring fresh intact specimens to see clearly). Taylor notes that the leaves of *U. resupinata* are septate, unlike those of *U. cornuta*.

3. **U. cornuta** Michaux   Fig. 123   Horned Bladderwort
    Map 355. A calciphile of fens (sometimes in sphagnum), margins of ponds and lakes, marshy shores, pools between dunes, marly areas; however, also in softwater lakes.

This is undoubtedly the bladderwort most often seen by travelers in Michigan, for it can form extensive bright yellow carpets in sandy interdunal hollows, borrow pits along highways, and wet peatlands. In unusually dry years for their habitat (whether caused by drought or lowering of Great Lakes water levels), flowers may be scarce, as they may be also in extraordinarily wet seasons. The flowers (sometimes mistaken for some kind of orchid) are very fragrant and attract insects, even large yellow tiger swallowtail butterflies.

Robust plants may be as tall as 35 cm. The tiny bladder traps of this so-called terrestrial species are, as in the preceding, buried with the branch and leaf system in the substrate; however, Lloyd and Taylor both aver that they are functional, no matter how reduced in size or number they may appear.

If one has a blooming *Utricularia* without basal parts, *U. cornuta* can be recognized by having tiny bracteoles narrower but slightly longer than the

355. Utricularia cornuta

356. Utricularia purpurea

357. Utricularia radiata

bracts of the inflorescence subtending them. No other of our species has such bracteoles.

4. **U. purpurea** Walter   Purple Bladderwort

Map 356. Unlike many of our species in this genus, not a calciphile, but usually occurs in shallow to often deep (3 feet) water of softwater lakes; sometimes, however, in fens; river margins, marshes.

Not as often seen in flower as most of our species, although foliage may be dense on a lake bottom; but the magenta flowers with yellow palate are relatively large and attractive. The species is easily recognized from a distance in the field, with bushy whorls of foliage separated by open internodes. The young leaves, especially, are conspicuously circinate at the tips and tend to have a basket-like aspect.

5. **U. radiata** Small

Map 357. Known in Michigan only from old intermittently wet interdunal ponds in the Grand Beach area (collected by Medley & Kohring in 1976, MOR; and later collections). Also known from adjacent Indiana, but otherwise only from Nova Scotia south mostly along the Coastal Plain to Mississippi, Tennessee, and Arkansas.

This species has often been included in the larger *U. inflata* Walter, of which it may be treated as var. *minor* Chapman.

6. **U. intermedia** Hayne   Fig. 127   Flat-leaved Bladderwort

Map 358. A circumpolar species, usually of alkaline waters: interdunal pools and wet flats, fens, marshes, ponds, rivers, lakes; may grow in water several feet deep, but thrives and flowers in shallow water.

The lower lip of the corolla is relatively flat (except at the base) compared with the lower lip in *U. vulgaris*, which has the sides turned down, giving it a strongly humped aspect. The spur in *U. vulgaris* is more conical (broad-based) and curved less strongly under the lip; the more slender spur in *U. intermedia* is ± closely pressed to the bottom of the lip—even into a concavity. (Cf. figs. 126 & 127.) Both species have ± auriculate bracts. These two common species, often growing together, can thus be easily identified without close examination when in bloom.

The bladders in *U. intermedia* are as long as 5 mm on some individuals, though usually they are 1–3.5 mm. Some plants, apparently deep-water ones, have the leaf segments distinctly narrower than most plants. In shallow water (or very recently exposed wet mud) the green leaves when dense often give a mossy appearance (though actually flat and fan-shaped in outline), while the bladder-bearing branches are buried in the substrate.

Michigan is included by Taylor in the range of the problematic *U. ochroleuca* R. W. Hartman, but I have seen no specimens so annotated or displaying the characters which supposedly distinguish that poorly understood and possibly hybrid taxon. In none of our flowering specimens is the spur as short as half as long as the lower lip, as it is in *U. ochroleuca*; the latter is also characterized by more narrowly acute ultimate leaf segments, with fewer spicules along the margin. Our material has mostly blunt tips on the ultimate leaf segments, with ca. 10–20 marginal spicules (or fascicles of 2–6 spicules).

7. **U. minor** L. Figs. 124, 125

Map 359. Another circumpolar and usually calciphile species, its range extending in some places, like *U. intermedia* and *U. vulgaris*, north of the Arctic Circle. Fens (sometimes in sphagnum), sedge mats, swales and marshes; wet open thickets and peaty lake margins; often in ruts and animal trails in such habitats.

In shape, the leaves resemble very miniature flat fronds of staghorn fern. This species and the next are both very slender and small-flowered, the corolla not over 9 mm long, and they often grow intermixed. However, the more flattened leaf segments of *U. minor* are nearly always easily seen under a lens. Small shoots (e.g., from turions) of *U. intermedia* may resemble *U. minor* but can be distinguished by the complete absence of bladders and the presence of marginal spicules on the leaf segments. The flowers of *U. minor*, being a rather pale dirty yellow, are not as conspicuous as the equally small but bright yellow ones of *U. gibba* and *U. geminiscapa*, apart from the more chubby and less linear shape of these two. Some fragmentary specimens which otherwise appear to be *U. minor* have unusually narrow, elongate, flat ultimate leaf segments with a few marginal spicules; I suspect they may be derived from expanded turions.

8. **U. gibba** L. Fig. 128

Map 360. Lakes, ponds, pools, marshes, fens, and other wet areas. While masses—sometimes quite large—of foliage of this species may drift in quiet waters, flowering occurs usually on damp, often recently exposed shores, so that the "terrestrial" habit mimics that of *U. cornuta* and *U. resupinata*.

U. gibba apparently does not produce turions, while the stems of *U. minor* often bear small compact ones up to 2 or even 4 mm broad.

9. **U. geminiscapa** Benj.

Map 361. Softwater lakes and bog pools.

Plants with either chasmogamous flowers (very similar to those of *U. gibba*) or cleistogamous ones, or both, are easily identified. Sterile fragments of foliage, on the other hand, have always given trouble to botanists. The extraordinarily delicate foliage (alga-like, going totally limp on removal from the water) and restriction to boggy habitat are helpful clues. A few minute spicules [use 30× lens] are usually present on the margins of the ultimate leaf segments, distinguishing this species from *U. gibba* (as does the greater number of forks in the branching). The segments, though usually hair-like, sometimes appear somewhat flattened, especially when dried, though narrower and more forked than usual for *U. minor*; but the distinction is not always as easy as one would like. Just as small fragments of *U. geminiscapa* may mimic *U. minor* or *U. gibba*, small branchlets of *U. vulgaris* may mimic *U. geminiscapa* (see below).

10. **U. vulgaris** L.   Figs. 126, 129   Common Bladderwort
    Map 362. Lakes of all kinds, interdunal (and other) ponds and swales, wet peatlands and marshes, rivers and streams. Often in water up to 6 feet deep; the deepest recorded on Michigan specimens is 10–15 feet.
    Plants of North America and eastern Asia are often treated as a distinct species (based on minor floral characters, not vegetative ones), *U. macrorhiza* Leconte, or at least as ssp. *macrorhiza* (Leconte) Clausen. The typical subspecies is Eurasian.
    After the main forking of the leaves next to the stem, those of robust *U. vulgaris* often have a more pinnate than dichotomous aspect, with a slightly zigzag and broader central axis and many forked lateral segments. The main segments on such large leaves are sometimes flattish (at least when dried).
    Fertile specimens are easily distinguished from other species, and so are those large coarse sterile ones, with leaves 3–5 cm or more long and up to 22 levels of forking, bladders nearly or quite 5 mm long, and massive turions as much as 3 cm or more across. The total diameter of a leafy branch may be 10 cm. Small fragments of depauperate growth, on the other hand, are often not safely distinguishable from sterile *U. geminiscapa*, although Haber (1979) indicates some tendencies, and experience with good fertile material of both species will then help one decide on the basis of characters not able to be quantified to the extent desirable in a key. *U. geminiscapa* is often said to have the ultimate segments of the leaves entire or with few spicules, while in *U. vulgaris* there are more numerous spicules. I have been unable to see a consistent difference in number of

358. Utricularia
     intermedia

359. Utricularia minor

360. Utricularia gibba

spicules (always few), but in *U. vulgaris* they do tend to be somewhat larger, arising at times from a tiny green tooth on the leaf margin. In turions of *U. vulgaris* (usually 5–30 mm broad), the crowded leaves are heavily provided with glistening apical and marginal spicules often giving a grayish as well as minutely prickly aspect to the structure (like those of *U. intermedia*). The turions of *U. geminiscapa* run smaller, perhaps less compact, and certainly more green and less prickly in aspect. *U. geminiscapa* is much more narrow in habitat and is never expected to develop a layer of marl on the foliage, whereas *U. vulgaris* when growing in alkaline water may develop (as do many other aquatics) a layer of carbonates.

ACANTHACEAE   Acanthus Family

This family is closely related to the Scrophulariaceae, differing in seed characters and in having an explosive capsule that forcibly ejects the seeds (which are on stalks specialized for this purpose). Acanthaceae are a large tropical family with numerous species grown as ornamentals. Representatives of the two largest genera barely reach southern Michigan, where all three are considered to be threatened species.

REFERENCE

Long, Robert W. 1970. The Genera of Acanthaceae in the Southeastern United States. Jour. Arnold Arb. 51: 257–309.

KEY TO THE GENERA

1. Leaves less than 3.5 times as long as broad, with at least minute strigose pubescence; calyx and stem pubescent; corolla (except on cleistogamous flowers) nearly regular, ca. 3–7 cm long, the lobes much shorter than the tube; stamens 4 . . . . . 1. **Ruellia**
1. Leaves ca. 7–15 times as long as broad, completely glabrous; calyx and stems glabrous; corolla bilateral, less than 1.5 cm long, the lobes about as long as the tube or longer; stamens 2 . . . . .2. **Justicia**

361. Utricularia
     geminiscapa

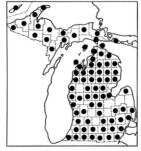

362. Utricularia vulgaris

363. Ruellia strepens

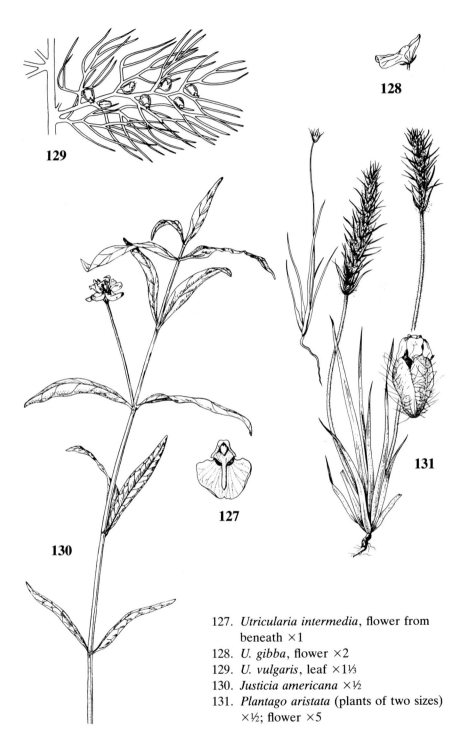

127. *Utricularia intermedia*, flower from beneath ×1
128. *U. gibba*, flower ×2
129. *U. vulgaris*, leaf ×1⅓
130. *Justicia americana* ×½
131. *Plantago aristata* (plants of two sizes) ×½; flower ×5

1. **Ruellia**   Ruellia; Wild-petunia

Both of our species have few but rather showy pink-violet flowers in nearly sessile or short-stalked terminal or axillary clusters. Flowers are usually self-pollinated—even the showy (as well as the common cleistogamous) ones.

REFERENCE

Long, Robert W., & Leonard J. Uttal. 1962. Some Observations on Flowering in Ruellia (Acanthaceae). Rhodora 64: 200–206.

KEY TO THE SPECIES

1. Calyx lobes lanceolate, ca. 3–6 times as long as broad; leaves with minute usually ± strigose pubescence . . . . .1. **R. strepens**
1. Calyx lobes narrowly linear, up to 1 mm broad at base and ca. 15–30 times as long; leaves with spreading hairs (much longer than any minute strigose ones), especially on main veins and margins . . . . .2. **R. humilis**

1. **R. strepens** L.

Map 363. Discovered at two sites in rich lowland woods in Lenawee County in 1957 (*Wagner 8468 & 8469*, MICH). No specimens to support listing by the First Survey (1838) have been seen.

2. **R. humilis** Nutt.   Plate 5-B

Map 364. Unlike the previous species, a plant of dry open ground including prairies.

2. **Justicia**

REFERENCE

Penfound, William T. 1940. The Biology of Dianthera americana L. Am. Midl. Nat. 24: 242–247.

1. **J. americana** (L.) Vahl   Fig. 130   Water-willow

Map 365. Spreads extensively by leafy stolons, forming large dense colo-

364. Ruellia humilis   365. Justicia americana   366. Littorella uniflora

nies, albeit rather local, along the Huron and Raisin rivers and nearby lakes and streams.

The flowers have white corollas with one or more lobes purple or streaked with purple and are crowded into small dense heads on axillary peduncles (4.5) 6–11 cm long.

PLANTAGINACEAE   Plantain Family

All of our species except one have the leaves in a basal rosette (or at least crowded at the base), usually arising (in the perennial species) from a strong woody caudex. All have 4-merous flowers (with 2 carpels) apparently adapted for wind pollination.

REFERENCES

Bassett, I. John. 1973. The Plantains of Canada. Canada Dep. Agr. Monogr. 7. 47 pp.
Rosatti, Thomas J. 1984. The Plantaginaceae in the Southeastern United States. Jour. Arnold Arb. 65: 533–562.
Tessene, Melvern F. 1968. Preliminary Reports on the Flora of Wisconsin No. 59. Plantaginaceae—Plantain Family. Trans. Wisconsin Acad. 56: 281–313.

KEY TO THE GENERA

1. Leaves terete (ca. 3 mm or less thick at the middle, thence tapering to apex), at most 1-veined, glabrous; flowers unisexual (the staminate long-stalked, the pistillate basal); fruit indehiscent; plant submersed or on damp shores . . . . 1. **Littorella**
1. Leaves flat, in most species with at least 3 prominent veins and/or pubescent; flowers perfect (in heads or spikes); fruit circumscissile; plant of dry or rarely wet habitats . . . . .2. **Plantago**

1. Littorella

REFERENCES

Fernald, M. L. 1918. The North American Littorella. Rhodora 20: 61–62.
Lakela, Olga. 1958. Distribution of Littorella americana in the Mid-Arrowhead Region of Minnesota. Rhodora 60: 33–37.
Voss, Edward G. 1965. Some Rare and Interesting Aquatic Vascular Plants of Northern Michigan, with Special Reference to Cusino Lake (Schoolcraft Co.). Michigan Bot. 4: 11–25. [*Littorella*, pp. 22–23.]

1. **L. uniflora** (L.) Asch.   Plate 5-C

Map 366. Sandy to mucky shores of softwater lakes and at depths (but sterile) up to 3 feet or more. First recognized in Michigan in 1964 (Voss 1965) and soon found to be in a considerable number of northern Michigan lakes—a history similar to that of its rediscovery in Minnesota (Lakela 1958).

Fernald (1918) described American plants as a new species distinct from the European one, but I share the reluctance of some other authors to follow that distinction and thus treat our taxon as var. *americana* (Fernald) Gleason. In all fresh plants that I have seen, from a dozen or more Michigan lakes, the leaves are terete, not flattened, rather yellow-green, thickest about the middle, tapering from there to the apex. The submersed rosettes are easily distinguished at a distance from those of associated rosette-formers like *Lobelia dortmanna, Isoëtes, Eriocaulon*, and *Juncus pelocarpus*. Fernald declared that American plants have "flattish" leaves (as indeed they do on herbarium sheets) and European ones "subterete or semi-cylindric." *Flora Europaea* states that the leaves are "semicircular in section, sometimes flat and wider." Some European specimens do look more robust than ours, but others are as small and delicate as most local plants.

Wholly underwater plants do not flower, but within days of emergence the wet to moist shore of a receding lake may be carpeted with blooming *Littorella*. The staminate flowers are more conspicuous, on stalks up to 2.5 cm long and with long filaments. The pistillate flowers are at the base of the plant and have an elongate stigmatic style. Tessene (1968) was able to induce flowering under long days in potted plants at the University of Michigan Botanical Gardens. However, the potted plants had long lax leaves and looked quite unlike field-collected ones.

Plants may form colonies by stolons at or near the surface as well as by deeper rhizomes. Pollination is by wind and Tessene suggested that plants of a clone are self-incompatible.

2. **Plantago**   Plantain

Although the reduced flowers seem well adapted for wind pollination, some are cleistogamous in certain species and some (e.g., *P. lanceolata*) are in part insect-pollinated (Mesler 1977).

REFERENCES

Allen, Gary M., & Michael J. Oldham. 1985. Plantago cordata Lam. (Heart-leaved Plantain) Still Survives in Canada. Plant Press (Ontario) 3: 94–97.

Cavers, P. B., I. J. Bassett, & C. W. Crompton. 1980. The Biology of Canadian Weeds. 47. Plantago lanceolata L. Canad. Jour. Pl. Sci. 60: 1269–1282.

Hawthorn, Wayne R. 1974. The Biology of Canadian Weeds. 4. Plantago major and P. rugelii. Canad. Jour. Pl. Sci. 54: 383–396.

Mesler, Michael R. 1977. Notes on the Floral Biology of Plantago in Michigan. Michigan Bot. 16: 73–83.

Shinners, Lloyd H. 1950. The North Texas Species of Plantago (Plantaginaceae). Field & Lab. 18: 113–119.

Stromberg, Julie, & Forest Stearns. 1989. Plantago cordata in Wisconsin. Michigan Bot. 28: 3–16.

Tessene, Melvern F. 1969. Systematic and Ecological Studies on Plantago cordata. Michigan Bot. 8: 72–104.

KEY TO THE SPECIES

1. Plant with several pairs of opposite linear cauline leaves, annual without a basal rosette, the stem often branched...................................1. **P. arenaria**
1. Plant with leaves all basal or nearly so, the stem obsolete and unbranched (though several scapes may arise from a rosette)
   2. Bracts of inflorescence pubescent or puberulent across the back, at least toward their base
      3. Mature bracts (at least the lower ones) (2) 3–10 (12) times as long as the calyx ....2. **P. aristata**
      3. Mature bracts scarcely if at all longer than the calyx
         4. Corolla lobes erect, appearing beak-like over the fruit; pubescence at middle of scape spreading, the hairs (as also in the spike) septate; leaves narrowly lanceolate to elliptic ....3. **P. virginica**
         4. Corolla lobes widely spreading at maturity; pubescence at middle of scape strongly ascending or appressed, the hairs (as in the spike) not septate; leaves linear-lanceolate ....4. **P. patagonica**
   2. Bracts of inflorescence glabrous or at most ciliate
      5. Bracts with broad scarious margins occupying more than half their area, ovate and at least the lower ones ± acute-acuminate; scapes deeply furrowed and ridged; calyx with the 2 sepals beneath the bract connate, the other 2 sepals distinct; leaves (including obscurely differentiated petiole) (2.5) 6–21 times as long as broad ....5. **P. lanceolata**
      5. Bracts with scarious margins occupying less than half their area, of various shape; scape terete or at most weakly ridged; calyx with all 4 sepals distinct; leaves less than 3 times as long as broad or if longer, then including a definite long petiole
         6. Leaf blades flat on the ground, pubescent on both surfaces, tapering into rather indistinct and short petioles; bracts ovate ....6. **P. media**
         6. Leaf blades ± ascending, glabrous or slightly pubescent, abruptly tapered to cordate at the base, on definite long petioles; bracts linear to ovate
            7. Older leaf blades often ± cordate, the principal longitudinal veins, although basal, closely paralleling the midvein before diverging (hence, with pinnate aspect); bracts broadly ovate; plant of wet places (typically a streambed in rich woods), blooming in May (–early June), completely glabrous; major roots large and fleshy (5 mm or more thick) ....7. **P. cordata**
            7. Older leaf blades broadly tapered to truncate at base, the principal longitudinal veins all clearly diverging from the base; bracts various; plant of dry to moist places, blooming June–August, often pubescent on scape and/or leaves; major roots fibrous (from stout caudex)
               8. Capsule narrowly ellipsoid-oblong, dehiscing distinctly below the middle; bracts narrowly lanceolate, ± sharply acute, ca. 2–3 times as long as broad (fig. 134) ....8. **P. rugelii**
               8. Capsule plumply ovoid (shaped like a fat acorn), dehiscing about the middle; bracts broadly ovate, nearly obtuse to rounded, less than twice as long as broad (fig. 135)...9. **P. major**

1. **P. arenaria** Waldst. & Kit.   Psyllium

Map. 367. A European species, local along railroads and sandy roadsides, in waste ground such as parking lots and dumps, and on shores. The oldest Michigan collection I have seen is from 1909, as a weed in alfalfa in Alcona County. No other collections antedate 1928 (Emmet Co.).

This species has often been called *P. psyllium* L. or *P. indica* L., but in

Flora Europaea the former is rejected as ambiguous and the latter as illegitimate (evaluations confirmed in Lejeunia 101: 47. 1980). The seeds have long been used for their laxative and stabilizing properties, having a very mucilaginous seed coat when wet—as do those of other species.

2. **P. aristata** Michaux   Fig. 131   Bracted Plantain
    Map 368. Perhaps not native east of the prairies and plains. The first Michigan collections are from 1885 (Ingham Co.), 1892 (St. Clair Co.), and 1896 (Kent Co.). Sandy fields, hillsides, and dunes; disturbed ground including railroads, roadsides, and gravel pits.
    Very depauperate plants with shorter bracts than usual may be confused with the next two species, but the bracts are still more prolonged and the pubescence is not densely silky (unlike *P. patagonica*); the corolla lobes are spreading (unlike *P. virginica*).

3. **P. virginica** L.
    Map 369. Considered a native species in eastern North America, but weedy in habit and not collected in Michigan before the 1890s (St. Clair Co.). Sandy hillsides, clearings, and fields (old or recently abandoned).
    The characteristic spreading septate hairs are sparse enough to reveal clearly the surface which bears them.

4. **P. patagonica** Jacq.
    Map 370. Like *P. aristata*, probably not native quite this far east in North America. The earliest Michigan collections are from the counties of Washtenaw (1928), Cheboygan (1929), and Delta (1934). Sandy roadsides, fields, and open ground; also pine barrens and river banks in oak woods.
    The silvery white (to tawny) hairs are usually so dense and long as to conceal the surface, and they give the plant a decidedly silky appearance. Eastern American plants have often been referred to *P. purshii* Roemer & Schultes.

5. **P. lanceolata** L.   Fig. 136.   Ribgrass; Buckhorn;
                                                  Narrow-leaved or English Plantain
    Map 371. Considered to be a Eurasian native. Known in Michigan since the First Survey and now in disturbed or waste ground everywhere: roadsides, railroads, parking lots, gravel pits, farmyards, filled land; lawns and fields; borders and clearings in woods and pine plantations; shores and river banks.
    The distinctive calyx of this species is usually easy to see on young spikes (before the corolla lobes have spread), when turning back a bract near the base of the spike will reveal 2 fused sepals forming a slightly bilobed structure with each green midvein evident. The leaves vary from nearly glabrous to densely villous. Robust plants may have a dozen or more

Page 274

scapes. Leaves may be as broad as 4 cm, although usually they are much narrower.

6. **P. media** L.   Fig. 132   Hoary Plantain
    Map 372. An Old World species, occasionally found in North America. Collected by Bradford in waste ground at Bay City in 1895 (MSC) and by Billington in the Cranbrook area of Oakland County in 1919 (MICH, BLH, WUD, WMU, GH; Rep. Michigan Acad. 21: 370. 1920).
    The flowers are said to be fragrant and visited by insects.

7. **P. cordata** Lam.   Plate 5-D   Heart-leaved Plantain
    Map 373. A large handsome native species that has become exceedingly rare (or local) in recent years, and the subject of much attention (see references). Long thought to have become extirpated in Michigan, until the first plants to be seen since 1925 were discovered in Hillsdale County in 1990 (*Wagner 90012 & Fritsch*, MICH) along (and in) an ephemeral stream in rich woods, a typical habitat. In 1995, rediscovered in Ionia County (*F. & R. Case*, MICH) in deep floodplain woods with *Saururus*. Older records are from brooks, small streams in woods, and ditches.
    Like many plants with large fleshy roots, those of this species are reputed to have great medicinal value.

367. Plantago arenaria

368. Plantago aristata

369. Plantago virginica

370. Plantago patagonica

371. Plantago lanceolata

372. Plantago media

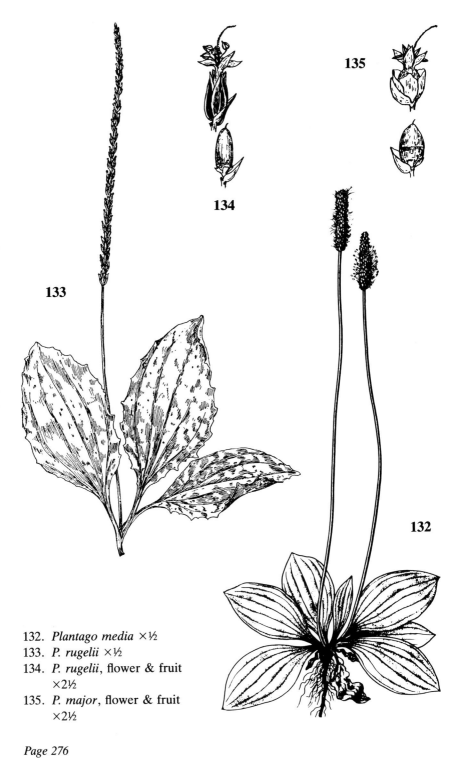

133

135

134

132

132. *Plantago media* ×½
133. *P. rugelii* ×½
134. *P. rugelii*, flower & fruit
 ×2½
135. *P. major*, flower & fruit
 ×2½

8. **P. rugelii** Decne.   Figs. 133, 134   Rugel's Plantain
Map 374. Although this is considered a native species, the only Michigan collection I have recorded from before 1890 is from Kent County in 1883. Besides the usual waste places and disturbed ground (vacant lots, roadsides, parking lots, fields and gardens, lawns, etc.), it inhabits damp shores, floodplains, river banks, woods (especially along trails and in clearings).

This species is almost as variable as the next and grows in much the same places except the driest and hardest packed soil. It can usually be reliably distinguished even before the fruit is ripe by the narrow acute bracts and nearly linear sepals, giving the flower a more elongate, streamlined appearance than the chubby ones of *P. major*. (Furthermore, the spikes of robust plants may be as long as 65 cm.) The petioles are nearly always pink to deep red-purple at the base, while in *P. major* they are green or at most a pale pink. The leaves are usually glabrous, sometimes sparsely puberulent, and usually have a few small marginal teeth (though *P. major* may occasionally have teeth and more often a somewhat undulate margin).

9. **P. major** L.   Fig. 135   Common Plantain
Map 375. Originally native to Eurasia, but like *P. lanceolata* now a thoroughly naturalized weed throughout much of the world. Collected in Washtenaw County in 1861, in Houghton County in 1878, in Wayne County in 1884, and in Keweenaw County in 1888, so probably well distributed throughout the state at an early date. Waste places and disturbed ground, including roadsides and railroads, parking lots, gravel pits and filled land, dumps; damp shores, woods, river banks, floodplains; spreading to mixed woods (including pine, aspen, birch), especially along trails and old roads.

A variable species in habit, size, and foliage. The leaves are usually pubescent on one or both surfaces (more so than in *P. rugelii*). The sepals are often broadly rounded and the fruit also shorter than in *P. rugelii*, as well as more rotund. The most striking variant is f. *rosea* (Decne.) Prahl, with 2–3 leaves on the scape just below the spike; it has been collected in Jackson, Kent, Keweenaw, Washtenaw, and Wayne counties.

373. Plantago cordata

374. Plantago rugelii

375. Plantago major

RUBIACEAE

Madder Family

This is a large, mostly tropical and subtropical family, estimated to include from 7,000 to 10,000 species, of which only relatively few are herbaceous (including the cosmopolitan genus *Galium*). Plants typically have entire, opposite, stipulate leaves or whorled leaves (the added leaves in a whorl representing transformed stipules), with flowers characterized by an inferior ovary and radial symmetry. The common name does not indicate a family more angry than another but comes from an Old World species of the type genus, *Rubia tinctorium* L., from which the roots formerly provided a well known red dye (madder), now largely supplanted by synthetics. Other Rubiaceae with some economic uses include coffee (*Coffea* spp.), the emetic ipecac (*Cephaëlis ipecacuanha* (Brot.) Tussac), quinine (*Cinchona* spp.), a few ornamentals (such as *Gardenia* spp.), and some other timber, dye, and drug species.

REFERENCES

Hauser, Edward J. P. 1964. The Rubiaceae of Ohio. Ohio Jour. Sci. 64: 27–35.
Rogers, George K. 1987. The Genera of Cinchonoideae (Rubiaceae) in the Southeastern United States. Jour. Arnold Arb. 68: 137–183. [Includes family characters and only three of our genera: *Cephalanthus* and *Houstonia/Hedyotis*.]
Urban, Emil K., & Hugh H. Iltis. 1958 ["1957"]. Preliminary Reports on the Flora of Wisconsin. No. 38. Rubiaceae—Madder Family. Trans. Wisconsin Acad. 46: 91–104.

KEY TO THE GENERA

1. Plant an erect woody shrub; flowers and fruits in dense, spherical, peduncled heads . . . . .1. **Cephalanthus**
1. Plant herbaceous (or at most a prostrate trailing subshrub); flowers and fruits nearly sessile or pediceled, but not in heads
  2. Leaves whorled (i.e., stipules, even if fused or further divided, closely resembling the leaves)
    3. Corolla tube much longer than the lobes; flowers in heads, the corolla pink to blue . . . . .2. **Sherardia**
    3. Corolla tube distinctly shorter than the lobes; flowers not in heads, the corolla white, greenish, yellow, or maroon . . . . .3. **Galium**
  2. Leaves opposite (with stipules united around the stem)
    4. Stem prostrate, rooting at the nodes; leaves evergreen, scarcely (rarely as much as 50%) longer than wide and usually ± variegated with the midrib broadly marked in pale green or whitish; fruit indehiscent, red, fleshy, formed by fusion of ovaries from the 2 adjacent flowers and hence with two perianth scars . .4. **Mitchella**
    4. Stem erect; leaves deciduous (or overwintering only in a basal rosette), the blades distinctly longer than broad, not variegated; fruit dehiscent, a dry capsule formed from 1 ovary
      5. Flowers axillary, sessile; stipules with fringe 2–7 times as long as body of the stipule, equalling or surpassing the flowers and fruit (fig. 146); fruit consisting of 2 1-seeded nutlets; stem pubescent . . . . .5. **Diodia**
      5. Flowers terminal (or on a scape), usually stalked; stipules with no fringe or a very short one; fruit a many-seeded capsule; stem glabrous or pubescent

6. Calyx lobes shorter than the inferior portion of the ovary (at least as fruit develops); capsule longer than wide; stipules of cauline leaves ciliate or sparsely fringed or with a prolonged bristle; leaves narrowly linear . . . . . . 6. **Hedyotis**
6. Calyx lobes, even in fruit, equalling or exceeding the inferior portion of the ovary (fig. 144); capsule no longer than wide; stipules of cauline leaves not ciliate or fringed (though may be pubescent at base); leaves often broader . . .
. . . . .7. **Houstonia**

1.   Cephalanthus

1. **C. occidentalis** L.   Fig. 137   Buttonbush
Map 376. Hardwood swamps, wet thickets, river margins, edges of marshes, swales, and shores; often in standing water or in very deep muck.
The leaves are opposite or in whorls of 3 and are typical Rubiaceous leaves, resembling those of coffee and many other members of the family.

2.   Sherardia

1. **S. arvensis** L.   Fig. 138   Field-madder
Map 377. An Old World native, still fairly local as a weed of waste ground in North America. Collected in 1920 at East Lansing (*Walpole*, BLH, MSC), in 1933 as a lawn weed on the campus of Western Michigan University (*Kenoyer*, WMU), and in 1993 as a lawn weed in Berrien County (*Dritz 1116*, MICH, MOR). Quite possibly overlooked elsewhere.
Plants resemble a low, creeping *Galium* with bristly ovaries and leaves, but the little heads of pink to blue flowers with long corolla tubes are distinctive.

3.   Galium   Bedstraw
This is a difficult genus, with a relatively high percentage of misidentifications in herbaria where the specimens have not been critically checked. The characters are sometimes subtle, and pressed specimens may lack the distinctive aspects of fresh material. Several of the smaller species can be especially troublesome to identify and, indeed, even more than in some groups it is necessary to use a good, well illuminated lens and to consider a balance of characters rather than relying on a single one to separate similar species. Expect exceptions! And read the comments offered here, as well as the keys, very carefully.
References to the pedicel are strictly to the stalk of an individual flower, not to any portion of the stem or inflorescence from which that (un-branched) stalk arises.

REFERENCES

Malik, N., & W. H. Vanden Born. 1988. The Biology of Canadian Weeds. 86. Galium aparine L. and Galium spurium L. Canad. Jour. Pl. Sci. 68: 481–499.

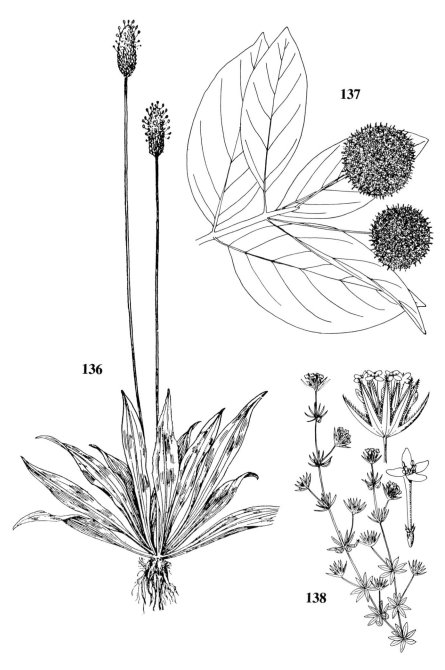

136. *Plantago lanceolata* ×½
137. *Cephalanthus occidentalis* ×⅔
138. *Sherardia arvensis* ×¼; inflorescence ×1; flower ×2

Page 280

Moore, R. J. 1975. The Galium aparine Complex in Canada. Canad. Jour. Bot. 53: 877–893.

Moore, R. J. 1988. Key to the Bedstraws (*Galium*) in Canada. Pl. Press (Ontario) 5: 21–23.

Puff, Christian. 1976a ["1975"]. Morphologie, Biologie und Abgrenzung von Galium L. sect. Aparinoides (Jord.) Gren. (Rubiaceae). Beitr. Biol. Pfl. 51: 17–40.

Puff, C. 1976b. The Galium trifidum Group (Galium sect. Aparinoides, Rubiaceae). Canad. Jour. Bot. 54: 1911–1925.

Puff, Christian. 1977. The Galium obtusum Group (Galium sect. Aparinoides, Rubiaceae). Bull. Torrey Bot. Club 104: 202–208.

Seaholm, John E. 1964. A Taxonomic Study of the Genus Galium in Minnesota. Proc. Minnesota Acad. 31: 99–104.

KEY TO THE SPECIES

1. Ovary and fruit spreading-bristly, or leaves with 3 prominent longitudinal veins, or (usually) plant with both these conditions (ovary smooth in some plants of *G. boreale*; leaves with only the midrib prominent at base of the blade in some others)
  2. Leaves all or mostly in whorls of 6–8 (9) with only one longitudinal vein (the midrib) prominent at base of the blade; plant ± reclining or trailing
    3. Stem and leaf *margin* rough-scabrous with retrorse (or at least divergent) broad-based barbs (may also have slender hairs at or above the nodes); leaves linear-oblanceolate (usually broadest beyond the middle, tapering more abruptly to apex than to base); plant annual . . . . .1. **G. aparine**
    3. Stem smooth or nearly so (occasionally with some spreading hairs) and leaf *margin* antrorse-scabrous; leaves narrowly (rarely broadly) elliptic (tapered ± equally to apex and base); plant perennial
      4. Inflorescence terminal; corolla funnel-shaped, the tube at least half as long as the lobes; plant a rare escape from cultivation .2. **G. odoratum**
      4. Inflorescences axillary as well as terminal; corolla flat, the tube less than a third as long as the lobes; plant a common native species . . . . .3. **G. triflorum**
  2. Leaves all in whorls of 4 (or fewer), often with a pair of weaker longitudinal veins besides the midrib at base of the blade; plant erect
    5. Branches of the inflorescence with lateral flowers (and fruits) mostly sessile or nearly so (pedicels if any less than 1 mm long); corolla yellow-green to deep maroon
      6. Principal leaves broadly ovate-elliptic, obtuse, mostly less than 2.5 times as long as broad; internodes and outside of corolla usually pubescent; corolla greenish . . . . .4. **G. circaezans**
      6. Principal leaves (not the lowest) broadly lanceolate (widest below the middle), acute, (2) 2.6–4.2 (5.5) times as long as broad; internodes and outside of corolla glabrous; corolla maroon . . . . .5. **G. lanceolatum**
    5. Branches of the inflorescence with all flowers (and fruits) distinctly pediceled; corolla white to whitish green
      7. Leaves linear-lanceolate, the principal ones mostly 4–14 times as long as broad, with 1–2 pairs of longitudinal veins (besides the midrib) prominent about half or more the length of the leaf; fruit with bristles none or if present not hooked at the tip (at most curved) . . . . .6. **G. boreale**
      7. Leaves broadly ovate-elliptic, less than 3 times as long as broad, with the lateral pair of longitudinal veins very weak or absent (conspicuous only in the rare northern *G. kamtschaticum*); fruit (and ovary) with bristles hooked at the tip
        8. Plants low, with only (2) 3–5 whorls of leaves, the upper ones the largest; stems glabrous; leaves not gland-dotted . . . . .7. **G. kamtschaticum**
        8. Plants taller, with numerous whorls of leaves, the upper ones smaller (or at least no larger) than the middle and lower ones; stems ± pubescent; leaves obscurely to distinctly gland-dotted beneath . . . . .8. **G. pilosum**

Page 281

1. Ovary and fruit smooth or at most roughened (with tubercles or a few appressed spicules) *and* leaves with only one prominent longitudinal vein (i.e., the midrib)
  9. Leaves sharply cuspidate or awned at apex, (5) 6–12 in a whorl
    10. Stems smooth, glabrous or puberulent, erect or ascending; leaves at main nodes 8–12 per whorl
      11. Corolla yellow, with acute but scarcely apiculate lobes; internodes puberulent (at least in inflorescence); leaves narrowly linear, strongly revolute, more densely pubescent on the nearly hidden lower surface than on the upper surface . . . . .9. **G. verum**
      11. Corolla white or greenish with at least some lobes often apiculate with prolonged tips; internodes glabrous (rarely puberulent); leaves usually ± oblanceolate, flat or very slightly revolute, nearly or quite glabrous (except on margins)
        12. Largest leaves less than 2 (2.5) cm long, stiff, densely antrorse-scabrous on the margins . . . . .10. **G. mollugo**
        12. Largest leaves 2.5–4 cm long, thin and membranous, sparsely scabrous on the margins . . . . .11. **G. sylvaticum**
    10. Stems smooth or retrorse-scabrous (not puberulent), reclining on or supported by other plants; leaves at main nodes (5) 6 in a whorl
      13. Margins of leaves smooth or weakly antrorse-scabrous; midrib smooth; leaves linear to very narrowly elliptic-oblong . . . . .12. **G. concinnum**
      13. Margins of leaves and midrib beneath strongly retrorse-scabrous (antrorse in the rare *G. verrucosum*); leaves narrowly elliptic to obovate
        14. Mature fruit ca. 3–4 mm long, densely tuberculate or papillose, on stout, scabrous, strongly recurved pedicels; plant an annual, sparsely if at all branched, a rare waif . . . . .13. **G. verrucosum**
        14. Mature fruit less than 2 mm long, smooth or nearly so, on slender, smooth, straight pedicels; plant a much-branched perennial, native throughout the state . . . . .14. **G. asprellum**
  9. Leaves merely acute to rounded (not cuspidate or awned) at apex, not over 4–6 in a whorl
    15. Flowers all or mostly (7) 10–25 per inflorescence (i.e., beyond the distal reduced but leaf-like bract); corolla ca. 2–4 mm broad, 4-lobed; nodes glabrous or nearly so . . . . .15. **G. palustre**
    15. Flowers 1–6 (8) per inflorescence; corolla less than 2 mm broad (3- or 4-lobed) *or* nodes pubescent
      16. Corolla 2–3 (4) mm broad (i.e., ca. 1 mm or more long), 4-lobed, the lobes longer than broad; nodes ± bearded with short hairs; leaf margins scabrous-ciliate with short, conical (broad-based, symmetrical) ± spreading hairs (or tiny prickles)

376. Cephalanthus occidentalis

377. Sherardia arvensis

378. Galium aparine

17. Leaves less than 2.5 mm broad (to 3.5 mm in shade forms), ± linear with revolute margins, all or mostly reflexed, the midrib beneath smooth or nearly so; mature fruit ca. 1–1.8 mm long, on pedicels 1–4 mm long; plant of wet peatlands . . . . 16. **G. labradoricum**
17. Leaves (at least the broadest) 2.5–6 (7) mm broad, narrowly elliptical, the midrib beneath usually nearly as or quite as bristly-ciliate as the margins; mature fruit ca. 2–2.8 mm long, the longest pedicels 3.5–7 (12) mm long; plant of swamps and mesic woods, especially on floodplains . . . . 17. **G. obtusum**
16. Corolla not over 2 mm broad (i.e., less than 1 mm long), usually 3-lobed (some flowers sometimes 4-lobed), the lobes no longer than wide; nodes glabrous; leaf margins smooth or usually scabrous with short retrorse deltoid barbs
18. Pedicels ± densely (but minutely) retrorse-scabrous, the longer ones (6) 7–17 (18) mm long, curved toward the tip at maturity, usually 2 or 3 subtended by a whorl of leaves at ends of branchlets, without a common peduncle (fig. 143) . . . . 18. **G. trifidum**
18. Pedicels smooth, not exceeding 8 mm long, and usually straight at maturity, often 2 or 3 on a peduncle (which may be scabrous)
19. Pedicels 0.5–3.5 (4.5) mm long and often curved at maturity, solitary or in pairs in leaf axils or at ends of branches but not on a common peduncle; corolla not over 1 mm wide; mature fruit 0.8–1.2 mm long; leaves mostly 2.5–7 mm long . . . . 19. **G. brevipes**
19. Pedicels (at least the longest) 3–8 mm long and nearly always straight at maturity, often on a peduncle; corolla ca. 1–1.8 mm wide; mature fruit 1–1.8 (2) mm long; leaves mostly 5.5–14 (22) mm long . . . . 20. **G. tinctorium**

1. **G. aparine** L.   Fig. 139   Goosegrass; Cleavers

Map 378. This is a circumpolar species of deciduous woods and thickets, especially in disturbed areas, apparently spreading northward in Michigan in recent years. There are very few pre-1900 collections, all from the southern Lower Peninsula (except for a lone Keweenaw County specimen from 1890) and none before 1879 (Kent Co.). There seems to have been rapid spread about the 1940s and subsequently.

Our plants do all seem to be the true *G. aparine* and not the very similar *G. spurium* L., which can be a serious weed of cultivated fields—as it is in some parts of Canada (Malik & Vanden Born 1988). *G. spurium* has sometimes been treated as, at best, a diploid (n = 20) variety or subspecies of *G. aparine* (which has been reported with several chromosome numbers from 22 to 88, with 64 and 66 the most common). Alleged morphological distinctions are not clearcut. *G. spurium* has smaller fruit and flowers, with the corolla greenish yellow rather than white. Corollas in Michigan material may be smaller but usually are ca. 1.5–2.8 mm across; insofar as collectors have noted color on their labels it is invariably white. Fruits range up to 4.8 mm long (excluding spiny bristles). Herbarium specimens are probably biased toward smaller (immature) fruit since the ripe ones detach readily and adhere to other objects.

The bristles on the fruit are hooked at the tips and thus it is easily dispersed on fur, feathers, and human clothing. The stems are sometimes rather smooth, with a good lens (or sensitive fingers) needed to be sure of the retrorse barbs on some internodes. The greater roughness of the upper surface of the leaves, as well as the retrorse (or at least spreading) barbs on the margins, are generally good distinctions from *G. triflorum*, also common in deciduous woods.

2. **G. odoratum** (L.) Scop.   Sweet Woodruff
Map 379. An Old World native, long a garden favorite and rarely escaping. Local in shaded thickets on the north side of Little Traverse Bay.

Long known as *Asperula odorata* L., under which name an established Michigan colony was first reported (Michigan Bot. 16: 137. 1977). The little lawn weed *Sherardia* might key here if one did not recognize that genus by its elongate corolla tube and pink to blue flowers in little heads.

3. **G. triflorum** Michaux
Map 380. Deciduous, coniferous, and mixed woods, even swamps (especially cedar if not too wet), fens, river and stream banks; spreading (or persisting) on shores and in clearings.

Although the stem is sometimes roughened (or even pubescent), this species is rather easily distinguished from *G. aparine* by the antrorse barbs or minute bristles on the leaf margins (usually best seen by viewing the leaf from beneath) and the more elliptical leaf shape. The mature fruits (excluding the hooked bristles) are up to 2 (occasionally 2.5) mm long and thus smaller than is usual in *G. aparine*.

4. **G. circaezans** Michaux
Map 381. Woods, ranging from dry oak-hickory to rich beech-maple (rarely in swampy or coniferous sites).

Sometimes difficult to distinguish from the next species, at least by leaf shape, but generally more pubescent, green-flowered, and with leaves more obtuse and shorter.

5. **G. lanceolatum** Torrey   Fig. 140
Map 382. Deciduous woods (beech-maple more often than oak-hickory or northern hemlock-hardwoods).

One of the few plants blooming in hardwoods in summer.

6. **G. boreale** L.   Northern Bedstraw
Map 383. Upland, ± open woods of oak, hickory, aspen, and/or pine; fields, meadows, prairie remnants; fens, tamarack swamps, thickets; banks of ditches, streams, and lakes.

This variable species has been a target of the "splitters," but most recent authors disregard the numerous named variants. The fruit may be bristly or smooth, the leaves narrowly linear or somewhat lanceolate (but always with 3 (5) prominent longitudinal veins at the base), the stems glabrous or occasionally pubescent, and the chromosomes tetraploid or hexaploid.

7. **G. kamtschaticum** Schultes & J. H. Schultes

Map 384. This is a decidedly boreal species, ranging from Alaska and northwestern Canada south to Washington, disjunct in the eastern Lake Superior area, and again disjunct from northern New York and the mountains of New England to Newfoundland. Known for many years from forests north of Sault Ste. Marie in Ontario, but first discovered in Michigan in 1995 (*M. Jaunzems*, MICH), in old-growth beech-maple forest ca. 3 mi. south of Lake Superior, north of Eckerman.

8. **G. pilosum** Aiton

Map 385. Dry sandy woodland with oak or jack pine, thriving in clearings; fields, grasslands.

The undersides of the leaves are minutely gland-dotted, a helpful character for distinguishing this species from *G. circaezans* when an inflorescence is lacking (or too immature to be sure of the distinct and thickish pedicels in *G. pilosum*).

9. **G. verum** L.   Yellow Bedstraw

Map 386. Native to Europe and the Middle East, escaping from cultivation. Fields, meadows, roadsides, railroads; several of our relatively few

379. Galium odoratum

380. Galium triflorum

381. Galium circaezans

382. Galium lanceolatum

383. Galium boreale

384. Galium kamtschaticum

collections are from the vicinity of golf courses and untended lawns, suggesting a possible origin in contaminated grass seed.

The dense, attractive, bright yellow flowers have a strong aroma. The distinctive needle-like leaves may be as many as 12 at a node.

10. **G. mollugo** L. White Bedstraw

Map 387. A Eurasian native, locally established along trails, roads, and fencerows; in other disturbed ground and waste places; sometimes a garden weed. First collected in Michigan in 1895 in Bay County.

Depauperate or lax plants might be confused with *G. concinnum*, but the latter never has more than 6 leaves in a whorl even at the principal nodes of the main stem, and the leaves are nearly or quite linear. Known to hybridize with *G. verum*, but such hybrids have not been detected in this state.

11. **G. sylvaticum** L.

Map 388. Another introduction from Europe. Collected in 1921 at Bay View by R. Ford (labeled by B. A. Walpole, BLH, ALBC).

The inflorescence is more diffuse than in *G. mollugo*; the leaves a little longer, distinctly thinner and pale beneath; and the corolla lobes are merely acute rather than apiculate.

12. **G. concinnum** T. & G.

Map 389. Deciduous woods, including beech-maple and oak-hickory; banks, swampy ground along streams.

See comments under *G. mollugo*, with which some specimens might be confused. An occasional specimen of *G. boreale* with very small leaves might be mistaken for *G. concinnum*, but the former has leaves all blunt-tipped and only 4 per whorl.

13. **G. verrucosum** Hudson

Map 390. A native of southern Europe, collected by Farwell (*2002*, BLH) in waste ground at Detroit in 1906 and not likely to be found again in the state.

385. Galium pilosum   386. Galium verum   387. Galium mollugo

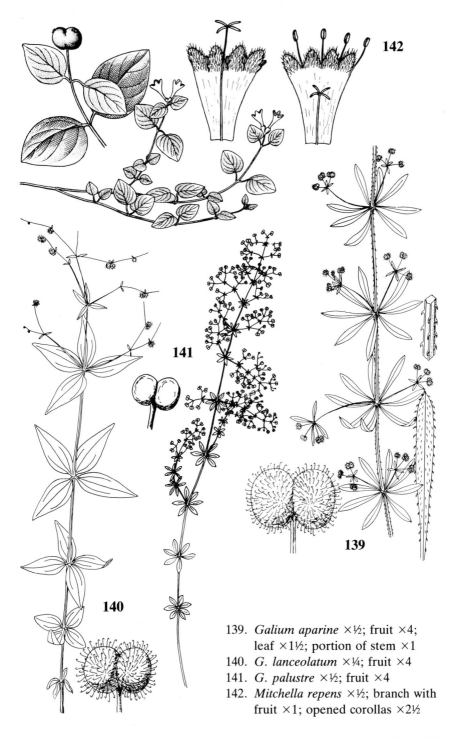

142

141

140

139

139. *Galium aparine* ×½; fruit ×4;
    leaf ×1½; portion of stem ×1
140. *G. lanceolatum* ×¼; fruit ×4
141. *G. palustre* ×½; fruit ×4
142. *Mitchella repens* ×½; branch with
    fruit ×1; opened corollas ×2½

This species has the sturdiest, harsh, thick-based prickles of any of our bedstraws—quite a rough, large-fruited, distinctive plant with remarkably thick pedicels curved into a U-shape compared with pedicels of other species. Farwell's specimen is a scrap with immature but clearly tuberculate fruit. He identified it as what is now called *G. tricornutum* Dandy, but the leaves are shorter (not over 1 cm) with margins antrorse-scabrous, whereas in the latter species the leaves are longer with retrorse-scabrous margins and the stem is even more densely prickly.

14. **G. asprellum** Michaux   Rough Bedstraw
    Map 391. Sprawling over and clinging to other plants in wet thickets, cedar swamps, borders of rivers and streams, marshes and wet meadows; clearings and disturbed moist ground.
    A rough, clambering plant, rather variable and some individuals smoother than others. Although the "prickles" are thick-based, they are not as coarse as in the preceding species. Rarely the ovaries and fruit are partly covered with very short appressed spicules (sometimes ± hooked).

15. **G. palustre** L.   Fig. 141   Marsh Bedstraw
    Map 392. A circumpolar species, ranging west in North America only to Lake Michigan and northern Ontario. Wet gravelly, rocky, or sandy shores and banks; low ground, meadows, swampy hollows.
    This is a rather showy species, as *Galium* goes, thanks to the numerous white flowers. Not only does it have more flowers in each ultimate cyme (distal to a bract), but there is also usually more than one such ultimate cyme at the end of a stem or branch, so the impression is of many more flowers in a terminal inflorescence than are seen in the next five species (of which, furthermore, only the next two have as large flowers as in *G. palustre*). The pedicels at maturity are ± strongly divergent, a pattern evident even before the fruit begins to ripen. In *G. obtusum* and other species, the pedicels may be divergent as the fruit nears maturity, but usually they are ± ascending. The leaves in *G. palustre* are generally 5–6 in well developed whorls, with margins spreading- or antrorse-scabrous.

388. Galium sylvaticum   389. Galium concinnum   390. Galium verrucosum

16.   **G. labradoricum** (Wiegand) Wiegand

Map 393. A northern (but not Arctic) species ranging westward in North America to Minnesota, Alberta, and the Northwest Territories. With us, it is a plant of fens, bogs, cedar and tamarack swamps, sedge meadows, and marshy ground along streams and lakes.

Sometimes difficult to recognize, especially if quite young. The leaves are not always reflexed (nor always revolute). Shade forms (e.g., in coniferous swamps) apparently can have broader, flatter leaves than plants in open bogs and fens. Stem internodes may be scabrous, while in *G. obtusum* they are smooth. Besides the habitat and morphological characters in the key, it may be helpful to note that in *G. labradoricum* the terminal inflorescence is often overtopped by lateral branches later in the season. The stems, though weak, are often quite erect (as they may also be in *G. obtusum*).

17.   **G. obtusum** Bigelow

Map 394. Deciduous woods (usually beech-maple, not oak-hickory), low ground (even floodplains) along rivers and streams, shrubby thickets, occasionally in grassy or disturbed ground.

This species and the preceding are compared with the next three in a helpful table by Puff (1977) except that his stated fruit length appears to be the combined breadth of the two mature carpels (as is clear from Puff 1976a, p. 26). Our plants are all ssp. *obtusum* as treated by Puff. See also comments above under *G. labradoricum*.

This species was once called (misidentified as) *G. tinctorium*, so older literature must be interpreted with caution.

18.   **G. trifidum** L.   Fig. 143

Map 395. Shores of lakes and ponds, river banks, fens, swamps, ditches, hummocks in marshes.

The very slender, even filiform arcuate pedicels, scabrous (seen under a good lens), make this species rather easy to recognize except when very young. The leaves are usually more linear than in *G. tinctorium*, but there

391. Galium asprellum

392. Galium palustre

393. Galium labradoricum

is not enough consistency in shape (or scabrousness) for identification of sterile material. The little retrorse barbules that make the leaf margins scabrous are short-conical, deltoid in aspect, like miniature rose thorns, and make these last three species rather easy to separate from *G. obtusum* and *G. labradoricum*, which have longer, straighter, and more slender bristly hairs along the leaf margins (as well as bearded nodes). Puff (1976b) refers all material from this part of North America to ssp. *trifidum*.

19.   **G. brevipes** Fernald & Wiegand

Map 396. Marshy ground, conifer thickets; especially characteristic of recently exposed calcareous shores and bottoms, interdunal hollows, and ditches.

Typically forms dense ± prostrate tangled mats (though other species may do the same if growing on open sites without vegetation to support them). The very small pedicels (usually ± recurved), fruits, corollas, and leaves—if all are displayed—indicate the distinctness of this species, which was long overlooked in Michigan, the first published reports being in 1977 (Michigan Bot. 16: 137).

20.   **G. tinctorium** L.

Map 397. Marshy ground, wet shores and thickets, cedar and tamarack swamps, peatlands, swampy woods and pond borders. More to be expected in peaty, mucky places than in sandy ones.

As in some other difficult genera, one is tempted to summarize, upon reaching the last species keyed, that if "none of the above" applies then this must be the species at hand. It *is* a variable one! Nevertheless, ssp. *tinctorium* is the only subspecies recognized by Puff in the Great Lakes region. This was once known as *G. claytonii* Michaux.

G. tinctorium is larger in some or all respects than *G. brevipes*, but it is not always easy to separate the two. A very few collections with some pedicels arcuate but (unlike *G. trifidum*) smooth, up to twice as long as in *G. brevipes*, are tentatively referred to *G. tinctorium*. The larger corolla and fruit (insofar as present) on these specimens would rule out *G.*

394. Galium obtusum   395. Galium trifidum   396. Galium brevipes

brevipes, and the pedicels are less slender than in *G. trifidum.* It should be noted that *G. trifidum* ssp. *subbiflorum* (Wiegand) Puff, of western North America, has arcuate smooth pedicels (and Puff includes in its synonymy *G. brandegei* A. Gray, with which some authors have combined *G. brevipes*!).

4. Mitchella

This genus consists of only two species (or varieties of one species), one in eastern Asia and the other in eastern North America. No other Rubiaceae display this pattern.

REFERENCE

Keegan, Christine R., Robert H. Voss, & Kamaljit S. Bawa. 1979. Heterostyly in Mitchella repens (Rubiaceae). Rhodora 81: 567–573.

1. **M. repens** L.   Fig. 142   Partridge-berry
Map 398. Deciduous woods with beech, maple, birch, aspen, and/or oak; often with cedar and hemlock or with pine, including hummocks in cedar swamps.

The flowers grow in pairs at the ends of trailing or slightly ascending stems, and are very fragrant. They are heterostylous, all those on any one plant having either long styles and short stamens or vice versa (fig. 142). The corollas are funnel-shaped, with lobes hairy above (suggestive of minia-ture *Menyanthes*—which is, however, 5- rather than 4-lobed). The "two-eyed" red fruit (derived from the two flowers of a pair) survives the winter and is conspicuous the following summer, if not hidden beneath leaves or consumed by animals.

5. Diodia

1. **D. teres** Walter   Fig. 146   Buttonweed
Map 399. Michigan is at the northern edge of range for this species,

397. Galium tinctorium   398. Mitchella repens   399. Diodia teres

which may in fact be merely adventive in the state. Its weedy habit makes origins obscure. There are no Michigan records before 1917 (railroad yard at Ypsilanti). Waste ground, usually sandy; disturbed areas on dunes.

Whatever its origin, the species at least can be fairly easily recognized by the bristles fringing the stipules, much exceeding the unfringed portion as well as the fruit.

6. Hedyotis

1. **H. nigricans** (Lam.) Fosberg
Map 400. An 1889 collection by Josephine M. Milligan, labeled only as from "Branch Co., Mich." (US) is somewhat suspicious, there being no other known Michigan specimens, although the species is locally common as close to this state as central Indiana and Ottawa and Erie counties, Ohio, on the Erie Islands and nearby mainland, especially on outcrops and in quarries. The preferred habitat of this species—rocky or gravelly ledges and cliffs, especially in limestone areas—does occur uncommonly in Branch County.

The ovary is about two-thirds inferior, whereas in the next genus it is about half inferior. The species has long been called *Houstonia nigricans* (Lam.) Fernald, but Terrell now places it in *Hedyotis*, although recognizing *Houstonia* for other species.

7. Houstonia   Bluets

The species of this genus have often been placed by recent authors—though usually with some doubt—in *Hedyotis*, but Terrell, who has studied the group intensively, believes strongly that *Houstonia* should be maintained.

REFERENCES

Terrell, Edward E. 1959. A Revision of the Houstonia purpurea Group (Rubiaceae). Rhodora 61: 157–180; 188–207.
Terrell, Edward E. 1991. Overview and Annotated List of North American Species of Hedyotis, Houstonia, Oldenlandia (Rubiaceae), and Related Genera. Phytologia 71: 212–243.

KEY TO THE SPECIES

1. Flowers solitary, much shorter than their tiny-bracted scapes . . . . .1. **H. caerulea**
1. Flowers in cymes, on short pedicels
  2. Cauline leaves mostly lance-ovate (broadest below the middle), with 3 definite longitudinal veins (the midrib most prominent) . . . . .2. **H. purpurea**
  2. Cauline leaves linear, narrowly oblong-elliptic, or oblanceolate (± parallel-sided or broadest beyond the middle), with only the midrib evident beneath

3. Basal leaves forming conspicuous rosettes, pubescent above (especially toward the edges) and with ciliolate margins; main stems with 4–6 internodes (including short lower ones but not branches of the cyme); median leaves oblanceolate (or some narrowly elliptic) . . . . .3. **H. canadensis**
3. Basal leaves none or if present usually glabrous; main stems with (5) 6–9 (10) internodes; median leaves linear (or some narrowly elliptic). . . . . .4. **H. longifolia**

1. **H. caerulea** L.   Bluets

Map 401. The status of this southern species in Michigan is uncertain. It is said to have been introduced deliberately in Tuscola County. It is not included in Allmendinger's 1876 "Flora of Ann Arbor and Vicinity" so older reports from Ann Arbor are presumably based on cultivated specimens or on misinterpreted data (e.g., labels of Mary Clark, who died in 1875, display "Ann Arbor" prominently printed—but as her address, even when the specimens came from elsewhere). The only possibly indigenous collection is from Edwardsburg, Cass County (*Hebert* in 1930, ND), with no habitat stated.

2. **H. purpurea** L.

Map 402. Presumably adventive from farther south. The only Michigan collections are by Farwell, from "low grounds" at Redford, in 1932 and 1933 (*9196*, MICH, BLH; Pap. Michigan Acad. 26: 19. 1941.)

Terrell confirms that the Michigan specimens can be referred to var. *calycosa* A. Gray.

3. **H. canadensis** Roemer & Schultes

Map 403. Sandy banks, fields, and woodlands in southern Michigan; rocky and gravelly shores and ridges in the limestone country of the northeastern Lower Peninsula.

Not always easy to distinguish from the next species, in which some authors have included it. (Others include both of these taxa in *H. purpurea*.) In Livingston County, intermediates have been found; and Terrell considers a St. Joseph County collection to be atypical, possibly a hybrid with *H. longifolia*, for the basal leaves are only sparsely ciliolate.

400. Hedyotis nigricans   401. Houstonia caerulea   402. Houstonia purpurea

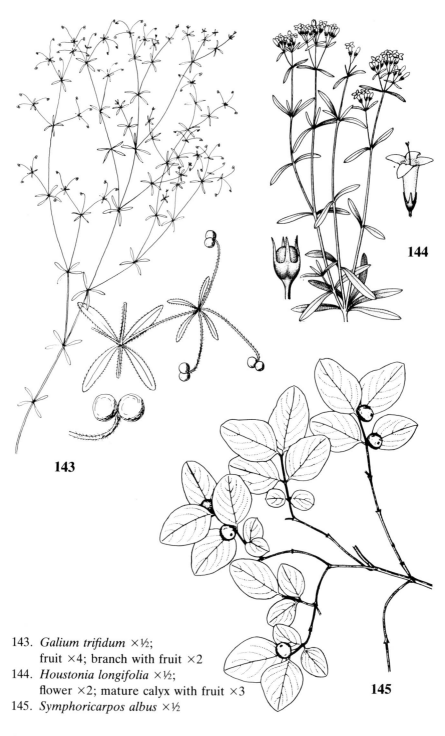

143. *Galium trifidum* ×½;
   fruit ×4; branch with fruit ×2
144. *Houstonia longifolia* ×½;
   flower ×2; mature calyx with fruit ×3
145. *Symphoricarpos albus* ×½

4. **H. longifolia** Gaertner   Fig. 144

Map 404. Sandy, often barren woodland with jack pine, oak, and/or aspen, both dry sites and damp areas; shallow soil over limestone; sandy fields.

Our commonest species of bluets. When there are basal leaves, it is often difficult to see the short lowest internodes among them. Occasional plants have pubescent leaves and are close to *H. canadensis*, but the leaves are all parallel-sided or narrowly elliptical, none of them oblanceolate.

CAPRIFOLIACEAE   Honeysuckle Family

This family, as traditionally circumscribed, has only about 5% as many species as the Rubiaceae, and they are mostly northern rather than tropical. Despite some superficial similarity, these two families are usually now classified in two different orders. For a convenient key character by which to separate them, Caprifoliaceae have opposite leaves without stipules, or with small stipules partly adnate to the petiole. In Rubiaceae, the stipules are present and free (though they may be united to each other around the stem between the petioles of the opposite leaves), or they are transformed into additional (and thus whorled) leaves. Further, the Caprifoliaceae may need to be defined as to exclude *Viburnum* and *Sambucus* but to include Valerianaceae and Dipsacaceae (see Harvard Pap. Bot. 5. 1994).

Except for some well known shrubs cultivated for the ornamental flowers or fruit, the family is of relatively little economic importance.

REFERENCES

Ferguson, I. K. 1966. The Genera of Caprifoliaceae in the Southeastern United States. Jour. Arnold Arb. 47: 33–59.
Hauser, Edward J. P. 1965. The Caprifoliaceae of Ohio. Ohio Jour. Sci. 65: 118–129.
Salamun, Peter J. 1980 ["1979"]. Preliminary Reports on the Flora of Wisconsin No. 68. Caprifoliaceae—Honeysuckle Family. Trans. Wisconsin Acad. 67: 103–129.

403. Houstonia canadensis   404. Houstonia longifolia   405. Triosteum perfoliatum

KEY TO THE GENERA

1. Leaves strictly entire (neither toothed nor lobed)
   2. Stem herbaceous, unbranched; leaves sessile, pubescent; fruit a dry drupe (with 3 pits) . . . . .1. **Triosteum**
   2. Stem woody, branched; leaves in most species at least short-petioled, glabrous or pubescent; fruit a fleshy drupe (with 2 pits) or several-seeded berry
      3. Corolla campanulate, less than 1 cm long, white to pink, regular or nearly so; fruit a white drupe with 2 pits (red to purple in 1 occasionally escaped species) . . . . .2. **Symphoricarpos**
      3. Corolla with prolonged tube, over 1 cm long, yellowish to red, bilaterally symmetrical; fruit a berry, yellow to red or blue to purple-black . . . . .3. **Lonicera**
1. Leaves (or leaflets) with margins lobed, toothed, crenate, finely crenulate, or at least with minute gland-like teeth
   4. Leaves pinnately compound, the leaflets closely and evenly toothed but not lobed; fruit fleshy with 3 (or more) seed-like pits . . . . .4. **Sambucus**
   4. Leaves simple, in many species crenate, crenulate, coarsely toothed, or lobed; fruit if fleshy with only 1–2 pits
      5. Calyx lobes up to 1.5 mm long and deltoid to broadly rounded or obsolete; fruit fleshy (a berry-like drupe)
         6. Calyx ± campanulate; fruit with 2 pits; leaf margins somewhat undulate or sinuous (neither sharply nor regularly toothed) . . . . .2. **Symphoricarpos**
         6. Corolla rotate (flat with very short tube); fruit with 1 pit; leaf margins usually sharply and/or regularly toothed (obscurely in *V. cassinoides*) . . . . .5. **Viburnum**
      5. Calyx lobes (1.6) 2–6.5 (7.5) mm long, linear or narrowly lanceolate; fruit dry (a nutlet or capsule, dehiscent or not)
         7. Plant with long-trailing stem from which arise short erect nearly herbaceous leafy shoots; leaf blades obtuse or rounded, shallowly toothed on apical half only; flowers pink, pendent, normally in pairs on long peduncles terminating the leafy shoots and much exceeding the leaves; ovary (i.e., its closely enveloping bractlets) and fruit densely glandular-bristly . . . . .6. **Linnaea**
         7. Plant an erect shrub; leaf blades acute to acuminate, toothed (even if minutely) below as well as beyond the middle; flowers yellow or pink, spreading but not pendent, usually more than 2 in an inflorescence, on peduncles shorter than the subtending leaves; ovary and fruit without bristles or with bristles not gland-tipped
            8. Corolla yellow, turning orange or even flushed with red; ovary and fruit glabrous (occasionally pubescent); leaf margin closely toothed throughout; plant a common native . . . . .7. **Diervilla**
            8. Corolla ground color pink; ovary and fruit with dense spreading bristly hairs not gland-tipped; leaf margins with remote and minute teeth appearing to be merely glandular bumps; plant a rare escape from cultivation . . . . .8. **Kolkwitzia**

1. **Triosteum**   Horse-gentian; Feverwort

These are hairy, coarse-looking plants, often with several to many stems arising from a crown. The corolla is reddish to maroon and the fruits are orange to red.

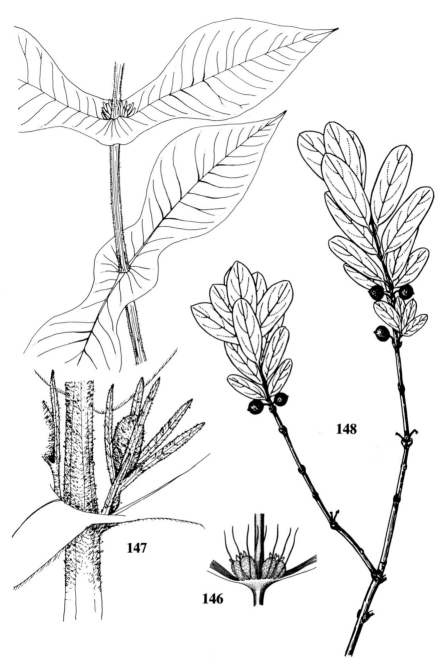

146. *Diodia teres*, node with fruit ×2
147. *Triosteum perfoliatum* ×⅓; node with fruit ×3
148. *Lonicera villosa* ×½

KEY TO THE SPECIES

1. Principal mid-cauline leaves both ca. 3–5 cm broad at the base *and* connate (upper leaves merely tapered basally); stem pubescence all or mostly of ± dense, short (0.5 mm or less), gland-tipped hairs; style exserted beyond the corolla . . . . . . 1. **T. perfoliatum**
1. Principal mid-cauline leaves tapered to narrow base (but upper leaves often narrowly connate); stem pubescence often chiefly of long (ca. 1–2 mm) eglandular hairs (usually with some short eglandular or gland-tipped hairs besides); style equalling or slightly shorter than the corolla (rarely exserted) . . . . 2. **T. aurantiacum**

1. **T. perfoliatum** L.   Fig 147

Map 405. In woods of beech and maple, oak and hickory, or pine; also in marshy ground and grassy areas.

Fully developed plants with strikingly broad connate-perfoliate leaves at the middle of the stem are easily recognized, but fragmentary specimens can be troublesome. See comments under the next species. *T. perfoliatum* blooms later than *T. aurantiacum*.

2. **T. aurantiacum** E. P. Bicknell

Map 406. In diverse habitats, from low swamp forest and rich deciduous woods to dry sandy or rocky woods with oak, aspen, sassafras, and/or pines, often at the edges.

Until this species was described in 1901, all Michigan reports used the name *T. perfoliatum*. The two are not always easily distinguished and some authors treat this as *T. perfoliatum* var. *aurantiacum* (E. P. Bicknell) Wiegand. A few specimens have styles clearly exserted (as much as 2 mm), but the leaf bases and stem pubescence are like those of *T. aurantiacum*. Pubescence characters are not as consistent as some keys suggest, but if many of the stem hairs are long and eglandular or if the pubescence is sparse, a specimen is pretty surely *T. aurantiacum*. If the hairs are dense, all or almost all short and mostly glandular, it is probably *T. perfoliatum*. Stems with short rather dense glandular hairs *and* a goodly number of longer hairs apparently occur in both species, especially on the upper internodes.

2. **Symphoricarpos**   Snowberry

A familiar genus in cultivation, especially in hedges, the principal taxon being a variety of *S. albus*. The white fruits in the fall make this species attractive to human eyes and to hungry birds and mammals. Even mature first-year stems can show the distinctly hollow pith characteristic of our common species.

REFERENCE

Jones, George Neville. 1940. A Monograph of the Genus Symphoricarpos. Jour. Arnold Arb. 21: 201–252.

KEY TO THE SPECIES

1. Corolla ca. 3–4 mm long; fruit red to purple; pith usually ± uniform (dense or loose but continuous—or even absent); flowers in dense axillary clusters of 6 or more (= 12 or more per node) . . . . .1. **S. orbiculatus**
1. Corolla ca. (4.5) 5–9 mm long; fruit white; pith of twigs hollow between the nodes; flowers various
  2. Stamens and style exserted at maturity; anthers 1.7–2.4 mm long; corolla ca. (5) 6–9 (11) mm long; flowers mostly 6 or more in axillary (and terminal) spikes . . . . .2. **S. occidentalis**
  2. Stamens and style included in the corolla; anthers ca. 1–1.5 mm long; corolla ca. (4.5) 5–6 (6.5) mm long; flowers when axillary 1 (–3) per axil (more when terminal and sometimes so in var. *laevigatus*) . . . . .3. **S. albus**

1. **S. orbiculatus** Moench   Coralberry
    Map 407. Native south of Michigan, but here apparently an occasional escape from cultivation to railroads, roadsides, and waste ground.
    The abundant fruits are said to be insipid and not appealing to birds.

2. **S. occidentalis** Hooker   Wolfberry
    Map 408. Native west of the Great Lakes. Our collections are all from disturbed places such as railroad embankments, dry fields, roadsides, and shores. The species also grows along railroads from the St. Mary's River around the north side of Lake Superior.
    The larger leaves on vigorous growth often have broadly toothed or crenate margins (a condition rarely seen in our other species).

3. **S. albus** (L.) S. F. Blake   Fig. 145   Snowberry
    Map 409. Usually in dry ± open sandy or rocky ground, in woodland of oak, aspen, and pine, as on old dunes, jack pine plains, and rock outcrops; river bluffs and shores (occasionally even in moist woods); often at borders of conifer thickets along shores. The introduced variety is found along roadsides and waste places, even spreading into woods.
    Our widespread native shrub is var. *albus* [including var. *pauciflorus* (A.

406. Triosteum   407. Symphoricarpos   408. Symphoricarpos
     aurantiacum   orbiculatus   occidentalis

Gray) S. F. Blake], with young twigs puberulent, leaves pubescent, fruit ca. 5–10 (12) mm in diameter and mostly solitary in leaf axils. Common in cultivation and occasionally escaped is the larger-leaved var. *laevigatus* (Fernald) S. F. Blake [sometimes recognized as a distinct species, *S. rivularis* Suksd.], native to the Pacific Northwest. It has young twigs and often the leaves glabrous (unlike *S. occidentalis*), with fruit at least 1 cm in diameter and mostly in clusters at the ends of branchlets. Intermediates do occur. The records mapped for Berrien, Cass, Gratiot, Kent, Oakland, and Washtenaw counties are solely var. *laevigatus*; both varieties are known from at least eight of the other counties.

3. **Lonicera**   Honeysuckle

Many honeysuckles are familiar garden plants, several species often too luxurious unless ruthlessly and frequently pruned. Except for the shrubby, weedy species (11–15 below), ours are more easily identified than the key might imply. An occasional exception to almost any character makes it necessary to evaluate several characters before naming some specimens.

The distinction based on pith in couplet 7 easily separates the weedy (and hybridizing) naturalized shrubs from a group of fine native species which have other good recognition characters that will, one by one, serve to identify them without resort to cutting a longitudinal section of twig to examine the pith; these characters can be reviewed in couplets 8–10 of the key.

Besides the more conspicuous bracts usually present at the summit of the peduncle in species 7–15, there are useful characters often in the bracteoles (bractlets), usually very short and ± appressed to the ovaries. In *L. villosa*, however, the bracteoles form a cupule around the united ovaries, and in *L. involucrata* they are greatly enlarged though not quite as long as the bracts themselves, being part of the distinctive showy involucre.

REFERENCES

Green, P. S. 1966. Identification of the Species and Hybrids in the Lonicera tatarica Complex. Jour. Arnold Arb. 47: 75–88.

Hauser, Edward J. P. 1966. The Natural Occurrence of a Hybrid Honeysuckle (Lonicera ×bella) in Ohio and Michigan. Michigan Bot. 5: 211–217.

Luken, James O., & John W. Thieret. 1995. Amur Honeysuckle (Lonicera maackii; Caprifoliaceae): Its Ascent, Decline, and Fall. Sida 16: 479–503.

Luken, James O., & John W. Thieret. 1996. Amur Honeysuckle, Its Fall from Grace. BioScience 46: 18–24.

KEY TO THE SPECIES

1. Leaves (1–2 pairs) below inflorescences connate around the stem; flowers in clusters at the ends of branchlets (or at nodes with connate leaves); plant a vine (twining or somewhat shrubby)
  2. Corollas (2.8) 3–5 cm long, the tube 1.7–4.2 cm; plant a rare escape from cultivation

3. Flowers in 1–3 whorls in a peduncled inflorescence; corolla nearly regular, the lobes ± equal and erect; leaves glabrous or at least slightly pubescent beneath. .
. . . . .1. **L. sempervirens**
3. Flowers sessile at bases of the connate leaves; corolla bilateral, 2-lipped, the lobes ± spreading; leaves glabrous . . . . .2. **L. caprifolium**
2. Corollas (1.2) 1.4–2.9 cm long, the tube (0.7) 0.9–1.8 cm; plant native, in woodlands and thickets
   4. Pair of connate leaves beneath inflorescence orbicular or nearly so, very glaucous above, broadly rounded or even shallowly notched at each leaf tip . . . . .3. **L. reticulata**
   4. Pair of connate leaves beneath inflorescence distinctly longer than broad, not (or very little) glaucous above, ± acute at each leaf tip
     5. Leaves with ± densely spreading-ciliate margins and at least sparsely strigose above; corolla tube (and usually branches of inflorescence) glandular-pubescent . . . . .4. **L. hirsuta**
     5. Leaves eciliate, glabrous above; corolla tube (and branches of inflorescence) glabrous or with fine long straight hairs (overtopping any tiny glandular hairs) .
. . . . .5. **L. dioica**
1. Leaves all distinct; flowers sessile (or ovaries even united) in pairs on axillary peduncles; plant an erect bushy shrub (slender vine only in *L. japonica*)
  6. Plant a trailing or climbing vine; young stems and at least midrib of leaves above hairy; corolla ca. [2.5] 3–5 cm long, the lobes nearly as long as the tube or longer; fruit purple-black, nearly sessile or on peduncles less than 10 [15] mm long . . . . . .
. . . . .6. **L. japonica**
  6. Plant an erect bushy shrub (at most, some of the branchlets slightly twisted); corolla less than 2.5 cm long; fruit and peduncles various
   7. Pith of twigs white and continuous; fruit red, blue, or deep purple-black; plant native, in woods or peatlands
    8. Peduncles, even in fruit, less than 1 cm long (rarely one to 1.5 cm); ovaries united (appearing as one, with 2 corollas), forming (in a cupule) a single blue-glaucous fruit per peduncle; corolla nearly regular, the lobes ± equal; bracts at base of ovary equalling or exceeding it at anthesis, but very narrow; pubescence of branchlets, petioles, lower leaf surface, and/or leaf margins mostly of straight ± brownish hairs or cilia . . . . .7. **L. villosa**
    8. Peduncles 1.1–3.5 (4.5) cm long; ovaries separate or ± united, forming separate red or dark purple-black fruits or a 2-lobed one; corolla regular or 2-lipped; bracts at base of ovary broadly foliaceous or shorter than the ovary; pubescence, if any, less extensive (straight brownish hairs in *L. canadensis*, but limited to cilia on leaves)
     9. Bracts apparently 4 at summit of peduncle, all foliaceous, covering ovaries and (until reflexing as a ruddy involucre) the 2 distinct purple-black berries; corolla nearly regular, the lobes ± equal and erect; leaves large (longest 9–13 cm or more), the apex short-acuminate or acute . . . . .8. **L. involucrata**
     9. Bracts 2 at summit of peduncle (not including minute bracteoles), shorter than the ovaries (rarely one longer) and than the red to purplish berries united at their bases; corolla various; leaves shorter (very rarely over 9 cm long in *L. canadensis*), the apex rounded to obtuse or acute (not at all acuminate)
      10. Leaf blades downy-puberulent beneath; leaf margin and petiole eciliate; the 2 ovaries in a pair partly to fully united but both ascending (not divergent); corolla 2-lipped (lower lobes spreading) . . . . .9. **L. oblongifolia**
      10. Leaf blades glabrous beneath or rarely with scattered long hairs (like the cilia); leaf margin and petiole ciliate; the 2 ovaries in a pair barely united at the base and strongly diverging (± 180°) in fruit; corolla nearly regular . . . .
. . . . .10. **L. canadensis**

7. Pith of twigs brown and hollow between the nodes; fruit orange to red; plants introduced, escaping to fields, borders of woods, etc.
    11. Peduncles distinctly shorter than the subtending petioles (flowers and fruits sessile or subsessile); leaves acuminate . . . . .11. **L. maackii**
    11. Peduncles much longer than the subtending petioles; leaves merely acute or obtuse
        12. Leaves and young stems completely glabrous or with a very few hairs . . . .
            . . . . .12. **L. tatarica**
        12. Leaves and at least young stems pubescent
            13. Ovary glandular; bracteoles glandular and pubescent on back and margins; filaments sometimes pubescent on distal half; leaf blades mostly broadest beyond the middle . . . . .13. **L. xylosteum**
            13. Ovary and bracteoles glandless; filaments glabrous except on basal half (or less); leaf blades broadest at or below the middle
                14. Bracteoles more than half as long as the ovary at anthesis . . . . . . .14. **L. morrowii**
                14. Bracteoles half or less as long as the ovary at anthesis . . . . .15. **L. ×bella**

1. **L. sempervirens** L.   Trumpet Honeysuckle
    Map 410. Native farther south, where evergreen; cultivated and rarely escaped northward. Roadsides, near old homesites, and woods.

2. **L. caprifolium** L.   Italian-woodbine
    Map 411. A species of Eurasia, cultivated and rarely escaping. Collected as naturalized in dryish open sandy or rocky ground around Marquette in 1916 (*N. M. Fairbanks* ex Dodge, MICH, BLH; Univ. Michigan Mus. Zool. Misc. Pap. 5: 40. 1918).

3. **L. reticulata** Raf.   Grape Honeysuckle
    Map 412. Generally considered to be native well south of the Keweenaw Peninsula, where Farwell collected specimens over many years (det. Rehder; Pap. Michigan Acad. 23: 133. 1938). Perhaps the Michigan material was escaped from cultivation.
    The circular disk of uppermost connate leaves, combined with very glaucous leaves (including upper surface of that disk) and tendency for the inflorescence to be separated into 2 or more whorls, make this a rather distinctive species among our climbing honeysuckles. Long known as *L. prolifera* (Kirschner) Rehder.

4. **L. hirsuta** Eaton   Hairy Honeysuckle
    Map 413. Woods and thickets, with cedar, fir, oak, aspen, pine, birch, and other trees; like the next species, particularly along borders, clearings, and banks; often in sandy or rocky ground, occasionally in swamps, especially cedar. The Ingham County collection seems quite out of range (*Bailey* in 1887, GH).
    The leaves are tapered to a sessile or subsessile base, which is generally densely pubescent, while in the next species it is glabrous.

5. **L. dioica** L.   Fig. 149   Glaucous Honeysuckle

Map 414. Woods and thickets both coniferous and deciduous; especially characteristic of borders and clearings, banks and rock outcrops, old dunes, fencerows; cedar swamps and sometimes other wetlands.

The foliage is often quite glaucous, especially beneath. Corollas range from yellow to deep maroon, more often at the red end of the spectrum than in *L. hirsuta*, which is most often yellow. The leaves range from glabrous to pubescent beneath; the ovaries and fruit are glabrous to densely glandular-pubescent. Variants have been named—even as distinct species—but seem too inconstant to recognize.

Two collections (from Baraga and Emmet cos.) have two-lipped corollas with tube and total length (over 3 cm) both in the range for *L. caprifolium* and distinctly too large for *L. dioica* or *L. hirsuta*. However, neither has the inflorescence sessile or the leaves glabrous beneath. Both have some cilia on the leaf margins. These may be hybrids involving *L. hirsuta* and some large-flowered species.

6. **L. japonica** Thunb.   Japanese Honeysuckle

Map 415. Introduced (with good intentions but disastrous results) from eastern Asia. Now in many places an aggressive vine that defies eradication, forming dense tangles that overwhelm the native (or other) vegeta-

409. Symphoricarpos albus

410. Lonicera sempervirens

411. Lonicera caprifolium

412. Lonicera reticulata

413. Lonicera hirsuta

414. Lonicera dioica

tion. The earliest Michigan records are from Washtenaw (1892) and Wayne (1904) counties; the dates of all other specimens I have seen are scattered from 1937 onwards. Usually in ± disturbed ground: roadsides, railroads, banks and bluffs, thickets.

The corolla is white, fading quickly to pale or dull yellow.

7. **L. villosa** (Michaux) Schultes   Fig. 148   Mountain Fly
Honeysuckle
Map 416. Cedar, tamarack, alder, and other swamps; fens, wet thickets.
A northern species, by some authors included in the Old World *L. caerulea* L.

8. **L. involucrata** Sprengel   Bracted Honeysuckle
Map 417. A boreal and western species barely ranging to Lake Superior. It is frequent along the north shore in Ontario. Spruce-fir woods and borders, rocky openings, stream banks.

Especially striking in fruit, the large glossy nearly black berries set on a shining red involucre.

9. **L. oblongifolia** (Goldie) Hooker   Fig. 150   Swamp Fly
Honeysuckle
Map 418. Peatlands, especially fens and cedar or tamarack swamps; calcareous marshy shores and stream margins.

Easily distinguished from *L. villosa* and *L. canadensis* by the absence of long (and ± tawny) cilia and the presence of downy pubescence on the undersides of the leaves. The 2-lipped aspect of the corolla is best displayed in fresh specimens. The fruits are sometimes rather purplish red.

10. **L. canadensis** Marshall   Plate 5-E   Fly Honeysuckle
Map 419. Woods of all kinds: beech-maple-hemlock, aspen, oak, mature red and white pine, spruce-fir, and mixed conifer-hardwoods; swamps, especially cedar.

The characteristic cilia on the leaves may be hard to see in very young

415. Lonicera japonica   416. Lonicera villosa   417. Lonicera involucrata

material and some of them may be eroded away on very old material. The two rather conic-cylindric berries generally diverge in completely opposite directions at the summit of the peduncle. Occasionally flowering peduncles are as short as 8 mm, but they are glabrous, as are the bracts, whereas in *L. villosa* the peduncles and bracts are usually puberulent, hairy, or ciliate.

The name has often been attributed to Bartram, who, however, never validly published it.

11. **L. maacki** (Rupr.) Herder   Fig. 151   Amur Honeysuckle
    Map 420. An Asian species, cultivated for the attraction its red berries have for birds, and becoming a pest like its relatives as a result of ready avian distribution. Woods (upland and swampy), thickets, banks, fencerows, often near a landscaped source. First collected out of cultivation in Michigan in 1963 at East Lansing (*L. C. Anderson 2369*, MSC). Luken and Thieret (1995, 1996) present thoroughly the history of its introduction and fall from grace.

12. **L. tatarica** L.   Tartarian Honeysuckle
    Map 421. Native to Eurasia, widely cultivated and naturalized along roadsides and railroads; thickets, shores, and along rivers; borders of woods and invading them; fields and waste places. Collected out of cultivation in Michigan as early as the 1890s (1891, Washtenaw Co,; 1897, St. Clair Co.).

This promiscuous species hybridizes freely with its relatives (see Green 1966)—in our area, particularly with *L. morrowii*. Plants with glabrous inflorescence and undersides of leaves, except at most for a very few scattered hairs (other than cilia, which are rarely present on leaf margins), I have referred to *L. tatarica*, which also rarely may have a few glands on the margins of the tiny, often nearly orbicular bracteoles. The distinction from *L. ×bella* is presumably obscured by backcrossing. *L. ×bella* also has short bracteoles at most half as long as the ovaries at anthesis, and the pubescence of leaves and branchlets is less dense than in *L. morrowii*, which has

418. Lonicera oblongifolia

419. Lonicera canadensis

420. Lonicera maackii

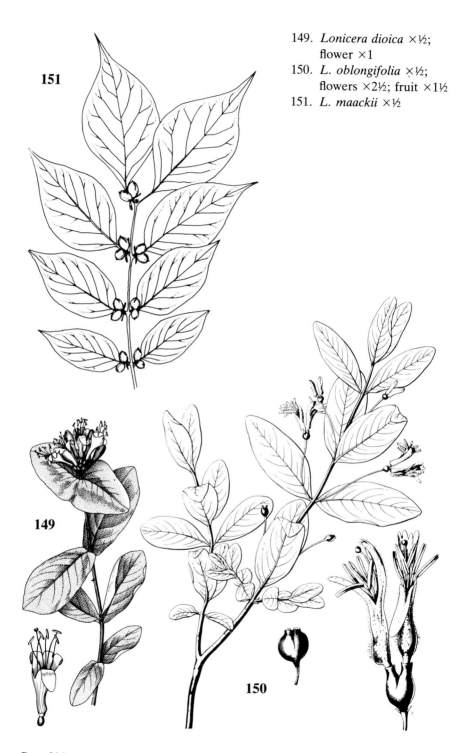

149. *Lonicera dioica* ×½; flower ×1
150. *L. oblongifolia* ×½; flowers ×2½; fruit ×1½
151. *L. maackii* ×½

the bracteoles more than half as long as the ovaries—often nearly or quite as long and tending to be oblong. The branches and undersides of the leaves in *L. morrowii* are rather densely pubescent. But, again, the distinction from *L. ×bella* is sometimes obscure.

The corolla is deep to pale pink or white, not tending to turn yellow with age.

13. **L. xylosteum** L.   European Fly Honeysuckle

Map 422. Another Eurasian species, less commonly escaped in this region than some of its relatives. Woods and thickets; swamps.

Hybrids may be expected, with very short-stipitate glands on ovaries and other parts as a clue for recognizing involvement of this parent.

The corolla is yellowish white or white turning to yellow the second day.

14. **L. morrowii** A. Gray   Morrow Honeysuckle

Map 423. A Japanese species, rather frequently escaped and hybridizing with *L. tatarica* to form *L. ×bella*. Along roadsides and railroads; thickets, banks, and shores; borders of woods and invading them. First collected out of cultivation in Michigan in 1939 in Kalamazoo County (*Rapp 3257*, MICH, WMU).

The corolla is white, turning yellow in age.

15. **L. ×bella** Zabel

Map 424. This is the result, apparently, of hybridization between *L. tatarica* and *L. morrowii* and is established in similar habitats. It has arisen in cultivation and probably spontaneously in the wild as well.

L. ×bella is a name applied to a swarm of variants and is distinguished by Green principally by the combination of pubescence and very short bracteoles (see comments under *L. tatarica* above). Corollas are colored as in *L. tatarica* from deep pink to white, but fade to a dull yellow as in *L. morrowii*. There seems to be no sharp line of separation from *L. morrowii* in either density of pubescence or length of bracteoles, and intermediate

421. Lonicera tatarica   422. Lonicera xylosteum   423. Lonicera morrowii

specimens are frequent. I have usually come down on the side of *L. ×bella* for such specimens (e.g., with very densely pubescent leaves but short bracteoles or with very lightly pubescent leaves but longer bracteoles), especially if such plants also have deep pink corollas. Because the hybrid is so common, it is keyed and mapped here to show its documented distribution (doubtless understating its abundance). A few individuals of intermediate nature may not truly be *L. ×bella* but of other parentage. Small corollas (less than 12 mm long) and/or somewhat obovate blades could derive from *L. ruprechtiana* Regel; obovate blades and glandular bracteoles would suggest *L. xylosteum*. These are all treated by Green and by Swink and Wilhelm (who have an even greater mess in the Chicago area than we do!).

4. **Sambucus** Elderberry

The pith occupies a large portion of the stem diameter and hence the distinguishing color is easy to see at all times of the year. However, the pith of the current year's growth cannot be relied upon; it may be light brown or "off-white" in both species. The corollas are rotate, i.e., flat with a very short tube, as in *Viburnum*.

KEY TO THE SPECIES

1. Pith white (sometimes a little brown at the edges); mature fruit purple-black; inflorescence distinctly broader than long and ± flat-topped, blooming in early to mid summer and fruiting in late summer to fall; lowest node of inflorescence with 4–5 (7) rays . . . . .1. **S. canadensis**
1. Pith tan or orange-brown on branches 1 year or more old; mature fruit red; inflorescence about as long as broad or up to 1.5 times as long, ± pyramidal or elongate, blooming in spring and fruiting in early to late summer (depending on latitude); lowest node of inflorescence with 3 rays (main axis and 2 branches) . . . .
. . . . .2. **S. racemosa**

1. **S. canadensis** L. Common Elder

Map 425. Wet ground, including swamps and floodplains, ditches, borders of woods, thickets, shores; scattered in marshes and meadows.

The fully ripe purple-black fruit in large flat-topped, umbelliform inflorescences can be used for making jelly, pie, or wine—just as the even more abundant (and glaucous) fruits of *S. cerulea* Raf. of the West Coast.

The new branchlets and branches of the inflorescence are glabrous or nearly so.

2. **S. racemosa** L. Fig. 152 Red-berried Elder; Red Elderberry

Map 426. Primarily in beech-maple woods, especially along borders, trails, and clearings, but also with conifers and in mixed woods, thickets, occasionally swamps.

The bright red fruit is very attractive to the eye (and to birds, as with the preceding species also), but is usually said to be inedible or even poisonous to humans. Some years ago a student whose motives I had no reason to distrust gave me a jar of fine red homemade jelly from the fruit of this species. It was tasty and I suffered no ill effects (perhaps buttered toast is the appropriate antidote).

This is a variable species, with several allied taxa in the West. Eastern American plants have often been called *S. pubens* Michaux, which is now usually treated as ssp. *pubens* (Michaux) House or var. *pubens* (Michaux) Koehne of the European *S. racemosa*, which is sometimes cultivated. Typical *S. racemosa* is essentially glabrous, while our plants are usually slightly to densely pubescent on new branchlets and in the inflorescence. There is also much variation in leaflet shape.

The uncommon yellow-fruited form [f. *xanthocarpa* House] has been collected in Mackinac and Washtenaw counties. A cut-leaved variant has rarely been found—sometimes as a branch on an otherwise normal plant.

5. **Viburnum**   Viburnum; Arrow-wood

This is a difficult genus with the lines between taxa in several complexes not clear or at least their taxonomic significance not agreed upon. The fruit of most species is popular with wildlife. The flowers (including large sterile ones in a few species), colorful autumn foliage, and red or purple-black fruit make several species, American as well as Asian, well known in cultivation. Reports of *V. alnifolium* Marshall [*V. lantanoides* Michaux] from Michigan are not confirmed—and not likely to be.

REFERENCES

Fernald, M. L. 1941. Viburnum edule and Its Nomenclature. Rhodora 43: 481–483.
McAtee, W. L. 1956. A Review of the Nearctic Viburnum. [Author], Chapel Hill. 125 pp. + 9 pl. + postscript. [Still useful, especially for seed and pit characters, although the author deliberately refused to make his nomenclature conform to the Code.]

424. Lonicera ×bella

425. Sambucus canadensis

426. Sambucus racemosa

KEY TO THE SPECIES

1. Leaves lobed (except sometimes the distal pair on a branch), ± palmately veined with 3 (–5) main veins (each bearing regular lateral veins) arising at base of blade
  2. Marginal flowers of inflorescence sterile and much larger than flowers toward the center; petioles with prominent large glands toward the summit and stipules at the base; undersides of leaves not dotted . . . . .1. **V. opulus**
  2. Marginal flowers similar to the others; petioles without large glands, and stipules sometimes absent; undersides of leaves ± sprinkled with red-orange dots
    3. Inflorescences only on lateral shoots bearing one pair of leaves (primary shoots ending in leaves, not cymes); mature fruit red or yellowish; leaves glabrous beneath or with only simple hairs . . . . .2. **V. edule**
    3. Inflorescences all or at least partly terminal on long leafy branches; mature fruit purple-black; leaves ± densely pubescent beneath with mostly stellate or tufted hairs . . . . .3. **V. acerifolium**
1. Leaves not lobed, pinnately veined (only the midrib with regular lateral veins)
  4. New growth and leaves (margins, lower surfaces, petioles) all ± strongly stellate-pubescent; winter buds naked (rudimentary leaves covered by no additional scales) . . . . .4. **V. lantana**
  4. New growth and leaves bearing only scattered if any stellate hairs (dense in young *V. plicatum*); winter buds covered with 1 or more scales that do not develop into leaves
    5. Leaf veins (and their branches if any) each extending to the tip of a prominent large marginal tooth
      6. Marginal (or all) flowers of an inflorescence sterile and much enlarged; new growth, especially youngest shoots, with ± dense stellate (or tufted) hairs; plant a rare escape from cultivation. . . . . .5. **V. plicatum**
      6. Marginal flowers similar to the others; new growth with few if any stellate hairs; plants widespread native species
        7. Enlarged base of style (persistent on fruit) antrorse-hairy; petioles 13–22 mm long, glabrous or nearly so, without stipules at base; leaf margins usually eciliate. . . . . .6. **V. dentatum**
        7. Enlarged base of style glabrous; petioles up to 8 (12) mm long, usually pubescent and usually stipulate; leaf margins ciliate . . . . .7. **V. rafinesquianum**
    5. Leaf veins anastamosing extensively before reaching the closely crenulate-toothed margin
      8. Teeth of leaf margin low and obtuse to broadly rounded or reduced to mere gland-like swellings; inflorescence on a single terminal peduncle (6) 8–17 (26) mm long . . . . .8. **V. cassinoides**
      8. Teeth of leaf margins acute; inflorescence sessile (at most a peduncle 2.5–4 mm long), several branches of it arising from the axils of a terminal pair of leaves
        9. Petioles scarcely if at all margined, the margins if any on petioles of sterile branchlets not crisped, revolute, or appearing toothed (may be broader and reddened on leaves beneath inflorescences on fertile branchlets); apex of leaf rounded or acute to very short-acuminate; shrub of southernmost Lower Peninsula . . . . .9. **V. prunifolium**
        9. Petioles with a green often revolute margin ± crisped, undulate, or even appearing toothed; apex of most if not all leaves with a distinctly prolonged acuminate tip usually ca. 1–1.5 cm long; shrub occurring throughout the state . . . . .10. **V. lentago**

1. **V. opulus** L.   Plate 5-F   Highbush-cranberry; Guelder-rose
    Map 427. Swamps (both hardwood and coniferous), borders of woods
and shores, wet roadsides and ditches; banks and thickets along rivers and
streams, fens; and other damp often open ground.
    This species, in the broad sense adopted here, is circumpolar. Opinions
of authors are quite divided on how distinct the American plants are from
the Eurasian. Our native tall shrub may be designated as var. *americanum*
Aiton. It is also often called *V. trilobum* Marshall or *V. opulus* ssp. *trilobum*
(Marshall) R. T. Clausen. The European plant is often cultivated, including a cultivar with all the flowers enlarged and sterile ("Snowball Bush"). I
am convinced that only very few of our Michigan specimens (other than
some explicitly from cultivation) represent the Old World taxon, which is
said sometimes to escape. All have at least some stipules enlarged or
knobbed at the tip—but so do some European specimens, although they
are supposed to have thin narrow tips on the stipules. Old World plants
supposedly have broader, sessile, saucer-like (concave), often more numerous glands toward the summit of the petiole; some of our apparently native
plants approach this condition, although often they have short-stalked
glands flat or convex at the end—and some of our apparently escaped
plants have small glands. At least one collection from Isle Royale lacks
both glands and stipules. As a number of authors have observed, the variation is too great—and too continuous—to make clear distinctions.
    The ripe fruit is red and about the size of a cranberry; it makes a tasty
jelly. The Old World taxon is said to have more bitter fruit. This species
and the next are the only Michigan viburnums to have the mature fruit red;
our other species, including the two escapes from cultivation, have purple-
black fruit when mature (although it may be red when it is green).

2. **V. edule** (Michaux) Raf.   Fig. 153   Squashberry; Mooseberry
    Map 428. A shrub almost entirely of the Boreal Forest region of North
America, in Michigan found only at Isle Royale, especially at borders,
openings, and thickets in spruce-fir forest and along shores.
    Readily distinguished from the preceding species and the next by the
lateral, even axillary, aspect of the inflorescences. The fruit is considered
edible.
    This species has sometimes been called, even in recent works, *V.
pauciflorum* Raf., a binomial not validly published by Rafinesque although
erroneously attributed to him in some indexes (see full explanation in
Fernald 1941).

3. **V. acerifolium** L.   Fig. 154   Maple-leaved Viburnum
    Map 429. Most often an undershrub in deciduous woods, both dry and
sandy with oak, pine, aspen, and/or sassafras and rich beech-maple forest;
often on slopes and hillsides, including old wooded dunes as well as river

banks; less often in somewhat swampy woods (but not conifer swamps). In Michigan, found only south of the known range of *V. edule*.

A rare form [f. *ovatum* Rehder] with leaf blades ± ovate and pinnately veined was collected by Farwell in Wayne County in 1922 and in Lenawee County in 1927. It can be distinguished from *V. dentatum* and *V. rafinesquianum* not only by the tiny red dots on undersides of the leaves but also by the stellate pubescence on petioles and blades beneath.

4. **V. lantana** L.   Wayfaring Tree

Map 430. A Eurasian species, widely cultivated and occasionally escaped to deciduous woods, thickets, roadsides, borrow pits, and waste places.

5. **V. plicatum** Thunb.   Japanese Snowball

Map 431. A native of eastern Asia, established as an escape from cultivation in a woodlot on the Michigan State University campus.

6. **V. dentatum** L.   Arrow-wood

Map 432. Most, perhaps all, of our records are probably escaped from cultivation or adventive from a little farther east or south. Usually in somewhat disturbed areas along roads and borders of trails and thickets in parks; fields, meadows, and shores; in both swampy and upland areas.

427. Viburnum opulus

428. Viburnum edule

429. Viburnum acerifolium

430. Viburnum lantana

431. Viburnum plicatum

432. Viburnum dentatum

Our few specimens of this mostly more southern species are apparently the glabrate var. *lucidum* Aiton, often recognized at specific rank as *V. recognitum* Fernald. The species in the broad sense is a variable and complex one. One of the collections from Berrien County has very short petioles but style and other characters of *V. dentatum*.

7. **V. rafinesquianum** Schultes   Downy Arrow-wood
Map 433. Typically in dry sandy or rocky woods with oak, hickory, pine, or other trees, sometimes beech and maple and along associated banks and openings; fencerows; sometimes in swampy ground and thickets along rivers.

Most of our specimens are var. *rafinesquianum*, with leaf blades ± downy beneath and petioles rather densely pubescent. The glabrate extreme, with leaf blades at most pubescent on the main veins and with the petioles glabrate to sparsely hairy, is var. *affine* (B. F. Bush) House. This variety is known mainly from the southeastern Lower Peninsula, but ranges as far to the northwest as Ontonagon County. Intermediates between this variable species and the preceding are said to occur where there is more overlap in range. Fortunately we seem to have the glabrate extreme of *V. dentatum* and mostly the pubescent extreme of *V. rafinesquianum*, so there is little difficulty in Michigan distinguishing the two. See also note above on *V. acerifolum* f. *ovatum*.

8. **V. cassinoides** L.   Wild-raisin
Map 434. A tall shrub of diverse usually wet ground: peatlands (fens, bogs, cedar swamps, boggy woods), forested tamarack, red maple, hardwood wetlands; moist areas or even drier sites in oak, pine, or jack pine woodland.

In much of Michigan, this is one of our commonest viburnums of damp places, but it barely ranges into northeastern Wisconsin and not at all into Minnesota. Of our three common wetland viburnums, this is the latest to

433. Viburnum
     rafinesquianum

434. Viburnum cassinoides

435. Viburnum
     prunifolium

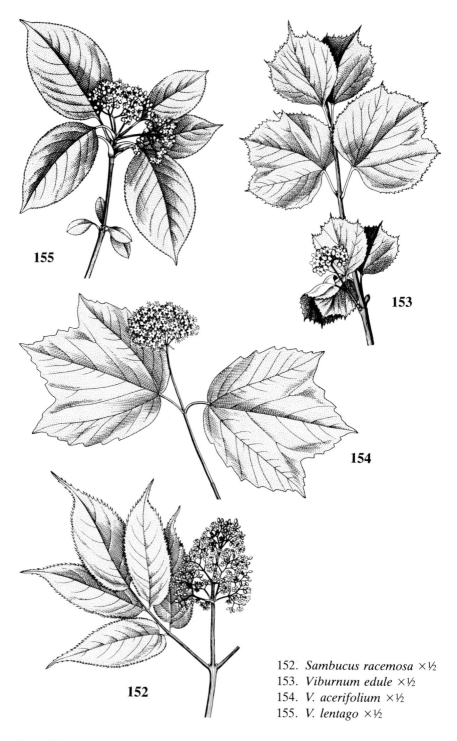

155

153

154

152

152. *Sambucus racemosa* ×½
153. *Viburnum edule* ×½
154. *V. acerifolium* ×½
155. *V. lentago* ×½

bloom, usually overlapping the last of *V. opulus*, which in turn may overlap the last blooming of *V. lentago*.

The immature fruit is red or whitish, though it may turn a handsome glaucous blue when dried. Leaf tips vary from acute to rounded to short-acuminate. The petioles and branches of the inflorescence are usually densely scurfy with rusty-orange scales. Such scales are sparse or absent in the next two species.

Some authors include this taxon in the southern and eastern *V. nudum* L., as var. *cassinoides* (L.) T. & G.

9. **V. prunifolium** L.   Black-haw

Map 435. Oak woods and other upland deciduous forest; also woods and thickets along streams, including river floodplain swamps.

This species barely ranges as far north as Michigan, and is not always easy to separate from the next. The leaves are very finely toothed, the teeth less—typically much less—than 0.4 mm long (measured perpendicular to the basic leaf margin); the lateral branchlets diverge widely and stiffly from the main branches. In *V. lentago*, the teeth are usually ca. 0.5 mm long and the lateral branches more ascending, lax, or less stiffly straight in aspect.

10. **V. lentago** L.   Fig. 155   Nannyberry

Map 436. Borders and banks of streams and rivers; fens, sedge meadows, tamarack swamps; shrubby swamps, swampy woods, thickets, fencerows.

Not always quickly distinguishable from the preceding species (see above). The mature stamens of *V. lentago* are said to protrude at least half their length from the corolla, while in *V. prunifolium* the stamens usually protrude about a fourth of their length. This is not an easy character to use, considering the shrinking of over-mature filaments and incomplete elongation of immature ones. But specimens which indeed have more than half of the stamen length extending beyond the corolla can probably be referred to *V. lentago*.

6. **Linnaea**

1. **L. borealis** L.   Fig. 156   Twinflower

Map 437. Woods and thickets, neither the wettest nor the driest but ranging from cedar swamp to aspen woods, usually with conifers (most often cedar); especially common at borders (as along shores or on old dunes), openings, and clearings; very local southwards in the state, often abundant northward.

This circumpolar plant was a favorite of the great Swedish naturalist Carl Linnaeus, and the genus was named for him before his own works were taken as the formal starting point of botanical nomenclature. Plants of Eurasia and parts of Alaska represent the typical subspecies or variety; ours may be called var. *longiflora* Torrey or ssp. *longiflora* (Torrey) Hultén.

The fruit, variously described as a 1-seeded tardily dehiscent capsule or as an achene, is rarely found mature in eastern North America. The flowers are very fragrant, and a carpet of blooming pink twinflower in early summer casting its fragrance upon the area is a worthy memorial to its namesake. Rarely the flowers are solitary or more than two. Occasional late-blooming individuals (September) tend to have an extra flower on the peduncle, below the usual pair.

7. **Diervilla**

REFERENCES

Hardin, James W. 1968. Diervilla (Caprifoliaceae) of the Southeastern U. S. Castanea 33: 31–36.

Schoen, Daniel J. 1977. Flora Biology of Diervilla lonicera (Caprifoliaceae). Bull. Torrey Bot. Club 104: 234–240.

1. **D. lonicera** Miller   Fig. 157   Bush-honeysuckle
Map 438. Dry woods, usually sandy or rocky, with aspen, birch, oak, or conifers, especially along borders (e.g., at shores, fields, roads, and clearings); thriving after some disturbance such as fire; old dunes, sandy bluffs, railroad embankments, thickets, fencerows.

The ovary is occasionally puberulent but usually glabrous; the leaves are sometimes ± densely pubescent beneath [f. *hypomalacum* Fernald] but usually glabrous or nearly so except on the midrib. The fruit of this genus and the closely related *Weigela*, often cultivated, is a slender many-seeded 2-valved capsule.

In Cheboygan County, Schoen found the species to be pollinated by bumblebees and the diurnal bumblebee hawkmoth, whose larvae also feed on the leaves.

436. Viburnum lentago   437. Linnaea borealis   438. Diervilla lonicera

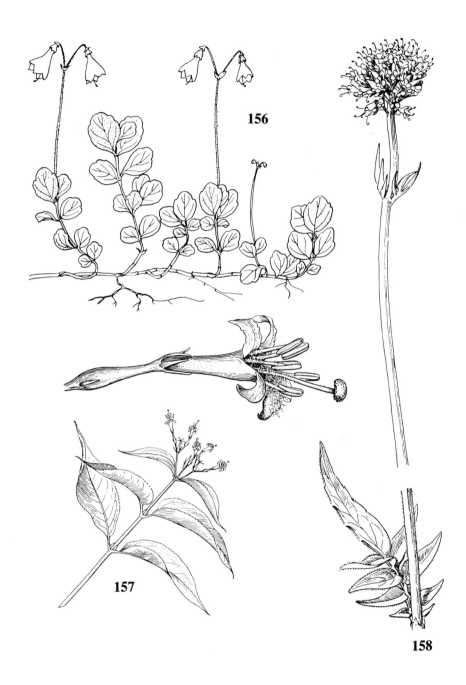

156

157

158

156. *Linnaea borealis* ×¾
157. *Diervilla lonicera* ×½; flower ×3
158. *Valeriana uliginosa* ×½

8. Kolkwitzia

1. **K. amabilis** Graebner   Beauty-bush
   Map 439. The only species in this genus, a native of China, is a showy
shrub often cultivated but only rarely escaping. Established at the edge of a
dry woods near Rogers City (*C. Newhouse 116* in 1972, MSC).
   Species of *Abelia*, often cultivated especially south of Michigan, might
key here but differ in lacking bristles on the ovary; the fruit in both genera
is an achene or nutlet.

VALERIANACEAE   Valerian Family

The flowers are small, white (occasionally pink), and ± crowded (at
least when young) into rather dense inflorescences. The corolla is slightly
bilateral, with 3 (or 4) stamens inserted on the tube, and the inferior ovary
is composed of 3 carpels of which only 1 develops a seed.

REFERENCES

Ferguson, I. K. 1965. The Genera of Valerianaceae and Dipsacaceae in the Southeastern
    United States. Jour. Arnold Arb. 46: 218–231.
Hauser, Edward J. P. 1963. The Dipsacaceae and Valerianaceae of Ohio. Ohio Jour. Sci. 63:
    26–30.

KEY TO THE GENERA

1. Cauline leaves deeply pinnately lobed or even compound; calyx developing into a
   conspicuous plumose pappus; plant a perennial with stout, strong-scented roots or
   rhizomes . . . . .1. **Valeriana**
1. Cauline leaves (like the basal) unlobed, entire or slightly toothed; calyx minute or
   obsolete even in fruit; plant a slender-rooted annual . . . . .2. **Valerianella**

1. Valeriana   Valerian
   A box or herbarium cabinet containing any specimens of valerian will
emit, when opened, a strong, long-persistent "valerianic odor" so characteristic of these plants, an odor also evident when one digs up the underground parts. Yet these pungent parts have been used as food when properly prepared, and they have a long reputation for medicinal value.
   The incurved calyx unfurls and elongates on the mature fruit, where it
resembles a pappus of plumose bristles, as in some Compositae. The flowers in *Valeriana* are perfect or unisexual, usually mixed on an individual and
with corollas of differing lengths. Corolla measurements in the key below
are based on *mature* corollas of the *larger* flowers (i.e., staminate or perfect, not pistillate ones, which run smaller).

REFERENCE

Meyer, Frederick G. 1951. Valeriana in North America and the West Indies (Valerianaceae). Ann. Missouri Bot. Gard. 38: 377–503.

KEY TO THE SPECIES

1. Rachis of cauline leaves broader than the lateral lobes; leaf margins very densely ciliate; plant with a large (sometimes branched) taproot; inflorescence becoming longer than broad; corollas ca. 2.7–4 mm long . . . . .1. **V. edulis**
1. Rachis of cauline leaves narrower than the lateral lobes; leaf margins moderately to sparsely ciliate; plant rhizomatous; inflorescence often as broad as long or broader; corollas various
   2. Basal leaves all or mostly simple, at most with one pair of small lobes; cauline leaves with 2–5 (7) pairs of lateral lobes (or leaflets), these with marginal cilia mostly spreading (if any); corolla (tube) ca. 4.5–8 mm long; plant native in fens and conifer swamps . . . . .2. **V. uliginosa**
   2. Basal leaves all deeply divided (similar to cauline leaves); cauline leaves with (6) 7–9 pairs of lateral lobes (or leaflets), these with marginal cilia mostly antrorse; corolla (tube) ca. 3–4.5 mm long; plant an escape from cultivation . . . . . .3. **V. officinalis**

1. **V. edulis** T. & G.
    Map 440. Fens, meadows, and wet prairies.
    Eastern American plants (centered around the southern Great Lakes region) are referred to var. *ciliata* (T. & G.) Cronquist, sometimes recognized as a distinct species from the typical variety, which is more widespread in the west.

2. **V. uliginosa** (T. & G.) Rydb.   Fig. 158   Swamp Valerian
    Map 441. Rather local in wet alkaline places: fens, tamarack and cedar swamps (sometimes with spruce and sphagnum), wet prairies. A tall and noticeable plant where found.
    Our plants of the eastern United States and adjacent Canada are sometimes treated as a variety or subspecies of the western *V. sitchensis* Bong. or of the more northern *V. sylvatica* S. Watson [= *V. septentrionalis* Rydb.]— which is itself sometimes included in the Old World *V. dioica* L.!

439. Kolkwitzia amabilis   440. Valeriana edulis   441. Valeriana uliginosa

3. **V. officinalis** L.   Common Valerian; Garden-heliotrope
Map 442. A Eurasian native, often cultivated for its very fragrant flowers and occasionally established as an escape to roadsides, ditches, shores, fields, and borders of woods.

2. Valerianella   Corn-salad; Lamb's-lettuce
Fruit morphology may vary on different individuals of the same species, depending on relative shape and size of the fertile locule and the 2 sterile locules.

REFERENCE

Ware, Donna M. Eggers. 1983. Genetic Fruit Polymorphism in North American Valerianella (Valerianaceae) and Its Taxonomic Implications. Syst. Bot. 8: 33–44.

KEY TO THE SPECIES

1. Bracts of inflorescence at least partly ciliate; fruit bearing a corky mass of nearly equal size; corolla lobes blue . . . . .1. **V. locusta**
1. Bracts completely eciliate; fruit without such a mass; corolla lobes white
  2. Fruit 2.7–4 (4.2) mm long; stems glabrous (except sometimes at nodes) and leaves glabrous (except sometimes toward base) . . . . .2. **V. chenopodiifolia**
  2. Fruit ca. 1.5–2 mm long; stems pubescent on angles and leaf margins ciliate (sometimes glabrate with age) . . . . .3. **V. umbilicata**

1. **V. locusta** (L.) Latterell
Map 443. A Eurasian native, sometimes cultivated as a salad plant, especially in Europe. Collected several times in 1929 and 1930 (ND, MICH) along the St. Joseph River at Bertrand, presumably as an escape.

2. **V. chenopodiifolia** (Pursh) DC.   Fig. 159
Map 444. Moist ground, especially river floodplains.
Cooperrider (1995) has noted that the fruit at the base of each inflorescence ripens very early, while other flowers of the same inflores-

442. Valeriana officinalis   443. Valerianella locusta   444. Valerianella chenopodiifolia

cence are still in anthesis; in the next species, such early-ripening fruits are absent.

3. **V. umbilicata** (Sull.) A. W. Wood

Map 445. Plants from several localities in southern Monroe County are from swamps and floodplains insofar as habitat is indicated.

Michigan plants are apparently all f. *intermedia* (Dyal) D. M. Eggers, with fruit ± pointed, longer than broad. The lower leaves of *V. umbilicata* are usually more narrowed to a petiolar base than those of *V. chenopodiifolia*.

DIPSACACEAE   Teasel Family

This is strictly an Old World family. All of our species are European natives, escaped from cultivation or otherwise naturalized. The flowers are in a dense involucrate head, but can be distinguished readily from those of the Compositae by the 4 corolla lobes and 4 conspicuous, separate stamens. The achenes in our species are hairy (glabrate in *Succisella inflexa*).

REFERENCES

Ferguson, I. K. 1965. The Genera of Valerianaceae and Dipsacaceae in the Southeastern United States. Jour. Arnold Arb. 46: 218–231.

Salamun, Peter J., & Theodore S. Cochrane. 1974. Preliminary Reports on the Flora of Wisconsin No. 65. Dipsacaceae—Teasel Family. Trans. Wisconsin Acad. 62: 253–260.

KEY TO THE GENERA

1. Involucral bracts with a prolonged spine at the tip, more than 4 times as long as wide; stem prickly with broad-based thorns; flowers uniform; receptacle with a spine-tipped bract subtending and exceeding each flower . . . . .1. **Dipsacus**
1. Involucral bracts without spine at the tip, the outer ones less than 3 times as long as wide (excluding marginal bristles); stem glabrous or hairy but not prickly; flowers various; receptacle merely hairy or with short spineless bracts
   2. Leaves (at least middle ones) pinnately lobed; stems pubescent; flowers toward outside of head distinctly larger (and showier) than central flowers; calyx bearing 8 (or more) prominent but deciduous awns; receptacle hairy but without bracts; achenes hairy, obscurely 4-ribbed . . . . .2. **Knautia**
   2. Leaves unlobed (± lanceolate and nearly or quite entire); stems (not peduncles) glabrous (or finely pubescent at the nodes); flowers nearly uniform in head; calyx very short, awnless; receptacle with bracts shorter than achenes; achenes glabrate, strongly 8-ribbed . . . . .3. **Succisella**

1. **Dipsacus**   Teasel

Teasels are biennial, producing from taproots conspicuous basal leaves that overwinter their first year. The dry fruiting heads are familiar and conspicuous, often being used in winter bouquets and various floral arrangements.

Heads of *D. sativus* (L.) Honckeny, fuller's teasel [often called *D. fullonum*; see below], have long been used to raise the nap on woolen fabrics, being less damaging than metal devices (Mullins 1951; Ryder 1996). Although this species has sometimes been collected around woolen mills, it has not been reported or collected in Michigan; it differs from *D. fullonum,* as that name is here applied, in the strongly recurved spiny tips of the receptacular bracts (which are straight in both species below).

REFERENCES

Ferguson, I. K., & George K. Brizicky. 1965. Nomenclatural Notes on Dipsacus fullonum and Dipsacus sativus. Jour. Arnold Arb. 46: 362–365.
Mullins, Donald. 1951. Teasel Growing—An Ancient Practice. World Crops 3: 146–147.
Ryder, Michael L. 1996. Is the Fuller's Teasel (Dipsacus sativus) Really a Distinct Species? Linnean 11(4): 21–27. [Continued use at woolen mills in England.]
Werner, Patricia A. 1975. The Biology of Canadian Weeds 12. Dipsacus sylvestris Huds. Canad. Jour. Pl. Sci. 55: 783–794.

KEY TO THE SPECIES

1. Cauline leaves ± deeply pinnatifid..................................1. **D. laciniatus**
1. Cauline leaves entire to regularly crenate-toothed.....................2. **D. fullonum**

1. **D. laciniatus** L.   Fig. 162   Cut-leaf Teasel
   Map 446. Roadsides and railroads, ditches, river banks, clearings and sandy disturbed ground. Spreading rapidly in recent years, although known from banks of the Red Cedar River in East Lansing as early as 1894.
   A conspicuous tall plant, often 2–3 m or even more in height, with strongly ascending branches. The connate bases of the cauline leaves form a conspicuous cup ca. 1.5–7 (13) cm deep around the stem at some nodes.

2. **D. fullonum** L.   Fig. 160   Wild Teasel
   Map 447. The oldest Michigan collections I have seen are dated 1844 (presumably Macomb or Oakland Co.) and 1861 ("rare" in Washtenaw Co.). Apparently not spread much in the state until the second decade of

445. Valerianella   446. Dipsacus laciniatus   447. Dipsacus fullonum
     umbilicata

the 20th century, and still very local in northern Michigan, while it is now common southward along roadsides and railroads, in fields and open disturbed ground, along trails and floodplain meadows.

Although the leaves are normally opposite, Farwell discovered a plant with 3 leaves in each whorl at Detroit in 1918.

This is the species long called *D. sylvestris* Hudson, while *D. fullonum* was applied to the fuller's teasel, now to be called *D. sativus*.

2. Knautia

1. **K. arvensis** (L.) Coulter   Blue-buttons; "Scabiosa"
Map 448. Locally well established as a weed in old fields and roadsides in the Upper Peninsula, since at least around 1950. A specimen collected in 1923 near Manchester bears no statement as to whether it was established or in cultivation; since the species was never included on subsequent unpublished lists for Washtenaw County and it was not collected elsewhere in the county until 1992 as definitely established, the status of that older collection remains dubious.

The foliage is quite variable, ranging from leaves entire or nearly so (toward base of plant) to deeply pinnatifid.

3. Succisella

1. **S. inflexa** (Kluk) Beck
Map 449. A native of southern and eastern Europe, locally established in northeastern North America, perhaps as an escape from cultivation although not generally listed as a horticultural species. Collected as abundant in a roadside ditch in southern Kent County in 1944 (*Bazuin 6194*, ALMA).

This species was long known as *Scabiosa australis* Wulfen.

448. Knautia arvensis   449. Succisella inflexa   450. Citrullus lanatus

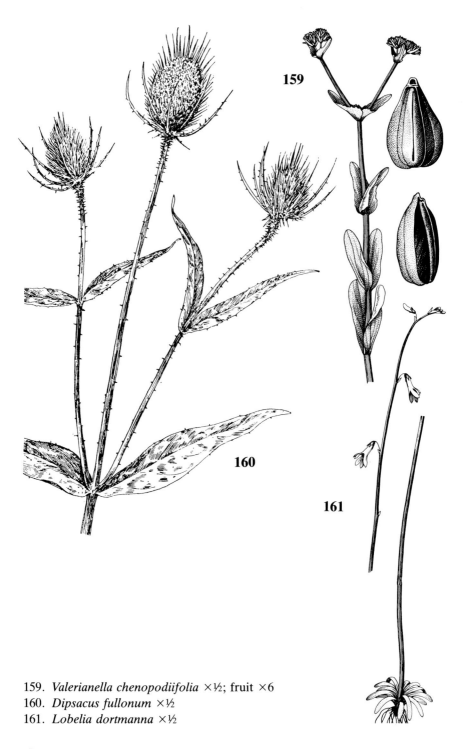

159. *Valerianella chenopodiifolia* ×½; fruit ×6
160. *Dipsacus fullonum* ×½
161. *Lobelia dortmanna* ×½

CUCURBITACEAE
Gourd Family

This family includes only two species native in Michigan. The others that have been collected or reported in the state are, at best, mere waifs. It has been suggested to me that the occurrence of cucurbits on shores and such places results not so much from picnickers at those sites as from the large seeds' surviving inadequate garbage and sewage treatment and hence washing up, still viable, far from their origins.

The flowers are unisexual, and in our species both sexes are on the same plant (monoecious). Modern systems of classification generally place the family near the Violaceae, in the subclass Dilleniidae, far from the old Englerian position where it appears (as here) by the Dipsacaceae, in the Asteridae.

The white-flowered gourd or "calabash" is *Lagenaria siceraria* (Molina) Standley and has been cultivated for ornamental and technological uses in both Old World and New for thousands of years.

Adequate herbarium specimens of garden species, with often very large leaves and bulky fruit, are sparse, for these do not inspire collectors to gather and press what is needed. Temporary waifs are doubtless much more common than the maps indicate, and species additional to those documented and included here may be found.

REFERENCES

Bates, David M., Richard W. Robinson, & Charles Jeffrey (eds.). 1990. Biology and Utilization of the Curcurbitaceae [sic]. Cornell Univ. Press, Ithaca. 485 pp.

Scholz, Hildemar. 1983. Ordnung (Reihe) Cucurbitales. Hegi Illustrierte Flora von Mitteleuropa (ed. 2). 6(2): A1–A36. [Includes both of our native species, which are locally naturalized in Europe, as well as the commonly cultivated ones.]

Whitaker, Thomas W., & Glen N. Davis. 1962. Cucurbits. Botany, Cultivation, and Utilization. Leonard Hill, London. 250 pp.

KEY TO THE GENERA

1. Leaves deeply pinnately lobed (usually at least halfway to the midrib) . . . . . . .1. **Citrullus**
1. Leaves palmately (if at all) lobed
  2. Corolla at least 5 cm long and 5 cm broad, yellow . . . . .2. **Cucurbita**
  2. Corolla less than 4 cm long and broad, yellow or greenish white
    3. Plant rarely escaped from cultivation, with large (ca. 1.5–3 cm long) yellow flowers and unbranched tendrils . . . . .3. **Cucumis**
    3. Plant a native vine, with small (less than 1 or very rarely 1.5 cm long) greenish white to cream flowers and branched tendrils
      4. Calyx and corolla 6-lobed; internodes glabrous; fruit solitary, ca. 3–5 cm long, 4-seeded, ellipsoid and broadly rounded at apex . . . . .4. **Echinocystis**
      4. Calyx and corolla 5-lobed; internodes pubescent; fruits each ca. 1.1–1.7 cm long, 1-seeded, ovoid and tapered to acute apex, but crowded into a ± globose cluster ca. 2.5–3.7 cm broad . . . . .5. **Sicyos**

1. Citrullus

1. **C. lanatus** (Thunb.) Matsum. & Nakai   Watermelon
   Map 450. A native of Africa, widely grown for the fruit, which produces more than enough seeds to account for the few collections as a waif on filled land, shores, and waste places; the Baraga County collection label states explicitly "Many escapes at L'Anse Sewage Plant" (*Bourdo* in 1960, MSC).
   Long known as *C. vulgaris* Schrader.

2. Cucurbita   Gourd
Species of this genus are originally native to the Western Hemisphere. The history of the cultivated species is complex. Some taxa have been grown for ornament (especially the hard-shelled gourds) and others for the large edible fruits (squashes and pumpkins)—part of the traditional corn–bean–squash agriculture of pre-Columbian civilizations in North and Central America.

REFERENCES

Bailey, L. H. 1937. The Garden of Gourds. Macmillan, New York. 134 pp.
Heiser, Charles B., Jr. 1979. The Gourd Book. Univ. Oklahoma Press, Norman. 284 pp.
Whitaker, Thomas W., & W. P. Bemis. 1975. Origin and Evolution of the Cultivated Cucurbita. Bull. Torrey Bot. Club 102: 362–368.

KEY TO THE SPECIES

1. Leaf blades triangular-ovate (clearly longer than broad), not deeply lobed, very stiff and rough on both sides; plant perennial from a large root, adventive in dry waste ground . . . . .1. **C. foetidissima**
1. Leaf blades ± orbicular to reniform (broader than long), unlobed to deeply palmately lobed, thin and flexible; plant an annual, rarely escaped from cultivation
   2. Fruiting peduncle ± terete, spongy, not expanded; leaves not lobed . . . . .2. **C. maxima**
   2. Fruiting peduncle ± angled, hard, expanded at junction with fruit; leaves usually lobed . . . . .3. **C. pepo**

1. **C. foetidissima** Kunth   Missouri Gourd
   Map 451. Native in the Southwest and adventive northeastward. Collected in 1973 in waste ground between railroad tracks near Niles (*Schulenberg & Kohout 73–432*, MOR).
   An ill-smelling plant with striped, gourd-like fruit.

2. **C. maxima** Lam.   Winter Squash; Turban Gourd; Pumpkin
   Map 452. Sandy disturbed ground at a dump near Mullett Lake (*Voss 15515* in 1982, MICH, UMBS).
   Buttercup and Hubbard squash are among those belonging to this species.

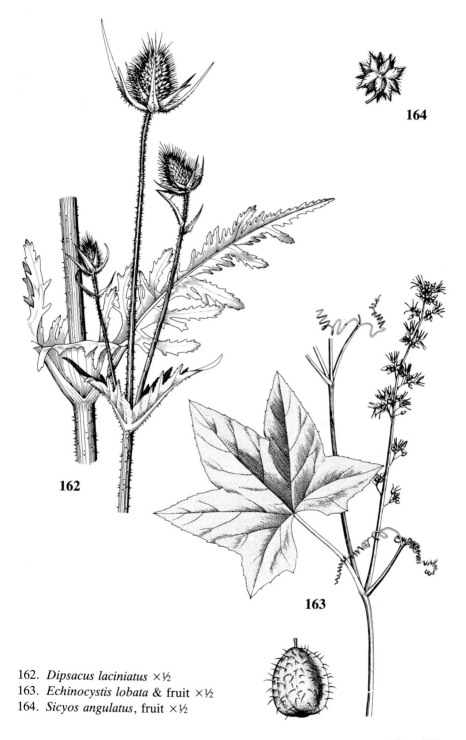

164

162

163

162. *Dipsacus laciniatus* ×½
163. *Echinocystis lobata* & fruit ×½
164. *Sicyos angulatus*, fruit ×½

3. **C. pepo** L. Summer Squash; Ornamental Gourd; Pumpkin
Map 453. Occasional around refuse piles and near cultivated ground.
Zucchini and acorn squashes are among the edible fruits of this species, of which other variants include bicolored and other diverse ornamental gourds. Cultivated pumpkins are mostly this species, but some are *C. maxima* and others are *C. mixta* Pangalo and *C. moschata* Poiret, which may be the ancestral species in the genus.

3. Cucumis

KEY TO THE SPECIES

1. Apex of leaf sharply acute or short-acuminate; fruit ± cylindrical, warty or tuberculate but glabrous .... 1. **C. sativus**
1. Apex of leaf obtuse to broadly rounded; fruit ± globose to ellipsoid, smooth to
reticulate, ± pubescent (at least when young) .... 2. **C. melo**

1. **C. sativus** L. Cucumber
Map 454. Probably originated in southern Asia, now well known in cultivation throughout the world but only rarely escaped. Collected in 1925 as seedlings on the sandy shore of Douglas Lake (*Gates 14196*, UMBS).
Michigan is the nation's top producer of cucumbers for pickling and ranks fourth in production of fresh cucumbers.

2. **C. melo** L. Melon
Map 455. Probably originated in Africa or southern Asia. Occasionally found in waste ground.
From this species have come the muskmelon, cantaloupe (in this country, a name largely applied to the muskmelon), and honeydew—among other melons.

4. Echinocystis

1. **E. lobata** (Michaux) T. & G. Fig. 163 Wild-cucumber
Map 456. Swamps (deciduous), floodplains, river banks, streamsides, marshy ground, thickets, borders of woods, fencerows, roadsides, railroad embankments, gravel pits, waste ground such as alleys and refuse areas. The species is sometimes cultivated as an annual vine, and it is often hard to tell whether a given stand is natural or the result of garden escape.
The thoroughly distinctive fruit is greatly inflated and watery, opening irregularly at the apex to release the seeds. Squeezing the fruit just before it is mature will eject the seeds. The contents are otherwise so unsubstantial (mostly air) that one is not tempted to use the fruit like a true cucumber, which it loosely resembles.

The calyx is very inconspicuous in this species and the next, and both have soft spines on the fruit. The corolla lobes tend to be more narrowly linear-lanceolate in *Echinocystis* than in *Sicyos*.

5. **Sicyos**

1. **S. angulatus** L.   Fig. 164   Bur-cucumber
   Map 457. In similar habitats to the preceding species and sometimes growing with it. Swamp forests, floodplains, river banks; wet thickets, edges of marshy shores; fencerows, alleys and other waste places.
   The fruits are dry and indehiscent—less of a curiosity than those of *Echinocystis*.

CAMPANULACEAE   Bellflower Family

This family includes two subfamilies in our area, so strikingly different from each other in floral symmetry that they have sometimes been recognized as two distinct families. However, both generally have milky sap, simple alternate leaves, an inferior ovary of 2 or 3 carpels (the flowers otherwise 5-merous), and stamens slightly if at all fused to the corolla but

451. Cucurbita
     foetidissima

452. Cucurbita maxima

453. Cucurbita pepo

454. Cucumis sativus

455. Cucumis melo

456. Echinocystis lobata

with anthers forming a tube into which the pollen is shed and through which it is then pushed by elongation of the style.

REFERENCE

Rosatti, Thomas J. 1986. The Genera of Sphenocleaceae and Campanulaceae in the Southeastern United States. Jour. Arnold Arb. 67: 1–64.

KEY TO THE GENERA

1. Corolla strongly bilaterally symmetrical, 2-lipped; carpels 2; anthers fused, not separating after anthesis; capsules opening by apical valves or irregularly .....1. **Lobelia**
1. Corolla radially symmetrical; carpels 3; anthers coherent, separating after anthesis; capsules opening by lateral pores
   2. Corolla bell-shaped (campanulate), with lobes shorter than the tube (except in the lax, scabrous, small-flowered *C. aparinoides*); flowers stalked (except in *C. glomerata*); plants perennial (biennial in *C. medium*) ....2. **Campanula**
   2. Corolla nearly flat (rotate) or broadly funnel-shaped, with lobes longer than the tube; flowers sessile or subsessile; plants annuals, winter-annuals, or biennials
      3. Flowers subtended by leaves or bracts linear or tapered to a petiolar base (not at all cordate); style at maturity declined basally and then curved up apically .....
      ....2. **Campanula (americana)**
      3. Flowers subtended by ± cordate-clasping sessile bract-like leaves; style straight...3. **Triodanis**

1. Lobelia   Lobelia

Several taxa are cultivated, and our wild species are at least as attractive as many of the exotic ones. The lower lip, 3-lobed, is anatomically the upper one, as the flower twists on its pedicel 180° in development (just as in most orchids). Unlike many plants with milky juice, lobelias last very well when cut and placed in water. However, they are ± poisonous if taken internally, and some species, e.g., *L. inflata*, were used medicinally in the past.

REFERENCES

Bowden, Wray M. 1982. The Taxonomy of Lobelia ×speciosa s.l. and Its Parental Species, L. siphilitica and L. cardinalis s.l. (Lobeliaceae). Canad. Jour. Bot. 60: 2054–2070.
McVaugh, Rogers. 1943. Campanulaceae (Lobelioideae). N. Am. Fl. 32A(1): 1–134.

KEY TO THE SPECIES

1. Principal cauline leaves (if any) linear to linear-lanceolate (or linear-oblanceolate), entire to remotely denticulate, the widest less than 0.4 (0.7) cm broad
   2. Leaves all basal (sometimes a few tiny bracts on stem), usually submersed, somewhat fleshy (consisting of 2 hollow tubes) ....1. **L. dortmanna**
   2. Leaves all or mostly cauline, thin (not tubular in construction)
      3. Calyx lobes at least twice as long as the calyx tube and inferior part of the ovary at anthesis; plant a rare escape from cultivation, diffuse or partly trailing; pedicels without bracteoles, much exceeding their subtending bracts ....2. **L. erinus**

3. Calyx lobes mostly less than twice as long as the calyx tube and inferior part of the ovary; plant a common native calciphile, ± stiffly erect; pedicels each with a pair of tiny bracteoles near the middle, often scarcely if at all exceeding their subtending bracts . . . . .3. **L. kalmii**
1. Principal cauline leaves elliptic to oblanceolate or obovate, ± toothed or crenulate (or at least denticulate), (0.7) 1–4.5 (6) cm broad
  4. Mature corolla 1.8–4 (4.3) cm long, with an open slit on each side near the base of the tube (black in fig. 169)
    5. Corolla bright red (except in albinos), at least 3 cm long; calyx lobes glabrous, not auriculate . . . . .4. **L. cardinalis**
    5. Corolla bright blue (except in albinos), less than 2.4 (2.6) cm long; calyx lobes bristly-ciliate, auriculate at base . . . . .5. **L. siphilitica**
  4. Mature corolla less than 1.3 cm long, without lateral slits
    6. Stem ± hairy throughout, with many of the hairs 0.5–1 mm long; calyx lobes usually less than twice as long as inferior ovary at anthesis; capsule wholly inferior, much inflated at maturity . . . . .6. **L. inflata**
    6. Stem mostly glabrous or glabrate, usually becoming minutely but densely pubescent toward the base with hairs not over 0.5 mm; calyx lobes at least twice as long as inferior part of the ovary (less as ovary expands); capsule only about half to three-quarters inferior, not inflated . . . . .7. **L. spicata**

1. **L. dortmanna** L.   Fig. 161   Water Lobelia
Map 458. A circumpolar aquatic species characteristic of northern softwater lakes with low pH.

The leaves are almost always submersed or on recently exposed wet sandy shores, and the flowers may bloom and set fruit under water as well as above it. Even a small fragment of leaf is identifiable in the field, as it consists of little more than 2 hollow tubes side by side, and these have milky juice. The dense rosettes of slightly flattened, parallel-sided, slightly curved, obtuse to round-tipped leaves are very distinctive from a distance, often growing with *Littorella, Eriocaulon, Isoëtes*, and other rosette plants. *Elatine* and *Myriophyllum tenellum* are other interesting and characteristic associates. (These are all distinguished vegetatively in Key A of the General Keys.)

2. **L. erinus** L.   Edging Lobelia
Map 459. A native of South Africa, popular in gardens, especially as a border plant, and occasionally escaping to lawns and waste places. Collected by Farwell at Lake Linden in 1914 (*3918*, BLH), with no data on habitat.

Fragmentary specimens may resemble the next species, but *L. erinus* has deep blue or violet corollas, slender even setaceous calyx lobes, and long pedicels without bracteoles; a pair of minute bracteoles is on each pedicel in *L. kalmii*.

3. **L. kalmii** L.   Fig. 165   Kalm's or Brook Lobelia
Map 460. A calciphile of moist, sandy, gravelly, or marly shores, ditches, meadows, marshes, interdunal hollows; clay banks, fens, cedar and tama-

rack swamps; rock crevices, ledges, and pools. Often locally abundant, forming carpets of colorful blue.

The corolla is blue with a white "eye," sometimes quite pale and occasionally all white [f. *leucantha* Rouleau]. Depauperate plants of *L. inflata* or *L. spicata* may sometimes be confused with this species or the preceding; for distinctions, see the individual characters in the key and notes in the text below.

4. **L. cardinalis** L.   Plate 6-A; fig. 169   Red Lobelia; Cardinal-flower

Map 461. Swamps and floodplain forests; thickets and open ground along rivers and streams; marshes, wet shores, ditches, and swales.

The rich red flowers with their long corolla tube are pollinated normally by hummingbirds, and *L. siphilitica*, by bees. However, bees may visit the flowers of *L. cardinalis*, gathering nectar through the slits on the sides of the corolla. Hybrids [*L.* ×*speciosa* Sweet] can be produced artificially, and rarely occur in nature (none reported from Michigan), suggesting that the two species are more closely related than first appears. Some very showy hybrids, including triploids and tetraploids, are in cultivation.

White-flowered plants occur occasionally throughout the range and are striking in appearance [f. *alba* (McNab) H. St. John]. Plants with the corollas an intermediate pink may be found with the red and white forms.

457. Sicyos angulatus

458. Lobelia dortmanna

459. Lobelia erinus

460. Lobelia kalmii

461. Lobelia cardinalis

462. Lobelia siphilitica

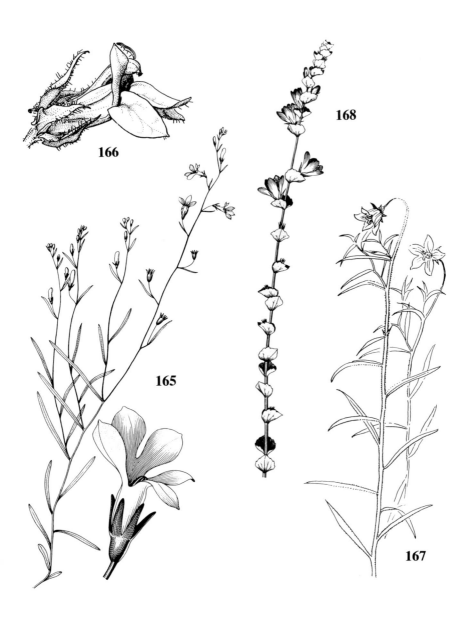

165. *Lobelia kalmii* ×½; flower ×3
166. *L. siphilitica*, flower ×2
167. *Campanula aparinoides* ×½
168. *Triodanis perfoliata* ×½

5. **L. siphilitica** L.   Fig. 166   Great Blue Lobelia

Map 462. Swamps (hardwood, cedar, tamarack), floodplains, river banks; fens, wet meadows and thickets, shores, ditches.

White-flowered forms [f. *albiflora* Britton] seem more rare than in the preceding species.

6. **L. inflata** L.   Indian-tobacco

Map 463. Usually in moist (sometimes dry), often disturbed, ground: roadsides, ditches, borrow pits; trails, openings, and utility line clearings in deciduous woods; fields and meadows; river banks, swamp borders.

Even young depauperate plants with leaves as narrow as in *L. kalmii* and *L. erinus* can be distinguished by the long, irregular hairs at the base of the stem. Furthermore, in *L. inflata* and *L. spicata* the minute bracteoles are at the base of the pedicel; in *L. kalmii* they are near the middle, and in *L. erinus* they are absent.

7. **L. spicata** Lam.   Pale Spiked Lobelia

Map 464. Low ground, often at least seasonally wet, including meadows, fields, shores, calcareous flats and rocky openings, edges of swamps and marshes, oak woodland, prairies. The only collections purporting to have come from Keweenaw County were made by Farwell in the 1880s.

Depauperate plants might be confused with *L. kalmii*, especially since the two may grow together, but the latter has relatively shorter calyx lobes, bracteoles near the middle of the pedicel, and linear glabrous cauline leaves. In *L. spicata*, even small leaves are elliptic and finely pubescent.

The corolla is often rather pale blue, sometimes nearly or quite white. The inflorescence is unbranched, a slender spike-like raceme. *L. inflata*, except in small plants, is branched. Both this species and *L. inflata* vary greatly in size and other characters.

2. **Campanula**   Bellflower

A number of species are cultivated, and some additional ones may turn up as waifs in Michigan besides those included below.

463. Lobelia inflata   464. Lobelia spicata   465. Campanula
                                                       americana

REFERENCES

Bailey, L. H. 1953. The Garden of Bellflowers in North America. Macmillan, New York 155 pp. [Keys and describes 137 cultivated species, including all those treated below.]

Shetler, Stanwyn G. 1962. Notes on the Life History of Campanula americana, the Tall Bellflower. Michigan Bot. 1: 9–14.

Shetler, Stanwyn G. 1963. A Checklist and Key to the Species of Campanula Native or Commonly Naturalized in North America. Rhodora 65: 319–337.

Shetler, Stanwyn G. 1982. Variation and Evolution of the Nearctic Harebells. Phan. Monogr. 11. 516 pp.

KEY TO THE SPECIES

1. Corolla ± rotate, the lobes much longer than the tube; flowers sessile or nearly so (sometimes on short axillary branches) . . . . .1. **C. americana**
1. Corolla campanulate, the lobes shorter than the tube (except in the small-flowered, lax, scabrous *C. aparinoiodes*); flowers stalked
  2. Corolla ca. [2.5] 3.5–5.5 cm long; calyx with large reflexed appendages at the base of the lobes; stigmas 5 . . . . .2. **C. medium**
  2. Corolla less than 3 (3.5) cm long; calyx without appendages; stigmas 3
    3. Plant lax, the weak 3-angled stem and leaves retrorsely scabrous, clinging to other plants; corolla (4) 5.5–10 (11) mm long, the lobes longer than the tube; leaves all less than 7 (very rarely 8) mm wide; plant of damp or wet habitat . . .3. **C. aparinoides**
    3. Plant (at least flowering stems) ± erect, the stem and leaves smooth (or if scabrous, not retrorsely so); corolla (9) 12–30 (35) mm long, the lobes shorter than the tube; leaves various; plant usually of dry or rocky habitat
      4. Ovary glabrous (very rarely puberulent); cauline leaves obscurely if at all petiolate, linear to narrowly lanceolate and entire to weakly denticulate at middle of stem, ranging to lanceolate, oblanceolate, or rotund and coarsely toothed at the base of the stem
        5. Capsules opening by pores near the summit; corollas [18] 22-30 (35) mm long; middle and upper cauline leaves sparsely denticulate; basal leaves oblanceolate; plant rarely escaped from cultivation .4. **C. persicifolia**
        5. Capsules opening by pores near the base; corollas (9) 12–22 (24) mm long; middle and upper cauline leaves entire; basal and juvenile leaves (often absent) suborbicular, ovate, or even deltoid; plant a common native (though also cultivated) . . . . .5. **C. rotundifolia**
      4. Ovary spreading-pubescent or retrorse-hispidulous (rarely glabrous); cauline leaves distinctly petioled toward lower part of stem, regularly to irregularly crenate to sharply toothed, lanceolate to ovate or deltoid at middle of stem, ± cordate
        6. Flowers all or mostly crowded in a terminal cluster . . . . .6. **C. glomerata**
        6. Flowers solitary or in an elongate or open inflorescence
          7. Plant completely glabrous; corolla spreading or flaring . . . . .7. **C. carpatica**
          7. Plant pubescent, at least on ovaries and/or petioles; corolla tubular-campanulate (sides ± parallel)
            8. Flowers nodding; calyx lobes usually spreading-reflexed at maturity; ovary retrorse-hispidulous (hairs less than 0.3 mm long), rarely glabrous . . . .
. . . . .8. **C. rapunculoides**
            8. Flowers erect or ascending; calyx lobes erect into maturity; ovary bristly with ± spreading hairs more than 0.5 mm long . . . . .9. **C. trachelium**

1. **C. americana** L.　　Fig. 170　　　　　Tall or American Bellflower

Map 465. Deciduous woods, both upland and floodplain, especially in openings and ± disturbed areas such as trails, edges of fields and railroads; marshy ground, stream banks.

This distinctive species, an annual, has been segregated by a few authors as *Campanulastrum americanum* (L.) Small.

2. **C. medium** L.　　　　　　　　　　　　　　　　Canterbury Bells

Map 466. A native of Europe, popular in cultivation and rarely escaped as a waif along roadsides.

The corolla varies from white to blue or pink.

3. **C. aparinoides** Pursh　　Fig. 167　　　　　　Marsh Bellflower

Map 467. Usually a calciphile and often associated with grasses and sedges including hummocks in wet habitats; marshes (if not too wet), ditches, swales; sedge and other fens, cedar and tamarack swamps; meadows, shores, stream and pond borders.

The corolla is pale blue, often nearly white. Smaller-flowered plants with narrowly elliptical leaves are typical var. *aparinoides*, a southern taxon barely reaching southwestern Michigan (Berrien, Kalamazoo, and Kent cos.). Our plants are mostly var. *grandiflora* Holz., long known as *C. uliginosa* Rydb., with flowers running a bit larger on average and the leaves more linear. The distinctions are not sharp and most authors no longer recognize the two as separate species.

4. **C. persicifolia** L.　　　　　　　　　　　　　Willow Bellflower

Map 468. A Eurasian species, popular in cultivation but only occasional outside the garden, spread to roadsides and dooryards.

The corollas are blue, pink, or white.

5. **C. rotundifolia** L.　　Fig. 171　　　　　　Bluebell; Harebell

Map 469. Sandy shores and dunes (old and young); dry, usually sandy, woodland and openings with oak, hickory, sassafras, and/or jack pine; limestone crevices and gravels around the Niagara Escarpment, and on other sedimentary and granitic rock outcrops and crevices elsewhere; prairies, dry meadows, and river banks, thickets; occasionally along roadsides.

A highly variable circumpolar complex, exhaustingly treated for North America by Shetler (1982), who did not formally recognize named infraspecific taxa. While the corolla is usually blue, white-flowered plants [f. *albiflora* E. L. Rand & Redfield in North America] are occasionally found, together with intermediate shades. One collection from Presque Isle County has all of the floral parts doubled. Plants with the vegetative parts and ovaries ± densely puberulent were described from Michigan (TL: Indian River, Cheboygan Co.) in 1878 as var. *canescens* E. J. Hill.

6. **C. glomerata** L. Clustered Bellflower

Map 470. A species of Eurasia, cultivated and rarely escaped. Collected by Farwell in 1934 (*9758*, BLH) at Lake Linden.

A distinctive-appearing bellflower, with short-pubescent foliage, the lower leaves petioled.

7. **C. carpatica** Jacq. Tussock Bellflower

Map 471. A central European species, widely cultivated. Collected by Farwell in 1921 as escaped at Eagle River (BLH; Pap. Michigan Acad. 2: 41. 1923).

The leaves (at least the lower ones) have distinct petioles as in the next two species, but the ovaries are glabrous and the corolla distinctly open. The capsule opens by subapical pores, unlike the sometimes nearly glabrous *C. rapunculoides*, and the pedicels are very long.

8. **C. rapunculoides** L. Roving or Creeping Bellflower

Map 472. Another Eurasian native, rather aggressively spreading underground to form persistent weedy colonies along roadsides and railroads, in dooryards and thickets, and in waste places such as dumping areas.

466. Campanula medium

467. Campanula aparinoides

468. Campanula persicifolia

469. Campanula rotundifolia

470. Campanula glomerata

471. Campanula carpatica

The blue flowers are in a ± one-sided raceme and the reflexed calyx lobes are distinctive; the capsules open by pores near the base, as in the next species, but unlike *C. persicifolia.*

9. **C. trachelium** L.   Nettle-leaved Bellflower
   Map 473. Yet another Eurasian species, rarely escaped from cultivation, as locally on Mackinac Island (*Overlease* in 1994, MICH).

3. **Triodanis**
   This genus has long been included by many authors in *Specularia.* Cleistogamous flowers occur at least at the lower nodes and often have only 3 or 4 short calyx lobes, whereas on the showy deep blue flowers, both calyx and corolla have 5 long lobes. The capsules open by neat little elliptical pores, one on the side of each locule.

REFERENCE

McVaugh, Rogers. 1945. The Genus Triodanis Rafinesque and Its Relationships to Specularia and Campanula. Wrightia 1: 13–52.

1. **T. perfoliata** (L.) Nieuwl.   Fig. 168   Venus' Looking-glass
   Map 474. Sandy, barren, open ground; dry fields and bluffs; sometimes in moister sites; woodland with oak, pine, sassafras, especially at clearings, trails, roads; occasionally a garden weed.
   Rarely the stems are lax and trailing rather than erect. The leaves are short-ovate or even nearly reniform, with a few coarse teeth, and are very different from the ovate-elliptic ± acuminate leaves of *Campanula americana,* which also has ± rotate corollas but grows in rich habitats and has larger flowers (corolla lobes ca. (7) 8–13 mm long, compared to 6–8 mm in *Triodanis*)

472. Campanula   473. Campanula   474. Triodanis perfoliata
     rapunculoides   trachelium

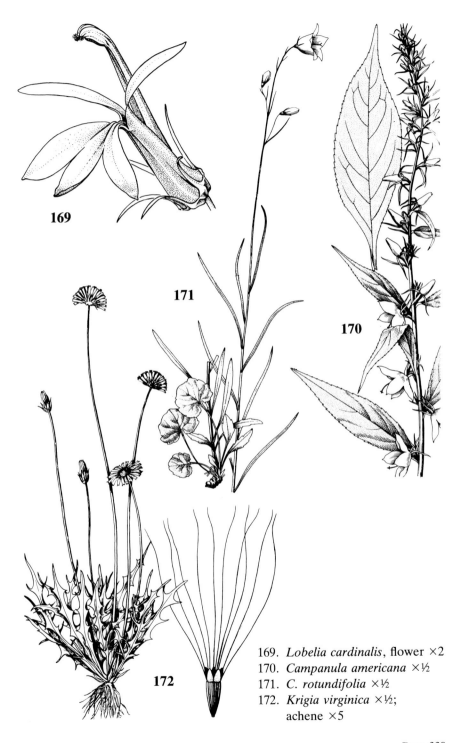

169. *Lobelia cardinalis*, flower ×2
170. *Campanula americana* ×½
171. *C. rotundifolia* ×½
172. *Krigia virginica* ×½;
     achene ×5

COMPOSITAE (ASTERACEAE)   Aster or Daisy Family

The composites are often considered to be the largest or second largest family of flowering plants in the world, with perhaps 20,000 or more species. Many are known as ornamentals. Only a few provide human food, e.g., in this part of the world, lettuce, artichokes, and sunflower (for oil). Low levels of toxins or at least bitter substances occur in many of these plants. Successfully obnoxious weeds abound. As Mabberley aptly concluded his summary of this family in *The Plant Book*, "With increasing clearance of native vegetation throughout the world these aggressive toxic plants will inherit it."

The family is easily recognized by its involucrate heads, and seldom will one be in doubt as to whether any plant belongs to it. Close beneath the clusters of flowers (*florets*) comprising a head are few to many bracts making up an *involucre* (which is to the head much as the calyx is to a single flower in other families). The individual bracts of the involucre are called *phyllaries* (or in some works, *tegules*). Radially symmetrical flowers, as massed in the center of a daisy or sunflower or as seen with long slender corolla lobes in a thistle, are *disk flowers* and make up the *disk*. Surrounding them in some species are petal-like flowers, bilaterally symmetrical, called *ray flowers*, and heads with both kinds of flowers can be called *radiate*; those lacking rays are *discoid*. Ray flowers are sometimes also called *ligulate* — especially in the tribe in which all heads consist entirely of such ligulate flowers. The central flowers in such heads, as the last to open, may appear radially symmetrical while still in bud, as in the common dandelion, but should not be mistakenly interpreted as disk flowers. In some species, the ray flowers are sterile (lacking a functional pistil); in others, they are pistillate. Disk flowers are usually perfect, but there are exceptions, especially in dioecious species (where, e.g., functionally staminate flowers may have only rudimentary styles).

The term "ray" generally refers to the *expanded*, petal-like part of a ray flower, and is to be so interpreted when ray measurements and colors are stated in the keys and text. However, since the number of rays necessarily equals the number of ray flowers, the term is also often used when a count of ray flowers is intended. "Double" forms in some species, with only rudimentary reproductive parts and all flowers converted to rays, such as the goldenglow of the garden, may be found occasionally as escapes from cultivation and may be falsely thought to belong in "Group A" — in which, however, the ligulate flowers are all perfect.

Measurements of the length of the involucre are from its base to the tip of the longest phyllaries. As in other circumstances, measurements should be made on mature structures.

The ovary of each individual flower is inferior. Surrounding the base of the corolla, or on the summit of the achene, may be a ring of tissue, a series of bristles, two or more scales or awns, or some combination of such structures. These are the *pappus*, of which the ancestral structure was apparently a calyx. Pappus characters are very helpful in classification and identification. Often the pappus is an important seed-dispersal device. On the *receptacle*, to which the flowers of a head are attached, there may be *chaff*: scales or bristles, short or long, also useful in classification and identification. Beginners must take care to remember that *chaff* is on the receptacle and the *pappus* is at the summit of the ovary (and of the achene into which the ovary ripens). The stamens are normally united in a ring around the style (in *Iva* and *Ambrosia*, of our flora, the anthers are separate). Beginners should also be warned not to confuse the ring of anthers inside or protruding from a disk corolla with the corolla itself.

The individual fruits are here called "achenes" although some authors prefer the specialized term *cypsela* for the peculiar fruit of the Compositae, derived from a 2-carpellate, inferior ovary, reserving "achene" for a fruit of similar texture but derived from a single superior carpel (as in many Ranunculaceae and Rosaceae, for example).

Just as spikelets in the grass family are usually more than single flowers, so also are heads in the Compositae. Nevertheless, it is often convenient, as in the grasses, to use the word *pedicel* to refer to the stalk of a single head (the florets of which are never stalked) and to use the regular terms that describe inflorescences when describing the arrangement of heads: e.g., spicate, racemose, corymbose (although variant terms like *corymbiform* or *paniculiform* can be used). When both "pedicel" and "peduncle" are used, the latter refers to the stalk of a group of heads. The term *scape* (or sometimes *peduncle*) is used to refer to a leafless stalk (above the basal leaves) bearing one or more heads at the summit.

As might be expected in so large and diverse a family, it is divided into a number of tribes, most but by no means all of which are rather easily recognized even though the characters that define them may include some pretty obscure anatomical, chromosomal, molecular, or chemical ones. (For an example of modern techniques for interpreting relationships, see Jansen, Michaels, & Palmer 1991; and also Bremer 1994.) It is not easy to prepare a practical identification key to the tribes, and I do not do so here. A good summary of traditional tribes, with details well illustrated, is in Solbrig (1963). For a thorough, fairly recent account of the family and its tribes worldwide, including many problem issues, one may consult the two volumes edited by Heywood, Harborne, and Turner (1977); for a detailed update on the tribe Anthemideae, the major paper by Bremer and Humphries (1993—appearing just after this text was drafted); and for what amounts to a major "progress report" detailing many problems yet remaining toward a full new classification, Bremer (1994).

The sequence of genera in the present work is, as always, the same as that in which the genera appear in the key. While not fully based on tribal classification, it is at least more likely than an alphabetical sequence to place similar and even related genera near one another. For example, "Group A" (genera 1–16) is identical with the tribe Lactuceae. Genera 24–29 include all our Anthemideae. The Cardueae are genera 67–72 plus 23 (which is a highly unusual one). A full synopsis of local tribes is in the Gleason and Cronquist *Manual* and other works.

There are many ways in which the family can be divided into manageable-sized groups easier for identification purposes than are the tribes. I use three main groups here (with acknowledgment to L. H. Shinners). It is no easier to write (or use) a key to 80+ genera and 300 species than it is to learn the students in a class of large size. The process takes time, patience, and work—but the result is gratifying.

I had hoped to prevail on Art Cronquist for assistance with the Compositae, but his untimely death early in 1992 prevented such personal consultation. However, his 1991 thorough revision of the Gleason and Cronquist *Manual* presented his views as a lifelong specialist in this family. In cases of doubt or lack of strong conviction to the contrary, I have accepted his circumscriptions of species and genera, as a convenience to users and a gesture of respect for what is, in effect, his posthumous advice. I must also acknowledge a great debt to the late Lloyd H. Shinners, who started to set me straight on composites in 1952 at the University of Michigan Biological Station and whose draft key to genera of that region has been most helpful over many years.

REFERENCES

These references deal in some way with the whole family or at least with a number of genera.

Beals, Edward W., & Ralph F. Peters. 1967 ["1966"]. Preliminary Reports on the Flora of Wisconsin No. 56. Compositae V—Composite Family V Tribe Inuleae. Trans. Wisconsin Acad. 55: 223–241.

Boivin, Bernard. 1972. Flora of the Prairie Provinces Part III (continued) 113. Compositae. Phytologia 23: 85–216 (also in Provancheria 4).

Bremer, Kåre, & Christopher John Humphries. 1993. Generic Monograph of the Asteraceae–Anthemideae. Bull. Nat. Hist. Mus. Bot. 23: 71–177.

Bremer, Kåre. 1994. Asteraceae: Cladistics & Classification. Timber Press, Portland, Oregon. 752 pp.

Fisher, T. Richard. 1989 ["1988"]. The Dicotyledoneae of Ohio Part 3. Asteraceae. Ohio State Univ. Press, Columbus. 280 pp.

Heywood, V. H., J. B. Harborne, & B. L. Turner. 1977. The Biology and Chemistry of the Compositae. Academic Press, London. 2 vol.

Jansen, Robert K., Helen J. Michaels, & Jeffrey D. Palmer. 1991. Phylogeny and Character Evolution in the Asteraceae Based on Chloroplast DNA Restriction Site Mapping. Syst. Bot. 16: 98–115.

Johnson, Miles F., & Hugh H. Iltis. 1964 ["1963"]. Preliminary Reports on the Flora of Wisconsin No. 48. Compositae I—Composite Family I (Tribes Eupatorieae, Vernonieae, Cynarieae, and Cichorieae). Trans. Wisconsin Acad. 52: 255–342.

Mickelson, Carol J., & Hugh H. Iltis. 1967 ["1966"]. Preliminary Reports on the Flora of Wisconsin No. 55 Compositae IV—Composite Family IV (Tribes Helenieae and Anthemideae). Trans. Wisconsin Acad. 55: 187–222.

Moore, R. J., & C. Frankton. 1974. The Thistles of Canada. Canada Dep. Agr. Monogr. 10. 112 pp.

Scott, Randall W. 1990. The Genera of Cardueae (Compositae; Asteraceae) in the Southeastern United States. Jour. Arnold Arb. 71: 391–451.

Solbrig, Otto T. 1963. The Tribes of Compositae in the Southeastern United States. Jour. Arnold Arb. 44: 436–461.

Soreng, Robert J., & Edward A. Cope. 1991. On the Taxonomy of Cultivated Species of the Chrysanthemum Genus-Complex (Anthemideae; Compositae). Baileya 23: 145–165.

Vuilleumier, Beryl Simpson. 1969. The Genera of Senecioneae in the Southeastern United States. Jour. Arnold Arb. 50: 104–123.

Vuilleumier, Beryl Simpson. 1973. The Genera of Lactuceae (Compositae) in the Southeastern United States. Jour. Arnold Arb. 54: 42–93.

KEY TO THE GROUPS

1. Heads entirely of flat ligulate flowers (these all perfect); sap milky; leaves never opposite (alternate or basal) ....**GROUP A** (p. 343)
1. Heads all or partly of regular disk flowers (sometimes thread-like, sometimes very deeply lobed); ligulate (ray) flowers sterile or pistillate and only around the margin of the head; sap watery; leaves various (opposite in some species)
   2. Pappus none or entirely of scales, teeth, flattened spines, or few (up to 8) stiff awns ....**GROUP B** (p. 376)
   2. Pappus all or primarily of numerous long soft hairs or bristles (or these rising from scale-like bases in *Dyssodia*, or 2 long bristles and numerous tiny ones in *Boltonia*) ....**GROUP C** (p. 442)

KEY TO THE GENERA (Group A)

1. Pappus none or a crown or entirely of scales (without separate bristles)
   2. Leaves all basal
      3. Pappus none; scapes gradually enlarged from base to summit; involucres, scapes, and leaves without distinctly longer hairs...........................1. **Arnoseris**
      3. Pappus present (only as scale-like crown on outermost achenes; of long hairs on inner achenes); scapes at most slightly enlarged beneath the head; involucres, scapes, and especially leaves with distinctly longer hairs minutely forked at the tip ....6. **Leontodon (taraxacoides)**
   2. Leaves all or partly cauline
      4. Flowers blue (rarely white or pink), in heads 3–4.5 cm broad; pappus of numerous tiny scales; cauline leaves sessile; phyllaries (at least the outer ones at maturity) nearly always with few to many gland-tipped hairs ....2. **Cichorium**
      4. Flowers yellow, in heads ca. 1 cm or less broad; pappus none; cauline leaves (below inflorescence) tapered to a petiolar base; phyllaries glabrous or nearly so ....3. **Lapsana**
1. Pappus at least partly of bristles or soft hairs
   5. Pappus composed of both scales and bristles (scales may be short & inconspicuous); involucre a single series of phyllaries ....4. **Krigia**
   5. Pappus composed entirely of hairs or bristles; involucre various
      6. Pappus at least partly of plumose bristles
         7. Plants with leaves in a basal rosette, the stems (simple or branched) at most with very reduced bracts

8. Receptacle with a ± membranous scale subtending each flower and achene and bearing an elongate bristle-like tip (evident even among the flowers, and equalling the pappus); achene with a slender beak longer than the body; involucre 11–16 mm long (or even 25 mm in fruit), the phyllaries glabrous or (usually) with few to many bristles on the midrib only . . . . .5. **Hypochaeris**
8. Receptacle without scales or chaff; achene beakless; involucre ca. 7–10 (14) mm long, even in fruit, the phyllaries glabrous or ± hairy all across . . .6. **Leontodon**
7. Plants with leaves of fertile stems all or partly cauline
  9. Leaves with prominent midrib and lateral veins; stems, leaves, and involucres with scattered stiff hairs; achenes beakless, the pappus readily deciduous; phyllaries imbricate . . . . .7. **Picris**
  9. Leaves parallel-veined (even the midrib scarcely if at all more prominent than other veins); stems, leaves, and involucres glabrous (or when young with a little loose tomentum); achenes very long-beaked, the pappus persistent with intertangled plumose bristles, the whole head globose at maturity; phyllaries in a single series . . . . .8. **Tragopogon**
6. Pappus entirely of simple bristles or hairs (at most barbed or scabrous)
  10. Heads solitary on completely naked scapes; leaves all in a basal rosette; scapes glabrous or with a little loose tomentum
    11. Phyllaries in 1 series or irregularly imbricate; achenes antrorsely hispidulous, with beak less than half as long as the body; leaves entire (or with an obscure occasional denticulation) . . . . .9. **Agoseris**
    11. Phyllaries in two definite series, the outer ones strongly spreading or usually recurved; achenes tuberculate or spiny toward summit of body, with elongate slender beak; leaves ± toothed or pinnatifid . . . . .10. **Taraxacum**
  10. Heads (1) 2–several on naked to leafy stems; leaves basal or cauline; scapes on 1-headed plants rather densely pubescent with stellate and/or long straight hairs
    12. Leaves all or mostly in a basal rosette at flowering time (or on stoloniferous basal shoots), entire and unlobed . . . . .11. **Hieracium**
    12. Leaves all or mostly cauline at flowering time, entire or toothed or lobed—or if largely basal, then toothed
      13. Involucres cup-shaped or bell-shaped, at least two-thirds as broad as long at anthesis; achenes scarcely if at all beaked
        14. Cauline leaves tapered to base or sessile but not auriculate-clasping; pappus light to darkish brown . . . . .11. **Hieracium**
        14. Cauline leaves (at least the upper ones) ± auriculate or clasping at the base; pappus pure white, copious
          15. Leaf margin not spiny; involucre 5.5–9.5 (10) mm long; achenes ± terete (but ribbed) . . . . .12. **Crepis**
          15. Leaf margin prickly with prolonged ± spiny-tipped teeth; involucre ca. 8–18 mm long at maturity; achenes at least somewhat flattened . . . . . . .13. **Sonchus**
      13. Involucres (at least at anthesis) cylindrical to urn-shaped, at least twice as long as broad; achenes long-beaked, short-beaked, or beakless
        16. Cauline leaves less (usually much less) than 1 cm broad, entire to denticulate, readily deciduous; body of achene spiny at apex (as in *Taraxacum*) and bearing a long slender beak . . . . .14. **Chondrilla**
        16. Cauline leaves broader or if less than 1 cm then entire and persistent; body of achene smooth or only ribbed, beakless or beaked
          17. Blades of cauline leaves (in outline) less than twice as long as broad, entire or with lobes 3 (5) and somewhat palmate in aspect (*or* involucre hairy); achenes neither flattened nor beaked . . . . .15. **Prenanthes**
          17. Blades of cauline leaves (in outline) over three times as long as broad, entire or with lobes 5–7 and clearly pinnate; involucres glabrous; achenes strongly flattened, beaked (usually) or not . . . . .16. **Lactuca**

Page 344

1. Arnoseris

1. **A. minima** (L.) Schweigger & Körte   Lamb-succory;
                                                        Dwarf-nipplewort
Map 475. A European species very locally adventive as a weed in North America. Collected in Michigan only in 1886 near Lansing (*Bailey*, GH).

2. Cichorium

C. endiva L. is endive, an Asian native better known than our weedy species as a plant cultivated for greens.

REFERENCE

Steiner, Erich. 1983. The Blue Sailor: Weed of Many Uses. Michigan Bot. 22: 63–67.

1. **C. intybus** L.   Fig. 176   Chicory; Blue-sailors
Map 476. An Old World species now a familiar weed of roadsides, fields, and waste ground. The earliest Michigan collection dates from 1840 (presumably Macomb or Oakland Co.); well established across the Lower Peninsula by 1911 but apparently slow to spread in the Upper Peninsula.

Sometimes cultivated for its roots, used to flavor or adulterate coffee or even as a substitute for it. The large blue heads are very colorful en masse, but the plant is generally rather coarse and unattractive. Occasionally some heads are on elongate peduncles, and they may be rosy or white [f. *album* Neuman] rather than blue.

3. Lapsana

1. **L. communis** L.   Fig. 173   Nipplewort
Map 477. A Eurasian native locally naturalized, becoming abundant in the 1970s and subsequently. The earliest Michigan collection I have seen is from Mackinac Island in 1910, but before 1920 it had also turned up in Oakland, Berrien, and Ingham counties. Like so many species it was doubtless largely overlooked until the population exploded (as in the similar case

475. Arnoseris minima   476. Cichorium intybus   477. Lapsana communis

of *Epipactis helleborine*). For example, *Lapsana* was collected in 1923 in Emmet County, but for over half a century was hardly noticed there again; now, it is truly abundant in newly disturbed ground. Especially characteristic of trails and roadsides in deciduous woods, but also in gardens, hedges, and a diversity of waste places.

This is a distinctive plant in its small, few-flowered heads of pale yellow flowers in a corymbiform or paniculiform arrangement.

4. **Krigia** Dwarf-dandelion

The ligules in both species tend toward orange or yellow-orange compared to the more clear yellow ones of many genera in this group.

REFERENCES

Kim, Ki-Joong, & Billie L. Turner. 1992. Systematic Overview of Krigia (Asteraceae—Lactuceae). Brittonia 44: 173–198.
Shinners, Lloyd H. 1947. Revision of the Genus Krigia Schreber. Wrightia 1: 187–206.

KEY TO THE SPECIES

1. Heads 1–5 (6) on stems bearing (usually) at least 1 well developed bract or leaf; pappus consisting of more than 5 each bristles and scales . . . . 1. **K. biflora**
1. Heads solitary on strictly naked scapes (often many scapes per plant); pappus consisting of 5 bristles and 5 scales (the latter outside and alternating with the bristles) . . . . 2. **K. virginica**

1. **K. biflora** (Walter) S. F. Blake

Map 478. Woodlands, especially oak or jack pine, sometimes spruce, often in moist ground and on banks and borders; fens, wet meadows, boggy ground.

The pappus bristles are much more numerous than in the next species, and in both they are strongly antrorse-scabrous. All of our specimens from the Upper Peninsula (except the single one alleged to be from Keweenaw County, *Farwell 1123b*, BLH) have gland-tipped hairs at the base of the involucre and on the pedicel below it. None of the Lower Peninsula specimens bear such glands. This form has been named f. *glandulifera* Fernald (TL: near Humboldt, Marquette Co.)—raised to subspecific rank by Iltis.

2. **K. virginica** (L.) Willd. Fig. 172

Map 479. Dry, open, sandy, ± barren ground; oak and pine woodland, especially in disturbed areas.

Leaves and scape range from nearly glabrous to, usually, with ± dense gland-tipped hairs. Plants start blooming in early summer (May–June) and continue (if not dried up) to produce new scapes till late in the summer— even September. Because of the long blooming season, it is usually possible to find flowering and fruiting scapes at the same time, the latter revealing the very characteristic pappus of few scales and few bristles.

5. Hypochaeris
A change in the International Code of Botanical Nomenclature (Art. 13.4) in 1981 settled the spelling of this generic name (no longer *Hypochoeris*).

REFERENCES

Aarssen, Lonnie W. 1981. The Biology of Canadian Weeds. 50. Hypochoeris radicata L. Canad. Jour. Pl. Sci. 61: 365–381.
Harriman, Neil A. 1980. Leontodon and Hypochaeris (Compositae) in Wisconsin. Michigan Bot. 19: 93–95.

1. **H. radicata** L.   Cat's-ear
Map 480. Native in Europe and neighboring regions. A weed of lawns, roadsides, vacant lots, fields; occasionally in prairie remnants and wooded areas. A lawn weed in Ingham County as early as 1886 but not collected again in Michigan until 1915 (Kent & Oakland cos.); started spreading quite rapidly in the 1960s.

The stem is usually branched, with a few yellow heads on long pedicels. In this respect, the plant resembles *Leontodon autumnalis*. The pappus consists of two rows of hairs, the outer shorter and sparsely if at all plumose (as in all but the outermost flowers of *L. taraxacoides*). The pappus of *L. autumnalis* consists of one row of plumose hairs, and the receptacle lacks the elongate, slender-tipped chaffy scales conspicuous in *Hypochaeris*.

6. **Leontodon**   False-dandelion; Hawkbit
Both our species are Old World natives, still sparingly introduced in Michigan. The heads are yellow. See also comments under *Hypochaeris radicata* above.

REFERENCE

Harriman, Neil A. 1980. Leontodon and Hypochaeris (Compositae) in Wisconsin. Michigan Bot. 19: 93–95.

478. Krigia biflora

479. Krigia virginica

480. Hypochaeris radicata

KEY TO THE SPECIES

1. Stem usually branching, with a few heads on long pedicels; hairs all simple, not forked; outermost achenes in head with pappus of one row of plumose hairs—as in the other achenes . . . . .1. **L. autumnalis**
1. Stem simple, the heads on naked scapes; hairs (especially those on the leaves) minutely 2 (–3)-forked at the apex [use strong lens]; outermost achenes in head with pappus solely of scales, the other achenes with pappus of plumose hairs surrounded by short simple hairs . . . . .2. **L. taraxacoides**

1. **L. autumnalis** L.   Fall-dandelion

Map 481. Primarily a lawn weed, collected in Michigan as early as 1881 (Ionia Co.) and 1897 (St. Clair Co.) but still very local.

The leaves tend to be more deeply pinnatifid compared with the next species (or with *Hypochaeris*), the deepest sinuses extending more than halfway to the midrib.

2. **L. taraxacoides** (Vill.) Mérat   Fig. 174

Map 482. A roadside and lawn weed, first found in Michigan at Detroit in 1930 (*Farwell 8730*, BLH), but not gathered again until the late 1950s (in the Upper Peninsula).

The distinctive minutely forked tips on many of the hairs are a feature shared with the leafy-stemmed *Picris hieracioides* and with *L. hispidus* L., not yet known from Michigan and differing in having also some short-stalked stellate hairs and all achenes with a pappus of the type occurring in the central flowers of *L. taraxacoides*.

7. **Picris**

1. **P. hieracioides** L.   Ox-tongue

Map 483. Collected in 1896 in East Lansing, where apparently introduced with grass seed from France (according to labels by Beal and Wheeler); Farwell found it "rare" at Detroit in 1901, and Swink collected it

481. Leontodon
autumnalis

482. Leontodon
taraxacoides

483. Picris hieracioides

in weedy ground in Berrien County in 1966. Obviously not yet a major weed in the state. I have been unable to find the basis for McAtee's report (Pap. Michigan Acad. 1: 65. 1923) from Crawford County in 1919.

This is another yellow-flowered Eurasian weed, even more sparing in occurrence than the two previous genera, and differing most conspicuously in having leafy stems.

8. **Tragopogon**   Goat's-beard

Our three species are all natives of Europe, whence they came to North America to become attractive weeds, even sometimes cultivated. Hybrids are known in Europe and may be expected wherever two or three species grow together. The first artificial experimental plant hybrid was apparently produced by Linnaeus in his garden in 1759, when he crossed *T. porrifolius* and *T. pratensis*.

One good site for hybrids in Michigan for over 40 years has been in the vicinity of the railroad tracks and station (now all gone) in Cheboygan, including nearby roadsides and vacant lots (see Michigan Bot. 16: 140. 1977). All three species and all their hybrids can be found blooming there in mid June, generally in the morning, for the heads close by mid day, especially on sunny days. The hybrids are rather easily recognized, especially those involving the purple *T. porrifolius*, for the heads are often a handsome red-orange or sometimes pale purplish; the bases of the ligules are yellow. If the phyllaries are reddish-margined, only equalling the ligules, and/or the leaf tips curled, *T. pratensis* is the yellow parent. If the phyllaries are entirely green and the leaf tips straight, *T. dubius* is the yellow parent. Hybrids between the two yellow species (plate 7-A) are less easily spotted, but close examination again reveals a mixture of characters, such as phyllaries exceeding the ligules or peduncle inflated (as in *T. dubius*), but phyllaries reddish-margined and/or leaf tips curled (as in *T. pratensis*).

The most frequent hybrid, *T. porrifolius* × *T. pratensis*, was early described from Michigan by Farwell, who named it *T.* ×*neohybridus*, having found it in Ann Arbor Township (Am. Midl. Nat. 12: 133–134. 1930). However, this hybrid had already been named *T.* ×*mirabilis* Rouy 40 years earlier in France. The hybrid between the two yellow species may be called *T.* ×*crantzii* Dichtl, named in Europe in 1883.

First-generation hybrids are generally sterile, but Ownbey described as species two derivatives rendered fertile by doubling of chromosomes (amphidiploids). Hybrids between these amphidiploids were later analyzed by Rieseberg and Warner.

The globose fruiting heads of goat's-beard resemble those of dandelion but are much larger and on leafy stems (besides having plumose rather than simple pappus-hairs); they can be used effectively in "winter bouquets" if sprayed to prevent disintegration.

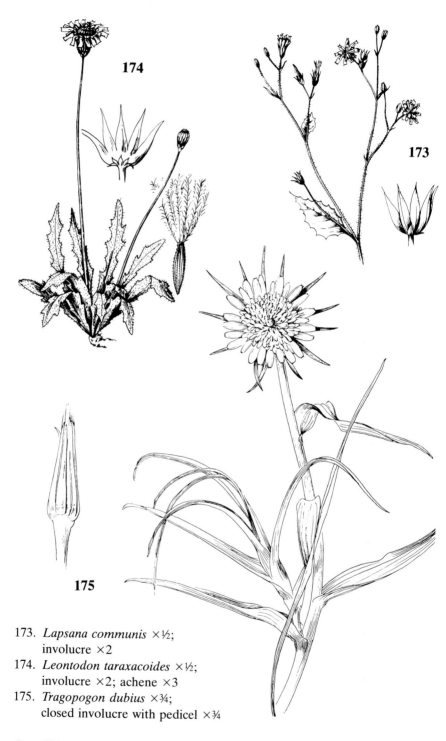

174

173

175

173. *Lapsana communis* ×½;
   involucre ×2
174. *Leontodon taraxacoides* ×½;
   involucre ×2; achene ×3
175. *Tragopogon dubius* ×¾;
   closed involucre with pedicel ×¾

REFERENCES

Ownbey, Marion. 1950. Natural Hybridization and Amphiploidy in the Genus Tragopogon. Am. Jour. Bot. 37: 487–499.

Rieseberg, Loren H., & Donn A. Warner. 1987. Electrophoretic Evidence for Hybridization between Tragopogon mirus and T. miscellus (Compositae). Syst. Bot. 12: 281–285.

Thompson, Paul W. 1952. The Genus Tragopogon in Michigan. Asa Gray Bull. n.s. 1: 58–60.

KEY TO THE SPECIES

1. Ligules purple; pappus brownish . . . . .1. **T. porrifolius**
1. Ligules yellow; pappus dingy whitish
  2. Margins of phyllaries green (or pale); leaf tips straight; ligules pale yellow, distinctly shorter than the longest phyllaries; pedicels much expanded beneath the heads . . . . .2. **T. dubius**
  2. Margins of phyllaries reddish purple (rarely green); leaf tips ± curled or curved; ligules bright yellow, as long as or longer than the phyllaries; pedicels scarcely if at all expanded . . . . .3. **T. pratensis**

1. **T. porrifolius** L.   Salsify; Vegetable-oyster

Map 484. Collected in Michigan as early as 1843 by Cooley (presumably Macomb or Oakland Co.) and by the mid 1880s in Ingham and Keweenaw counties. Roadsides and railroads, fields, vacant lots and other waste places. Doubtless originally an escape from cultivation, for the species has long been grown for its taproot, which is considered edible.

As in *T. dubius*, the phyllaries exceed the ligules and the pedicels are inflated. Collectors who press specimens lacking ligules or who fail to note their color on the labels waste space with specimens of these two species and their hybrids, which are often unidentifiable in such a condition.

2. **T. dubius** Scop.   Fig. 175

Map 485. Not collected in Michigan until 1911, in Berrien County (*O. E. Lansing 3214*, GH), and spreading since then. Roadsides, railroads, fields, fencerows; vacant lots and other waste places; clearings in woods, shores.

The achenes (including beak but not pappus) are ca. (2.5) 3–3.5 cm long at maturity.

484. Tragopogon porrifolius

485. Tragopogon dubius

486. Tragopogon pratensis

3. **T. pratensis** L.   Plate 7-A (hybrid)
Map 486. Our earliest collections are from Kent and Van Buren counties in 1880, but other counties in the 1880s include Mecosta, Ingham, and Ionia, so the species must have been well established by then. Like the other members of the genus, inhabits roadsides and railroads, fields, fencerows, clearings, waste places, and shores.

The achenes are smallest in this species, running at maturity ca. 1.8–2.5 cm long (excluding pappus).

9.   Agoseris

REFERENCE

Mustard, Timothy S. 1982. The Distribution and Autecology of Pale Agoseris, Agoseris glauca, in Michigan. Michigan Bot. 21: 205–211.

1. **A. glauca** (Pursh) Raf.   Plate 6-B   False Dandelion
Map 487. Local on sandy glacial outwash in the open jack pine plains, especially in grassy areas, near the corner where four counties meet (Otsego, Montmorency, Oscoda, Crawford). Although first collected in 1918, not reported from Michigan until 1957 (Brittonia 9: 96), and widespread search has not documented any larger range for this disjunct occurrence of a species wide-ranging in grasslands and grassy clearings from Hudson Bay and western Minnesota to Arizona and British Columbia.

This species has a long blooming season, from May to September.

10.   Taraxacum   Dandelion
Everyone knows a dandelion, but sometimes it is hard to say *which* dandelion it is, especially if one tries to understand the many hundreds of microspecies that have been named in Europe. Most of these reproduce asexually as apomictic polyploids and thus maintain their little identities just as do the similar entities in *Crataegus, Rubus*, and *Hieracium*. Whether the evolution or the nomenclature of *Taraxacum* is proceeding faster may be open to question. Even if one adopts a *sensu lato* approach, as I do here, it is not clear what names are correct for our common North American species. Some of the nomenclatural problems are described by Kirschner and Štěpánek (1987), and some aspects of the disgraceful behavior of the genus are described by Mogie and Ford (1988) and by Hughes and Richards (1989). Richards (1972) shows an example of what all this can mean in a local flora: over 130 species in the British Isles alone. While analysis of accurately named microspecies might be a profitable phytogeographc exercise, providing evidence for the introduction and spread of these taxa, most of the available herbarium material is inadequate and would have to be ignored.

Besides the two traditional broadly defined species treated below, Michigan botanists should watch for the marsh dandelion, *T. palustre* (Lyons) Symons sens. lat. [*T. turfosum* (Sch. Bip.) Soest], which has been found in wet sites, including seasonally flooded highway ditches with other halophytes, in Ontario and New York. In addition to the habitat, it differs in having the outer series of phyllaries straight (not recurved) and the leaves only shallowly lobed or toothed.

REFERENCES

Brunton, Daniel F. 1989. The Marsh Dandelion (Taraxacum section Palustria; Asteraceae) in Canada and the Adjacent United States. Rhodora 91: 213–219.

Hughes, Jane, & A. J. Richards. 1989. Isozymes, and the Status of Taraxacum (Asteraceae) Agamospecies. Jour. Linn. Soc. Bot. 99: 365–376.

Kirschner, Jan, & Jan Štěpánek. 1987. Again on the Sections in Taraxacum (Cichoriaceae) (Studies in Taraxacum). Taxon 36: 608–617.

Mogie, Michael, & Henry Ford. 1988. Sexual and Asexual Taraxacum Species. Jour. Linn. Soc. Biol. 35: 155–168.

Richards, A. J. 1972. The Taraxacum Flora of the British Isles. Watsonia 9 Suppl. 141 pp.

KEY TO THE SPECIES

1. Body of achene red-brown, ca. 2.5–3 mm long, usually spiny from apex to middle or below; leaves deeply pinnatifid, cut nearly to midrib; phyllaries (some, not all) often with a small protuberance just below the apex .... 1. **T. erythrospermum**
1. Body of achene gray to yellow-green or pale brown, ca. 3–3.5 (3.7) mm long, spiny above the middle; leaves various (from deeply pinnatifid to nearly entire); phyllaries without a protuberance .... 2. **T. officinale**

1.   **T. erythrospermum** Besser   Fig. 177   Red-seeded Dandelion
Map 488. A Eurasian native. I have seen no Michigan specimens from the 1800s, but Farwell collected this species at Detroit in 1900. By 1919, E. A. Bessey collected it in Gogebic County, so it must have been scattered locally throughout the state early in the 20th century. Roadsides, railroads, fields, clearings (often gravelly or rocky), shores; farmyards, lawns, parking lots, and other disturbed or waste places.

487. Agoseris glauca

488. Taraxacum erythrospermum

489. Taraxacum officinale

Specimens with deeply pinnatifid (lacerate) leaves but lacking both fruit and protuberances on the phyllaries cannot safely be identified unless the terminal lobe of the leaf is distinctly larger than the lateral lobes, in which case the specimen is presumably *T. officinale* (although in the shade, red-fruited plants can have a relatively large terminal lobe).

This taxon is also widely known as *T. laevigatum* (Willd.) DC.

A number of specimens referred here were named in 1985 by Reinhard Doll, a current European specialist in *Taraxacum*, as *T. scanicum* Dahlst.— a common microspecies in dry, especially sandy, places in Europe. Doll identified one specimen (from Manitou Island, Keweenaw Co.) as *T. disseminatum* G. E. Haglund and one (from Washtenaw Co.) as *T.* aff. *fulvum* Raunk. (A detailed monograph of over 100 species in the "Section Erythrosperma" was published by Doll in vols. 83 & 84 of *Feddes Repertorium*, 1973.) However, some of the specimens mapped here as this species were considered unidentifiable (at least as microspecies) by Doll or were not seen by him at all.

2. **T. officinale** Wiggers   Fig. 178   Common Dandelion
Map 489. Another Eurasian species, collected by the First Survey in 1838 (Kalamazoo Co.) and now thoroughly naturalized in most habitats except the wettest: lawns, roadsides, railroads, fields, dunes; woods, especially disturbed areas; often on dry sand or rock outcrops; occasionally in wet ground; meadows, river banks, shores.

Even when broadly treated, as here, this species is sometimes hard to separate from the preceding in the absence of mature fruit (see notes above). However, some specimens from throughout the state have been confirmed by Doll as *T. officinale*, and it is clear that both species, sens. lat., can be expected anywhere in Michigan.

11. **Hieracium**   Hawkweed
Traditionally a difficult genus, with many problems in both native and introduced species. The latter, particularly, are mostly in groups where hybridization is common and hundreds of variants have been dignified with scientific names (as well as unprintable, unscientific ones). Even the native species are split more finely by some authors. More broadly defined species, as here, are quite variable morphologically, especially in regard to leaf shape and arrangement and to pubescence. Several species, therefore, will run at more than one place in the key—and the arrangement is in no way intended to imply any taxonomic distinctions related to more narrowly defined characters. Exceptions may occur to almost any character, whether due to mere variability or introgression, and a good deal of judgment is often needed to place puzzling specimens even in the broadly defined species or species complexes recognized below. Most hairs, especially on the stems, are pale when fresh but after a few years turn a rich orange-brown or tawny shade.

176. *Cichorium intybus* ×½; achene ×5

177. *Taraxacum erythrospermum* ×½; achene ×2

178. *T. officinale* ×½; achene (minus beak & pappus) ×8

The annotations of Mark Garland on many specimens of species 6–9 and 13 have been very helpful. E. Lepage earlier annnotated species 14–15 and their allies. Species 1–5 were briefly dealt with by Voss and Böhlke (1978).

REFERENCES

Lepage, Ernest. 1960. Hieracium canadense Michx. et ses Alliées en Amérique du Nord. Nat. Canad. 87: 59–106.

Lepage, Ernest. 1971. Les Épervières du Québec. Nat. Canad. 98: 657–674.

Voss, Edward G., & Mark W. Böhlke. 1978. The Status of Certain Hawkweeds (Hieracium subgenus Pilosella) in Michigan. Michigan Bot. 17: 35–47.

KEY TO THE SPECIES

1. Leaves of erect flowering stems all or mostly in a basal rosette, or densely crowded very near the base of the stem (leafy stolons may also be present)
   2. Heads 1 (–3); leaves with numerous to dense stellate hairs beneath
      3. Leaves beneath with a dense felt of stellate hairs (white when fresh, reddish brown after several years); involucre ca. 7–8 (9.5) mm long; heads solitary . . . . .1. **H. pilosella**
      3. Leaves beneath with numerous but not densely felted stellate hairs; involucre ca. 9–12 mm long; heads mostly 2–3 . . . . .2. **H. flagellare**
   2. Heads several to many; leaves with sparse stellate hairs (or none) beneath (rarely such hairs numerous in *H. caespitosum*)
      4. Ligules red-orange; plants stoloniferous (or with shallow rhizomes) . . . . .3. **H. aurantiacum**
      4. Ligules yellow; plants stoloniferous or not
         5. Achenes 1.5–2 mm long; heads ± crowded in a corymbiform inflorescence; well developed cauline leaves 0–3 (and smaller than rosette leaves); rosette leaves ca. (3) 6–20 times as long as broad
            6. Leaves glabrous or unevenly and sparsely hairy above, ± glaucous; mature involucres usually ca. (4) 5–7 (7.5) mm long; pedicels with stellate hairs none or usually sparse (occasionally dense) . . . . .4. **H. piloselloides**
            6. Leaves hairy above (many hairs as close together as 1 mm) and not glaucous; mature involucres ca. (6.5) 7–8 (9) mm long; pedicels densely stellate-pubescent . . . . .5. **H. caespitosum**
         5. Achenes over 2 mm long; heads usually in a ± wide open or elongate inflorescence; well developed cauline leaves usually more than 3 (and equal to rosette leaves) (usually none in *H. venosum*, with rosette leaves up to 3 (4) times as long as wide)
            7. Basal leaves with soft minutely barbellate hairs, these at least partly curled and/or tangled; principal leaves usually coarsely toothed; achenes truncate, not narrowed toward apex . . . .[go to couplet 15]
            7. Basal leaves with prominent long setose hairs at least on margin (and often upper surface), these mostly straight and not tangled; principal leaves entire (or minutely denticulate); achenes usually at least slightly constricted or narrowed toward apex at maturity
               8. Principal leaves all basal or nearly so (at most a few small cauline bracts), ca. 3 (4) times as long as broad (or shorter), usually with veins outlined in reddish purple; stem glabrous or at most finely pubescent toward base; mature achenes (2.2) 2.5–3 mm long . . . . .6. **H. venosum**

8. Principal leaves partly cauline (as large as basal leaves), 3–8 times as long as broad, the veins not outlined; stem long-hairy at least toward base; mature achenes 3–4 mm long
   9. Longest hairs of stem and leaves at least 8 mm; involucre 8.5–10.5 (11.5) mm long . . . . .7. **H. longipilum**
   9. Longest hairs ca. 4–6 mm; involucre 6.5–9 mm long . . . . .8. **H. gronovii**
1. Leaves of erect flowering stem all or mostly cauline (not basal)
  10. Longest hairs toward base of stem and on lower leaves 3–10 mm (or even more)
    11. Flowering heads large (ca. 1.5–2.5 cm long including ligules); leaf blades elliptic, usually coarsely toothed, at least the lowest on clearly defined petioles . . . .
    . . . .[go to couplet 15]
    11. Flowering heads smaller; leaf blades elliptic to oblong or oblanceolate, usually sessile or tapering into indistinct petioles
      12. Leaves distinctly paler beneath than above (glaucous when fresh); heads with fewer than 20 florets, on very slender (ca. 0.2–0.3 mm—almost filiform) elongate pedicels nearly or quite glabrous (or often sparsely glandular-stipitate) . .
      . . . . .9. **H. paniculatum**
      12. Leaves not glaucous, little if at all paler beneath; heads many-flowered, on stouter and ± densely pubescent and/or stipitate-glandular pedicels
        13. Achenes truncate (not narrowed at apex) at maturity; pedicels ± densely stipitate-glandular; longest hairs at base of plant less than 7 mm . . . .13. **H. scabrum**
        13. Achenes narrowed toward apex; pedicels only sparsely if at all stipitate-glandular *or* longest hairs at least 8 mm long . . . .[go to couplet 9]
  10. Longest hairs toward base of stem and on lower leaves less than 3 mm (or none)
    14. Leaf blades elliptic, usually coarsely toothed, at least the lowest clearly petioled; flowering heads large (ca. 15–25 mm long, including ligules)
      15. Leaves streaked and blotched with red-purple . . . . .10. **H. maculatum**
      15. Leaves green throughout
        16. Cauline leaves (2) 4–7; leaf blades tapering narrowly into the petiole . . . .
        . . . . .11. **H. lachenalii**
        16. Cauline leaves 0–1 (or 2 near base); leaf blades broadly rounded to subcordate at base . . . . .12. **H. murorum**
    14. Leaf blades oblong to elliptic or oblanceolate, toothed or entire, sessile (or if the lower ones clearly petioled, the heads less than 1.5 cm long); flowering heads 8–25 mm long (including ligules)
      17. Involucres and pedicels with ± dense dark gland-tipped hairs; leaves entire or barely and remotely denticulate, tapered more gradually to the base than to the apex (lower leaves even into petioles); flowering heads 8–13 mm long . . . .
      . . . . .13. **H. scabrum**
      17. Involucres and pedicels without dark gland-tipped hairs; leaves almost always with evident but remote teeth, all sessile or nearly so, tapered about equally to base and apex or more abruptly to the base; flowering heads ca. (12) 15–25 mm long
        18. Leaves strongly roughened with dense, stout, broad-based, conical (but very short) prickle-like "hairs," especially above and toward the margins; involucre glabrous (at most slightly pubescent at the very base) . . . . .14. **H. umbellatum**
        18. Leaves without such projections on the surface (at most slightly roughened to the touch); involucre often at least sparsely strigose and/or stellate-pubescent . . . . .15. **H. kalmii**

1. **H. pilosella** L.   Mouse-ear Hawkweed

Map 490. A species of Europe and western Asia, sometimes planted but spreading aggressively once it gets out of control (which is not often in

Michigan). Lawns and roadsides; known as a lawn weed in Benzie County since 1861.

Plants are quite short and strongly stoloniferous.

2. **H. flagellare** Willd.

Map 491. An introduction from Europe. Old fields, roads, railroads, and other disturbed ground; lawns and edges of woods. First collected in Michigan in 1969 (Sanilac Co.) and 1971 (Ingham Co.).

Averaging only slightly taller than the preceding species and also strongly stoloniferous. Few-headed plants of *H. caespitosum* might be thought to belong here, but that species has shorter involucres and the leaves are both more acute at the apex and more tapered at the base than in *H. flagellare*. Intermediate plants with ± numerous stellate hairs on the leaves beneath may represent hybrids between these two species (see below).

3. **H. aurantiacum** L.   Fig. 179   Orange Hawkweed;
                                                        Devil's-paintbrush

Map 492. Almost everywhere: can be colorfully abundant along road-sides and in dry fields, but also invading logged areas, woods, shores, moist pastures, marshy ground, and cedar swamps. Apparently introduced from Europe as a garden ornamental in Vermont by 1875, and soon spread throughout northern New England and beyond. The first Michigan collec-tion is from Alpena in 1895, and it spread rapidly in that part of the state. However, there are also reports from Lenawee County in 1897 and Kent County in 1901, so the species may have been introduced ± simultaneously at more than one place in Michigan (see Voss & Böhlke 1978).

Hybrids with *H. piloselloides* occur occasionally where both parents are found and can be recognized by the intermediate shade of the ligules and/ or by bicolored ligules (red apically, yellow basally). The bicolored aspect is, if anything, intensified in herbarium specimens several years after they were dried. Hybrids with ligule color closer to one of the parents are more likely to be keyed as that parent unless the collector has carefully described

490. Hieracium pilosella

491. Hieracium flagellare

492. Hieracium
     aurantiacum

the situation on the label. Besides, *H. aurantiacum* ligules tend to be paler in the shade. Morphological characters are not very helpful in confirming hybrids. Specimens at least temporarily identified as hybrids are known from Alcona, Alger, Benzie, Cheboygan, Delta, Kalamazoo, Keweenaw, Lenawee, Luce, Mackinac, Schoolcraft, and Washtenaw counties. A collection from Iosco County with bicolored ligules may be *H. aurantiacum* × *H. caespitosum*, with very broad, hairy, denticulate leaves.

This species has small achenes, like those of the next two closely related species.

4. **H. piloselloides** Vill.   King Devil; Yellow Hawkweed

Map 493. Like the preceding species, almost everywhere, especially along roadsides and in fields; railroads, gravel pits, disturbed and waste places of all kinds; invading dry woods, plantations, sandy and rocky openings, shores, even wet ground.

This is another European species, introduced a little more recently than the preceding and now about equally abundant in northern Michigan, where the two often dominate old fields and roadsides. The first Michigan collection is from Houghton County in 1914, and it spread rapidly in the Copper Country. I have seen no collection from south of the Straits of Mackinac before 1935 (Mackinaw City) but the species was doubtless ahead of its collectors at all times (see Voss & Böhlke 1978).

Long known under the name *H. florentinum* All. in American manuals, and here including also local plants sometimes ascribed to Michigan under names of other species, such as *H. floribundum* Wimmer & Grab. and *H. praealtum* Gochnat. Our plants are quite variable and sometimes distinguished only with difficulty from *H. caespitosum*, which generally is a much hairier species on stems and leaves, with larger heads and darker, gland-tipped hairs among the dense stellate pubescence, especially on pedicels. However, some plants referred to *H. piloselloides* have large heads on rather densely stellate pedicels, but with leaves and stems nearly glabrous. These and assorted other ambiguous plants (such as a very few hairy ones referred to *H. caespitosum* despite smaller heads than usual for that species) could result from hybridization. It could even be that species once more distinct are now swamped by introgression and losing their identity. Until some convincing evidence becomes available as to what is really going on, these two closely related species are maintained here—albeit with some reservations (Voss & Böhlke 1978). Hybrids with the previous species are mentioned above.

5. **H. caespitosum** Dumort.   King Devil; Yellow Hawkweed

Map 494. A less common immigrant than the previous two species, evidently in the state only since the 1930s (Cheboygan Co.), and spreading since the 1960s. Roadsides, fields, clearings, waste ground; dry woodland, meadows, banks.

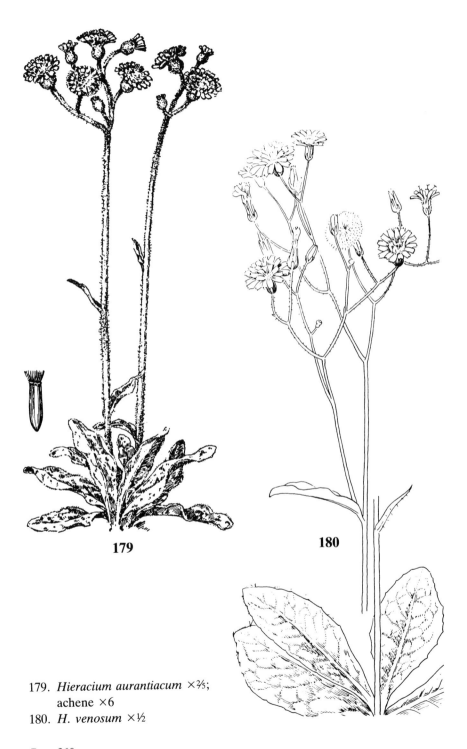

179. *Hieracium aurantiacum* ×⅖;
  achene ×6
180. *H. venosum* ×½

The ± glaucous nature of leaves of *H. piloselloides* is often not evident on dried specimens, making separation more difficult, especially from equally large-headed specimens occasional in the preceding species. The conspicuous long, often stiff hairs on leaves and stem, often larger leaves, and darker inflorescence of larger heads give *H. caespitosum* [long known as *H. pratense* Tausch] a reasonably distinct aspect hard to express in a key and open to exception on any one character. The leaves are often sparsely stellate-pubescent beneath (in addition to long hairs), but are more acute at the apex than the ± obtuse to rounded ones of *H. flagellare*.

A few collections from Allegan, Berrien, Calhoun, Kalamazoo, Lenawee, Luce, Newaygo, and Oceana counties appear intermediate between this species and *H. flagellare*; and one from Iosco County appears intermediate with *H. aurantiacum*.

6. **H. venosum** L.   Fig. 180   Rattlesnake-weed; Veined Hawkweed
    Map 495. Sandy plains and banks, especially jack pine, oak, and aspen woodlands; does well after fire.
    Rarely the base of the stem has some scattered long ± straight hairs, though these are less dense than in the next two species. Occasionally a plant will lack the conspicuous and distinctive reddish purple borders on the leaf veins. And a very few plants have the setose hairs around the margin of the leaves sparse or lacking. The heads are fewer-flowered (not over 25) than in many species. The inflorescence at maturity is very broadly paniculate, with ± divaricate branching (as in fig. 180). See also comments under *H. gronovii*, which may have basal leaves of similar shape (although usually not all leaves are basal).

7. **H. longipilum** Hooker   Prairie or Long-bearded Hawkweed
    Map 496. Dry open often sandy plains and small hills, prairies, oak or pine savana, fields.
    This is our shaggiest appearing hawkweed, with hairs toward the base of the plant at least 8–10 mm long and often twice that. As in other species, these are pale when fresh and deep tawny or rusty in color when several years old and dry. The phyllaries are mostly stipitate-glandular throughout, while in the next species they are stipitate-glandular, if at all, only on the basal two-thirds.
    This species was originally described—very casually—from Michigan in 1833 (TL: River St. Clair, United States [St. Clair Co.]).

8. **H. gronovii** L.
    Map 497. Prairie–savana and open woodland with oak, pine, sassafras, and/or aspen; openings in beech-maple woods; dry open fields and roadsides.
    The achenes tend to be a little longer than those of *H. scabrum*, as well as tapered toward the apex. The hairs toward the base of the plant are

sometimes soft and curled, but such plants, even if the leaves are all or mostly basal, can be distinguished from *H. venosum* by the abundance of stem pubescence, absence of reddish purple veins, larger achenes, and often elongate-cylindrical panicle of numerous heads. *H. longipilum* has longer leaves, mostly 5–8 times as long as broad, sometimes densely crowded near the base, as well as longer hairs.

From time to time (including Farwell, Rep. Michigan Acad. 20: 194. 1918), the application of this name has been challenged, with its being placed in the synonymy of *H. venosum*. The latest suggestion (Garland) is that *H. floridanum* Britton is the correct name for our plant. In view of some confusion over typification, I tentatively retain the familiar *H. gronovii*, as was done by Beaman for Mexico (Syst. Bot. Monogr. 29: 42. 1990).

9. **H. paniculatum** L.

Map 498. Sandy woods, especially with oak, as on old dunes along Lake Michigan.

Although seemingly difficult to accommodate in the key, this species is actually unusually distinctive, with its very small heads on extremely slender pedicels and with the leaves thin, nearly or quite glabrous, and pale-glaucous beneath.

10. **H. maculatum** Sm.   Spotted Hawkweed

Map 499. A European species, occasionally cultivated. Well established for over 40 years at the borders of calcareous mixed woods on a bluff west of St. Ignace.

Sometimes included in the next species, to which it is very similar except for the conspicuously spotted and streaked leaves.

11. **H. lachenalii** C. C. Gmelin

Map 500. Another European species, at borders, trails, and openings in dry rocky woods; yards and hedgerows.

493. Hieracium   494. Hieracium   495. Hieracium venosum
    piloselloides   caespitosum

Like the preceding, quite attractive with its large heads and relatively few, large, toothed leaves. This is the species long called *H. vulgatum* Fries by American authors.

12. **H. murorum** L.
Map 501. Along trails and borders of woods, clearings, shores.
This is a third European species in Section *Vulgata*, not always easily separated from the preceding.

13. **H. scabrum** Michaux
Map 502. In similar habitats to those of *H. gronovii* and sometimes growing with it where their ranges overlap. Dry sandy or rocky woods with oak, sassafras, aspen, pines, etc., including jack pine plains, clearings, oak savana, mixed conifer-hardwoods, sand dunes; in old fields and rarely damp sites.
An intermediate plant from dry sandy soil in Schoolcraft county (*Henson 1269* in 1981, MICH) has been confirmed by Garland as perhaps *H. scabrum* × *H. venosum*; a very few other odd specimens could conceivably also be hybrids.

496. Hieracium longipilum

497. Hieracium gronovii

498. Hieracium paniculatum

499. Hieracium maculatum

500. Hieracium lachenalii

501. Hieracium murorum

14. **H. umbellatum** L.   Northern Hawkweed

Map 503. Often in dune areas, at edges of woods and in grassy openings, and in other sandy ± open sites; rock crevices and forest borders at Lake Superior; shores, thickets, and mixed woods. Several mixed and simultaneous collections of this species and the next indicate that the two frequently grow together.

American plants of this species have sometimes been segregated as *H. scabriusculum* Schwein. The inflorescence of both this species and the next is often umbelliform, although the heads may also be in an elongate panicle, as in *H. scabrum*. The characteristic dense little cone-shaped projections along the leaf margins are generally ca. 0.2–0.3 mm long.

15. **H. kalmii** L.   Canada Hawkweed

Map 504. Dry sandy or rocky woods and thickets with oak, pine (sometimes other conifers), aspen, birch—as in jack pine plains and old dunes; grassy clearings, sedge meadows, burned sites, shores.

Hybrids between this species and each of the previous two are known (and often suspected). Some plants have roughish margins on the leaves, but not as rough (i.e., the projections not as long nor as dense) as in *H. umbellatum*; the line is hard to draw. A few plants from the Keweenaw Peninsula and Isle Royale, with some very short-stalked (and not black) glands at the base of the involucre, have been referred by Lepage to *H. canadense* Michaux, but the distinction appears slight. In view of the varied application and interpretation of that name it is not taken up here, although long used for *H. kalmii* (an older name and hence to be adopted if the two are not distinguished).

Lepage has named a few specimens from Keweenaw County (and one from Oakland Co.) as *H. ×grohii* Lepage (*H. canadense* × *H. lachenalii*). Although these plants tend to be a little more leafy than many *H. kalmii*, I am not convinced that most of them show any significant difference from that species; a duplicate of the type of Lepage's *H. ×grohii* nm. *farwellii* (TL: Copper Harbor, Keweenaw Co.) was in fact cited by him elsewhere in the same paper as *H. canadense*.

502. Hieracium scabrum   503. Hieracium   504. Hieracium kalmii
                            umbellatum

12. Crepis   Hawk's-beard

REFERENCE

Najda, H. G., A. L. Darwent, & G. Hamilton. 1982. The Biology of Canadian Weeds. 54. Crepis tectorum L. Canad. Jour. Pl. Sci. 62: 473–481.

KEY TO THE SPECIES

1. Achenes pale or tan; longer phyllaries (like the shorter) glabrous within . . . . .1. **C. capillaris**
1. Achenes dark purple-brown (rarely paler); longer phyllaries antrorse-strigose within
    2. Plant ± rough-pubescent on leaves and stems; achenes yellow or yellow-brown, ca. 13–25-ribbed, over 4 mm long . . . . .2. **C. biennis**
    2. Plant smooth or nearly so; achenes dark purple-brown, 10-ribbed, ca. 4 mm long . . . . .3. **C. tectorum**

1. C. capillaris (L.) Wallr.

Map 505. A native of Europe, sparsely introduced and locally weedy in lawns, gardens, cemeteries, fields, parking lots, and such places. First collected in Michigan in 1915 in Antrim County.

The pale achenes run smaller than in the other two species, and are 10-ribbed.

2. C. biennis L.

Map 506. This European weed was collected 1897–1900 at East Lansing, introduced in a meadow sown with grass seed from France and then spreading along a railroad.

3. C. tectorum L.   Fig. 183

Map 507. A Eurasian species of local occurrence. A lawn weed in Ingham and St. Clair counties in the 1880s, but except for a 1935 collection from Ann Arbor, not gathered again in the state until 1978; since, apparently spreading (or being discovered) rapidly along roadsides and in lawns, fields, plantations, vacant lots, logged areas, gravel pits, and other disturbed ground.

Good magnification (20–30×) and illumination at the right angle are often necessary to see, on the inside of the phyllaries, the distinctive tiny hairs which will identify this species and the preceding if achenes are not mature.

13. Sonchus   Sow-thistle

All three of our species are native in Europe but are now widespread as weeds around the world. Leaf shape is excessively variable in the genus, ranging from unlobed to deeply pinnatifid.

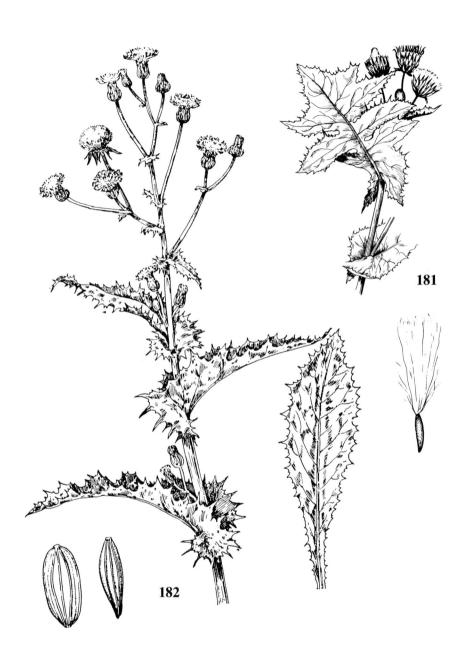

181. *Sonchus oleraceus* ×½; achene ×5
182. *S. asper* ×½; achenes (minus pappus) ×8

REFERENCES

Hutchinson, Ian, Joseph Colosi, & Ralph A. Lewin. 1984. The Biology of Canadian Weeds. 63. Sonchus asper (L.) Hill and S. oleraceus L. Canad. Jour. Pl. Sci. 64: 731–744.
Lemna, Wanda K., & Calvin G. Messersmith. 1990. The Biology of Canadian Weeds. 94. Sonchus arvensis L. Canad. Jour. Pl. Sci. 70: 509–532.

KEY TO THE SPECIES

1. Flowering heads ca. 2.5–4.5 (5) cm broad; largest mature involucres mostly (13) 14–18 mm long; plant a deep-rooted perennial, spreading by horizontal roots . . . . .1. **S. arvensis**
1. Flowering heads less than 2.5 cm broad; largest mature involucres 8–13 mm long; plant a tap-rooted annual (or winter annual)
  2. Leaf bases with ± acute lobes (fig. 181); mature achenes densely papillate-tuberculate, slightly flattened . . . . .2. **S. oleraceus**
  2. Leaf bases with conspicuously rounded lobes (fig. 182) — often so large and curved around as to suggest a nautilus shape; mature achenes essentially smooth except for 3 (5) prominent ribs on each face, very strongly flattened . . . . .3. **S. asper**

1. **S. arvensis** L.   Field or Perennial Sow-thistle
    Map 508. Roadsides, railroads; meadows, shores, disturbed wetlands and dikes; fields, weedy among crops and in gardens; waste ground, including fill, parking lots, dumps. Collected at Lansing as long ago as 1867, but not gathered again until the 1890s in several counties (Bay, St. Clair, Tuscola, Wayne); now all too common throughout the state.
    The sow-thistles are not always easy to distinguish, especially using dry specimens. Many collections lack indication of underground parts, the sure feature to recognize the deep-rooted perennial *S. arvensis*, which one cannot pull easily from the ground. Plants with the biggest heads and longest involucres are this species. The achenes are similar to those of *S. oleraceus*, with a prominently papillate-tuberculate surface, but they seem often not to develop in *S. arvensis*. The ligules are paler yellow in *S. oleraceus* than in the deep (even orange-) yellow heads of *S. arvensis*.
    In *S. arvensis* var. *arvensis*, the involucres and pedicels bear numerous stiff hairs tipped with dark glands. More common in Michigan is var.

505. Crepis capillaris   506. Crepis biennis   507. Crepis tectorum

glabrescens Günther, Grab., & Wimmer, which lacks the glandular hairs and may average slightly smaller involucres. This variety is sometimes recognized as a distinct species, *S. uliginosus* M. Bieb. or as *S. arvensis* ssp. *uliginosus* (M. Bieb.) Nyman.

2. **S. oleraceus** L.   Fig. 181   Common Sow-thistle
    Map 509. A persistent weed of yards and gardens, cultivated fields, city streets, parking areas, hedges, foundations of buildings, railroad yards, bulldozed land, and waste places generally. Michigan records go back at least to 1877 (Detroit) with a fragmentary specimen probably this species from 1846 (Macomb Co.).
    Specimens of this species with the largest involucres and those of *S. arvensis* with the smallest involucres are sometimes hard to assign, especially if underground parts and ligule shade are not observed or recorded on the label (see above). The often large and ± acute auricles on the leaves of *S. oleraceus* may help; in *S. arvensis* the auricles run smaller and are rounded (and are also usually smaller and less prickly than in well developed *S. asper*).

3. **S. asper** (L.) Hill   Fig. 182   Prickly Sow-thistle
    Map 510. Roadsides, railroads; a weed of cultivated fields and gardens; waste ground, including gravel pits, construction sites; deciduous woodland, especially along trails and recent clearings; shores, marshy ground, river banks. Collected by the First Survey (1838) in Lenawee County and by 1888 collected in Keweenaw County, so obviously long established in Michigan.
    This species seems to set fruit well in this area, so the distinctive very compressed and smooth achenes are often available for positive identification. The large almost coiled auricles are often striking, especially as they, like the rest of the leaf margin, have prickles that tend to be longer, stiffer, and more numerous than in the other two species. Occasional plants with gland-tipped hairs as in *S. arvensis* var. *arvensis* may be called f. *glandulosus* Beckh.

508. Sonchus arvensis

509. Sonchus oleraceus

510. Sonchus asper

14. **Chondrilla**

1. **C. juncea** L.   Fig. 184   Skeleton-weed

Map 511. A stout-rooted perennial introduced from Europe and spreading across southern Michigan in recent years along sandy roadsides, railroads, and dry fields. First found in the state in 1934 in Kalamazoo County.

The pinnatifid basal leaves wither before the plant flowers and the linear cauline leaves are also mostly shed early. But this species is easily recognized by its dandelion-like achenes, much-branched rush-like stems, and narrow heads with yellow florets.

15. **Prenanthes**   Rattlesnake-root

KEY TO THE SPECIES

1. Involucres and pedicels conspicuously hairy; cauline leaves only very rarely lobed . . . . .1. **P. racemosa**
1. Involucres and pedicels glabrous (or pedicels puberulent); cauline leaves often ± deeply lobed
   2. Inner (i.e., longer) phyllaries 4–5 (6); florets (& achenes) (4) 5 (6) per head; mature pappus pale brown or buff . . . . .2. **P. altissima**
   2. Inner (longer) phyllaries 6–8; florets (& achenes) 8–12 (13) per head; mature pappus dark cinnamon-brown (pappus pale when immature) . . . . .3. **P. alba**

1. **P. racemosa** Michaux

Map 512. Calcareous sandy or rocky shores and interdunal flats, including conifer thickets on them; fens, wet prairies, damp hollows; rock crevices on Lake Superior.

An easily recognized, rather glaucous plant with clasping leaf bases and an elongate, cylindrical, hairy inflorescence that strongly nods when in bud but becomes erect as the heads open. The stem and leaves below the inflorescence are glabrous. A form with pinnatifid lower leaves has been collected in St. Clair County (*Hayes 542 & 555* in 1961, WUD) as well as near Windsor, Ontario.

Two uncommon species, reported from northern Ohio and northern Indiana, may turn up some day in southern Michigan. In *P. aspera* Michaux, the stem and cauline leaves are ± rough-hairy. In *P. crepidinea* Michaux, the cauline leaves are petioled and usually coarsely dentate, the heads with more flowers and in a more open-branched inflorescence. Otherwise, these resemble *P. racemosa*.

2. **P. altissima** L.

Map 513. Deciduous woods and borders, thickets, swamp forests, and often at transitions between upland woods and wetlands.

The leaves of this species and the next vary tremendously from unlobed to variously lobed or parted; although names have been published for some of the expressions of leaf shape, they seem pointless.

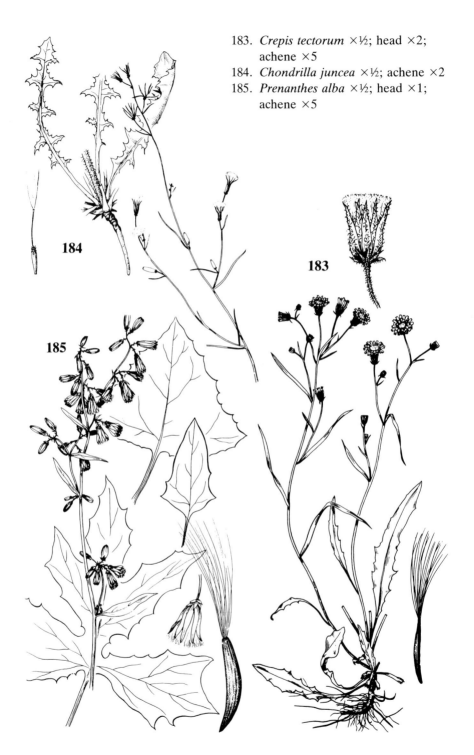

183. *Crepis tectorum* ×½; head ×2;
     achene ×5
184. *Chondrilla juncea* ×½; achene ×2
185. *Prenanthes alba* ×½; head ×1;
     achene ×5

184

183

185

3. **P. alba** L.　　Fig. 185

Map 514. Deciduous woods (beech-maple, northern hardwoods, oak, aspen) especially along borders and clearings; thickets, damp woods, river banks, occasionally even floodplain forests; rock outcrops, jack pines.

16. **Lactuca**　　　　　　　　　　　　　　　　　　Lettuce

The common garden lettuce, *L. sativa* L., is a familiar salad plant in all its forms, probably originating in or near the Middle East area and derived from *L. serriola*. It has apparently been collected only once as possibly out of cultivation in Michigan: on Belle Isle in 1892 (*Farwell 1315*, BLH). Farwell referred to it as occasional in waste places there and on the mainland (Asa Gray Bull. 3: 22. 1895; Rep. Michigan Acad. 2: 68. 1902). However, the evidence is too sparse to include this species in the state's flora. It can be recognized by its very broad leaves with smooth midrib; otherwise it is much like *L. serriola*.

REFERENCE

Vries, I. M. de, & C. E. Jarvis. 1987. Typification of Seven Linnaean Names in the Genus Lactuca L. (Compositae: Lactuceae). Taxon 36: 142–154. [Includes *L. saligna*, *L. sativa*, & *L. serriola*.]

KEY TO THE SPECIES

1. Heads with blue to white ligules; beak (if any) of achene stout and distinctly less than half as long as the body
  2. Mature involucres (12) 13–19 (21) mm long; middle and upper cauline leaves entire, unlobed; pappus white; plant a deep-rooted perennial . . . . .1. **L. pulchella**
  2. Mature involucres (9) 9.5–12 (12.5) mm long; middle and upper cauline leaves all or mostly toothed and lobed; pappus dark (white in one rare species); plant annual or biennial
    3. Pappus white; inflorescence open, the branches widely spreading . . . . . . .2. **L. floridana**
    3. Pappus brownish or dirty gray; inflorescence various, but often elongate-cylindrical with crowded heads . . . . .3. **L. biennis**
1. Heads with yellow ligules (drying blue in nos. 5 & 6); beak of achene distinct, filiform, when fully mature nearly half or more as long as the body (shorter in *L. muralis*)

511. Chondrilla juncea

512. Prenanthes racemosa

513. Prenanthes altissima

4. Florets 5 per head; longer phyllaries 5 or fewer; beak of achene ca. 0.5–1 mm long . . . . .4. **L. muralis**
4. Florets and phyllaries more numerous; beak much longer (at least when fully mature)
  5. Mature achenes with several strong nerves on each face; leaves either prickly on midrib beneath or linear-lanceolate and entire
    6. Leaves smooth on the midrib (rarely a few prickles toward base), linear-lanceolate (ca. 12–30 times as long as broad), entire and unlobed . . . . . . .5. **L. saligna**
    6. Leaves prickly along the midrib beneath, ± oblong or oblanceolate in outline (less than 4 times as long as broad), prickly-margined and often ± lobed . . . . . .
    . . . . .6. **L. serriola**
  5. Mature achenes with at most 1 strong nerve on each side (sometimes a weak pair besides); leaves neither prickly nor all linear-lanceolate and unlobed
    7. Involucres 15–21 mm long at maturity; mature achenes (7.5) 8–9 mm long (including beak), with pappus 8–10 (14) mm long .7. **L. hirsuta**
    7. Involucres (8) 9–12.5 mm long at maturity; mature achenes 4.5–6 mm long, with pappus ca. 5–7 mm . . . . .8. **L. canadensis**

1.   **L. pulchella** (Pursh) DC.   Blue Lettuce
Map 515. A large-headed species ranging mostly northwest of the Great Lakes. Apparently native on ridges at Isle Royale but adventive in waste ground elsewhere.
Sometimes treated as a variety or subspecies of the Asian *L. tatarica* (L.) C. A. Meyer.

2.   **L. floridana** (L.) Gaertner   Woodland Lettuce
Map 516. Our few specimens are from a sandy oak woods and a disturbed cottonwood thicket.

3.   **L. biennis** (Moench) Fernald   Fig. 186   Tall Blue Lettuce
Map 517. Borders of diverse woods and roadsides, trails and recent clearings; often in moist ground, even on floodplains and at edges of swamps; fields and disturbed places.
Robust plants, as often seen at recently disturbed roadsides, logged woods, or building sites, may tower over a person and have a massive ± cylindrical inflorescence. Leaf lobing is variable, as in other species in the genus.

4.   **L. muralis** (L.) Gaertner   Wall Lettuce
Map 518. An invader from Europe, locally abundant on calcareous shores, especially on islands, in the Straits of Mackinac area and spreading into cedar swamps and deciduous forests; also on a sandy river bank in Emmet County. First collected in Michigan at Mackinac Island in 1924 (*Hunnewell 9387*, GH).
By some recent authors, segregated into a separate genus as *Mycelis muralis* (L.) Dumort. Easily recognized by the large, diffuse panicle of

numerous very slender, few-flowered heads and the large, triangular or 3-lobed terminal portion of the pinnatifid leaves.

5. **L. saligna** L.   Willow-leaved Lettuce

Map 519. Another European weed, first collected in Michigan in 1935 near Ypsilanti (*Hermann 6922 & 6972*, MICH, MO, MSC). Still local along roadsides, in oak-hickory woods, and on river banks, but likely to spread.

The ligules are light yellow when fresh but turn blue on drying — illustrating the wisdom of always recording the color of fresh corollas on one's specimen labels! Our collections are all the nomenclaturally typical form [once called f. *ruppiana* (Wallr.) Beck] with unlobed, entire, linear leaves. A form in which the leaves are pinnatifid and rather oblong in overall outline, but otherwise entire or nearly so, may yet invade Michigan from farther south.

6. **L. serriola** L.   Fig. 187   Prickly Lettuce

Map 520. Another European weed, along roadsides and railroads; in fields, gravel pits, alleys, and waste ground; gardens and nurseries; sometimes in low ground. During the decade 1884–1894, collected in six counties across the southern Lower Peninsula; not in the Upper Peninsula until 1933 and still very local there.

514. Prenanthes alba

515. Lactuca pulchella

516. Lactuca floridana

517. Lactuca biennis

518. Lactuca muralis

519. Lactuca saligna

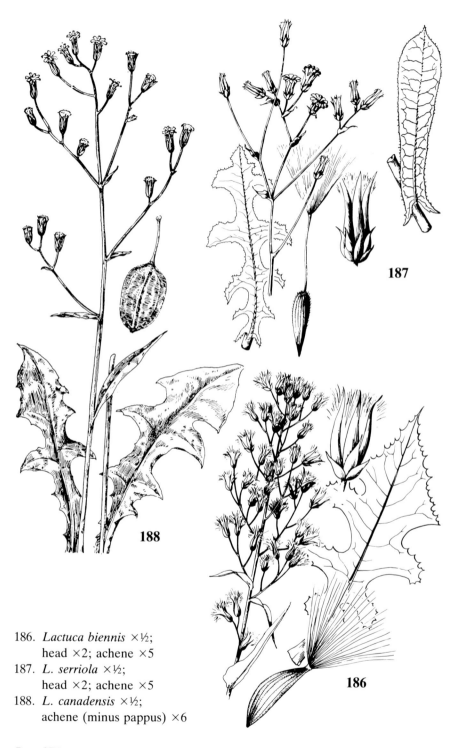

186. *Lactuca biennis* ×½;
     head ×2; achene ×5
187. *L. serriola* ×½;
     head ×2; achene ×5
188. *L. canadensis* ×½;
     achene (minus pappus) ×6

The stiff prickles on the midrib are unique among the species of *Lactuca* thus far known in Michigan (though some others may have hairy midribs) and will also distinguish this species from any *Sonchus* (e.g., *S. asper*) which may have prickly margins but smooth midribs. *L. biennis* may have a few bristles on the midrib beneath.

The name *L. scariola* L. is a later synonym for the same species, not (as often stated) merely a variant spelling.

7. **L. hirsuta** Nutt.

Map 521. Dry ± open ground, including oak-pine woodland on old dunes, clearings among jack pines, sandy bluffs and banks, prairie-like areas.

Our plants are all the glabrous (or very nearly so) phase of this species, which is sometimes hairy on the lower internodes.

8. **L. canadensis** L.   Fig. 188   Wild Lettuce

Map 522. In much the same diverse habitats as *L. biennis*: borders and trails in woods; promptly invading disturbed sites and clearings; sandy pine plains, conifer swamps, rocky ground, shores, fields; roadsides and railroads.

This species and *L. biennis* are the two widespread common wild lettuces in Michigan, and they can be easily separated by pappus color: silky white in *L. canadensis* and dark in *L. biennis*. The two species are known to hybridize but presumed hybrids, with intermediate characters, are apparently quite scarce in Michigan.

Some plants have many of the leaves unlobed and ± lanceolate, but usually the leaves are all or mostly lobed and the lobes tend to be narrowly lanceolate in contrast with the broad lobes in *L. biennis*. The foliage is amazingly variable in both species.

520. Lactuca serriola

521. Lactuca hirsuta

522. Lactuca canadensis

KEY TO THE GENERA (Group B)

1. Involucre of just 2 series of phyllaries distinctly different in size, texture, and/or orientation; heads at least 1 cm broad; cauline leaves opposite or whorled
   2. Leaves all or mostly aquatic (submersed), apparently whorled, fully dissected with filiform segments (usually at least 1 pair of simple, opposite, toothed leaves on emersed stem beneath head); heads radiate, 1–2 at the end of long leafy (submersed) branches; pappus awns elongating to 2.1 [4] cm in fruit . . . . 17. **Megalodonta**
   2. Leaves not aquatic (if finely divided, nevertheless terrestrial); heads various; pappus awns, if any, less than 1 cm long
      3. Phyllaries of inner series (longer and easily seen) connate at least a third of their length; ray flowers none . . . . .18. **Thelesperma**
      3. Phyllaries slightly or not at all connate; ray flowers present or not
         4. Pappus of distinct barbed awns
            5. Ray flowers red to white or orange; leaves deeply pinnately lobed into narrow or even filiform segments; plants rarely escaped into waste places . . . . . . .19. **Cosmos**
            5. Ray flowers none or clear yellow; leaves various (but narrowly lobed in only one species of wet places); plants native, common .20. **Bidens**
         4. Pappus of 2 teeth or scales or wholly obsolete
            6. Leaves all cauline, unlobed, sessile and clasping; outer phyllaries equalling or exceeding the inner, the latter resembling the broad membranous receptacular chaff; disk flowers pubescent just above the ovary .21. **Guizotia**
            6. Leaves cauline and/or basal, unlobed (narrowly lanceolate), lobed, or compound, not clasping; outer phyllaries shorter than the firm inner ones; disk flowers glabrous
               7. Outer phyllaries various (often less than half as long as inner), at most barely connate at the very base; leaves various: simple, or palmately lobed or compound, or if pinnately lobed at least the central lobe usually much larger than the lateral ones; achenes strongly flattened, not beaked, in most species winged . . . . .22. **Coreopsis**
               7. Outer phyllaries more than half as long as inner, clearly connate at the base (for ca. 0.5–1 mm above the receptacle); leaves deeply pinnatisect with uniformly narrowly lanceolate to filiform lobes; achenes little if at all flattened, beaked, without wing . . . . .19. **Cosmos**
1. Involucre imbricate, with phyllaries of more than 2 lengths, or with a single series of phyllaries, or heads less than 1 cm broad; cauline leaves various
   8. Corollas very deeply lobed, the narrow linear lobes at least 4 mm long and about equalling the tube or longer; phyllaries (the middle ones if not all) with prominent fringed or lacerate margins, at least at their tips
      9. Heads apparently globose (1-flowered heads in globose secondary heads!); pappus a distinct fringed crown; leaves all sessile and clasping, densely white-woolly beneath, spiny-margined . . . . .23. **Echinops**
      9. Heads less than hemispherical; pappus none or of a few linear scale-like bristles; leaves neither clasping nor white-woolly beneath, without spines or with spiny tips only at ends of narrow segments . . . . .68. **Centaurea**
   8. Corollas with shorter lobes if any; phyllary margins at most merely ciliate
      10. Margins of phyllaries with a well defined hyaline or scarious border at least around the tip; leaves alternate
         11. Receptacle—and hence the disk—conical or high-hemispherical, the disk (including florets) at least 6 mm broad at the base; leaves all deeply dissected
            12. Heads rayless; bruised plant with fruity (pineapple) aroma . . . .
            . . . . .24. **Matricaria (discoidea)**
            12. Heads with ray flowers; bruised plant with aroma, or unpleasant odor, or no odor

13. Ray flowers yellow . . . . .25. **Anthemis (tinctoria)**
13. Ray flowers white
  14. Receptacle without chaff; leaf segments mostly revolute-filiform, essentially
     glabrous, less than 0.5 mm broad . . . . .24. **Matricaria**
  14. Receptacle with chaff (except at the border in one species); leaf segments
     and axis pubescent, often all or mostly ca. 0.5–2 mm broad, flat . . . .25. **Anthemis**
11. Receptacle flat or slightly convex, the disk in some species less than 5 mm
  broad; leaves dissected or not
  15. Leaves unlobed (but closely and regularly toothed), glabrous or only slightly
     silky when mature
    16. Rays ca. 4–6 mm long; leaves not over 8 mm wide . . . . .27. **Achillea (ptarmica)**
    16. Rays over 10 mm long or absent; leaves over 10 mm wide . . . .
     . . . . .26. **Chrysanthemum** (couplet 2)
  15. Leaves lobed or dissected (or at least deeply toothed near base) or if unlobed
     then entire and densely tomentose at least beneath
    17. Rays present, white
      18. Rays 4–50 (65) mm long; receptacle without chaff . . . . .26. **Chrysanthemum**
      18. Rays 1–3 mm long; receptacle chaffy with an elongate scale at the base of
        each floret . . . . .27. **Achillea (millefolium)**
    17. Rays absent or yellow
      19. Disk conspicuous (yellow), 5–22 mm broad (if less than 10 mm, then the
        leaves nearly glabrous); inflorescence ± corymbiform . . . . .28. **Tanacetum**
      19. Disk inconspicuous, usually less than 5 mm broad (if 5–10 mm, then the
        leaves densely tomentose beneath); inflorescence various
        20. Leaves deeply pinnatifid, the segments toothed; heads in a dense convex
          corymbiform inflorescence . . . . .27. **Achillea (filipendulina)**
        20. Leaves finely dissected; heads in an elongate paniculiform inflorescence
          . . . . .29. **Artemisia**
10. Margins of all or most phyllaries not hyaline or scarious (or if so, the leaves
  opposite); leaves opposite or alternate or all basal
21. Heads without ray flowers
  22. Lower surface of leaves densely white-woolly (stem never spiny); corollas
     present . . . . .30. **Adenocaulon**
  22. Lower surface of leaves glabrate or, if densely pubescent, the hairs neither
     white nor woolly (if whitened in *Xanthium spinosum*, the stem spiny); corollas
     often absent on pistillate flowers
    23. Leaves all alternate; involucre of pistillate heads covered with hook-tipped
      spines, forming a beaked bur in fruit; staminate heads crowded immediately
      above the pistillate heads, with separate phyllaries . . . . .31. **Xanthium**
    23. Leaves (at least the middle and lower) opposite; involucre of all heads with-
      out hooked spines (at most with horn-like protuberances); staminate (or all)
      heads in a branched cymose inflorescence or in elongate raceme-like or
      spike-like inflorescences and the phyllaries united into a cup
      24. Inflorescence branched (cymose); phyllaries ca. 5 mm or more long, sepa-
        rate . . . . .34. **Polymnia**
      24. Inflorescence elongate, raceme-like or spike-like; phyllaries less than 5 mm
        and united into a cup
        25. Staminate and pistillate flowers in the same heads (pistillate at the mar-
          gin) . . . . .32. **Iva**
        25. Staminate and pistillate heads separate, the former on elongate branches,
         the latter at the bases of these branches (fig. 206) (rare plants unisexual) .
         . . . . .33. **Ambrosia**
21. Heads with small to conspicuous ray flowers

26. Phyllaries (not to be confused with adjacent broad receptacular bracts) in a single series, ± lance-ovate and green, only slightly overlapping at base; branches of inflorescence and at least margins of phyllaries with septate hairs and/or stipitate glands; leaves (except sometimes the uppermost) opposite, large (ca. 1–2 dm or even broader), and lobed ....34. **Polymnia**

26. Phyllaries various, usually in more than 1 series, or if in 1 series, the pubescence neither septate nor stipitate-glandular (except in *Madia*); leaves alternate, basal, or opposite (if the latter, less than 1 dm broad)

  27. Rays white to pink or rose-purple

    28. Heads solitary (one per stem or at end of a branch); rays showy, conspicuously longer than the involucre

      29. Plants small (less than 2 dm tall), scapose; rays ca. 1 cm or shorter ....35. **Bellis**

      29. Plants tall (well over 2 dm), leafy-stemmed; rays ca. 3–6 cm long ....
      ....36. **Echinacea**

    28. Heads several per stem or branch; rays inconspicuous, shorter than the involucre (but may protrude beyond it)

      30. Leaves alternate ....37. **Parthenium**

      30. Leaves opposite

        31. Phyllaries ca. 10–12; leaves lanceolate or lance-elliptic, with obscure petiole if any; pappus none or an obscure crown; receptacular bracts hairlike ....38. **Eclipta**

        31. Phyllaries ca. 5; leaves with ovate blade and definite petiole; pappus of narrow, flat, fringed scales (fig. 207); receptacular bracts flat, narrow scales ....39. **Galinsoga**

  27. Rays yellow to orange, sometimes mostly red-purple with yellow bands or tips

    32. Disk flowers sterile, the ovary smaller than on fertile ray flowers, and the style undivided; achenes flattened parallel to the phyllaries, wingmargined ....40. **Silphium**

    32. Disk flowers fertile, the ovary at least as large as on any fertile ray flowers, and the mature style forked; achenes thick and angled or ± flattened at right angles to the phyllaries, wingless (except in *Verbesina*)

      33. Leaves opposite, at least toward base of plant (fragmentary specimens lacking lower part of the stem may be inadequate for identification)[†]

        34. Leaves less than 4 mm wide; rays less than 3 mm long and fewer than 4; phyllaries 4 or fewer ....41. **Madia**

        34. Leaves wider; rays longer and more numerous; phyllaries more than 4

          35. Ray flowers fertile, with style; achenes 3–4-angled; leaves all opposite; phyllaries rounded or blunt at apex ....42. **Heliopsis**

          35. Ray flowers sterile, without style; achenes ± compressed but not angled; leaves of some plants alternate on upper part of stem; phyllaries in most species acute or acuminate ....43. **Helianthus**

      33. Leaves all alternate

        36. Phyllaries sticky or gummy (appearing strongly varnished when dry), the outer ones with prolonged narrow recurved tip; leaves sessile, clasping, toothed, and with shiny glandular dots on both surfaces; receptacle without chaffy bracts (but long pappus bristles evident) ....44. **Grindelia**

        36. Phyllaries not gummy, without recurved tip; leaves various but without above combination; receptacle with or without chaff

[†]**Hymenoxys herbacea** (Greene) Cusick (in Rhodora 93: 238–241. 1991), Lakeside Daisy, has leaves all basal, narrow, gland-dotted; solitary yellow heads; very hairy achenes. It was collected in 1996 (*Case et al.*, MICH) in Mackinac Co.; otherwise known only from Bruce Peninsula and Manitoulin Island in Ontario and Marblehead area in Ohio.

37. Stem (at least upper half) with prominent green wings decurrent from leaf bases the full length of the internode
  38. Rays cuneate, broadest at the 3 (–4)-lobed apex, gland-dotted at least beneath; receptacle without chaff; leaves gland-dotted on both surfaces; achenes angled and ribbed, the pappus of several thin awn-tipped scales . . . . .45. **Helenium**
  38. Rays ± oblanceolate, broadest near or shortly beyond the middle, 2-lobed at apex, not gland-dotted; receptacle with distinct chaff; leaves not gland-dotted; achenes strongly flattened, with 2 persistent rigid pappus-awns . . . . .46. **Verbesina (alternifolia)**
37. Stem not at all winged
  39. Receptacle and disk flat or nearly so
    40. Ray flowers sterile; achenes wingless; chaff of receptacle consisting of elongate flattish bracts easily seen among the disk flowers; plants common (native or introduced), annual or perennial . . . . .43. **Helianthus**
    40. Ray flowers pistillate; achenes winged *or* chaff none; plants rare waifs, annual
      41. Leaves less than 3 mm wide (very numerous); pappus of 5 translucent scales narrowed to an awn tip . . . . .45. **Helenium (amarum)**
      41. Leaves much broader; pappus of 2 awns or none
        42. Receptacular chaff elongate; achenes (of disk florets) flattened, winged, with 2 pappus awns; leaves (at least middle and lower ones) with definite petioles and coarsely toothed ± lanceolate to deltoid blades . . . . .46. **Verbesina (encelioides)**
        42. Receptacular chaff none; achenes boat-shaped, wingless, without pappus but strongly curved and spiny down the back; leaves tapered to a sessile, ± clasping base, entire or nearly so . . . . .47. **Calendula**
  39. Receptacle (and hence disk) conspicuously dome-shaped, conical, or even cylindrical (figs. 217–220)
    43. Rays strongly drooping; achenes ± flattened; receptacle with chaff subtending both ray and disk flowers, with the bracts curved (like a hood) and densely hairy at the apex, concealing the florets in bud; cauline leaves very deeply pinnately lobed or compound, with narrow segments (or leaflets) . . . . .48. **Ratibida**
    43. Rays mostly spreading; achenes plump, angled; receptacle with chaff none or only on the disk, of various nature but not hood-like; cauline leaves variously lobed or usually unlobed (in some *Rudbeckia* with few broad palmate or pinnate lobes, in *Gaillardia* sometimes pinnately lobed and toothed)
      44. Receptacle with irregular setae not subtending individual florets; pappus of several awn-tipped scales; achenes long-hairy, especially near base; rays mostly 3–5-lobed or -toothed at apex . . . . . . .49. **Gaillardia**
      44. Receptacle with a chaffy scale subtending each disk floret (and partly surrounding the achene); pappus none or a tiny crown; achenes glabrous; rays 2-lobed or -toothed at apex . . . . .50. **Rudbeckia**

17. Megalodonta

This genus consists of a single species, often included in *Bidens* but differing not only in its thoroughly aquatic habitat and dissected leaves but also in chromosome number and several morphological characters, most noticeably the extraordinary length attained by the mature pappus awns (which are often 5). Mature fruit, however, is apparently quite uncommon.

REFERENCE

Roberts, Marvin L. 1985. The Cytology, Biology and Systematics of Megalodonta beckii (Compositae). Aquat. Bot. 21: 99–110.

1. **M. beckii** (Sprengel) Greene   Plate 6-C; fig. 189   Water-marigold
Map 523. Lakes, ponds, rivers, and creeks. While the deepest recorded depth in Michigan is 15 feet, this species flowers (and fruits) best in relatively quiet and shallow water.

The heads closely resemble those of the *Bidens* species with ray flowers, but the submersed foliage is often superficially confused with that of other dissected-leaved aquatics. The submersed leaves (opposite but branching at the base so appearing whorled) differ from those of *Myriophyllum* and *Armoracia* in lacking a central axis and in being much branched; from those of *Ranunculus, Utricularia,* and *Armoracia* in being opposite (or seemingly whorled), without bladders. The bushy appearance of the leaves surrounding the stem at the nodes, with bare internodes between, gives sterile pieces, which are often seen washed up on shores, a distinctive tiered aspect (shared only with *Utricularia purpurea*). The simple, opposite, toothed emersed leaves are very different.

Like *Utricularia*, *Megalodonta* has clearly retained the aerial pollination habit of its terrestrial ancestors, while its foliage is for the most part thoroughly adapted for aquatic existence.

18.   Thelesperma
One species native to open places from the Great Plains westward has been collected as a waif in Michigan. It has no ray flowers and deeply pinnatisect leaves with linear or filiform segments. The report of Hanes and Hanes (1947) of *T. filifolium* (as *T. trifidum*) from Kalamazoo County is based on a collection of *Coreopsis tinctoria*.

REFERENCE

Alexander, Edward Johnston, in Earl Edward Sherff. 1955. Thelesperma. N. Am. Fl. II(2): 65–69.

1.   **T. megapotamicum** (Sprengel) Kuntze
Map 524. Collected in 1902 by Emma Cole (MICH) from a farm in Paris Tp., where other adventives from the southwest were also found, reportedly originating in refuse from a felt-boot factory (see Cole 1901, p. 160). In addition to the *Thelesperma*, the herbarium sheet includes a fruiting piece of *Schkuhria multiflora* Hooker & Arn. (not otherwise in the Cole collection) and basal parts of some other plant (with abundant alternate nodes but no leaves). It is hard to imagine that Miss Cole mixed three such different plants, and the mixture could have resulted from some later error in labeling or mounting. (The *Schkuhria* has strongly quadrangular, elon-

gate, obconic achenes hispid especially at the base and glandular; a pappus of membranous scales; and puberulent, distinct phyllaries.)

19. Cosmos   Cosmos

Both of our species are Mexican natives familiar in cultivation and occasionally escaped. They are annuals with deeply pinnatisect leaves, the segments narrowly lanceolate to filiform, and with very showy heads. Most cultivated plants lack the pappus awns normally characteristic of the genus.

REFERENCE

Sherff, Earl Edward. 1955. Cosmos. N. Am. Fl. II(2): 130–146.

KEY TO THE SPECIES

1. Ray flowers red to pink or white .... 1. **C. bipinnatus**
1. Ray flowers orange .... 2. **C. sulphureus**

1. **C. bipinnatus** Cav.
Map 525. Shores, farmyards, and waste places.

2. **C. sulphureus** Cav.
Map 526. Collected in 1942 by the Haneses along a roadside west of Schoolcraft.

The leaf segments run wider in this species than in the preceding (± lanceolate vs. filiform).

20. Bidens   Beggar-ticks

This is a conspicuous genus in moist habitats in the fall, several species often growing together on shores and recently exposed mudflats (our species all annual or biennial). The abundant heads of the radiate species may

523. Megalodonta beckii

524. Thelesperma
megapotamicum

525. Cosmos bipinnatus

offer a sea of color in such habitats—and the abundant fruits with usually retrorse-barbed awns offer abundant nuisance as they embed themselves in socks and other garments. Achenes of *B. cernuus* have been recorded as causing damage in the gill arches of young salmon, leading to fungus infection, at the Platte River Trout Rearing Station of the Michigan Department of Natural Resources (Allison 1967).

The late E. E. Sherff, long the authority on this genus, very kindly examined several hundred Michigan specimens for me in 1957 — after his retirement—and his annotations have been of great help. The distinctions emphasized in the key are largely those utilized by him.

American authors, in particular, have long treated the name *Bidens* as feminine (as did Linnaeus and as do an increasing number of European writers). However, the word is Latin for a 2-toothed hoe and the Code has long called for retention of classical gender of words used as generic names, even if not followed by Linnaeus. (Sherff's defense of the feminine was that *bidens* may be considered an adjective modifying the feminine word *herba* or *planta*—but the Code clearly declares that generic names are substantives [nouns] or treated as such.)

REFERENCES

Allison, Leonard N. 1967. Beggar-ticks Cause Mortality among Fingerling Coho Salmon. Progr. Fish-cult. 29: 113.
Sherff, Earl Edward. 1955. Bidens. N. Am. Fl. II(2): 70–129.

KEY TO THE SPECIES

1. Leaves simple, the middle and lowest sometimes cleft but without a distinctly petiolulate terminal leaflet (except in a rare variety of *B. connatus*); achenes normally 3–4-awned
  2. Leaves sessile, at least the larger ones somewhat connate basally; rays (rarely absent) large and showy, larger than the disk; heads ± nodding at maturity; achenes 4-angled and 4-awned, retrorsely barbed on angles and awns . . . . . 1. **B. cernuus**
  2. Leaves mostly narrowed to distinct (sometimes winged) petioles; rays (very rarely present) shorter than disk; heads erect at maturity; achenes various
    3. Achenes antrorsely barbed (often sparsely so!) at the very base (in one form, up to and including awns), ± flattened, usually rather warty, but at least the central ones in a head 4-angled (i.e., ribbed on 2 faces) and 4-awned at summit when mature (the median awns generally shorter) . . . . .2. **B. connatus**
    3. Achenes retrorsely barbed on the margins their entire length (smooth and glabrous on the faces), strongly flattened with at most 1 distinct midrib, 3- (rarely 2- or 4-) awned at summit . . . . .3. **B. comosus**
1. Leaves definitely compound or so deeply lobed that the terminal leaflet has a distinct petiolule; achenes normally 2-awned
  4. Heads with well developed rays; outer phyllaries small, scarcely if ever exceeding the rays (usually shorter than the disk); margins (and usually awns) of achenes ± antrorsely barbed or ciliate
    5. Achenes narrowly cuneate (± straight-sided), less than 3 mm wide, not thin-margined . . . . .4. **B. coronatus**
    5. Achenes strongly convex-sided, ca. 3 mm or more wide, with distinct thin margin

6. Outer (green) phyllaries 10 or fewer, scarcely if at all exceeding the inner (brownish) ones; sparsely to densely ciliate with slender hairs ....5. **B. aristosus**
6. Outer phyllaries more than 10, much exceeding the inner ones, densely ciliate with coarse broad-based hairs ....6. **B. polylepis**
4. Heads discoid (rarely with rays under 4 mm long); outer phyllaries mostly large and foliose, much exceeding the disk; achenes various
7. Outer phyllaries mostly 4 (3–5), not ciliate; achene faces, margins, and awns antrorsely pubescent ....7. **B. discoideus**
7. Outer phyllaries mostly 6–10 or more, very sparsely to densely ciliate, at least basally; awns of achenes (except in an uncommon form) retrorsely barbed (their margins and faces mostly antrorsely pubescent or ciliate, or glabrate)
8. Outer phyllaries 10 or more; faces of achenes glabrous or nearly so ....8. **B. vulgatus**
8. Outer phyllaries mostly 6–8 (very rarely 10); at least one face of achene antrorsely pubescent ....9. **B. frondosus**

1. **B. cernuus** L.   Plate 6-D; fig. 190   Nodding Beggar-ticks
Map 527. Damp to wet ground on shores (sandy or mucky), mudflats, mucky bottomland, depressions in woods, sedge meadows, fens and bogs, cedar swamps, streamsides, ponds, ditches, and marshes; often growing in abundance with other species of *Bidens* and species of *Cyperus* on recently exposed shores.

Distinctive in the heads' turning as they mature (fig. 190). Depauperate plants flowering when as small as a few cm tall are often frequent in this species (though they occur also in some others), and look quite different at first glance.

2. **B. connatus** Willd.   Fig. 191
Map 528. In the same habitats as the preceding.
A variable species, represented in Michigan by five named varieties and forms (which probably do not deserve such recognition). In f. *anomalus* (Farw.) E. G. Voss (TL: Detroit [Belle Isle, Wayne Co.]), the awns are antrorsely barbed (known also from Monroe, Charlevoix, and Mackinac cos.). In f. *ambiversus* (Fassett) E. G. Voss, the awns have both antrorse and retrorse barbs (collected once at Detroit in 1930). Our common plant

526. Cosmos sulphureus   527. Bidens cernuus   528. Bidens connatus

has been called var. *petiolatus* (Nutt.) Farw., with petioles narrowly margined and the principal leaves mostly undivided. In var. *gracilipes* Fernald, the petioles are very narrowly winged and the leaves often have as many as 5 lobes; this is supposedly a Coastal Plain disjunct, occurring in Muskegon and Kalamazoo counties and also along the Atlantic Coast from Maine to Connecticut.

3. **B. comosus** (A. Gray) Wiegand

Map 529. In the usual habitats for this genus, including shores, ditches, mudflats, floodplains, and other low ground.

Dry specimens are sometimes difficult to distinguish from the preceding, but the species has a distinctive aspect in the field. *B. comosus* is said to have the anthers pale, included in the pale yellow disk corollas, in contrast to blackish anthers in *B. connatus*, exserted beyond the orange disk corollas. The outer phyllaries in *B. comosus* are large and foliaceous, usually much exceeding the disk; in *B. connatus* they tend to be smaller, fewer, and more linear, but this distinction is poorly defined.

This species and the preceding are sometimes included in the Eurasian *B. tripartitus* L., and sometimes only one or the other of them is so included. The relationships are obviously close.

4. **B. coronatus** (L.) Britton   Plate 6-E   Tickseed-sunflower

Map 530. In the same habitats as *B. cernuus*, but more frequently than that species in peatlands (bogs, fens, tamarack swamps).

The awns and, more sparsely, the margins of the achenes are antrorsely ciliate in this species, which is a very showy plant.

5. **B. aristosus** (Michaux) Britton

Map 531. In low ground, but rare in Michigan, having been collected only twice (1930 & 1978), both times in saline areas.

6. **B. polylepis** S. F. Blake

Map 532. Collected in 1994 (*Higman et al. 901*, MICH) along a two-

529. Bidens comosus   530. Bidens coronatus   531. Bidens aristosus

track in an old field. This species and the preceding are perhaps only adventive in Michigan from farther south and west.

Some recent authors include *B. polylepis* in *B. aristosus*, on the basis of extensive intergradation in the heart of their range. Both are much too local in Michigan for evaluation of intergradation here and they are tentatively treated as separate to call attention to their existence in this state. In both taxa, forms exist in which the awns of the achenes are much reduced or absent.

7. B. discoideus (T. & G.) Britton

Map 533. Swamps, often on hummocks and mossy logs; muddy shores, usually in shaded places.

Even the larger plants of this species are rather small-headed, especially in contrast to the next.

8. B. vulgatus Greene

Map 534. Swampy and marshy ground, fields, ditches, along rivers and streams; more weedy than most species (except *B. frondosus*), its habitats including railroads, roadsides, yards, and vacant lots.

In f. *puberulus* (Wiegand) Fernald, the backs of both the inner and outer phyllaries, as well as much of the upper stem and leaves, are strongly puberulent. It has been collected rarely in the state.

9. B. frondosus L.   Fig. 192

Map 535. Moist ground of shores (sandy, gravelly, rocky), swamps (cedar, black ash), depressions in woods, stream and river banks, ponds, fens; more weedy than most other *Bidens*, in wet waste ground, clearings, trails, alleys, barnyards, moist fields, and railroad embankments.

In f. *anomalus* (Fernald) Fernald, the awns are antrorsely barbed (Emmet Co. & Drummond Is.). Depauperate plants may resemble *B. discoideus*, but differ in the awns and at least sparsely ciliate phyllaries.

532. Bidens polylepis   533. Bidens discoideus   534. Bidens vulgatus

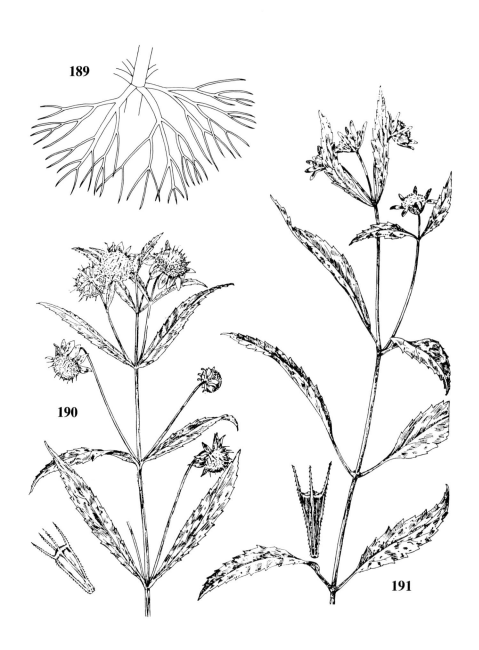

189. *Megalodonta beckii*, leaf ×1½
190. *Bidens cernuus* ×½; achene ×4
191. *B. connatus* ×½; achene ×5

21. Guizotia

1. **G. abyssinica** (L. f.) Cass.   Niger-seed

Map 536. A native of tropical Africa, this species is cultivated for its oily seeds, a component of food mixes for caged birds. The first report of its establishment outside of cultivation in the United States was by Farwell (Pap. Michigan Acad. 23: 133. 1938), who collected it on the banks of the River Rouge in 1930 (*8786*, MICH, GH). In 1976, collected at a parking lot in Richland, Kalamazoo County (*Gillis 13473*, MSC, A).

The ray and disk flowers are both yellow. The heads rather resemble those of a *Bidens*, but the foliage could be confused only with that of *B. cernuus*. The heads, however, do not nod and there are no awns or other pappus.

22. **Coreopsis**   Coreopsis; Tickseed

While the species have reasonably different aspects, it is hard to apply any one key character; and so, as in many composite genera, exceptions to any single feature in the key have to be tolerated. Most species may be either glabrous or pubescent, and of course these and other variants have names published for them.

C. nuecensis A. Heller, a Texas endemic, was collected by Farwell in 1925 (*7455*, BLH, as *C. coronata*; Am. Midl. Nat. 10: 45. 1926). It was a waif, with *C. tinctoria*, in a field by a railroad at Flat Rock, Wayne County. Both species are small annuals, with the lower portions of the rays a deeper yellow or marked with reddish, but the disk flowers of *C. nuecensis* are 5-lobed, the outer phyllaries longer, and the achenes winged. It seems not otherwise to be known as a waif in this part of the continent and is extremely unlikely to turn up in Michigan again. Furthermore, there is some question as to whether seed had been planted at the site.

535. Bidens frondosus   536. Guizotia abyssinica   537. Coreopsis tinctoria

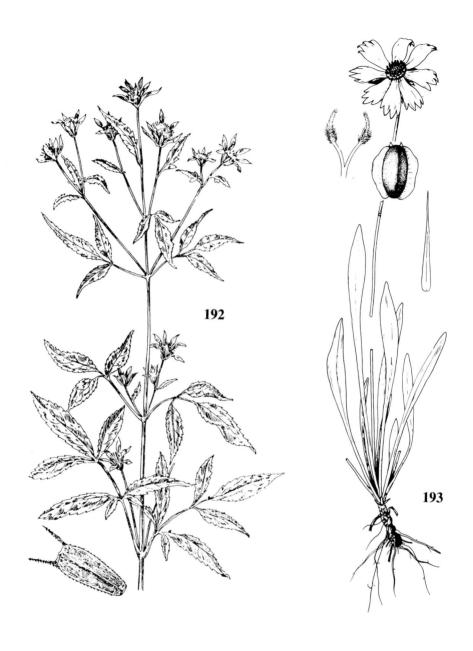

192. *Bidens frondosus* ×½; achene ×4
193. *Coreopsis lanceolata* ×½; achene ×5; style ×10;
    receptacular bract (chaff) ×3

REFERENCES

Sherff, Earl Edward. 1955. Coreopsis. N. Am. Fl. II(2): 4–41.

Smith, Edwin B. 1976. A Biosystematic Survey of Coreopsis in Eastern United States and Canada. Sida 6: 123–215.

KEY TO THE SPECIES

1. Phyllaries of outer series all or mostly less than 2.5 mm long; disk flowers 4-merous, red-purple; ray flowers less than 1.5 cm long, usually red-purple on basal half, otherwise yellow, and 3-lobed apically; achenes wingless . . . . .1. **C. tinctoria**
1. Phyllaries of outer series (2) 2.5–6 mm long; disk flowers 5-merous, yellow (becoming red-brown in *C. tripteris*); ray flowers ca. 1–2.6 (3) cm long, yellow, entire or (3) 4–5-toothed apically; achenes winged
   2. Leaves (except often the uppermost) deeply palmately lobed or trifoliolate (the leaflets sometimes deeply lobed); rays ± elliptical to narrowly obovate, entire or nearly so at apex
      3. Leaves apparently sessile or nearly so (petiole short and winged), with 3 (–5) narrow linear-lanceolate lobes; outer phyllaries nearly as long as the inner . . . . .
         . . . . .2. **C. palmata**
      3. Leaves petioled, compound (3 lance-elliptic leaflets or with central one deeply lobed); outer phyllaries usually about half as long as the inner, or shorter . . . . . .
         . . . . .3. **C. tripteris**
   2. Leaves unlobed (or some with narrow pinnate lobes); rays cuneate, with (3) 4–5 ± irregular apical teeth
      4. Stem with 4–6 nodes, these (and hence pairs of leaves) extending above middle of the plant (the peduncles, from heads to uppermost leaves or bracts, therefore mostly shorter than the rest of the stem); leaves (at least some of them) with (1) 2–4 (7) pairs of lateral lobes . . . . .4. **C. grandiflora**
      4. Stem with 2–4 (5) nodes, these (and hence pairs of leaves) all on lower half of plant (the peduncles therefore longer than rest of the stem); leaves unlobed . . . .
         . . . . .5. **C. lanceolata**

1. **C. tinctoria** Nutt.   Plains Coreopsis; Calliopsis
   Map 537. Native west of the Great Lakes, but a common garden annual sporadically escaped eastward. Occasional along roadsides and railroads.
   The principal leaves are deeply pinnatifid into very narrowly lanceolate, linear, or even filiform segments.

2. **C. palmata** Nutt.   Fig. 194   Finger or Prairie Coreopsis
   Map 538. Prairies and associated roadsides and railroads.
   The leaves are unlike those of any other of our species, with a "birdfoot" appearance.

3. **C. tripteris** L.   Tall Tickseed; Tall Coreopsis
   Map 539. Dry to wet prairies, meadows, marshes; oak woods, especially borders and clearings; fields, roadsides, railroads.
   Another species with distinctive leaves that will readily place it even if there are exceptions to the other key characters (e.g., unusually long outer

phyllaries). Plants are sometimes pubescent, sometimes glabrous, but neither monographer cited above thinks there is any merit in naming these forms. In 1937, L. H. Bailey collected a plant (*933*, BH) near Albion with an extra pair of leaflets, rendering the leaves pinnately 5-foliolate.

4. **C. grandiflora** Sweet
    Map 540. Native in the southeastern United States, locally established as an escape from cultivation. Railroads, fields, and roadsides; sand banks and clearings.
    Not always easily distinguished from the next. Both have achenes which may be granular-papillose on the surface and which usually have a large callus at each end on the inner (concave) face. Some of the variability may represent hybridization with *C. lanceolata*.

5. **C. lanceolata** L.   Fig. 193
    Map 541. Rather local (sometimes common) on sand dunes along Lake Michigan and Lake Huron and adjacent dry to moist shores and borders of woods; open sandy banks, grasslands, roadsides, oak-pine woodland; banks and bluffs. Some populations may represent escapes from cultivation

538. Coreopsis palmata

539. Coreopsis tripteris

540. Coreopsis grandiflora

541. Coreopsis lanceolata

542. Echinops
     sphaerocephalus

543. Matricaria discoidea

(the garden plants themselves transplanted from the wild!), for this showy native does well in gardens and spreads readily.

23. Echinops

REFERENCES

Moore, R. J., & C. Frankton. 1962. Cytotaxonomic Studies in the Tribe Cynareae (Compositae). Canad. Jour. Bot. 40: 281–293. [*Echinops* on pp. 289–290.]

Moore, R. J., & C. Frankton. 1974. The Thistles of Canada. Canada Dep. Agr. Monogr. 10. 112 pp. [*Echinops*, pp. 8–11.]

1. **E. sphaerocephalus** L. Globe-thistle
    Map 542. A native of Eurasia, grown as a garden ornamental and occasionally a conspicuous escape from cultivation. Roadsides and waste places, including abandoned railroads, filled land, grassy sites.
    The corollas, bracts, and anthers all contribute a decidedly bluish look to the large spherical heads, which may be 3–5 cm in diameter. The spiny foliage suggests a true thistle (*Cirsium*), but the heads are very different, not only in their shape but also in their structure. The globose compound heads consist actually of numerous 1-flowered heads, each with an involucre of several spine-tipped phyllaries surrounded by abundant stiff hairs or setae around the base.

24. Matricaria

The nomenclature in this genus is unusually complicated. I follow essentially the conclusions of Rauschert except that *Chamomilla* is not here given separate generic status. (But see Bremer & Humphries 1993.)

REFERENCE

Rauschert, Stephan. 1974. Nomenklatorische Probleme in der Gattung Matricaria L. Folia Geobot. Phytotax. 9: 249–260.

KEY TO THE SPECIES

1. Ray flowers none; disk corollas 4-lobed; bruised plant with pleasant fruity (pineapple) odor . . . . .1. **M. discoidea**
1. Ray flowers present, white; disk corollas 5-lobed; bruised plant aromatic or not
  2. Involucres ca. 4–5.5 mm long; pappus present (though minute); achene ribs 3, strongly thickened or wing-like; plant nearly or quite odorless . . . . .2. **M. perforata**
  2. Involucres ca. 2–3.2 mm long; pappus none; achene ribs 5, ± weak; plant aromatic when bruised . . . . .3. **M. recutita**

1. **M. discoidea** DC. Plate 6-F Pineapple-weed
    Map 543. Usually considered native to western North America, but has migrated eastward in historic times. First collected in Michigan by Farwell in 1921 (Houghton Co.; Pap. Michigan Acad. 2: 43. 1923) and 1925 (Oakland Co.; Am. Midl. Nat. 10: 45. 1926). In the 1930s, six widely dispersed

counties were added to the documented distribution, and the species is still spreading extensively. Usually in relatively bare disturbed ground: roadsides, railroads, cultivated fields and gardens, lawns, cut-over woods, and all sorts of waste ground including vacant lots, dumps, gravel pits, parking lots, fill. This is one of the few weedy American plants that has apparently spread to (rather than from) Europe.

Sometimes segregated in the genus *Chamomilla*, in which case the correct name is *C. suaveolens* (Pursh) Rydb. Long known as *M. matricarioides*, but that name is based on an illegitimate one which under the Code may not be used.

2. **M. perforata** Mérat   Scentless Chamomile
Map 544. A European native, locally established as an adventive. Collected as a lawn weed near Lansing in 1886 (*Bailey*, BH) and not again in the state until 1961 and 1962, when Bourdo found it to be a common weed of waste places in Baraga County. Rarely collected since, at weedy sites and shores (Saginaw River and Lake Superior at Isle Royale).

Often included in *M. maritima* L. (which differs in closer ribs on the achene, elongate resin glands, and more consistently perennial habit). It is currently often segregated in the genus *Tripleurospermum*. The illegitimate name *M. inodora* L. has also been applied to this species.

3. **M. recutita** L.   False Chamomile
Map 545. A Eurasian native, collected in Keweenaw County in the 1880s but like the preceding only very local with us. Dry roadsides and railroads and in other waste places.

When included in the genus *Chamomilla*, this species is called *C. recutita* (L.) Rauschert; in *Matricaria*, the name *M. chamomilla* L. was long misapplied to it.

25. **Anthemis**
A. nobilis L., a perennial now often segregated as *Chamaemelum nobile* (L.) All., is known as chamomile and (although one cannot be certain of

544. Matricaria perforata   545. Matricaria recutita   546. Anthemis tinctoria

such things) is presumably what flavored the chamomile tea given by his mother to Peter Rabbit. Other species in this group of genera are also called chamomile — although not all are so pleasantly fragrant. The three species included here are retained in *Anthemis* even in recent generic realignments within the tribe.

The receptacular chaff is usually quite visible as the slender bristle-like tips extend well up among or beyond the tips of the mature disk flowers.

KEY TO THE SPECIES

1. Ray flowers yellow . . . . .1. **A. tinctoria**
1. Ray flowers white
  2. Receptacle completely chaffy (all the way to the marginal disk flowers); ray flowers pistillate; involucre ca. (4) 4.5–6 (6.5) mm long; achenes ribbed but otherwise smooth; plant nearly or quite odorless. . . . . .2. **A. arvensis**
  2. Receptacle not chaffy around the marginal flowers of the disk; ray flowers sterile; involucre ca. 3–4.2 mm long; achenes warty-tuberculate; plant bad-smelling . . . .
    . . . . .3. **A. cotula**

1.   **A. tinctoria** L.   Yellow Chamomile; Golden Marguerite
    Map 546. A Eurasian species, cultivated and occasionally escaped with garden debris or otherwise. Roadsides, refuse areas, railroads, vacant lots, fields, and shores.

The ray flowers are fertile and retain their cheerful strong yellow color when dry.

Plants of *Tanacetum huronense* with a domed receptacle might be keyed here, but can easily be distinguished by lack of chaff on the receptacle and very short rays (less than 4 mm). *A. tinctoria* has a chaffy receptacle and long rays.

2.   **A. arvensis** L.   Fig. 195   Corn Chamomile
    Map 547. Probably originally native only in southern Europe, but now a widespread naturalized weed. Roadsides, railroads, fields, parking lots, and other waste ground; a weed in lawns, gardens, and nurseries. The earliest Michigan collection I have seen is from Ann Arbor in 1860; apparently collected soon afterwards in Flint (but specimen undated); well established across the state in the 1890s.

3.   **A. cotula** L.   Fig. 196   Mayweed; Dog-fennel;
                                                        Stinking Chamomile
    Map 548. Another Eurasian weed, now cosmopolitan. Although the species was listed by the First Survey, the earliest collection extant seems to be from 1843 (presumably Macomb or Oakland County). Roadsides, parking lots, dumps, farmyards, fields, lawns and gardens, and other disturbed ground.

A. arvensis and two species of *Matricaria* are often mistaken for this species, and the keys should be carefully checked.

194. *Coreopsis palmata* ×½;
     achene ×5
195. *Anthemis arvensis*, achenes ×5
196. *A. cotula* ×½
197. *Chrysanthemum balsamita* ×½

26. **Chrysanthemum**   Chrysanthemum

The common garden chrysanthemums (of which there are several species and hybrids) do not escape from cultivation. They are mostly of Asian origin and are prominent in oriental art and lore.

Generic limits have not been completely agreed upon. Some authors include *Tanacetum* in *Chrysanthemum*. Some segregate *Leucanthemum, Pyrethrum, Balsamita,* and others not in our flora. (See Soreng & Cope 1991 or Bremer & Humphries 1993 for full details.)

KEY TO THE SPECIES

1. Leaves pinnatifid (at least near base) or bipinnatifid
  2. Heads solitary on long stems, 3–5 (6.5) cm broad; foliage scentless, the leaves shallowly once-pinnately lobed . . . . .1. **C. leucanthemum**
  2. Heads several to many, less than 2 cm broad, ± corymbose; foliage strong-scented, the leaves bipinnatifid. . . . . .2. **C. parthenium**
1. Leaves undivided (though sharply or regularly and closely toothed)
  3. Disk less than 1 cm broad; rays none; foliage sweet-smelling; heads numerous on very short pedicels . . . . .3. **C. balsamita**
  3. Disk and conspicuous white rays each over 1 cm; foliage not aromatic; heads 1 to several on each stem or branch
    4. Leaves glandular-punctate, with large acute teeth; heads several on each stem or branch . . . . .4. **C. serotinum**
    4. Leaves not glandular-punctate, with low rounded teeth; heads 1–2 on long peduncles . . . . .5. **C. ×superbum**

1. **C. leucanthemum** L.   Fig. 198   Ox-eye Daisy
Map 549. A Eurasian native widely naturalized in North America. The oldest Michigan collections seen are from Wayne County in 1870 and from several counties in the 1880s (Van Buren, Keweenaw, Kent, Leelanau [N. Manitou], and Mackinac [Island]). So the species was evidently not with us from the earliest days but was widespread and well established by late in the 19th century. Now thoroughly naturalized throughout. Roadsides, fields, railroads, waste ground, shores, clearings and trails in woods; rare in wetlands.

547. Anthemis arvensis

548. Anthemis cotula

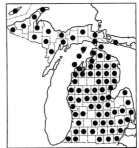

549. Chrysanthemum
     leucanthemum

The leaves ordinarily have 12 or fewer largish teeth and narrow little lobes on each side. Dwarf plants ca. 10–12 cm tall, with heads hardly 1.5–2 cm broad including rays 4–8 mm long, have been collected on the open rocky summit of East Bluff, Keweenaw County (*Richards 3734* in 1950, MICH). Ordinarily the rays are (9) 11–20 (24) mm long in this species, and only 4–6.5 mm long in *C. parthenium*.

C. leucanthemum and its allies (including *C. ×superbum*, below) are often segregated into the genus *Leucanthemum*, in which case (to avoid an inadmissible tautonym), our common roadside daisy becomes *L. vulgare* Lam.

2. **C. parthenium** (L.) Bernh.   Feverfew

Map 550. A European species, often cultivated but tending to be persistently weedy. Roadsides, shores, pastures, yards, trails, dumps; often near present or former sites of human habitation.

This is a bushy, leafy, aromatic plant, long reputed to have medicinal properties. It is also the source of an insecticide, although the silvery-silky *C. cinerariifolium* Trevir. is the usual commercial source of pyrethrin. Both species are sometimes placed in the segregate genus *Pyrethrum* and sometimes in *Tanacetum*.

3. **C. balsamita** L.   Fig. 197   Costmary; Mint-geranium

Map 551. A pleasantly aromatic Eurasian species, occasionally escaped from cultivation and locally well established along roadsides and railroads; fields, banks, meadows; often but by no means always near old homesites.

A form with very short rays is known but has not yet been collected in Michigan. This species is sometimes segregated into a separate genus, *Balsamita*, in which case the name becomes *B. major* Desf.—assuming that the names are treated as taxonomic synonyms. Some authors recognize the plants with ray flowers as a species distinct from those lacking ray flowers, and then the latter must apparently be called *Chrysanthemum majus* (Desf.) Asch. (and those with ray flowers, if placed in the genus *Balsamita*, would require an epithet other than *balsamita* to avoid a tautonym). The nomenclature is complicated by serious tautonym and homonym problems; e.g., *C. balsamita* L. of 1763, adopted here for a broad species concept, renders illegitimate any later combination based on *Tanacetum balsamita* L. (1753—also recognized in 1763), which is the same species in a broader concept but not a nomenclatural synonym. Furthermore, some recent authors have included this (or these) species in the genus *Pyrethrum*, but the most recent assignment is to accept *Tanacetum balsamita* L. (see Soreng & Cope 1991 or Bremer & Humphries 1993).

4. **C. serotinum** L.   Giant Daisy

Map 552. A native of Europe, reported [as *C. uliginosum* (Willd.) Pers.] from out of cultivation at several places in New England, Quebec, and

Ontario (Rhodora 47: 389–390. 1945), but not included in relevant manuals. Collected as an escape at Amasa in 1967 (*Bourdo*, MSC).

Sometimes considered a species of *Pyrethrum*, but now usually placed in a separate genus as *Leucanthemella serotina* (L.) Tsvelev.

5. **C. ×superbum** J. W. Ingram   Shasta Daisy
Map 553. This is a well known horticultural plant, presumed to be a hybrid between two European species; quite variable, with numerous cultivars (see Baileya 19: 167–168. 1975). Apparently collected only once out of cultivation in Michigan, along a sluggish stream in marly flats on Little Traverse Bay (*McVaugh 9622* in 1948, MICH).

The heads are similar to those of the weedy ox-eye daisy, but they run larger and the leaves are merely toothed rather than lobed, well developed toward the middle of the stem (rather than toward the base). Despite the distinctions, this taxon and its presumed parents are sometimes treated as variants of *C. leucanthemum*.

27. **Achillea**   Yarrow

REFERENCES

Gervais, Camille. 1977. Cytological Investigations of the Achillea millefolium Complex (Compositae) in Quebec. Canad. Jour. Bot. 55: 796–808.

Tyrl, Ronald J. 1975. Origin and Distribution of Polyploid Achillea (Compositae) in Western North America. Brittonia 27: 187–196.

Warwick, S. I., & L. Black. 1982. The Biology of Canadian Weeds. 52. Achillea millefolium L. s.l. Canad. Jour. Pl. Sci. 62: 163–182.

KEY TO THE SPECIES

1. Corollas yellow; leaves pinnatifid, with toothed segments . . . . .1. **A. filipendulina**
1. Corollas white (or pink); leaves either finely dissected or merely toothed
  2. Leaves simple, linear-lanceolate, sharply and closely (but finely) toothed; rays ca. 4–6 mm long. . . . . .2. **A. ptarmica**
  2. Leaves finely dissected; rays (1) 1.2–3 mm long. .3. **A. millefolium**

550. Chrysanthemum   551. Chrysanthemum   552. Chrysanthemum
      parthenium   balsamita   serotinum

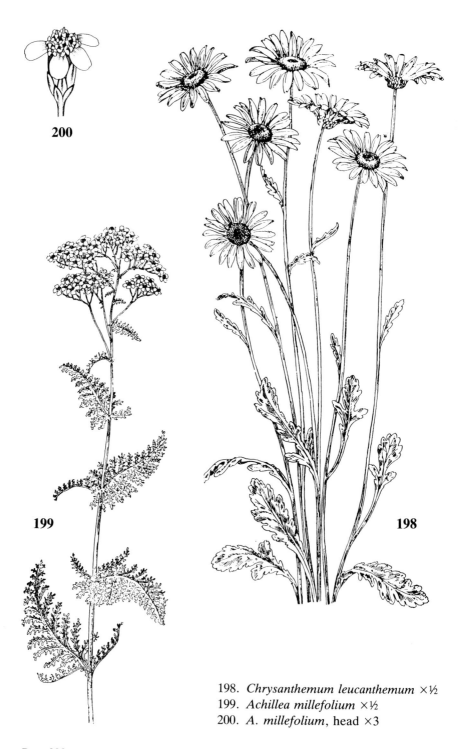

198. *Chrysanthemum leucanthemum* ×½
199. *Achillea millefolium* ×½
200. *A. millefolium*, head ×3

1. **A. filipendulina** Lam.   Fern-leaf Yarrow
   Map 554. Native of eastern Europe and Asia Minor, frequently cultivated and collected rarely in Michigan as an escape in waste ground.

2. **A. ptarmica** L.   Sneezewort; Sneezeweed
   Map 555. A Eurasian species sometimes grown in gardens (usually a
form with excess rays) and occasionally escaping to roadsides and railroads,
near old dwellings, in fields and waste places.

3. **A. millefolium** L.   Figs. 199, 200   Yarrow; Milfoil
   Map 556. Almost everywhere, especially dry roadsides, railroads, fields,
and waste places; rock shores and cliffs, dry woodlands, grasslands; less
often in damp ground (sometimes peaty), shores, river banks, and meadows.
   Although weedy in habit, most of our plants are presumably native. *A.
millefolium*, as treated here and in most other recent works, is a polyploid
complex of native and introduced plants that hybridize and intergrade with
one another.
   For a while, it appeared that native American plants [*A. lanulosa* Nutt.
or *A. millefolium* var. *occidentalis* DC.] were tetraploid, distinguishable
from the rarely introduced Old World hexaploid, *A. millefolium* sens. str.
A northern hexaploid with phyllaries very dark-margined, *A. borealis*
Bong. [which has sometimes been treated as including *A. millefolium* var.
nigrescens E. Meyer, a tetraploid], was supposed to range south to Lake
Superior. More recent work (see references) has revealed that the situation
is even more complex, with native hexaploids and pentaploid hybrids
occurring—as well as the occasional pink- to red-rayed forms found in wild
variants and escaping as cultivars of the Old World *A. millefolium*. Warwick and Black (1982) present a good summary and bibliography.
   Cytological studies have not been done on Michigan material, but morphological variation is rampant (as early noted by Farwell, Am. Midl. Nat.
11: 267–268. 1929). There is an apparent gradient from north to south in
decreasing incidence of dark-bordered phyllaries (with various shades of

553. Chrysanthemum
     ×superbum

554. Achillea filipendulina

555. Achillea ptarmica

intermediate and light brown also occurring). Plants with rays light to deep pink are rather frequently found. Their correct name at rank of form will vary depending on what classification if any is adopted at the rank of subspecies and/or variety. Some of the deepest magenta shade are doubtless cultivars of the Eurasian species and some may well be (as documented in Quebec) pentaploid hybrids. Native tetraploids tend to have a lighter rosy shade.

Yarrow has a strong aroma, especially when bruised, though less intense than that of *Tanacetum*.

28. **Tanacetum**   Tansy

This genus is sometimes included in *Chrysanthemum*—and sometimes circumscribed to include certain species often placed in *Chrysanthemum* (or *Pyrethrum*). Both of our species are strong-scented.

KEY TO THE SPECIES

1. Heads 5–10 mm broad; leaves glabrous or nearly so . . . . .1. **T. vulgare**
1. Heads (12) 13–19 (22) mm broad; leaves ± hairy .2. **T. huronense**

1. **T. vulgare** L.   Common Tansy

Map 557. A Eurasian species, much cultivated and widely naturalized, forming large colonies from strong rhizomes. Roadsides, fields, meadows, ditches, shores; vacant lots, farmyards, old building sites, fill, dumps, and other waste places; sometimes spreading into woodlands and clearings, but not doing well in shade. In Michigan at least in the 1860s and widespread by the 1890s.

Like other aromatic plants, this species has been used medicinally. The leaves are ± densely glandular-punctate as well as glabrous and less finely divided than in the next species.

2. **T. huronense** Nutt.   Frontispiece   Lake Huron Tansy

Map 558. Thrives on fairly active sand dunes along the Great Lakes; relic (usually without flowers) on old well stabilized dunes; on upper sandy or even cobbly beaches, withstanding some wave action at times of high water; rarely in crevices of limestone.

The type locality is the sandy shores of Lake Huron at the Straits of Mackinac, where it was found by Thomas Nuttall in 1810. *T. huronense* is remarkably restricted in its distribution on the Great Lakes, being found on the Door Peninsula of Wisconsin, northern Michigan shores, and adjacent Ontario shores of Lake Superior (see Michigan Bot. 2: 103–104. 1963). One would expect a plant of such northern affinity to occur farther north and west on sandy beaches of Lake Superior (as does *Elymus mollis*). The species also occurs in Maine and the Maritime region of Canada, north to Newfoundland, as well as in the Hudson Bay area and on beaches of

Lake Athabaska. The same or very similar species grow in Alaska and the Pacific Northwest. Hultén (who includes *Tanacetum* in *Chrysanthemum*) treats *huronense* as a subspecies of the similar *C. bipinnatum* L., which is also in Alaska. Plants of ocean dunes in the Pacific Northwest have been referred by different authors to *T. huronense* or *T. douglasii* DC. There is not agreement on the significance (if any) of the varieties that have been named in *T. huronense*. Whatever the best taxonomy of this complex may be, there is no doubt that the name applies at least to plants of the Great Lakes, whence it was described.

The leaves are usually not conspicuously glandular-dotted, but are hairy and more finely divided than in *T. vulgare*. The heads run twice as large as in that species, but are fewer, usually 3–12 (22) on a major stem, compared with 10–100 or more in *T. vulgare*. The aroma is similar. The rhizomes of *T. huronense* may help in stabilizing sand dunes, although very large colonies of the plant are not seen. The species usually has very small rays (to 3.5 mm), while these are absent in *T. vulgare*.

29. **Artemisia**   Wormwood

A rather large genus of mostly aromatic plants, some of which provide well known spices or flavorings, such as tarragon, absinth, and the bitter wormwood. Several species have been used as medicine, vermifuges, and insecticides. Others are cultivated for their ornamental foliage. *A. tridentata* Nutt. is the common sagebrush of the West, a bushy shrub.

Most species have ± deeply pinnatifid or at least lobed leaves. Sometimes there is a pair of lobes at the very base of the leaf, superficially suggesting stipules (and if these are the only lateral lobes, as is sometimes the case, the leaf may be misinterpreted as simple).

REFERENCE

Maw, M. G., A. G. Thomas, & A. Stahevitch. 1985. The Biology of Canadian Weeds. 66. Artemisia absinthium L. Canad. Jour. Pl. Sci. 65: 389–400.

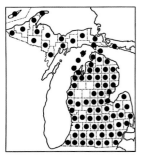
556. Achillea millefolium

557. Tanacetum vulgare

558. Tanacetum huronense

KEY TO THE SPECIES

1. Principal leaves unlobed or pinnately lobed (the axis and/or lobes mostly 4 mm or more broad); leaves strongly tomentose (at least beneath)
    2. Apex of leaf (and of lobes) rounded; heads large (for *Artemisia*), with involucres 5.5–7 mm long . . . . .1. **A. stelleriana**
    2. Apex of leaf (and of lobes if any) acute; heads smaller, with involucres (2.5) 3–4.2 (4.5) mm long
        3. Stem (at least upper part) and upper surface of leaves at least somewhat tomentose; leaves unlobed or at most once-pinnate with entire lobes . . . .
        . . . . .2. **A. ludoviciana**
        3. Stem above and upper surface of leaves glabrous or nearly so; leaves ± bipinnately lobed . . . . .3. **A. vulgaris**
1. Principal leaves finely dissected or bipinnatifid (the axis and lobes less than 4 mm broad); leaves tomentose or otherwise pubescent or glabrous
    4. Leaves on both surfaces and stems glabrous to puberulent or loosely hairy—not densely tomentose or silky
        5. Leaves bipinnatifid or pinnatifid with toothed segments, the segments all flat; plants annual or biennial, from a taproot
            6. Involucres ca. 1–1.8 mm long; heads all on short pedicels in an open, much-branched inflorescence; leaves sweetly aromatic, much reduced in the in-florescence . . . . .4. **A. annua**
            6. Involucres (at least the largest) ca. (1.8) 2–2.5 mm long; heads mostly sessile or subsessile, in tight spike-like clusters; leaves only weakly if at all aromatic (like faint *A. absinthium*), well developed in the inflorescence, usually exceeding the spike-like clusters of heads . . . . .5. **A. biennis**
        5. Leaves 1–3 times pinnatifid, often with ± revolute linear-filiform entire segments mostly less than 1.2 mm broad; plants biennial or (usually) perennial
            7. Stem and leaves (at least when young) ± closely puberulent; stem woody below, usually branched; plant a rare aromatic escape from cultivation . . . . .6. **A. abrotanum**
            7. Stem and leaves glabrous or with at least scattered long hairs; stem herbaceous, unbranched (except in the inflorescence); plant a common nearly scentless native on dunes and other sandy ground . . . . .7. **A. campestris**
    4. Leaves densely silky or tomentose at least beneath
        8. Heads with long white hairs on the receptacle, among the florets; leaf segments (or unlobed leaves in the inflorescence — except for *A. frigida*) clearly flat, mostly 1.3–3 mm broad
            9. Leaf segments linear-filiform, less than 1 mm broad; plant relatively small and slender (but woody at base), ca. 3.5 dm or less tall, mildly aromatic . . . . .8. **A. frigida**
            9. Leaf segments flat, over 1 mm broad; plant coarse and spreading, usually much taller, strongly aromatic . . . . .9. **A. absinthium**
        8. Heads with naked receptacle; leaf segments narrowly linear-filiform, mostly less than 1.2 mm broad
            10. Plant a common native of sandy places, biennial or perennial from a taproot or caudex; involucre usually glabrous; leaves 1–3 times pinnatifid, the ultimate segments many times as long as broad . . . . .7. **A. campestris**
            10. Plant a rare introduction, perennial from slender rhizomes; involucre ± densely tomentose; leaves once-pinnatifid and/or the ultimate segments appearing as prolonged teeth ca. 2–4 times as long as broad
                11. Leaves once-pinnatifid (or entire or with only a basal pair of lobes) . . . .
                . . . . .10. **A. carruthii**
                11. Leaves bipinnate, the ultimate segments ca. 2–4 times as long as broad (fig. 202) . . . . .11. **A. pontica**

1. **A. stelleriana** Besser   Dusty-miller; Beach Wormwood
   Map 559. A native of beaches in northeastern Asia, cultivated for its felt-like foliage and rarely escaped to beaches on the Great Lakes (also on the Atlantic coast), rarely to other habitats.
   This is one of the few non-aromatic species of *Artemisia*.

2. **A. ludoviciana** Nutt.   Western Mugwort
   May 560. Roadsides, railroads, fields, sandy open ground, waste places. Usually considered native on prairies and dry soils from Wisconsin and Illinois westward. A collection (*Henson 3015* in 1989, MICH, UMBS) from a prairie-like community along the Menominee River may well be native (as also, without habitat, an 1839 Menominee County collection by Douglass Houghton). Most of our occurrences are surely adventive from the West.
   This is an even more variable species farther west, but our specimens differ considerably in lobing (if any) of leaves. Most plants have narrow, ± lance-linear leaves, referable to what is often called var. *pabularis* (A. Nelson) Fernald, which is interpreted by Boivin (Phytologia 23: 91. 1972) as equal to the older var. *gnaphalodes* (Nutt.) T. & G.

3. **A. vulgaris** L.   Mugwort
   Map 561. A Eurasian species, locally established as an escape from cultivation along roadsides and railroads; in fields, sandy open ground, and waste places.

4. **A. annua** L.   Sweet Wormwood
   Map 562. A native of Asia and eastern Europe, sparingly established in waste ground such as vacant lots and railroad yards, and in other disturbed ground.
   The foliage looks much more delicate than in *A. biennis*, and the large terminal panicle of tiny heads is distinctive.

5. **A. biennis** Willd.   Fig. 201   Biennial Wormwood
   Map 563. A native of the northwestern United States, but spreading

559. Artemisia stelleriana   560. Artemisia ludoviciana   561. Artemisia vulgaris

eastward as a weed and early established in Michigan. Collected in the 1860s in Berrien, Washtenaw, and Wayne counties; said by Dodge in 1896 to be "becoming common" at Port Huron. Waste places, roadsides, railroads, agricultural land and gardens; shores, rocky ground, often in moist sites.

6. **A. abrotanum** L.   Southernwood
    Map 564. Another cultivated Eurasian species, rarely escaped to roadsides and waste places.

7. **A. campestris** L.   Wild Wormwood
    Map 565. Sand beaches and dunes, gravelly and rocky shores; sandy oak and jack pine woodlands; sandy fields, roadsides, railroad ballast, river banks, and prairies.
    This variable and common native species is now generally considered to include *A. caudata* Michaux, *A. canadensis* Michaux, and *A. borealis* Pallas. There is great diversity in size of heads, nature and distribution of pubescence on the leaves and involucres, amount of lobing of the leaves (although the lobes are always very long and narrow), and the nature of the inflorescence—often a large terminal panicle with hundreds of heads.
    The foliage is scarcely if at all aromatic, and variable as to pubescence. Collections with both involucres and leaves densely silky have been noted from Antrim and Keweenaw counties.

8. **A. frigida** Willd.
    Map 566. Native to dry open places west of the Great Lakes, but collected as probably a railroad waif by Farwell near Dearborn in 1917 (*4527*, BLH; Rep. Michigan Acad. 20: 192. 1919). The species is known from areas in both northern and southern Ontario adjacent to Michigan and is likely to appear in the state again.
    The slender habit and short stature (as this genus goes), with very narrow, densely silky leaf segments, characterize this species, which has long hairs on the receptacle as in the next one.

562. Artemisia annua   563. Artemisia biennis   564. Artemisia abrotanum

9. **A. absinthium** L.   Absinth; Common Wormwood

Map 567. A Eurasian species, frequently established as an escape from cultivation to roadsides and adjacent fields, fencerows, dooryards, and waste ground.

This is a spicy-aromatic plant, the bitter principle long in use medicinally although now generally illegal in beverages (as allegedly hallucinogenic and addictive).

10. **A. carruthii** Carruth   Kansas Mugwort

Map 568. A Great Plains species, rarely adventive eastward. What appears to be this was collected in 1902 on a farm near Grand Rapids (*Cole, MICH*)—the same site from which a number of other western plants were found as waifs from refuse. (See *Thelesperma megapotamicum*.)

11. **A. pontica** L.   Fig. 202   Roman Wormwood

Map 569. A European native, grown for its foliage and occasionally established along roadsides, banks, and other sandy open disturbed ground.

The lacy blue- or ashy-green foliage is attractive; the plant rarely forms flowers, but the feathery shoots are distinctive in the short segments of the dissected leaves, which are ± tomentose on both surfaces.

565. Artemisia campestris

566. Artemisia frigida

567. Artemisia absinthium

568. Artemisia carruthii

569. Artemisia pontica

570. Adenocaulon bicolor

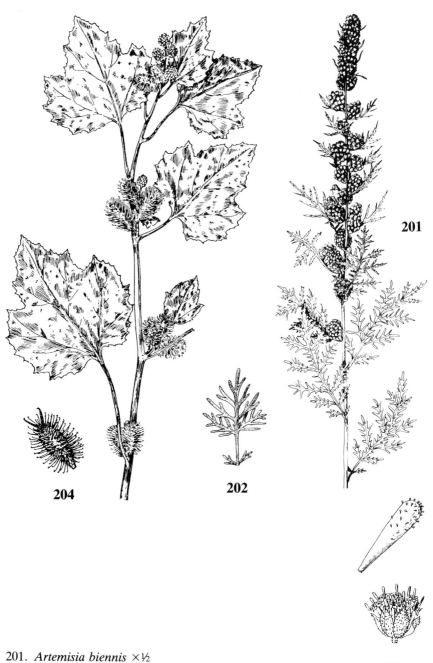

201. *Artemisia biennis* ×½
202. *A. pontica*, leaf ×1
203. *Adenocaulon bicolor*, head & achene ×4
204. *Xanthium strumarium* ×½

30. Adenocaulon

Besides the single species in North America, this genus is represented in Chile and in eastern Asia—a specialist in disjunct distribution.

1. **A. bicolor** Hooker   Plate 6-G; fig. 203   Trail Plant

Map 570. Deciduous and mixed woods, especially hemlock-hardwoods, often in damp ravines and along trails. Occurs in North America from southern British Columbia south to central California and east to northwestern Montana. It is found again in the Black Hills of South Dakota and is then disjunct to the upper Great Lakes region. In Ontario, it was collected on the Bruce Peninsula in 1895 and, 99 years later, near Owen Sound. Otherwise its only documented occurrence around the Great Lakes is in the Michigan counties bordering Lake Superior (although it may grow considerably inland in those counties). (See Michigan Bot. 20: 61–62. 1981.) There are good stands locally in the Porcupine Mountains State Park, Huron Mountains, Pictured Rocks National Lakeshore, and Tahquamenon Falls State Park.

The bicolored leaves (green above and white beneath) are distinctive at all times, crowded near the base of the stem, with long petioles and ± deltoid blades sinuate or shallowly lobed. The leaves are the source of the apt epithet *bicolor* and also the common name, for as one hikes in woods where the plant is common, the disturbed leaves with bottoms exposed reveal one's trail. The short gland-tipped hairs on the stem (whence the name *adeno-caulon*), inflorescence, and achenes are even more distinctive. The small white disk flowers are in a little button-like head, with the elongate achenes looking rather out of scale.

31. Xanthium   Cocklebur

REFERENCES

Cronquist, Arthur. 1945. Notes on the Compositae of the Northeastern United States. II. Heliantheae and Helenieae. Rhodora 47: 396–403.

Löve, Doris, & Pierre Dansereau. 1959. Biosystematic Studies on Xanthium: Taxonomic Appraisal and Ecological Status. Canad. Jour. Bot. 37: 173–208.

KEY TO THE SPECIES

1. Stems with trifid spines ca. 1 cm or more long at the nodes; leaf blades tapering to a short petiole . . . . .1. **X. spinosum**
1. Stems spineless; leaf blades ± cordate, mostly about 1–2 times as long as their petioles . . . . .2. **X. strumarium**

1. **X. spinosum** L.

Map 571. A species of questionable nativity but probably originally from South or Central America, now a widespread weed throughout the world. However, collected only twice in Michigan: in waste ground at Detroit in 1893 (*Farwell 1396*, BLH, MSC) and on the shores of Little Traverse Bay at Bay View in 1896 (*Fallass*, ALBC, MICH).

2. **X. strumarium** L.   Fig. 204

Map 572. Like the preceding species, a nearly cosmopolitan weed, perhaps in part native to North America but probably only an undocumented alien in Michigan. As in many species, the early records correspond quite well to the early distribution of botanists in the state. It was collected as early as 1838 by the First Survey on a dry roadside in St. Joseph County. Much later it was gathered in the counties of Wayne (1874, by A. B. Lyons), Ingham (1885, by L. H. Bailey), and Van Buren (1885, by Bailey); and it continued to spread (or at least to be collected): Gratiot (1890, by C. A. Davis), St. Clair (1892, by C. K. Dodge), Kent (1893, by E. J. Cole).

The species is represented in our area by two varieties of dubious merit, which tend to run into each other: var. *canadense* (Miller) T. & G., with burs hairy toward the base of the spines; and var. *glabratum* (DC.) Cronquist, with burs either glabrous or bearing only puberulence and/or glandular dots on lower portions of the spines and surface of the bur. Numerous species have been segregated in the past but are now generally recognized as part of the variable *X. strumarium*. Whatever their nativity, the burs are no more fun than those of burdock (*Arctium*) to have engaged in fur or clothing.

32. **Iva**   Marsh-elder

KEY TO THE SPECIES

1. Heads sessile in the axils of leafy bracts that exceed them . . . . .1. **I. annua**
1. Heads mostly short-pedicellate, bractless . . . . .2. **I. xanthiifolia**

1. **I. annua** L.

Map 573. Like the next, an annual weed, native southwest of Michigan. Collected only once, in 1916, as a waif in railroad yards on the west side of Detroit (*Billington*, MICH, WUD, BLH, WMU; & *Chandler*, MSC).

2. **I. xanthiifolia** Nutt.   Fig. 205

Map 574. Native to the central part of North America, adventive eastward and seldom seen in Michigan, although first collected in the Upper Peninsula as long ago as 1888 (Keweenaw Co.) and in the Lower Peninsula first in 1893 (Ingham Co.). Waste ground in dumps, roadsides, vacant lots, and around barnyards and old buildings.

33. **Ambrosia**   Ragweed

Were it not for *Ambrosia*, this Flora would probably never have been written. Grandparents seeking relief from hay-fever settled on the "water-washed" air of the Straits of Mackinac as a safe place for them—and me— to spend the summers. So it was the result of trying to escape a plant that led me to the botanical riches of my adopted state.

Of course, there *is* some ragweed in northern Michigan, although not a great deal, and there is even less still farther north. Ragweed is a short-day plant, initiating its flower buds (which ultimately lead to seeds) after the days begin to shorten in mid-summer. Days of the required brevity arrive later and later the farther north one goes, and soon there is insufficent time for seeds to set before plants are killed by early frost. So ragweed and many other short-day plants (including *Xanthium*) do not reproduce well north of the latitudes for which they are programmed.

The involucre encloses the fruit, which is thus dispersed as a hard, nut-like package.

Ambrosia, Iva, and *Xanthium,* with their copious wind-dispersed pollen and very reduced flowers, are fairly closely related and are generally placed as advanced (i.e., reduced) members of the tribe Heliantheae, as subtribe Ambrosiinae, less often at tribal rank as Ambrosieae (see Payne 1970). *Parthenium* (genus 37 below) also belongs to this group. All can cause hayfever, but tend to avoid some of the blame by blooming inconspicuously at the same time of year as more colorful but innocent genera, e.g., *Solidago* (goldenrod), which have showy heads and insect-dispersed pollen that cannot waft into one's nostrils.

REFERENCES

Bassett, I. J., & J. Terasmae. 1962. Ragweeds, Ambrosia Species, in Canada and their History in Postglacial Time. Canad. Jour. Bot. 40: 141–150.

Bassett, I. J., & C. W. Crompton. 1975. The Biology of Canadian Weeds. 11. Ambrosia artemisiifolia L. and A. psilostachya DC. Canad. Jour. Pl. Sci. 55: 463–476.

Bassett, I. J., & C. W. Crompton. 1982. The Biology of Canadian Weeds. 55. Ambrosia trifida L. Canad. Jour. Pl. Sci. 62: 1003–1010.

Payne, Willard W. 1970. Preliminary Reports on the Flora of Wisconsin No. 62. Compositae VI. Composite Family VI. The Genus Ambrosia—the Ragweeds. Trans. Wisconsin Acad. 58: 353–371.

Wagner, W. H., Jr., & T. F. Beals. 1958. Perennial Ragweeds (Ambrosia) in Michigan, with the Description of a New, Intermediate Taxon. Rhodora 60: 177–204.

Wagner, W. H., Jr. 1958. The Hybrid Ragweed, Ambrosia artemisiifolia × trifida. Rhodora 60: 309–316.

571. Xanthium spinosum   572. Xanthium strumarium   573. Iva annua

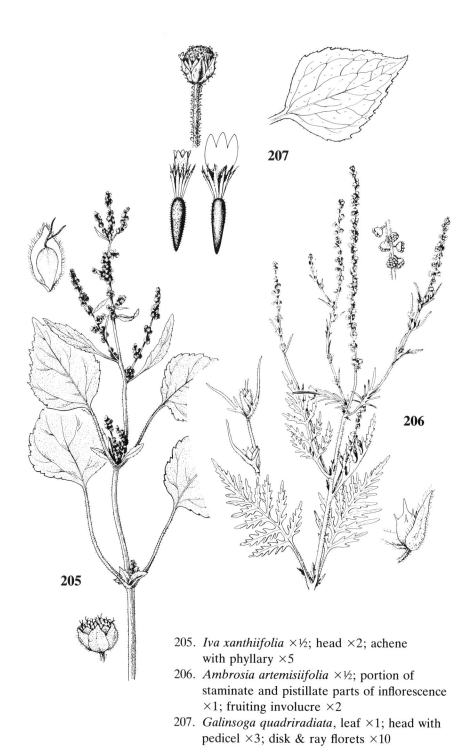

205. *Iva xanthiifolia* ×½; head ×2; achene
with phyllary ×5
206. *Ambrosia artemisiifolia* ×½; portion of
staminate and pistillate parts of inflorescence
×1; fruiting involucre ×2
207. *Galinsoga quadriradiata*, leaf ×1; head with
pedicel ×3; disk & ray florets ×10

KEY TO THE SPECIES

1. Leaves 3 (–5)-lobed (± palmate) or simple, all opposite . . . . .1. **A. trifida**
1. Leaves once or more pinnatifid, the upper ones tending to be alternate
  2. Leaves once-pinnatifid, rough above with ± appressed stiff hairs; plants perennial,
    with horizontal roots . . . . .2. **A. psilostachya**
  2. Leaves once- to thrice-pinnatifid, smooth to scabrous above with tiny appressed
    hairs, any longer hairs spreading; plants annual, from a taproot . . . . .3. **A. artemisiifolia**

1. **A. trifida** L.   Fig. 208   Giant Ragweed
    Map 575. Considered a native species, typical of river floodplains, but weedy in habit. In richer, moister soil than our other species, including low ground near lakes and streams, fens, meadows; roadsides, farmyards, gardens, alleys, vacant lots, and other disturbed ground. Collected by the First Survey in 1837 (St. Joseph Co.), but the relative recency of many collections, especially in the Upper Peninsula, suggests considerable expansion of range in historic time.
    Plants 2 m tall are not unusual; our other two species never come close to such a height. The fruiting involucres are ca. 6–11 mm long, including a short beak below which are several stout, short, thorn-like protuberances. In our other species, the fruits are less than 5 mm long.

2. **A. psilostachya** DC.   Western Ragweed
    Map 576. Usually considered native west of Michigan; not collected in the state until 1902 (Emmet and Kent cos.) and 1903 (St. Clair Co.); it was already on Isle Royale by 1910. So western ragweed must have spread rapidly once it reached the state, apparently around the turn of the century.
    The fruits lack thorn-like protuberances below the beak, or at least some of them are poorly developed and blunt. The longer, denser hairs on the leaves give the plant a decidedly gray-green aspect, quite noticeable in the field, contrasting with the darker green aspect of *A. artemisiifolia*.
    This species has sometimes gone under the name of *A. coronopifolia* T. & G.

574. Iva xanthiifolia   575. Ambrosia trifida   576. Ambrosia
                                                            psilostachya

Hybrids with the next species were reported by Wagner and Beals to be widespread in Michigan, and have been named *A.* ×*intergradiens* W. H. Wagner (TL: Benzie Co.). Also perennial, the hybrid often closely resembles *A. psilostachya*, but tends to be intermediate in several respects, as detailed by Wagner and Beals. Intermediate pubescence on the leaves is perhaps the most easily recognized feature, although sometimes that is close to *A. artemisiifolia*. In the field, the hybrid usually exists in a clone near the parent species, making it more readily distinguished.

3. **A. artemisiifolia** L.   Fig. 206   Common Ragweed
     Map 577. Although weedy in habit, considered native and known from Michigan as early as 1838 (Cass Co.). By the 1880s, documented from Detroit to Keweenaw County—as good a record as for many other unquestionably native species. Favored habitats include roadsides, railroads, fields, farmyards, cultivated ground, logged areas, and other disturbed sites; also in oak woodland, on stream banks and even floodplains, rocky ground.
     Plants can be quite large (even to 1 m tall) and bushy, whereas in *A. psilostachya* they are usually little if at all branched. The fruits have a ring of well developed thorn-like protuberances below the base of the beak.
     Besides the hybrid with *A. psilostachya* mentioned above, a much less common hybrid with *A. trifida* is known, *A.* ×*helenae* Rouleau. This has been recognized thus far in Michigan only in Cheboygan and Washtenaw counties (Wagner 1958).

34. **Polymnia**   Leaf-cup
     Both species are at the northern edge of their range in Michigan. They are both large-leaved, conspicuous plants but as they are relatively rare they are seldom noticed.

577. Ambrosia
      artemisiifolia

578. Polymnia uvedalia

579. Polymnia canadensis

REFERENCES

Wells, James R. 1965. A Taxonomic Study of Polymnia (Compositae). Brittonia 17: 144–159.
Wells, James R. 1967. The Genus Polymnia (Compositae) in Michigan. Michigan Bot. 6: 94–96.

KEY TO THE SPECIES

1. Pedicels and phyllaries stipitate-glandular, without septate hairs; leaves (at least the lower) palmately lobed; rays yellow, ca. 1.3–2 cm long ....1. **P. uvedalia**
1. Pedicels and phyllaries with septate hairs (sometimes stipitate-glandular besides); leaves deeply pinnately lobed; rays white or pale yellow, minute to ca. 1 mm long or absent ....2. **P. canadensis**

1. **P. uvedalia** (L.) L.   Fig. 209
   Map 578. Very rare and local in rich woods and moist borders of mixed hardwoods.

2. **P. canadensis** L.
   Map 579. Dry to swampy deciduous woods and occasionally in conifer swamps.

35. Bellis

1. **B. perennis** L.   English Daisy
   Map 580. An introduction from Europe, locally well established as a weed of lawns, golf courses, and rarely waste places. First collected in Michigan in 1897 at Port Huron, but very rarely gathered again until the 1960s.
   Resembles a miniature ox-eye daisy (*Chrysanthemum leucanthemum*) in its white rays and yellow disk, but the spatulate leaves are all crowded at the base of the stem.

36. Echinacea   Purple Coneflower

These plants are sometimes cultivated for their handsome large heads with pale to deep reddish purple rays and dark (maroon to purple or brown) disk. The receptacular bracts have conspicuous slender rigid tips that extend to and beyond the tips of the disk flowers.

The familiar garden *Zinnia violacea* Cav. [long known as *Z. elegans* Jacq.], a Mexican native, has been reported from Michigan, but without documentation or assurance that the plants were not actually in cultivation. The genus would key here, but differs from *Echinacea* in having opposite rather than alternate leaves (and the cultivated zinnias also generally are "double"—with extra ray flowers).

REFERENCE

McGregor, Ronald L. 1968. The Taxonomy of the Genus Echinacea (Compositae). Univ. Kansas Sci. Bull. 48: 113–142.

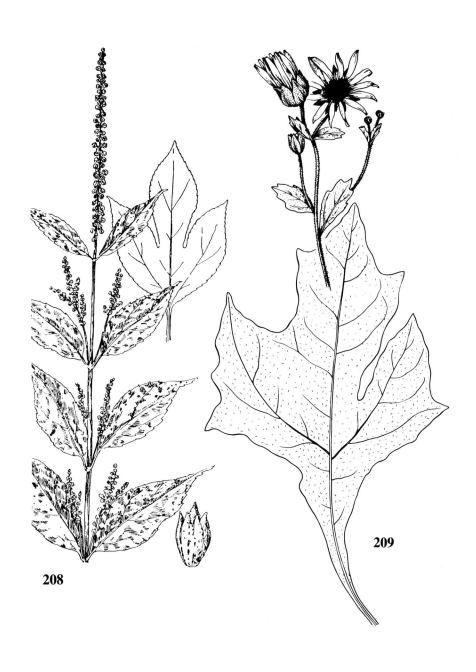

208. *Ambrosia trifida* ×½; fruiting involucre ×2
209. *Polymnia uvedalia* ×½

KEY TO THE SPECIES

1. Leaf blades ovate to ovate-lanceolate (ca. 2–4 times as long as broad, on distinct petioles), with at least a few teeth . . . . .1. **E. purpurea**
1. Leaf blades oblanceolate to long-elliptical (ca. 8–15 times as long as broad, tapering into the petiole), entire . . . . .2. **E. pallida**

1. **E. purpurea** (L.) Moench

Map 581. Collected by the First Survey in 1838 at Gull Prairie, Kalamazoo County, and also in St. Joseph County (no stated locality); these may well have been native occurrences for this species, but other counties must surely represent adventives or escapes from cultivation in fields and along roadsides.

2. **E. pallida** (Nutt.) Nutt.

Map 582. Chiefly in disturbed ground: along railroads (and adjacent prairie), fields, rocky clearings, edge of a gravel pit (Beaver Island). Like the preceding, native south and west of Michigan. Even our southern collections are from along railroad tracks and it is doubtful whether they represent native stands. One of the best known populations in the state is in a clearing along the highway at the site of the old lumbering town of Kenneth, northwest of Moran in Mackinac County, where there are thousands of plants blooming in July. The citing of this species from Baraga County by McGregor resulted from incorrect assignment of material labeled as from Keweenaw Point, where Farwell collected it in 1886 and again in 1941.

The rays in this species are paler than in the preceding, and at maturity are decidedly drooping.

37. **Parthenium**

A shrubby species in this genus, native to the Southwest and adjacent Mexico, is known as guayule [*P. argentatum* A. Gray] and has sufficient latex to have been the subject of investigations as a source of rubber. Our species are herbaceous and lack latex.

580. Bellis perennis   581. Echinacea purpurea   582. Echinacea pallida

REFERENCE

Rollins, Reed C. 1950. The Guayule Rubber Plant and Its Relatives. Contr. Gray Herb. 172. 73 pp. [Treats the entire genus.]

KEY TO THE SPECIES

1. Leaves deeply pinnatifid or bipinnatifid; plants annual; pappus of 2 scales ....
   ....1. **P. hysterophorus**
1. Leaves merely toothed or crenate; plants perennial; pappus of 2–3 weak awns
   2. Upper leaves clearly auriculate-clasping, with spreading hairs on nerves beneath; heads with disk over 7 mm broad ....2. **P. hispidum**
   2. Upper leaves sessile but not clasping, with mostly appressed hairs on nerves beneath; heads with disk at most 7 mm broad ....3. **P. integrifolium**

1. **P. hysterophorus** L.   Santa Maria
   Map 583. Native to Latin America, but adventive northward. Collected in 1917 at a construction site on the University of Michigan campus in Ann Arbor (*Ehlers 652*, MICH, GH).

2. **P. hispidum** Raf.
   Map 584. Native from Kansas and Missouri southward and westward. A railroad waif collected in Kalamazoo County from 1936 to 1969.
   Sometimes treated as a variety of the next species [var. *hispidum* (Raf.) Mears].

3. **P. integrifolium** L.   American Feverfew
   Map 585. Native in prairie remnants in adjacent Indiana (and beyond), so should be sought in southwestern Michigan. However, our only collections, presumably as waifs or escapes from cultivation, are from three sites in Keweenaw County 1939–1950.

38. Eclipta

1. **E. prostrata** (L.) L.   Yerba de Tajo
   Map 586. Apparently a native American plant of weedy habit in moist ground, but not collected in Michigan until 1949. Now, at least, locally frequent in the Lake Erie area on muddy banks and shores, railroad ballast, and grassy hollows.
   The species has sometimes been known as *E. alba* (L.) Hassk.

39. Galinsoga   Peruvian Daisy; Quickweed

REFERENCES

Canne, Judith M. 1977. A Revision of the Genus Galinsoga (Compositae: Heliantheae). Rhodora 79: 319–389.
Warwick, S. I., & R. D. Sweet. 1983. The Biology of Canadian Weeds. 58. Galinsoga parviflora and G. quadriradiata (= G. ciliata). Canad. Jour. Pl. Sci. 63: 695–709.

1. Pappus scales (of disk flowers) lacking an awn; ray flowers with pappus obsolete; gland-tipped hairs of pedicels less than 0.5 mm long; teeth of leaf margin obscure, broadly rounded or little more than thickened bumps . . . . .1. **G. parviflora**
1. Pappus scales, or many of them, with a slender terminal awn distinguishable from the cilia; ray flowers with pappus well developed; gland-tipped hairs of pedicels mostly 0.5 mm or more long; teeth of leaf margins usually definite, acute in outline . . . . .2. **G. quadriradiata**

1. **G. parviflora** Cav.

Map 587. Probably originally native to the southwestern United States and Central America, but now a nearly worldwide weed, especially in gardens and cultivated fields (although very local in Michigan). All definite records from the state out of cultivation are from 1938 and later.

2. **G. quadriradiata** Cav.   Fig. 207

Map 588. Native to Central and South America but like the preceding species now a nearly cosmopolitan weed of temperate and subtropical areas. Thrives in populated areas: poorly kept lawns; along hedges, sidewalks, and gardens; roadsides and parks; and in greenhouses. First col-

583. Parthenium
   hysterophorus

584. Parthenium hispidum

585. Parthenium
   integrifolium

586. Eclipta prostrata

587. Galinsoga parviflora

588. Galinsoga
   quadriradiata

lected in Michigan in Ingham County (1892), but soon (1901) found in Detroit; widespread across southern Michigan and even in Gogebic County before 1920.

Most authors, following Canne, now include *G. ciliata* (Raf.) S. F. Blake (as our plants were long known) in *G. quadriradiata*, an Andean species originally described from Peru.

40. Silphium   Rosin-weed

These are strikingly large plants, with big heads and usually very large leaves. Both surfaces of the leaves (or chiefly the upper in *S. integrifolium*) are strongly roughened with small, pale, hard papillae, these sometimes with a tiny sharp tip.

REFERENCE

Fisher, T. Richard. 1966. The Genus Silphium in Ohio. Ohio Jour. Sci. 66: 259–263.

KEY TO THE SPECIES

1. Principal leaves all opposite
   2. Leaf bases connate around a strongly 4-angled stem . . . . 1. **S. perfoliatum**
   2. Leaf bases sessile (or even slightly clasping) but not connate, on a ± terete stem
   . . . . 2. **S. integrifolium**
1. Principal leaves alternate or basal
   3. Leaves alternate, deeply pinnately lobed . . . . 3. **S. laciniatum**
   3. Leaves basal, unlobed (occasionally one or more much-reduced cauline leaves present) . . . . 4. **S. terebinthinaceum**

1. S. perfoliatum L.   Fig. 210   Cup Plant

Map 589. River banks and floodplain woods; plants of fields, an abandoned orchard, railroad embankments, and such places are probably not native there. The species is native south and west of Michigan, barely entering the state, and is perhaps in part only adventive here (e.g., in Cass, Kalamazoo, and Lenawee cos.).

2. S. integrifolium Michaux

Map 590. Michigan is at the northeastern edge of the range of this species of prairies and open woods. Our collections are from dry and wet prairies, sedge meadows, and damp open ground at borders of woods; gathered once from a Scots pine plantation.

See comments below under *Heliopsis* on possible confusion with that genus.

3. S. laciniatum L.   Compass Plant

Map 591. Another plant ranging, like *S. perfoliatum*, mostly to the south and west. At least in part probably adventive in Michigan along railroads (e.g., Jackson, St. Clair, Washtenaw, and Wayne cos.)—although railroad rights-of-way and depauperate prairies are indeed about all we have left of its proper habitat.

The common name refers to the tendency of the lower leaves to orient themselves vertically in a north-south direction when exposed to full sun (thus escaping the full radiation that a horizontal leaf would receive).

Plants having leaves intermediate with those of the next species are presumably hybrids and have been collected in Menominee and Wayne counties.

4. **S. terebinthinaceum** Jacq.   Plate 7-B   Prairie-dock
Map 592. This species does not range as far west into the prairies as the others, but does seem at home in southern Michigan, especially in calcareous places. Prairies and similar grassy habitats (even conspicuous along roadsides), fens, railroad embankments.

The very large, vertical leaves (blade up to 3 dm broad on long petioles) make this a very visible plant, recognizable from a distance even without the tall (at least to 3.5 m) stem with large heads. The large woody taproot is likewise impressive.

41. **Madia**

1. **M. glomerata** Hooker   Mountain Tarweed
Map 593. Native in the far West, rarely adventive this far east. Collected in 1929 by Louise Bach and Bruno Gladewitz in River Rouge Park (*Farwell 8570*, BLH; Am. Midl. Nat. 12: 73. 1930).

The phyllaries and inflorescences are stipitate-glandular, and the plant smells of tar. It is a slender, nondescript, glandular-pubescent annual with lower leaves opposite and the upper ones alternate.

589. Silphium perfoliatum   590. Silphium   591. Silphium laciniatum
                                   integrifolium

42. Heliopsis

REFERENCE

Fisher, T. Richard. 1958. Variation in Heliopsis helianthoides (L.) Sweet (Compositae). Ohio Jour. Sci. 58: 97–107.

1. **H. helianthoides** (L.) Sweet   Fig. 213   False Sunflower
   Map 594. Prairies, fens, meadows, river banks, and other low ground; woodlands and thickets (especially at borders and clearings) with aspen, birch, and other trees; evidently spreading along roadsides and railroads.

   This species is frequently mistaken for a *Helianthus* (see couplet 35 in key to genera of Group B for a reminder of the differences). It could also be hastily confused with *Silphium integrifolium* if one recognized a sunflower-like plant but with fertile ray flowers as not, therefore, being *Heliopsis*; but the *Silphium* has the upper leaves entire or nearly so and essentially sessile, while *Heliopsis* has the leaves rather coarsely toothed and clearly petioled.

   This is a variable species, but there is not agreement on what, if anything, to call the intergrading variants. Fisher recognized three subspecies: ssp. *helianthoides*, with leaves smooth above and stems glabrous, ranging north to southern Michigan; ssp. *occidentalis* T. R. Fisher, with leaves and upper stems glabrous, ranging east from the prairies into Michigan and to New England; ssp. *scabra* (Dunal) T. R. Fisher, also scabrous, restricted to Missouri, neighboring Illinois, and southwest into Texas. Others treat these as varieties, and most recent authors have included *occidentalis* in *scabra* [once recognized as a species, *H. scabra* Dunal]. Boivin (1972) called the northern scabrous plant var. *scabra* (Dunal) Fernald and renamed the more narrow-leaved scabrous plant from farther southwest.

43. **Helianthus**   Sunflower
   This is a notoriously difficult genus native to North America, with hybridization obscuring the differences between a number of species. Some speci-

592. Silphium terebinthinaceum

593. Madia glomerata

594. Heliopsis helianthoides

mens will not work in this (or any other) key. A perennial question is how to balance hybridization vs. natural variation within species. It is not easy to know where to place the blame! As Heiser (1969) stated: "The identification of hybrids is, obviously, facilitated if the other species which occur in the area are known." (He also noted that perhaps the greatest contribution of his treatise "is an explanation as to why sunflowers are difficult.") Long and Heiser examined a great many of our specimens 1959–1960, and other specialists in certain groups of species examined some earlier. Their annotations have been most helpful in understanding where they draw the lines; but the naming of more recent accessions in herbaria is less authoritative. Incomplete specimens (e.g., lacking basal parts or adequate notes on the label) are simply not satisfactory for identification in many instances, should never have been collected, and have sometimes had to be ignored rather than risk serious misidentification.

Many characters used in this genus (at least in the key) are difficult to apply in the absence of both options before the user. Experience with the genus will help to interpret such non-quantitative terms as "densely" pubescent or "nearly" glabrous stems. It is usually best to judge stem pubescence at the middle of the stem. Some species or individuals with hairy stems may have them glabrous toward the base (or even merely harsh, from persistent bases of worn-off hairs, at the middle); and some with stems glabrous at the middle may have short, usually antrorse, hairs toward the summit and on branches of the inflorescence. Even defining (and thus measuring) a petiole can be puzzling, as it is usual for at least a narrow strip of green leaf tissue to border the petiole on each side, making it "winged." In general, I have considered the distal end of the petiole to be where the blade ± abruptly expands and the sides of the winged petiole are therefore no longer nearly parallel. (I have not adopted the position of E. E. Watson, an early student of the genus, who defined the distal end of the petiole as the point of confluence of the lower pair of lateral veins with the midrib.)

In several species some of the upper leaves are usually or consistently alternate—perhaps 2 leaves or perhaps all but the lowermost 2 or 3 pairs. Depauperate plants (e.g., in *H. annuus*) tend to have more opposite leaves as well as looking in other ways quite different from plants grown under optimal conditions. Leaves in some species (e.g., *H. giganteus, H. strumosus, H. tuberosus*) may have scattered to dense glandular dots beneath. Some species thrive in moist ground and others on dry sites, but most at least tolerate a wide range of habitat and readily spread into disturbed ground, obscuring the source of their nativity.

"Helianthus *is a genus of infinite and bewildering variability, and its individuals must be studied in their entirety. Plant collecting is not an adjunct to an afternoon stroll; it is a serious but not very well paid occupation.*"

—E. E. Watson, 1929

"Doubtless sometimes you wish to call someone a mean name. Well I have found it. Just call him a sunflower. That combines all that is needed. The brutes have no principles, guided by no laws, and seem to be free for alls."

—C. C. Deam to Paul Weatherwax, May 1939

REFERENCES

Baker, Francis J. 1987. Flower of the Sun. Michigan Nat. Resources 56(4): 34–41.

Clevenger, Sarah, & Charles B. Heiser, Jr. 1963. Helianthus laetiflorus and Helianthus rigidus—Hybrids or Species? Rhodora 65: 121–133.

Dodge, Chas. K. 1895. Helianthus tuberosus in Eastern Michigan. Asa Gray Bull. 3: 17–18.

Heiser, Charles B., Jr., with Dale M. Smith, Sarah B. Clevenger, & William C. Martin, Jr. 1969. The North American Sunflowers (Helianthus). Mem. Torrey Bot. Club 22(3). 218 pp.

Heiser, Charles B., Jr. 1976. The Sunflower. Univ. Oklahoma Press, Norman. xxvi + 198 pp.

Jeffrey, C. 1981. Tab. 822. Helianthus pauciflorus. Bot. Mag. 183: 115–118 + pl.

Long, Robert W. 1961. Biosystematics of Two Perennial Species of Helianthus (Compositae), II. Natural Populations and Taxonomy. Brittonia 13: 129–141.

Swanton, C. J., P. B. Cavers, D. R. Clements, & M. J. Moore. 1992. The Biology of Canadian Weeds. 101. Helianthus tuberosus L. Canad. Jour. Pl. Sci. 72: 1367–1382.

KEY TO THE SPECIES

1. Plants annual, from a taproot; disk (i.e., lobes of disk corollas) brown to purple (very rarely yellow); leaves with blades ± broadly ovate on definite petioles, usually mostly alternate (except at lowest nodes)
    2. Leaf blades distinctly deltoid, less than 1.3 times as long as broad, sharply and regularly toothed (resembling leaves of *Populus deltoides*); receptacular chaff glabrate to merely ciliate at tip; phyllaries glabrous to sparsely hispidulous on back . . . . .1. **H. debilis**
    2. Leaf blades longer and/or irregularly (if at all) toothed; receptacular chaff with tuft of long white hairs at tip or phyllaries with long stiff hairs on back
        3. Receptacular chaff at center of disk tipped with fringe of prominent long white hairs; phyllary margins and back scabrous or with stiff bristles less than 0.3 mm long . . . . .2. **H. petiolaris**
        3. Receptacular chaff at most all uniformly ciliate; phyllary margins and backs with longer stiff hairs . . . . .3. **H. annuus**
1. Plants perennial; disk (i.e., lobes of disk corollas) yellow (except in *H. pauciflorus*); leaves with blades variable, in many species sessile or nearly so and/or narrowly elliptic, in some species all opposite
    4. Phyllaries all rounded to obtuse (or nearly so) at the apex; disk corollas red-purple (yellow in the hybrid *H.* ×*laetiflorus*, often with less obtuse phyllaries) . . . . .4. **H. pauciflorus**
    4. Phyllaries (except sometimes a few outer ones) acute to acuminate at the apex; disk corollas yellow
        5. Leaves mostly crowded at the base of the plant (only 1 or 2 pairs of reduced leaves or bracts on upper half of stem); phyllaries ± imbricate (of different lengths) and appressed in the involucre . . . . .5. **H. occidentalis**
        5. Leaves present along full length of stem; phyllaries looser, not overlapping much on apical half, often with prolonged spreading tips
            6. Principal mid-cauline leaves (if not all) sessile or subsessile (with petioles, if any, less than 4 mm long)

7. Leaves clasping at the base and (like the stem and phyllaries) ± densely soft-pubescent . . . . .6. **H. mollis**
7. Leaves with blades tapered or rounded at base but not clasping
  8. Widest part of leaf near or shortly below the middle, the base ± tapered and often including a short, indistinct, winged petiole
    9. Leaves as scabrous beneath as they are above, linear to narrowly elliptic-lanceolate, usually ± folded lengthwise, arcuate, and entire (rarely toothed); stem with ± appressed hairs, the upper part (or peduncles) usually with evident dense white antrorse pubescence; phyllaries with margins very rarely bearing cilia as long as 1 mm and at least some with tip prolonged into a soft, non-green bristle; plant of dry, open habitats . . . . .7. **H. maximilianii**
    9. Leaves less densely scabrous or smooth beneath (although softer pubescence may be present), narrowly elliptic to ovate-lanceolate, flat, ± distinctly though shallowly toothed; stem with spreading hairs, the upper part (or peduncles) seldom with appressed white pubescence; phyllaries with marginal cilia mostly 1 mm or more long and with tip acute or attenuate but hardly bristle-like; plant of moist habitats . . . . .8. **H. giganteus**
  8. Widest part of leaf at base of blade, which is broadly obtuse or rounded or even subcordate, with at most a tiny petiole
    10. Stem at least sparsely hairy (or scabrous from persistent bases of worn hairs); lowest lateral veins of leaf joining midrib slightly above base of blade (i.e., green tissue present below junction of the veins) . . . . .9. **H. hirsutus**
    10. Stem glabrous and smooth, even glaucous, at least below the inflorescence (except at nodes); lowest lateral veins of leaf usually joining midrib at base of blade (i.e., at summit of the extremely short "petiole") . . . . .10. **H. divaricatus**
6. Principal mid-cauline leaves with petioles (± winged) over 5 mm long
  11. Stems with ± spreading, sometimes scattered, hairs or at least scabrous from bases of worn hairs
    12. Phyllaries with marginal cilia all or mostly less than 0.8 mm long; leaf blades ovate-elliptic, the largest at least (3) 3.5 cm broad, on conspicuously winged petioles ca. 1.5 cm or longer . . . . .11. **H. tuberosus**
    12. Phyllaries with many or most cilia 1 mm or more long; leaf blades lance-elliptic, less than 3 cm (very rarely 4.5 cm) broad, on shorter petioles . . . . . .
      . . . . .8. **H. giganteus**
  11. Stems glabrous or essentially so, even glaucous, especially toward the base (not always near the heads)
    13. Heads (including rays) ca. 3 cm broad or smaller, the disk less than 1 cm broad and the rays ca. 0.8–1.6 cm long .12. **H. microcephalus**
    13. Heads at least 4 cm broad, the disk ca. 1.5–2.3 cm broad and the rays ca. 1.8–4.3 cm long
      14. Leaves only slightly scabrous to the touch above (the tiny pustulate-based projections appressed), usually not over 3 (rarely 5.5) cm broad, the blades ca. 4–8 times as long as broad, alternate on upper part of stem . . . .
        . . . . .13. **H. grosseserratus**
      14. Leaves very scabrous to the touch above (the tiny sharp projections more erect or spreading), usually over 3 cm broad, the blades ca. (2.2) 2.5–3.8 times as long as broad, opposite throughout (or sometimes 1–2 pairs alternate below heads)
        15. Leaf blades often coarsely and evidently toothed, glabrate or at most sparsely pubescent beneath; phyllaries with prolonged and loose tips distinctly longer than the disk (though some often reflexed) . . . . .14. **H. decapetalus**
        15. Leaf blades shallowly toothed to entire, ± densely pubescent beneath; phyllaries usually only slightly if at all exceeding the disk . . . . .15. **H. strumosus**

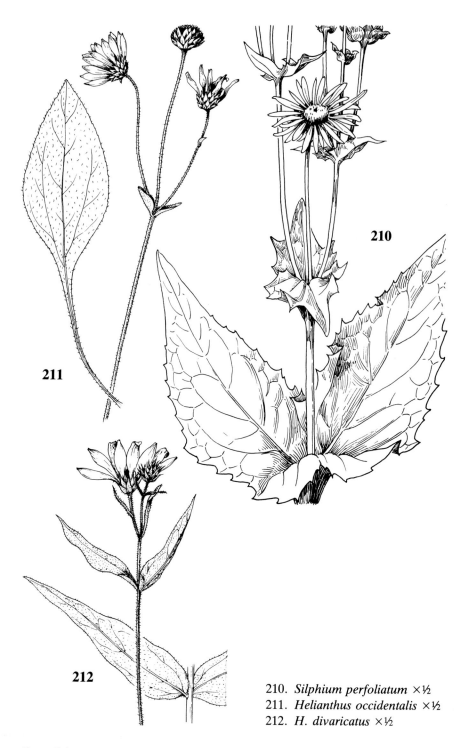

210. *Silphium perfoliatum* ×½
211. *Helianthus occidentalis* ×½
212. *H. divaricatus* ×½

1. **H. debilis** Nutt.

Map 595. Collected as a waif at River Rouge, Detroit, by Farwell in 1912 (*2888*, BLH) and along a railroad near East Lansing by Watson in 1923 (MSC). This rare escape from cultivation is var. *cucumerifolius* (T. & G.) A. Gray, a native of Texas.

The leaves are quite distinctive, with blades less than 8 (9) cm long, and plants never attain the robust size of which *H. annuus* is capable.

2. **H. petiolaris** Nutt.   Plains Sunflower

Map 596. Like many of our species, adventive from farther west, or spread from cultivation, along railroads and roadsides and in waste places near habitations. All our collections are from the 1890s (Kent Co.) or (usually) much later.

3. **H. annuus** L.   Common Sunflower

Map 597. Along railroads and roadsides; in dry fields and on river banks; shores (even among driftwood on Lake Superior); refuse heaps and other waste places. Presumably native to western North America, but long cultivated and early introduced to Europe.

Well known as a garden ornamental and also as a crop plant for the oil from its seeds, which are also popular in themselves as nutritious food for humans, birds, and other creatures. Hence, the species frequently escapes from cultivation as a weed. The waifs are generally much smaller than carefully selected and nurtured parents. Among many references on sunflowers, one may consult Baker (1987) and Heiser (1976).

4. **H. pauciflorus** Nutt.   Prairie Sunflower

Map 598. Presumably once, at least, native to Michigan prairies, but recent collections (over the past 75 years) are from railroads, fields, and roadsides in situations that suggest they are adventive from farther west or escapes from cultivation.

The leaves are usually ± reduced on the upper part of the stem, although not as much so as in the next species. The blades are ± elliptic, tapering into short or indistinct petioles.

595. Helianthus debilis   596. Helianthus petiolaris   597. Helianthus annuus

The name used here has priority over *H. rigidus* and includes *H. subrhomboideus* Rydb. (Jeffrey 1981). *H.* ×*laetiflorus* Pers. is now interpreted to apply to a hybrid between *H. pauciflorus* and *H. tuberosus*. It differs from the former in having yellow disk corollas and in general somewhat more acute phyllaries. It is often cultivated, and the sporadic occurrences probably represent escapes (Allegan, Benzie, Berrien, Ingham, Kalamazoo, Kent, Lenawee, Marquette, Midland, and Washtenaw cos.).

5. **H. occidentalis** Riddell   Fig. 211   Western Sunflower
Map 599. Despite the name, this sunflower ranges less far west than many, and is in fact concentrated in Michigan and Wisconsin, being less common southward. It is characteristic of our dry, very open, usually sandy woodlands, such as jack pine and oak plains. It has also spread along railroads and in fields and is found in dry prairies.

This is a very easily recognized species with the few leaves concentrated at the base of the plant, nearly naked stem, and colonial habit resulting from slender, shallow, pale rhizomes which usually produce an abundance of leaf rosettes in a colony.

6. **H. mollis** Lam.   Plate 8-A   Ashy or Downy Sunflower
Map 600. Degraded prairies, railroads, open sandy ground. This prairie species is probably native at some sites in Michigan although others are dubious. A large colony between the highway and an abandoned railroad near the Emmet/Charlevoix county line, for instance, is presumably adventive, as is a 1900 collection from near a railroad in Manistee County.

An intermediate specimen from Ypsilanti (*Walpole* in 1920, BLH), with leaves pubescent as in *H. mollis* but not clasping, has been determined by Heiser as a hybrid with *H. giganteus*.

7. **H. maximilianii** Schrader   Maximilian Sunflower
Map 601. Roadsides, railroads, fields; woodland of jack pine, oak, and aspen. Like so many of our sunflowers, the native range is obscure, but this is likely adventive from farther west (or escaped from gardens, as it is a showy species). Our earliest collections (like many later ones) are from along railroads, and none were made before 1890.

This is a fairly easily recognized species, especially if the typically folded (and arcuate) leaves are present. The upper part of the stem (or peduncles) usually has evident dense white pubescence, but such can also be seen in some plants of *H. giganteus*, which may also have nearly entire leaves (though not folded or arched). The habitat will help to separate these two if necessary. The rays in our specimens run 1.7–3.3 cm long; they are reported longer elsewhere. On richer soils the plants may approach 10 feet in height, but they are usually less, often much less, than a third that tall.

8. **H. giganteus** L. Tall Sunflower

Map 602. This is one species surely native in Michigan, where it favors moist sites: wet prairies, fens, sedge meadows, tamarack swamps; river banks and floodplain woods; borders of upland forests and wet depressions; marshes, ditches.

H. giganteus is an immensely variable species, Farwell alone having dignified with names at varietal rank five minor variants in southeastern Michigan. The leaves are rarely opposite or whorled, and may be densely gland-dotted beneath. The roots may be thickened or tuberous. Watson recorded plants as tall as 11 feet.

This species hybridizes with *H. grosseserratus*, producing *H.* ×*luxurians* E. E. Watson. Long identified this hybrid from Cass, Clinton, Genesee, Gratiot, Ingham, Kent, and Washtenaw counties; it is said to differ from *H. giganteus* in having petioles 2–3 cm long and stems green, yellow, or light red (rather than dark red or purple)—although one or both of these characters is lacking in many of the specimens examined. It differs from *H. grosseserratus* in having stems pubescent (rather than glabrous) and phyllaries usually pubescent and with marginal cilia (not glabrous or with short scattered hairs). The hybrid with *H. divaricatus* is mentioned under that species below.

598. Helianthus
     pauciflorus

599. Helianthus
     occidentalis

600. Helianthus mollis

601. Helianthus
     maximilianii

602. Helianthus giganteus

603. Helianthus hirsutus

213. *Heliopsis helianthoides*
     ×¼; section of
     head ×½
214. *Helianthus
     decapetalus* ×¾

213

214

9. **H. hirsutus** Raf. — Hairy Sunflower

Map 603. Our few specimens referred to this woodland species are mostly lacking in habitat data other than thickets along a railroad embankment. Presumably native in southern Michigan, as its range is usually stated to be from Pennsylvania to Minnesota and southward. It may stray into the Upper Peninsula from Wisconsin, but the Menominee County material is a little dubious as to identity.

The position of the lateral leaf veins is not as consistent a key character as one would like. Both this species and the next may have extremely short petioles (a sign of introgression in *H. divaricatus*?); these tend to run a trifle longer or more definite in *H. hirsutus*. Some authors describe the junction of lowest lateral veins with the midrib in terms of distance from *base* of petiole. In *H. hirsutus*, then, the branching is more than 2 mm from the petiole base. However, this is sometimes the condition also in individuals otherwise referable to *H. divaricatus*.

10. **H. divaricatus** L.   Fig. 212. — Woodland Sunflower

Map 604. This is one of the species clearly native to Michigan. It does not occur very far west of the Great Lakes except at latitudes much farther south. Typically in dry ± open woods of oak, pine, or aspen, especially at edges; fencerows, fields, roadsides, and railroads.

Some plants with distinct but short petioles and other intermediate characters are considered to be hybrids. Hybrids with *H. giganteus* have been named *H.* ×*ambiguus* (T. & G.) Britton; specimens so determined by Long (from Bay, Clinton, Ingham, Jackson, Kalamazoo, Kent, Livingston, St. Joseph, and Washtenaw cos.) and a similar one from Hillsdale County have leaves resembling *H. divaricatus* or intermediate in shape and sparsely hairy stems. The somewhat dubious Menominee County material referred to *H. hirsutus* could conceivably be this hybrid or something approaching *H. strumosus*. Other intermediate plants, but with glabrous stems, have been named *H.* ×*divariserratus* R. W. Long, the other parent being *H. grosseserratus*; specimens so identified by Long are from Berrien, Ingham, Lenawee, Livingston, St. Clair, and Shiawassee counties.

A plant with leaves in whorls of 3 was collected by Cooley in Macomb County in 1847.

11. **H. tuberosus** L. — Jerusalem-artichoke

Map 605. A native species of river banks and floodplains in our area (see Dodge 1895), but also escaped from cultivation and widespread along railroads and roadsides, in fields and fencerows, at old homesites and vacant lots, and in other disturbed places; found in both wet and dry ground.

This sunflower was once grown commonly for its crisp tubers, produced on the rhizomes, which can be eaten raw or cooked. While the species can be a serious weed, it does have a number of economic uses (see Swanton et al. 1992). Fragmentary specimens lacking information on habit of underground parts are easily confused with *H. annuus*, but the latter has larger heads (on all but very small plants) and a tendency to relatively broader leaves.

12. **H. microcephalus** T. & G.   Small-headed Sunflower
     Map 606. Found south of Michigan and barely west of the Mississippi. Like *H. divaricatus* it is a woodland species. The only Michigan collection (*Pepoon* in 1903, MSC) from "dry wood borders" gives no clue as to whether it should be considered indigenous here.
     H. divaricatus may have disks as small as in this species, but the rays are longer and the leaves nearly or quite sessile.

13. **H. grosseserratus** M. Martens   Sawtooth Sunflower
     Map 607. Low places, including swamps and marsh borders; disturbed ground, railroad yards. Near the edge of its range in southern Michigan, and much less common here than its close relative, *H. giganteus,* with which it hybridizes (see above).
     The specimens from Benzie County have been placed by Heiser as nearest this species, but the leaves are unusually scabrous. Watson recorded Michigan plants to 11 feet tall.

14. **H. decapetalus** L.   Fig. 214   Thin-leaved Sunflower
     Map 608. Native in shaded woodlands in Michigan, as elsewhere in eastern United States. Our records are from beech, maple, and red oak woods, especially at borders and openings; also river banks and floodplain woods.
     The leaves tend to be thinner in texture than the thick ones of the next species, but that is a comparative and qualitative character unsuited for a key. The petioles in both these species may be as short as 7 mm but in *H. decapetalus* they can be as long as 35 mm while in *H. strumosus* they do not, in our material, exceed 13 (or very rarely 15) mm. In regard to the

604. Helianthus
     divaricatus

605. Helianthus tuberosus

606. Helianthus
     microcephalus

lower surface of the leaves, the line between "sparsely" and "± densely" pubescent is admittedly tenuous.

Specimens of *Heliopsis helianthoides* are occasionally misidentified as this species, for there is some similarity in leaf shape and general pubescence characters. However, one should remember that in *Heliopsis* the outer phyllaries are rounded or only somewhat acute, and the ray flowers are fertile; while in *Helianthus decapetalus* the phyllaries are prolonged-acuminate and (as in other *Helianthus*) the ray flowers are sterile.

15. **H. strumosus** L.

Map 609. Wide-ranging in the eastern United States and native at least in part in Michigan. Habitats recorded for it are as diverse as its morphology; usually in dry sandy ground, including oak woods, but also in damp ground; roadsides, prairies, clearings, fields, river banks, and other places. Recorded to 6 feet tall.

Here at the end is the "none of the above" sunflower. Even the sunflower expert, Charles Heiser, has written: ". . . we are left with what may in a sense be described as a 'wastebasket' species which serves to accommodate all the eastern polyploid sunflowers that cannot be placed in any other species. . . . this is and will likely continue to be the 'rubbish heap' of the genus." *H. strumosus*—whatever that is—seems to intergrade with others and to be quite plastic in leaf shape and other characters. For example, several specimens with leaves densely pubescent beneath and other characters of this taxon have the long phyllaries of the preceding. Others seem to approach *H. divaricatus*.

44. **Grindelia**

1. **G. squarrosa** (Pursh) Dunal   Fig. 215   Gumweed
Map 610. Native to the prairies and beyond, west of the Great Lakes. Adventive eastward and first collected in Michigan 1901–1905 in Kent, Marquette, St. Clair, and Wayne counties; by 1920, recorded for six more

607. Helianthus
       grosseserratus

608. Helianthus
       decapetalus

609. Helianthus strumosus

counties. Roadsides, railroads, fields, sandy and gravelly banks, and waste ground including vacant lots, gravel pits, and filled land.

A very distinctive plant, often bushy-branched, with attractive yellow heads and very resinous, curly-looking phyllaries. See also comments under genus 61, *Haplopappus*.

45. **Helenium** Sneezeweed

The common name comes not from any pollen allergy that hay-fever victims might suffer (for wind-pollination is not characteristic of plants with showy flowers), but from use of the dried, pulverized heads to make a snuff that promotes sneezing.

KEY TO THE SPECIES

1. Leaves very numerous, crowded, less than 2 mm broad; stems not winged (may be angled or ridged); plant annual . . . . .1. **H. amarum**
1. Leaves not crowded, broader; stem with green wings the length of the internodes; plant perennial
  2. Disk corollas with 4 brown or purple lobes; ray flowers sterile . . . . .2. **H. flexuosum**
  2. Disk corollas with 5 yellow lobes; ray flowers pistillate . . . . .3. **H. autumnale**

1. **H. amarum** (Raf.) H. Rock

Map 611. Native in the southern states and a rare waif this far north. Collected by Dodge along a railroad near Port Huron in 1899 (MSC) and by Farwell in waste ground at Detroit in 1919 (GH, BLH).

2. **H. flexuosum** Raf.

Map 612. This species is probably not native in Michigan, but adventive from farther south. Habitats recorded for our specimens include marshy ground, roadsides, ditches in a conifer swamp, seasonally wet fields and sandpits.

3. **H. autumnale** L. Fig. 216

Map 613. Banks of streams and rivers, thickets, and floodplains; wet meadows, marshes, fens, tamarack swamps, edges of cedar swamps, and prairies; wet fields and shores.

46. **Verbesina**

KEY TO THE SPECIES

1. Stem winged; leaf with petiole obscure (and winged), tapering into blade; plant perennial . . . . .1. **V. alternifolia**
1. Stem wingless; leaf with well distinguished blade and petiole; plant annual . . . .
. . . . .2. **V. encelioides**

1. **V. alternifolia** (L.) Kearney   Wing-stem
   Map 614. Banks and riverbottom floodplain swamps; fens and thickets.
   This species has sometimes been placed in the segregate genus
   Actinomeris. The achenes may be winged narrowly, broadly, or not at all;
   both winged and wingless achenes may occur in the same head.

2. **V. encelioides** (Cav.) A. Gray   Golden Crownbeard
   Map 615. Native mostly in the Southwest, but occasionally adventive
   eastward. Collected in 1895 (first noted in 1892) in the same refuse area
   that produced *Thelesperma megapotamicum* (see genus 18 above) and
   other adventives introduced with wool from New Mexico. The *Verbesina*
   seeded itself and remained established for about a decade, at least.

47. **Calendula**

1. **C. officinalis** L.   Pot-marigold
   Map 616. A European native, well known in gardens, where a double
   cultivar is often grown (i.e., one with disk flowers mostly transformed to
   ray flowers). Collected by Farwell in Houghton (1894) and Wayne (1936)
   counties.

610. Grindelia squarrosa

611. Helenium amarum

612. Helenium flexuosum

613. Helenium autumnale

614 Verbesina alternifolia

615. Verbesina encelioides

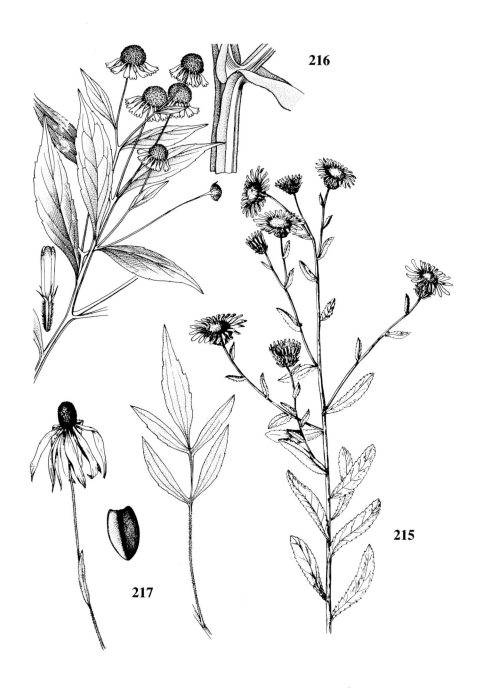

215. *Grindelia squarrosa* ×½
216. *Helenium autumnale* ×½; portion of stem ×2½; disk floret ×5
217. *Ratibida pinnata* ×½; achene ×5

Page 434

48. **Ratibida**   Prairie Coneflower

REFERENCES

Dress, William J. 1961. Notes on the Cultivated Compositae 6. The Coneflowers: Dracopis, Echinacea, Ratibida, and Rudbeckia. Baileya 9: 67–83.
Richards, Edward Leon. 1968. A Monograph of the Genus Ratibida. Rhodora 70: 348–393.

KEY TO THE SPECIES

1. Disk hemispherical (or even subglobose) to short-cylindrical (fig. 217), much shorter than the rays; rays ca. 2.5–3.5 (4.5) cm long; pappus none . . . . .1. **R. pinnata**
1. Disk cylindrical (fig. 218), longer than the rays; rays less than 2 [2.5] cm long; pappus of 1–2 tiny teeth . . . . .2. **R. columnifera**

1. **R. pinnata** (Vent.) Barnhart   Fig. 217   Yellow Coneflower
    Map 617. Apparently native in southern Michigan and Menominee County in or near prairie remnants (including roadsides and fencerows), at margins of swampy woods, and in dry open ground. Most northern occurrences represent garden escapes or other waifs along roadsides or in rocky fields.
    Specimens of *Rudbeckia laciniata* with drooping rays might be mistaken for this species, but differ in the essentially glabrous stems and peduncles and the greenish to brownish disk. Other *Rudbeckia* species with purplish disk and pubescent stem may have to be checked carefully for absence of chaff at the base of the ray flowers if they are puzzling; see also note under *Rudbeckia*.

2. **R. columnifera** (Nutt.) Wooten & Standley   Fig. 218
    Map 618. Native to prairies and open ground west of the Great Lakes, adventive eastward chiefly along railroads.
    A form with most or all of each ray deep red-purple or brown [f. *pulcherrima* (DC.) Fernald] has not yet been found in Michigan.

616. Calendula officinalis   617. Ratibida pinnata   618. Ratibida columnifera

49. Gaillardia   Gaillardia; Blanket-flower

It is often not easy to identify our gaillardias, all of which are presumably escapes from cultivation. The chief character usually given is the annual habit of *G. pulchella*, but as that species is often quite "woody" at the base, determining its duration is difficult. Of other characters given in the literature, only achene length appears to discriminate consistently. Color patterns can be less reliable than stated in the key.

REFERENCES

Biddulph, Susann Fry. 1944. A Revision of the Genus Gaillardia. Res. Stud. St. Coll. Washington 12: 195–256.
Stoutamire, Warren P. 1960. The History of Cultivated Gaillardias. Baileya 8: 12–17.

KEY TO THE SPECIES

1. Plants annual; rays deep red-purple more than half (basal half) their length; body of achenes ca. 2–2.5 mm long . . . . .1. **G. pulchella**
1. Plants perennial; rays yellow at least on the apical half; body of achenes ca. (3.5) 4– 5 (6) mm long . . . . .2. **G. aristata**

1. **G. pulchella** Foug.

Map 619. Native well south of the Great Lakes, west to New Mexico. Roadsides, railroad ballast, dooryards, near cemeteries.

2. **G. aristata** Pursh

Map 620. Native west of the Great Lakes, here escaped to roadsides, fields, fencerows, vicinity of old gardens, and waste ground.

A tetraploid hybrid between this species and the preceding arose in cultivation in Belgium and was named *G. ×grandiflora* Van Houtte. It is now grown in preference to the less vigorous *G. aristata* (to which it will key here). Some collections mapped as *G. aristata* may in fact be the hybrid. We have good collections of the hybrid at least from waste ground in Emmet and Schoolcraft counties; on these, the tip of the rays is barely 2-toothed or shallowly 2–3-lobed, the lobes less than 20% of the total length

619. Gaillardia pulchella

620. Gaillardia aristata

621. Rudbeckia laciniata

of the ray. Whether the small sample available is representative, I cannot say. In our *G. aristata*, the lobes are more numerous and more than 20% of the ray.

50. **Rudbeckia**   Coneflower; Rudbeckia

Individuals with lobed leaves are sometimes superficially confused with *Ratibida*, particularly *R. pinnata*, which does not have the distinctive cylindrical receptacle and disk of *Ratibida columnifera*. The usually bicolored rays of *Rudbeckia triloba* may help, as will the glabrous stem and peduncles and the greenish to brownish disk of *Rudbeckia laciniata*. Sometimes characters of chaff (location and shape) and pappus cannot be avoided. There is no pappus in *Ratibida pinnata*; in those rudbeckias with lobed leaves it is a very short crown.

The rays of *Rudbeckia* are yellow, but often with some shade of red, orange, or purple on the basal half.

REFERENCES

Dress, William J. 1961. Notes on the Cultivated Compositae 6. The Coneflowers: Dracopis, Echinacea, Ratibida, and Rudbeckia. Baileya 9: 67–83.
Perdue, Robert E., Jr. 1957. Synopsis of Rudbeckia subgenus Rudbeckia. Rhodora 59: 293–299.

KEY TO THE SPECIES

1. Leaves (at least the lower ones) ± deeply lobed (or even compound)
    2. Disk yellow or greenish yellow (brownish when dry); stems and pedicels (except immediately below the heads) glabrous or essentially so; largest leaves (often low on plant and seldom collected) 5–7 (9)-lobed . . . . .1. **R. laciniata**
    2. Disk deep purple-red; stems and pedicels pubescent; largest leaves 3-lobed
        3. Receptacular chaff ± obtuse or rounded and glandular-puberulent at apex; rays yellow throughout . . . . .2. **R. subtomentosa**
        3. Receptacular chaff (conspicuous among the disk florets) acuminate to a sharp point, glabrous throughout; rays usually with orange base . . . . .3. **R. triloba**
1. Leaves all unlobed
    4. Receptacular chaff acuminate to a sharp awn-like point, glabrous throughout; leaves ± ovate, toothed; pappus a tiny crown . . . . .3. **R. triloba**
    4. Receptacular chaff acute to rounded (but not prolonged to a sharp point) and pubescent, ciliolate, or glabrous at the tip; leaves various; pappus various
        5. Stems and leaves sparsely pubescent with scattered soft hairs; mid-cauline leaves with narrowly to broadly ovate blades and distinct petioles; pappus a tiny crown; tips of receptacular chaff glabrous or sparsely ciliolate . . . . .4. **R. fulgida**
        5. Stems and leaves coarsely pubescent with stiff, dense hairs; mid-cauline leaves ± lanceolate to oblanceolate, sessile or tapering into an often winged petiole; pappus none; tips of receptacular chaff pubescent . . . . .5. **R. hirta**

1. **R. laciniata** L.   Fig. 219   Tall or Cutleaf Coneflower

Map 621. River banks and floodplains, thickets and low woods, swamps (including cedar), wet ditches in (or by) woods and marshy ground.

Tall individuals may be 6–8 feet in height. In addition to the wild plants, a garden cultivar known as goldenglow, *R. laciniata* cv. Hortensia, derived from this species by conversion of all or most disk flowers to ray flowers, occasionally escapes. The dots on the map for Alpena, Ogemaw, and Schoolcraft counties and for the Manitou Islands represent only the cultivar, which is also documented as established in some of the counties from which the native plant has been collected.

2. **R. subtomentosa** Pursh   Sweet Coneflower
    Map 622. The only known Michigan collection is a fragment scarcely 15 cm long, bearing one head and two small unlobed leaves (*Davis* in 1894, US), obviously torn from a larger specimen. Identification was confirmed by R. E. Perdue; the rather dense pubescence on the stem and pedicel, and the short, stiff, appressed hairs on both surfaces of the leaves, help to eliminate any other possible species. The date and the unlikely habitat ("swamp") are the same as for some of Davis' collections of *R. fulgida*, also from near Alma, and one cannot help but wonder whether some mixture of specimen or data occurred long ago. The species ranges otherwise southwest of Michigan, but is known close enough in northwestern Indiana that it might be discovered at a more likely site in southwestern Michigan.
    The common name derives from a licorice-like aroma the involucre is said to possess.

3. **R. triloba** L.
    Map 623. Wet prairies and borders of floodplain woods; also in ± disturbed ground on shores and edges of marshy or swampy ground. The northernmost records, at least, and probably some others, may represent escapes from cultivation although the species has naturally weedy tendencies.
    The leaves are often all unlobed. Even when lobed leaves are present, the plant looks quite different from *R. subtomentosa*. Besides the characters in the key, the pubescence on stems and leaves consists of scattered long hairs, while in *R. subtomentosa* it consists of short dense hairs.

622. Rudbeckia   623. Rudbeckia triloba   624. Rudbeckia fulgida
    subtomentosa

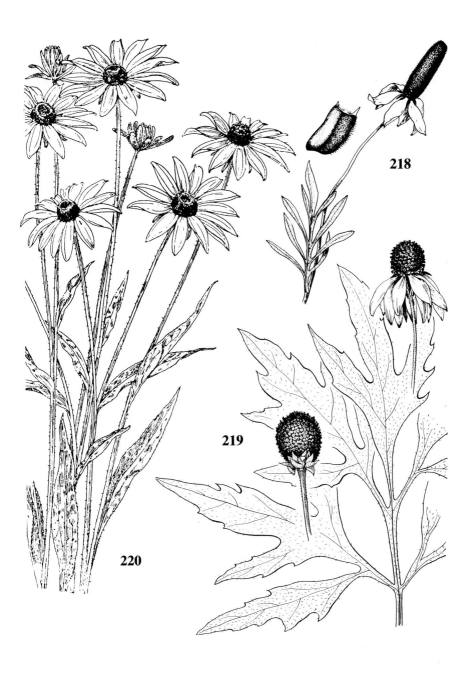

218. *Ratibida columnifera* ×½; achene ×5
219. *Rudbeckia laciniata* ×½
220. *Rudbeckia hirta* ×½

4. **R. fulgida** Aiton   Showy Coneflower
    Map 624. Fens, sedge meadows, calcareous springy banks, riverside
swamps, meadows, and other wet (sometimes rocky) ground.
    Our plants seem all to be var. *speciosa* (Wender.) Perdue, as recognized
by Cronquist in the 1991 *Manual*, with the best-developed rays (2.5) 2.8–
4.5 cm long [including var. *sullivantii* (Boynton & Beadle) Cronquist, recognized by Perdue as distinct, based on reduction of the uppermost leaves—
and which is what, in the narrowest sense, we have].

5. **R. hirta** L.   Fig. 220   Black-eyed Susan
    Map 625. Fields, roadsides, railroads, clearings, fencerows; woodland
(with jack pine, aspen, oak), especially in sandy or rocky openings; prairies, fens, sedge meadows, shores; gravel and borrow pits, other disturbed
ground.
    This is a complex species consisting, according to some authors, of as
many as 4 or 5 intergrading varieties. It is also cultivated, especially as large
tetraploid cultivars such as 'Gloriosa Daisy.' Most of our plants are the
rather weedy var. *pulcherrima* Farw. (TL: Detroit [Wayne Co.]), sometimes recognized at the rank of species as *R. serotina* Nutt. A few are
apparently var. *hirta*, with broader, elliptic-ovate lower leaves—an eastern
variety characteristic of undisturbed habitats. Determining the nativity of
any given population in Michigan is largely a futile task.
    The size and color of the rays, the shape and toothing and size of the
leaves, and the amount of pubescence on the receptacular chaff are all
quite variable. The rays are usually not bicolored. A rare form lacking the
ray florets was collected in Bay County in 1989 (*Penskar & Crispin 1112*,
MICH) and one with double heads (extra ray florets) in Gogebic County in
1958 (*Voss 7826*, MICH).

625. Rudbeckia hirta

626. Mikania scandens

627. Dyssodia papposa

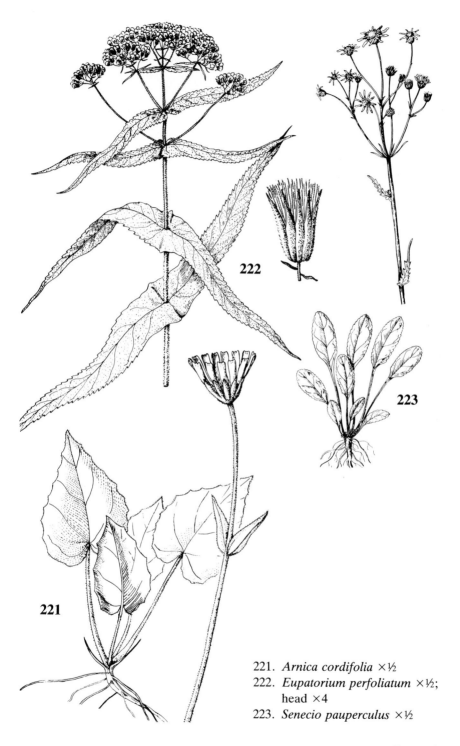

221. *Arnica cordifolia* ×½
222. *Eupatorium perfoliatum* ×½;
    head ×4
223. *Senecio pauperculus* ×½

KEY TO THE GENERA (Group C)

1. Leaves all or partly opposite or whorled
   2. Plant a vine, with twining, climbing stem; leaves with distinct wingless petiole and cordate blade . . . . .51. **Mikania**
   2. Plant a ± erect herb, the stem not twining or climbing; leaves various (but with petioles [winged] and blades cordate only at middle and lower nodes in *Arnica*)
      3. Leaves pinnatifid-pectinate; plant a malodorous annual with large (ca. 0.5–1 mm) glands on the involucre; pappus bristles united in groups at the very base . . . . .52. **Dyssodia**
      3. Leaves unlobed; plants perennial, with glands (if any) tiny on the involucre; pappus bristles not arising from scale-like bases
         4. Rays bright yellow, showy, greatly exceeding the involucre (ca. 14–25 mm long); leaf blades ± cordate or subcordate, the middle and lower on winged petioles . . . . .53. **Arnica**
         4. Rays none; leaf blades tapered to petiole or sessile (or even connate around stem)
            5. Leaves mostly alternate (occasionally a few opposite or subopposite); achenes with at least 10 ribs and plumose pappus bristles; disk cream . . . . .54. **Kuhnia**
            5. Leaves all or mostly opposite or whorled (occasionally the uppermost alternate); achenes with 5 ribs or angles and pappus bristles at most barbellate; disk white, purple, blue, or red-purple . . . . .55. **Eupatorium**
1. Leaves all alternate or basal
   6. Ray flowers present, evident (even if small) and ± bright yellow or orange (disk flowers also yellow)
      7. Phyllaries in a single series (all or essentially all ± equal in length and overlapping at most a little toward the base, but sometimes a few short bracteoles at base of involucre)
         8. Cauline leaves (at least the middle and lower) with blades, usually ± deeply lobed; stems with more than 1 head; disk flowers fertile (setting fruit), similar in number to the ray flowers . . . . .56. **Senecio**
         8. Cauline leaves all reduced to entire bracts; stems with solitary heads; disk flowers sterile, few, surrounded by ray flowers many times that number . . . . .57. **Tussilago**
      7. Phyllaries imbricate (of various lengths, overlapping)
         9. Heads small, the disk (receptacle) less than 5 mm broad
            10. Leaves narrowly linear-lanceolate, entire, usually punctate with shiny dots, best developed on the middle of the stem; heads all or mostly sessile or subsessile in little clusters of 2 or more, in a ± corymbiform inflorescence . . . . .58. **Euthamia**
            10. Leaves various but if entire and linear-lanceolate, then not punctate and/or best developed at the base and/or the inflorescence not corymbiform and/or the heads on distinct pedicels . . . . .59. **Solidago**
         9. Heads large, the disk (receptacle) ca. 7 mm or more broad
            11. Leaves entire (or very nearly so) . . . . .60. **Chrysopsis**
            11. Leaves toothed
               12. Teeth of leaves with prolonged spiny tips . . . . .61. **Haplopappus**
               12. Teeth of leaves blunt or rounded . . . . .62. **Inula**
   6. Ray flowers none (though marginal disk flowers may be enlarged and deeply lobed) or if present, not yellow (white, blue, pink, or purple)
      13. Rays present and showy, extending beyond the involucre at least as great a distance as the length of the involucre

Page 442

14. Heads ca. 4–6 cm broad, solitary at the ends of stems or branches; phyllaries fringed with long septate hairs; plant an annual, very rarely escaped from cultivation, with broad ovate leaves irregularly and deeply toothed . . . .
. . . . .63. **Callistephus**
14. Heads less than 4 cm broad, usually clustered; phyllaries ciliate or not; plants of most species perennial, native or not, with broad to narrow leaves, at least the cauline ones entire or regularly toothed
  15. Pappus of 2–4 (6) long (ca. 0.5–1.5 mm) and several much shorter bristles; achenes flat and winged . . . . .64. **Boltonia**
  15. Pappus of numerous silky hairs; achenes not winged
    16. Phyllaries all or almost all ± equal in length
      17. Cauline leaves consisting of modified broad multi-veined petiolar bracts with at most a rudimentary blade, on a thick (often over 5 mm diameter) stem . . . . .81. **Petasites**
      17. Cauline leaves not so modified, on a stem not over 5 (6) mm thick
        18. Pedicels and phyllaries with ± dense stalked glands . . . . . .66. **Aster** (couplet 13)
        18. Pedicels and phyllaries eglandular or at most with obscure glands
          19. Phyllaries at least sparsely pubescent on back, with no distinct expanded apex on central stripe; rays usually white to pink (showy on the few-headed *E. pulchellus*) . . . . .65. **Erigeron**
          19. Phyllaries glabrous on back, with central narrow green stripe expanded into a ± diamond-shaped green area below the tip; rays bright blue or purple . . . . .66. **Aster** (species 10 & 28)
    16. Phyllaries of several lengths, the involucre clearly imbricate
      20. Inflorescence compact, elongate, cylindrical; rays fewer than 10 . . . .
. . . . .59. **Solidago (bicolor)**
      20. Inflorescence ± open, corymbiform or paniculate; rays often more than 10
        21. Pappus hairs thickened (slenderly clavate) toward apex; rays white, 12–18; achenes glabrous; leaves ± oblanceolate . . . . .59. **Solidago (ptarmicoides)**
        21. Pappus hairs not thickened (if obscurely so, achenes pubescent); leaves various (but not oblanceolate if rays white and 12–18) . . . . .66. **Aster**
13. Rays none or very inconspicuous (extending beyond the involucre, if at all, less than its length), although in some species the marginal disk flowers may be much larger than the others
  22. Phyllaries (at least the outer ones) spine-tipped (hooked in *Arctium*) or strongly fringed or lacerate-margined, or the lobes of the leaves spine-tipped, or both conditions (phyllaries & leaves) present; corollas usually very deeply lobed
    23. Spine at tip of phyllary hooked; leaves very large and ovate, not spiny . . . . . .
. . . . .67. **Arctium**
    23. Spine (if any) at tip of phyllary straight; leaves various
      24. Phyllaries lacerate or fringed at the tip (including a terminal spine or not); leaves without spines . . . . .68. **Centaurea**
      24. Phyllaries with a terminal spine or none (not otherwise fringed except spiny basally in *Silybum*); leaves with spine-tipped lobes or spiny margins
        25. Principal cauline leaves at least (6) 8 cm broad, shallowly if at all lobed; plants either with outer phyllaries at least 1 cm broad at base or with stems broadly winged (the wings averaging over 5 mm broad and densely cobwebby-tomentose); plants very rare waifs
          26. Stem not winged; phyllaries expanded at base to over 1 cm wide; receptacle bristly; leaves variegated (at least the main veins bordered with whitish) . . . . .69. **Silybum**
          26. Stem with broad cobwebby-tomentose wings; phyllaries less than 3 mm wide; receptacle without bristles among the florets (but deeply pitted); leaves not variegated . . . . .70. **Onopordum**

Page 443

25. Principal cauline leaves less than 6 cm broad, or deeply lobed (sinus more than halfway to midrib); stem wings, if any, not cobwebby-tomentose and phyllaries not over 6 mm wide; plants (at least the genera) widespread, often common

    27. Pappus bristles simple (at most slightly barbed); internodes of stem continuously or intermittently winged its entire length . . . . 71. **Carduus**

    27. Pappus bristles plumose (feathery with very long fine lateral branches); internodes of most species not (or only slightly) winged . . . . 72. **Cirsium**

22. Phyllaries and leaves all without spines; corolla various

  28. Middle phyllaries wholly scarious or with prominent pale to brownish scarious tips and/or margins totalling a fourth or more of the length or width of the phyllary; corollas (except in staminate *Antennaria* and *Centaurea repens*, with pilose phyllaries) thread-like (i.e., extremely slender, ca. 0.1–0.3 mm broad)

    29. Mature leaves glabrous or nearly so, glandular-punctate; pappus bristles (the longest) short-plumose; tip of phyllaries pilose with dense straight hairs . . . .

    . . . . 68. **Centaurea (repens)**

    29. Mature leaves lightly to densely tomentose on one or both surfaces, not punctate; pappus bristles at most minutely barbed; tip of phyllaries glabrous or tomentose

      30. Plants mostly stoloniferous with leafy rosettes; cauline leaves much reduced, ± remote; pappus bristles (at least in pistillate or perfect flowers) united in a ring at the base

        31. Phyllaries brownish (or purplish) throughout; heads with perfect flowers centrally, pistillate ones outwards; inflorescence elongate, spicate . . . .

        . . . . 76. **Gnaphalium (purpureum)**

        31. Phyllaries white (or pinkish) at least apically; heads unisexual (plants dioecious); inflorescence ± corymbiform . . . . 73. **Antennaria**

      30. Plants annual or rhizomatous, with neither stolons nor basal rosettes; cauline leaves numerous, much overlapping; pappus bristles separate in all the common species

        32. Phyllaries tomentose nearly or quite to the tip; receptacle chaffy except at the middle (the chaff grading into the sparse phyllaries, but glabrate) . . . .

        . . . . 74. **Filago**

        32. Phyllaries glabrous on at least the apical half; receptacle without chaff

          33. Phyllaries (except at base) pure pearly white, appearing distinctly longitudinally striate (from tiny creases); leaves smooth and glabrous above or with loose white tomentum (rarely with a few tiny gland-tipped hairs hidden in the tomentum); plant rhizomatous and (fresh or dry) without sweetish odor . . . . 75. **Anaphalis**

          33. Phyllaries off-white to brownish, not appearing striate from tiny creases; leaves at least in common species with short gland-tipped hairs or at least roughened above; plants tap-rooted (rhizome only in *G. sylvaticum*) and the common species (fresh or dry) with sweetish (brown sugar?) odor especially when crushed . . . . 76. **Gnaphalium**

  28. Middle phyllaries with scarious margin none or very narrow (⅙ the length of phyllary or less — rarely approaching ¼ in *Liatris*); corollas in most species more than 0.3 mm broad

    34. Flowers (except for an occasional albino) purple or pink

      35. Rays present, tiny

        36. Involucre up to 4 mm long . . . . 80. **Conyza**

        36. Involucre at least 5 mm long

          37. Pedicels and often phyllaries glandular . . . . 65. **Erigeron (acris)**

          37. Pedicels and phyllaries eglandular . . . . 66. **Aster (subulatus)**

35. Rays none
  38. Corollas less than 0.2 mm broad, barely toothed at apex; phyllaries with septate hairs, at least on margins . . . . .77. **Pluchea**
  38. Corollas broader, deeply lobed (sinuses ca. 1 mm or deeper); phyllaries usually glabrous (or minutely ciliate) but in most species glandular punctate
    39. Phyllaries in a single series, of ± equal length (tiny bracteoles may be at base), smooth; cauline leaves reduced to broad bracts (at most a tiny blade at apex); basal leaves ca. 1–3 dm broad . . . . .81. **Petasites (hybridus)**
    39. Phyllaries imbricate, ± glandular-punctate; cauline leaves normal; basal leaves much less than 1 dm broad
      40. Pappus dark purple or brown, of 2 series: one of long, capillary, nearly smooth bristles and the other (outer) of short, somewhat flattened bristles; inflorescence ± open and corymbiform . . . . .78. **Vernonia**
      40. Pappus usually white, often becoming pink or dingy, of uniform long, plumose or strongly barbed bristles; inflorescence usually ± elongate or spicate . . . . .79. **Liatris**
34. Flowers white, cream, or yellow
  41. Phyllaries in 2 or more series of ± equal length or imbricate
    42. Pappus bristles plumose; leaves, involucre, and achenes all pubescent . . . . .54. **Kuhnia**
    42. Pappus bristles simple; leaves and involucre glabrous or pubescent (see below); achenes glabrous (usually) or pubescent
      43. Involucre 2.7–4 mm long
        44. Pubescence mostly of appressed antrorse hairs 0.5 mm or shorter; pappus double, with an outer series of minute scale-like bristles . . . . .65. **Erigeron (strigosus)**
        44. Pubescence mostly of spreading, longer hairs; pappus of a single series . . . . .80. **Conyza**
      43. Involucre at least 5 mm long
        45. Pedicels and often phyllaries with minute gland-tipped hairs; leaves lanceolate to oblanceolate; achenes glabrous . . . . .65. **Erigeron (acris)**
        45. Pedicels and phyllaries eglandular; leaves narrowly linear (less than 4 mm wide); achenes pubescent . . . . .66. **Aster (brachyactis)**
  41. Phyllaries in a single series of ± equal length, in some species with bracteoles very much shorter and usually narrower at base of involucre
    46. Flowers yellow; cauline leaves deeply pinnatifid, at least toward the base . . . . .56. **Senecio**
    46. Flowers white to cream; cauline leaves at most toothed
      47. Cauline leaves reduced to broad petiolar (parallel-veined) bracts with at most a rudimentary blade at the apex . . . . .81. **Petasites**
      47. Cauline leaves with well developed blades [if plant does not fit well here, try couplet 43]
        48. Leaves all cauline (crowded and overlapping), sharply toothed, pinnately veined with 1 midrib; pappus bristles smooth; plants annual . . . . .82. **Erechtites**
        48. Leaves cauline (remote and few) and basal, long-petioled, coarsely and broadly toothed, shallowly lobed, or entire, with several main palmate or longitudinal veins; pappus bristles scabrous or minutely antrorse-barbed; plants perennial . . . . .83. **Cacalia**

51. Mikania

This is a large, mostly tropical genus, of which only one species ranges as far north as our latitudes.

REFERENCES

Holmes, W. C. 1981. Mikania (Compositae) of the United States. Sida 9: 147–158.
McLouth, C. D. 1896. Mikania scandens, Willd. Asa Gray Bull. 4: 68.

1. **M. scandens** (L.) Willd. Climbing Hempweed

Map 626. Ranges north to Massachusetts and northern New York in the East, but in the interior widely scattered north of southern Illinois into Michigan. In 1894 the species was brought to the attention of C. D. McLouth, an accomplished botanist of Muskegon. He attempted to learn all he could about its history in what is presumably now part of the Muskegon State Game Area, and determined that the plant had apparently been abundant there since at least the middle of the 19th century. The area has long been popular with waterfowl, and the *Mikania* may have been dispersed to Michigan by such animals, if not by wind, before or during historical times. In any event, there is no reason to suppose that it was deliberately introduced. Modest efforts to rediscover it in 1994 were unsuccessful.

52. Dyssodia

This genus was long placed in the same tribe (Anthemideae) as other strong-smelling ones, such as *Anthemis, Chrysanthemum, Matricaria,* and *Tanacetum,* but differs strikingly from these in its usually well developed pappus bristles — especially in *D. papposa.* It is now considered, like the next genus, to be more closely allied to the tribe Heliantheae.

1. **D. papposa** (Vent.) Hitchc. Stinking-marigold

Map 627. Collected in 1902 by Emma Cole on the same farm that provided *Thelesperma* and other introductions from the West. To be expected elsewhere, as it became common along interstate highways near Chicago in the 1970s.

The rays are yellow but inconspicuous for they barely exceed the involucre.

53. Arnica Arnica

An additional species, *A. lonchophylla* Greene, is known in northeastern Minnesota and adjacent Ontario; it might be discovered on cold calcareous rocks in the Lake Superior region of the Upper Peninsula. It is disjunct in that region from its main range (Alaska to the Black Hills) and from the Maritime region of Canada (see Michigan Bot. 20: 72. 1981). From *A. cordifolia* it differs most obviously in the more elongate leaves, not at all cordate.

This genus has long been placed in the ragwort tribe (Senecioneae), from which it differs conspicuously in having opposite leaves. However, chemical and other evidence indicates that it is more closely allied to the Heliantheae, despite the showy white pappus (see Nordenstam in Heywood et al. 1977, pp. 822–823).

REFERENCE

Ediger, Robert I., & T. M. Barkley. 1978. Arnica. N. Am. Fl. II(10): 16–44.

1. **A. cordifolia** Hooker   Fig. 221

Map 628. Quite local but attractive where found, in and along the borders (including roadsides) of mixed or deciduous, often rocky or open woods. This is a western species occurring from Alaska south in the mountains to New Mexico and also in the Black Hills of South Dakota and the Riding Mountains of Manitoba. Its disjunct occurrence on the Keweenaw Peninsula (from Copper Harbor to Eagle Harbor) was discovered in 1849 by young W. D. Whitney, later to become a distinguished philologist at Yale. It was named for him in 1935 as an endemic new species, *A. whitneyi* Fernald (TL: Copper Harbor, Keweenaw Co.), but is now generally included in *A. cordifolia*. It is also known from the Sibley Peninsula on the northwest side of Lake Superior in Ontario (see Michigan Bot. 20: 62. 1981).

The hairy achenes are 5.5–7.5 mm long on our material. The pappus hairs are distinctly longer, slightly plumose with antrorse barbs.

54. **Kuhnia**

1. **K. eupatorioides** L.   False Boneset

Map 629. Sandy fields and prairies, along with associated bluffs and roadsides.

Kuhnia has been included in *Brickellia* by some authors, in which case this species becomes *B. eupatorioides* (L.) Shinners.

628. Arnica cordifolia   629. Kuhnia eupatorioides   630. Eupatorium
                                                            maculatum

55. Eupatorium

A diverse genus as treated here. Certain of our species are sometimes dispersed into other genera (even in different subtribes).

REFERENCES

Clewell, Andre F., & Jean W. Wooten. 1971. A Revision of Ageratina (Compositae: Eupatorieae) from Eastern North America. Brittonia 23: 123–144.

Duffy, David Cameron. 1990. Land of Milk and Poison. Nat. Hist. 99(7): 4, 6, 8.

Lamont, Eric E. 1995. Taxonomy of Eupatorium section Verticillata (Asteraceae). Mem. New York Bot. Gard. 72. 67 pp.

Moseley, Edwin Lincoln. 1941. Milk Sickness Caused by White Snakeroot. Author & Ohio Acad. Sci., Bowling Green. 171 pp.

Oldham, M. J. 1988. Tall Thoroughwort (Eupatorium altissimum L.) in Ontario. Pl. Press (Ontario) 5: 16–19.

Wooten, Jean W., & André F. Clewell. 1971. Fleischmannia and Conoclinium (Compositae, Eupatorieae) in Eastern North America. Rhodora 73: 566–574.

KEY TO THE SPECIES

1. Leaves whorled; heads pale pink to reddish or purple (very rarely albino)
  2. Florets mostly (8) 9–20 (22) per head; inflorescence flat-topped or of flat-topped subunits; stem distinctly spotted with red-purple or continuously red-purple . . . . .1. **E. maculatum**
  2. Florets 3–7 per head; inflorescence ± dome-shaped (convex); stem greenish (except at nodes) in common species (purplish in the other)
    3. Corollas 5–6 (6.7) mm long at maturity; middle internodes less than 5 (8) mm thick, with solid pith (or a very slender hollow), greenish; leaves 3–5 per whorl . . . . .2. **E. purpureum**
    3. Corollas 5 mm long or less; middle (if not all) internodes ca. (7) 10–20 mm or more thick, hollow over half their diameter, ± reddish throughout or spotted with elongate red dashes; leaves mostly 6 or more per whorl . . . . .3. **E. fistulosum**
1. Leaves opposite (sometimes a few upper ones alternate); heads white (except in *E. coelestinum*)
  4. Leaf bases (except sometimes the uppermost) connate around the stem . . . . .4. **E. perfoliatum**
  4. Leaf bases entirely free and separate from the stem
    5. Leaves sessile or nearly so; florets fewer than 8 (usually 5) per head
      6. Stem glabrous on middle internodes (or sparsely puberulent); leaves truncate to broadly rounded at the base, with only 1 long vein (the midrib) prominent beneath . . . . .5. **E. sessilifolium**
      6. Stem densely puberulent on middle internodes; leaves tapered to narrow base (or short-winged petiole), with 3 prominent longitudinal veins (at least on basal half) . . . . .6. **E. altissimum**
    5. Leaves with distinct petioles; florets 8–50 (or more) per head
      7. Florets blue or blue-purple, over 30 per head; receptacle ± conical . . . . .7. **E. coelestinum**
      7. Florets white, fewer than 25 per head; receptacle flat
        8. Leaf blades lanceolate, ca. 2–4 times as long as broad, pubescent beneath; phyllaries imbricate, the outer ones much shorter than the inner . . . .8. **E. serotinum**
        8. Leaf blades ovate, less than twice as long as broad (rarely 2.5 times), glabrous beneath or at most puberulent on main veins; phyllaries all the same length . . . . .9. **E. rugosum**

1. **E. maculatum** L.   Joe-pye-weed
Map 630. Marshes, meadows, swales, wet prairies and fields, shores, bogs and fens, cedar swamps, thickets, moist clearings and roadsides, damp hollows, borders of lakes and streams.

The heads are rarely white [f. *faxonii* Fernald]; such plants have been collected in Bay and Ontonagon counties. In this variable species, Lamont (1995) recognizes as occurring in Michigan two varieties in ssp. *maculatum* and also the western ssp. *bruneri* (A. Gray) G. W. Douglas (Keweenaw mainland and Isle Royale).

2. **E. purpureum** L.   Green-stemmed Joe-pye-weed
Map 631. Usually in ± shaded places, including floodplain and rich upland forests and borders; river and stream banks.

The heads, especially the involucres, are pale in this species, unlike the pink to deep rose-colored ones of the preceding and next species.

3. **E. fistulosum** Barratt   Hollow-stemmed Joe-pye-weed
Map 632. Generally a more southern and eastern species, collected in Michigan only once, near Union Pier, with red maple, green ash, American elm, touch-me-not, groundnut, common elder, and other plants of damp or wet ground (*Wilhelm & Wetstein 19805* in 1991, MICH, MOR).

Lamont (1995) does not recognize any difference between this species and the preceding in corolla length, as noted by Steyermark and apparently applicable in this area if not throughout the range.

4. **E. perfoliatum** L.   Fig. 222   Boneset
Map 633. In essentially the same habitats as *E. maculatum*, with which it often grows; sometimes weedy in its companions. Marshes, swales, wet fields, shores, fens, conifer swamps, thickets, low clearings, river and stream banks.

Ordinarily a very easily recognized plant with its connate pairs of leaves, abundant crinkly septate hairs, and white heads, but quite variable in ro-

631. Eupatorium
      purpureum

632. Eupatorium
      fistulosum

633. Eupatorium
      perfoliatum

Page 449

bustness and other features. Very rarely a plant will have at least several upper pairs of leaves merely sessile and truncate at the base, but not connate [f. *truncatum* (A. Gray) Fassett]; and the flowers are rarely pink-tipped [f. *purpureum* Britton]. A plant from Cass County (*Rogers 13011*, WUD) with leaves ± tapered to a sessile base and achenes shriveled could be a hybrid with *E. serotinum* [*E.* ×*polyneuron* (F. J. Herm.) Wunderlin]; it was apparently growing with the putative parents.

5. **E. sessilifolium** L.   Upland Boneset
    Map 634. Oak woods, shaded thickets and gravelly banks; evidently quite local but known from the state since 1840 and presumably native.
    Pubescent plants otherwise resembling this species are the rare *E. perfoliatum* f. *truncatum*.

6. **E. altissimum** L.   Tall Boneset
    Map 635. Not collected in Michigan before 1965 (Berrien Co.), suggesting that here, as in Ontario (Oldham 1988) and the Chicago area, it is largely if not entirely adventive from a little farther south along railroads and roadsides, including adjacent prairies and sedge meadows.

7. **E. coelestinum** L.   Mist-flower; "Ageratum"
    Map 636. Native south of Michigan (not in northern or northwestern Ohio), but rarely escaped from cultivation along roadsides.
    By some authors placed in a separate genus, as *Conoclinium coelestinum* (L.) DC.

8. **E. serotinum** Michaux   Late Boneset
    Map 637. Presumably adventive from farther south. Collected in Wayne County in 1930 (*Farwell 8771*, BLH). Not until the 1960s and 1970s did it show up in other counties. Along railroads, disturbed roadsides and dooryards, thickets on old dunes.

634. Eupatorium sessilifolium

635. Eupatorium altissimum

636. Eupatorium coelestinum

9. **E. rugosum** Houtt. White Snakeroot

Map 638. Wet woods and thickets, as in ravines, floodplains, and rich hardwoods; openings and trails, cedar swamps, river banks.

By some authors placed in a separate genus, in which case the name becomes *Ageratina altissima* (L.) R. M. King & H. Rob. (in *Eupatorium* it may not be called *E. altissimum* because of the prior *E. altissimum*, applying to the tall boneset).

A toxic substance in this plant can cause "trembles," a fatal disease of cattle which have browsed on it and transmittable to humans, in whom the consequent "milk sickness" caused many deaths on the 19th century frontier in this country (Duffy 1990; Moseley 1941).

56. **Senecio** Ragwort; Groundsel

In the broad sense, this is an enormous genus worldwide, variously estimated as including from one to three thousand species—woody, herbaceous, viny, in varying habitats. Some are cultivated as ornamentals. Though we suffer rather few species, all herbaceous, in this part of the world, they sometimes run into one another and are not always easy to distinguish. Where their ranges and habitats overlap, the species often intergrade as the result of hybridization. More than in many groups, exceptions can be expected to one or more of the characters given in the key.

The annotations of T. M. Barkley on many specimens in 1960 and on particularly puzzling ones since have been very helpful, as have his comments on an early draft of the manuscript.

REFERENCES

Bain, John F. 1991. The Biology of Canadian Weeds. 96. Senecio jacobaea L. Canad. Jour. Pl. Sci. 71: 127–140.

Barkley, T. M. 1963a ["1962"]. A Revision of Senecio aureus Linn. and Allied Species. Trans. Kansas Acad. 65: 318–408.

Barkley, T. M. 1963b. The Integradation of Senecio plattensis and Senecio pauperculus in Wisconsin. Rhodora 65: 65–67.

637. Eupatorium serotinum

638. Eupatorium rugosum

639. Senecio congestus

Barkley, T. M. 1964 ["1963"]. Preliminary Reports on the Flora of Wisconsin No. 49. Compositae II — Composite Family II. The Genus Senecio — the Ragworts — in Wisconsin. Trans. Wisconsin Acad. 52: 343–352.

Barkley, T. M. 1978. Senecio. N. Am. Fl. II(10): 60–139.

Barkley, T. M. 1988. Variation Among the Aureoid Senecios of North America: A Geohistorical Interpretation. Bot. Rev. 54: 82–106.

KEY TO THE SPECIES

1. Upper half of unbranched portion of stem nearly or quite as leafy as the lower half (i.e., leaves ± equally numerous and of similar size and shape, or only gradually reduced upwards); plants usually annual, without basal rosette of persistent leaves
    2. Heads with conspicuous rays clearly exceeding the involucre
        3. Leaves entire to coarsely toothed; involucre ca. (5) 7–10 mm long at maturity; plant of cold wet shores and conifer swamps, annual . . . . .1. **S. congestus**
        3. Leaves mostly ± bipinnatifid; involucre ca. 4–5 mm long; plant a waif of disturbed ground, perennial . . . . .2. **S. jacobaea**
    2. Heads with rays minute or none
        4. Bracteoles at base of involucre minute and unicolorous; rays present but minute; plant sparsely glandular-viscid . . . . .3. **S. sylvaticus**
        4. Bracteoles small but greenish with conspicuous black tip; rays none; plant without glandular-viscid hairs (glabrate to lightly tomentose) . . . . .4. **S. vulgaris**
1. Upper half of unbranched portion of stem less leafy than the lower half (i.e., leaves sparser and distinctly smaller); plants biennial or perennial, often with crowded basal leaves
    5. Blades of basal leaves cordate or ± truncate (abruptly contracted to the petiole)
        6. Rays none or less than 6.5 mm long; blades of basal leaves ± truncate at base; branches of inflorescence often all umbelliform; plant restricted to Lake Superior region . . . . .5. **S. indecorus**
        6. Rays present, (6) 6.5–9 (12) mm long when fully mature; blades of at least some basal leaves cordate at the base; branches of inflorescence often somewhat corymbiform; plant throughout the state . . . . .6. **S. aureus**
    5. Blades of basal leaves ± obtuse or tapering at the base
        7. Basal leaf blades ± broadly obovate, tapering into a petiole narrowly winged to its base; phyllaries broadest above the middle, thence rather abruptly tapered to the apex; plant with slender stolons giving rise to a rosette of leaves and glabrous or sparsely tomentose on lower leaves, near the nodes, and in the inflorescence . . . . .7. **S. obovatus**
        7. Basal leaf blades ± lanceolate to oblanceolate or narrowly elliptic, the petioles not winged; phyllaries usually tapered gradually from the middle or below; plant glabrous to tomentose, but with stolons only if tomentose
            8. Plant ± persistently lightly to densely tomentose (including the stems and in the inflorescence); stem solitary from erect base, with well developed lower cauline leaves (blade usually nearly or quite as large as basal leaves, but ± pinnatifid); habitat in dry prairie-like or jack pine sites . . . . .8. **S. plattensis**
            8. Plant glabrous or nearly so at maturity (occasionally ± tomentose even on internodes); stems often clumped at base and strongly curved basally, with cauline leaves usually much reduced; habitat often moist or even wet (but sometimes dry, especially with jack pines) . . . . .9. **S. pauperculus**

1. **S. congestus** (R. Br.) DC.   Marsh Groundsel

Map 639. A boreal circumpolar species, ranging south in the Great Lakes region sporadically to Minnesota, northern Wisconsin, and Michigan, where it has been collected only once: on the shore of Lake Michigan northwest of Harbor Springs (*Swift* in 1934, ALBC; see Pap. Michigan Acad. 42: 33. 1957). I discussed the location with the collector many years ago but was unable to relocate any plants.

This is a stout-stemmed species perhaps loosely resembling *Erechtites hieraciifolia*, but with at least the upper leaves ± clasping and the heads radiate (with light yellow rays).

2. **S. jacobaea** L. Tansy Ragwort

Map 640. An Old World species, widely established across northern North America but collected in Michigan only once, along a roadside north of Bergland (*Henson 1882* in 1985, MICH).

The species is poisonous to livestock (Bain 1991).

3. **S. sylvaticus** L.

Map 641. Another Old World waif, local in North America, collected in 1905 on an iron ore pile at Bessemer (*Davis*, MICH, NY).

A very similar species, likewise local in northern North America, but densely glandular-viscid, is *S. viscosus* L. This is to be expected in the state, for it has been collected in Ontario as close as Echo Bay (east of Sugar Island in the St. Mary's River) and Thunder Bay.

4. **S. vulgaris** L.   Fig. 224 Common Groundsel

Map 642. This is yet another — and with us the commonest — Old World, pinnatifid-leaved, weedy groundsel. It has evidently been long established here, having been collected by the First Survey as early as 1838 from a roadside in Calhoun County. Now a local weed in gardens, lawns, nurseries, and cultivated fields; in waste ground, including sidewalk cracks, parking lots, bulldozed areas, shores.

The involucre in this species and the preceding one are ca. (4) 6–7 mm long — generally a bit longer than in *S. jacobaea* and about the same length as in *S. eremophilus* Richardson, which has bracteoles almost as long as the primary phyllaries, has conspicuous rays, and has been collected at Thunder Bay, Ontario.

640. Senecio jacobaea   641. Senecio sylvaticus   642. Senecio vulgaris

5. **S. indecorus** Greene

Map 643. Coniferous and mixed, often rocky, woods and openings; cedar swamps. More often in rocky, easily drained soils than the permanently wet, rich soils inhabited by *S. aureus*.

Some or all heads in most Michigan collections are radiate [f. *burkei* (Greenman) Fernald], as they are also from the Lake Superior region of Ontario. Only radiate plants are reported from Wisconsin as well. Most keys will accommodate this species only in its discoid expression.

Most specimens mapped have been confirmed by Barkley as this species, which is easily confused with the next, especially if basal leaves are lacking. In the areas where their ranges overlap, one may not be sure of inadequate specimens. If the rays are long and/or the basal leaves are clearly cordate, one has *S. aureus*. If the rays are very short or absent, one has *S. indecorus*. (Rayless plants of *S. aureus*, known elsewhere, have apparently not been found in Michigan.) Farther west (as noted in some detail by Boivin 1972), plants allied to *S. aureus* but with basal leaves not clearly cordate are more common.

6. **S. aureus** L. Golden Ragwort

Map 644. Low woods and depressions in upland woods: floodplains, hardwood and conifer swamps (especially in openings, on hummocks, and in springy places); creek margins, sedge meadows, fens, moist thickets, ditches.

See comments under the preceding species, to which this one is rather similar. The branches of the inflorescence in *S. aureus* are more often umbelliform (arising from much the same point) than one might assume from descriptions. Blades of the basal leaves can range from 0.5 to 10 cm broad.

7. **S. obovatus** Willd. Round-leaved Ragwort

Map 645. Rich wooded hillsides, banks, oak woodland, dry to moist grassy sites.

Southernmost Michigan is along the northwest edge of the range of this

643. Senecio indecorus   644. Senecio aureus   645. Senecio obovatus

species, which occurs to the south and east of a line roughly from central New England to eastern Kansas and Texas. Where it meets the eastern edge of the range of *S. plattensis*, Barkley notes that it may intergrade with the latter, which is occasionally stoloniferous and somewhat tomentose. The collections mapped from Berrien and St. Joseph counties are somewhat tomentose and suggest such intermediacy.

8. **S. plattensis** Nutt.   Prairie Ragwort

Map 646. Dry sandy ± open prairies, fields, and barrens; woodland with oak, hickory, sassafras, aspen, and/or jack pine. Ranges from the prairies and plains of central North America east to Lake Huron (and, farther south, to Virginia).

Like some other species, this one tends to intergrade with its neighbors, especially with *S. pauperculus*, with which some authors have lumped it. However, as Barkley (1963b) has pointed out: "Combining *S. plattensis* and *S. pauperculus* into a single species could not stop there; it would be necessary to include *S. obovatus*, . . . plus the other taxa with which these in turn intergrade. . . . The one resulting species would contain so many and such diverse entities, and would encompass so much variation, as to be ridiculous. . . . The necessity for maintaining these entities as species becomes apparent only when all of the related senecios are considered."

This is typically a rather stout, stiffly erect plant, with broader leaves than the next and retaining pubescence longer. It is here defined rather strictly (see below). Although some specimens from Otsego, Montmorency, Crawford, and Oscoda counties have been referred by Barkley to the *S. pauperculus* side of this pair, it seems more natural to include them with the *S. plattensis* from immediately south — a natural range in jack pine barrens, where there are other species of prairie affinity.

9. **S. pauperculus** Michaux   Fig. 223   Northern Ragwort

Map 647. Generally a calciphile of damp sandy or gravelly (limestone) shores, fens, cedar swamps, thin soil over limestone (alvar); also in dry jack pine, aspen, and oak woodland (especially in low areas); meadows and marshy ground.

Robust plants may resemble *S. plattensis* but are often distinctly less pubescent at maturity. Small, slender plants are occasionally nearly as tomentose as *S. plattensis*, but tend to arise from a curved rather than erect base. Some of the plants from Oscoda, Crawford, Montmorency, Otsego, Mackinac, Schoolcraft, Iron, and Keweenaw counties appear close to *S. plattensis* and may represent hybrids. The advice of Ted Barkley is to include these in the wide-ranging and variable *S. pauperculus*, although probably "contaminated" with genes from *S. plattensis* — perhaps long ago when the latter was presumably more widespread in grasslands expanded during a period of warmer, drier climate (but see above).

224. *Senecio vulgaris* ×½;
 head ×2
225. *Tussilago farfara* ×½
226. *Euthamia remota* ×½;
 head ×2

57. Tussilago
Early reports of *Tussilago* from northern Michigan presumably resulted from misidentification of the closely related *Petasites frigidus*.

REFERENCE

Morton, Gary. 1978. Tussilago. N. Am. Fl. II(10): 174.

1. **T. farfara** L.   Fig. 225   Coltsfoot
Map 648. A Eurasian species, only very locally naturalized, usually in wet disturbed places, in our region. First collected in Michigan by Dennis Cooley in 1840 (presumably in Macomb or Oakland Co.); found in Macomb County in 1922 and 1923 by Farwell, in a tamarack swamp and along a railroad embankment, and in 1993 by E. Saxon, in flowing water of a canal. L. H. Bailey collected it near Lansing in 1886, and it was still (or again) a weed on the Michigan State University campus in 1978. First collected in Berrien County in 1964; in Jackson and Oakland counties, in 1978. Its history suggests that the species will spread and persist in additional places.

This rather woolly plant blooms early in the spring, before development of the long-petioled leaves with broad, ± round-cordate blades.

58. Euthamia
This genus has often been included in *Solidago*, but is sufficiently distinct that there is a growing practice to recognize it. However, the species are far from distinct and have been variously interpreted by authors, with named taxa sometimes combined with one and sometimes with another species, even with no evidence that types of the names have been examined. So application of some names may still be uncertain. I have tried for 40 years without success to recognize more than two species in Michigan— and even these two intergrade where their ranges overlap, including all three regions in Michigan where *E. remota* is known.

646. Senecio plattensis   647. Senecio pauperculus   648. Tussilago farfara

REFERENCE

Sieren, David J. 1981. The Taxonomy of the Genus Euthamia. Rhodora 83: 551–579.
[See also references under *Solidago*.]

KEY TO THE SPECIES

1. Principal cauline leaves with 3 strong longitudinal veins, the widest blades (3) 3.5–8 (13) mm broad and often with an additional pair of weaker veins; leaves and upper part of stem usually short-hairy to scabrous-hirtellous; uppermost leaves and leafy bracts dull in aspect, the glandular dots often obscure . . . . 1. **E. graminifolia**
1. Principal cauline leaves with 1 strong longitudinal vein (i.e., the midrib) and sometimes with a pair of weak veins, the widest blades less than 3 (3.5) mm broad; leaves and upper part of stem glabrous; uppermost leaves and leafy bracts glistening as if varnished and with conspicuous, dense, dark glandular dots . . . . 2. **E. remota**

1. E. graminifolia (L.) Nutt.

Flat-topped, Bushy, or Grass-leaved Goldenrod

Map 649. Although seldom considered a truly aquatic species, typical of damp sandy or gravelly (rarely clay) shores, ditches, fields, interdunal flats and beachpools, exposed lakebeds, clearings, borrow pits and other excavations, usually at or near the water table; fens, wet prairies, conifer swamps, lowland forest, meadows; rock crevices along Lake Superior.

A common, widespread, and variable species. The most pubescent extreme has been called var. *nuttallii* (Greene) Sieren. The heads are all (with only a rare exception) sessile in little clusters of 2 or more on peduncles, while in the next species at least an occasional head is solitary on its own pedicel. This is generally a more robust plant than the next species, with larger leaves and more florets per head, but the involucres are about the same length. The phyllaries are ± green-tipped. The sessile and somewhat broader heads give a more dense or crowded look to the inflorescence at the ends of the branches.

2. E. remota Greene   Fig. 226

Map 650. Moist or marshy sandy to mucky shores, exposed lakebeds, interdunal swales, and occasionally other damp depressions; often abundant in a distinct zone on the recently exposed shore of a softwater lake with fluctuating water levels.

Sieren includes this name in the synonymy of *E. gymnospermoides* Greene, although he agreed with all other authors in considering that species to have relatively long involucres (4–6 mm — most authors say 5–6 mm). I have seen no Michigan *Euthamia* with involucres as long as 5–6 mm. The plants referred here to *E. remota* have involucres ca. (3) 3.5–5 mm. A few have ± reflexed leaves and a very bushy aspect with well developed sterile axillary shoots, as attributed to *E. caroliniana* (L.) Porter & Britton [= *E. tenuifolia* (Pursh) Nutt. (see Taxon 40: 505–508. 1991)], the older name and hence to be used if the two species should be united.

Page 458

Indeed, they appear to be very close. *E. caroliniana* sens. str. is a species of the East Coast, and *E. remota* is endemic to the Great Lakes region (originally described from near Lake Michigan, whereas *E. gymnospermoides* was described from Oklahoma). Taken together, *E. caroliniana* and *E. remota* form a distribution pattern similar to that of other Coastal Plain disjunct species, and the habitat of *E. remota* in Michigan is usually that of such disjuncts.

Plants from the southwestern Lower Peninsula are most confidently referred to *E. remota* (or *E. caroliniana* sens. lat.). Those from the southeastern Lower Peninsula lack the sterile axillary shoots which are at least somewhat developed in *E. remota*. They are presumably equivalent to *Solidago moseleyi* Fernald, described from Lake Erie but later sunk by its author and others into the synonymy of *E. remota* (or, by Sieren, included in *E. gymnospermoides* despite the original description of the involucre as 3–4 mm long). The plants mapped from the Upper Peninsula are a little more puzzling, especially phytogeographically, but I see no tangible difference from Lower Peninsula specimens.

In all parts of the state, plants with the largest leaves a little broader than in *E. remota*, strongly 3- (or even weakly 5-) veined, but the foliage glutinous on the upper portion of the plant, appear to be intermediate between the two species recognized. Less common intermediate plants have dull foliage with obscure dots, but 1-veined leaves and sometimes a hint of pubescence. Often, intermediates are found on the same shores as less ambiguous *E. remota*, supporting the oft-claimed assumption of ready hybridization in *Euthamia*. The name *E. media* Greene and combinations based upon it have sometimes been applied to such intermediates, but that is apparently a misapplication if *media* (which was described from near the Mississippi River in western Illinois) belongs in the synonymy of *E. gymnospermoides*.

Although this species (or complex) is often said to average fewer than 20 florets per head and *E. graminifolia* to have more, there are too many plants of both species with counts at or even well beyond these limits to make them very useful for identification. The *remota* complex tends to

649. Euthamia
    graminifolia

650. Euthamia remota

651. Solidago rigida

have obconic heads, but in part, at least, the shape is a function of number of florets and their maturity, fully ripe heads being more campanulate.

59. **Solidago**   Goldenrod

The preceding genus, *Euthamia*, has often been included in this one, in which case the two species recognized are known as *S. graminifolia* (L.) Salisb. and *S. remota* Greene. On the other hand, *S. ptarmicoides* was long placed in the genus *Aster* but is now recognized as better included with the goldenrods despite its white rays.

Only one species, *S. virgaurea* L. ["golden rod" — literally], is native in Europe; it is similar to our *S. simplex*. In eastern North America alone there are about 50 species, easily illustrating the adage that "familiarity breeds contempt." Some of ours are valued as garden ornamentals in Europe (e.g., *S. canadensis, S. altissima,* and *S. gigantea*), while here they are often treated as weeds and even blamed for things they do not do, such as cause "hay-fever." Goldenrod pollen is transported by the many insects that can be seen busy at its showy heads. The heavy, sticky pollen is not designed for wind transport, which brings trouble to the nostrils of seasonal allergy sufferers. The bright beauty of goldenrods attracts attention at the same season as less conspicuous plants, like ragweed, shed their copious wind-dispersed pollen.

Goldenrods, especially *S. altissima* and much more rarely *S. gigantea* and *S. rugosa*, often have conspicuous swellings, or galls, on their stems. Most often noticed are the nearly spherical galls caused by a small fly, *Eurosta solidaginis* Fitch, whose growing larva is sheltered therein. The common elongate gall similarly harbors the larva of a tiny moth, *Gnorimoschema gallaesolidaginis* (Riley). The natural history of this gall-former, along with notes on some others, is described and illustrated by Miller (1963). Many additional insects also form galls on stems, leaves, and even roots of *Solidago* and *Euthamia*.

This genus has not been a favorite for identification of species, but its difficulties have inspired a number of cytological and other investigations. Most of the species are easily recognized when seen in their appropriate habitat and with typical inflorescences and leaves present. In a few species the inflorescences are axillary, subtended by ordinary-sized foliage leaves (fig. 230). Such a habit grades into a strictly terminal inflorescence, of which there are three principal shapes in our species. A few have flat-topped or corymbiform inflorescences (as also in *Euthamia*), with a somewhat dome-like (convex) or sunken (concave) outline across the top, the outer (lower) branches being longer than the central (upper) branches (plate 7-C). Nesom (1993) recognizes these four species in a separate genus, *Oligoneuron*. Some species have an elongate ± cylindrical inflorescence, a terminal "wand" or "rod" (plate 7-E & F). A number of conspicuous species have a pyramidal inflorescence broadest at or just above the base and tapering to the apex, which may nod a little; the lower branches in some species are ± recurved with the heads one-sided (i.e., oriented on top of the branches; plate 7-D).

In many species the total inflorescence varies considerably, sometimes compact, sometimes with wide-spreading branches, sometimes very depauperate. Plants growing in unaccustomed shade or stressful conditions may have very unusual inflorescences, such as abnormally small or condensed ones at the summit of the stem. The change, or lack of it, in leaf size from the base of the plant upwards is often very helpful for identification and complete specimens should always be collected. In some species, separate basal rosettes are formed, with leaves similar to (or even larger than) the lowest cauline leaves, which sometimes wither by flowering time. (Species with the mid-cauline leaves no smaller than the lower ones do not produce separate rosettes.) Incomplete specimens and extreme growth forms are not likely to work well in the key.

In determining length of an involucre, measure from its base to the tip of the longest phyllaries in *mature* heads. Likewise, look for *mature* achenes to determine their pubescence, for unlike the heads of some people, some achenes seem to become more hairy as they age. On the other hand, stem pubescence may abrade, especially on sand dunes, so that only scars may be visible on older stems.

As in so many genera, one should carefully evaluate alternatives presented in the key and comments, trying to place troublesome specimens where they seem best to fit but remembering that some hybrids do occur and a few individuals simply will not fit. The keys in most local treatments do not cover the same suite of species that we have in Michigan, but those in Salamun (1964) and Deam's *Flora of Indiana* have proved very helpful. The treatment by Semple (1992) is more up to date and reviews cytological data. Additional species may yet be discovered in Michigan (some have been reported but the documenting specimens, if any, prove to be other species). *S. macrophylla* Pursh, for example, comes as far south in Ontario as Batchawana Bay, not far north of Sault Ste. Marie. It has axillary inflorescences and large leaves, resembling those of *S. flexicaulis*, but also very large involucres (ca. 7.5–10 mm long) and glabrous achenes.

REFERENCES

Croat, Thomas. 1972. Solidago canadensis Complex of the Great Plains. Brittonia 24: 317–326.

Friesner, Ray C. 1933. The Genus Solidago in Northeastern North America. Butler Univ. Bot. Stud. 3: 1–64.

Melville, Mentor R., & John K. Morton. 1982. A Biosystematic Study of the Solidago canadensis (Compositae) Complex. I. The Ontario Populations. Canad. Jour. Bot. 60: 976–997.

Miller, William E. 1963. The Goldenrod Gall Moth Gnorimoschema gallaesolidaginis (Riley) and Its Parasites in Ohio. Ohio Jour. Sci. 63: 65–75.

Morton, J. K. 1979. Observations on Houghton's Goldenrod (Solidago houghtonii). Michigan Bot. 18: 31–35.

Nesom, Guy L. 1993. Taxonomic Infrastructure of Solidago and Oligoneuron (Asteraceae: Astereae) and Observations on their Phylogenetic Position. Phytologia 75: 1–44.

Pringle, James S. 1982. The Distribution of Solidago ohioensis. Michigan Bot. 21: 51–57.

Salamun, Peter J. 1964 ["1963"]. Preliminary Reports on the Flora of Wisconsin, No. 50. Compositae III — Composite Family III. The Genus Solidago — Goldenrod. Trans. Wisconsin Acad. 52: 353–382.

Semple, John C. 1992. The Goldenrods of Ontario: Solidago L. and Euthamia Nutt. Rev. ed. Univ. Waterloo Biol. Ser. 36. 82 pp. [Includes references to many previous publications.]

Werner, Patricia A., Ian K. Bradbury, & Ronald S. Gross. 1980. The Biology of Canadian Weeds. 45. Solidago canadensis L. Canad. Jour. Pl. Sci. 60: 1393–1409.

KEY TO THE SPECIES

1. Heads in a terminal ± flat-topped (somewhat domed to convex) corymbiform inflorescence
  2. Blades of middle and upper cauline leaves ovate to elliptical (less than 3 times as long as broad), densely pubescent on both surfaces . . . . .1. **S. rigida**
  2. Blades of middle and upper cauline leaves linear to lanceolate or oblanceolate (over 10 times as long as broad), glabrous or nearly so
    3. Rays 12–18, white, 4.5–8 mm long; pappus hairs slightly but clearly thickened (slenderly clavate) toward tip; upper cauline leaves slightly oblanceolate (broadest beyond the middle) . . . . .2. **S. ptarmicoides**
    3. Rays 10 or (usually) fewer, yellow, not over 4.5 (7) mm long; pappus hairs not thickened (or some thickening scarcely visible in *S. houghtonii*); upper cauline leaves broadest at or below the middle
      4. Heads larger: rays 3–4.5 (7) mm long and involucre ca. 5–7 mm long; pedicels scabrous-hirtellous . . . . .3. **S. houghtonii**
      4. Heads smaller: rays 1.5–3 mm long and involucre ca. 3.5–5.5 (6.5) mm long; pedicels smooth and glabrous or rough-hirtellous
        5. Pedicels and inflorescence branches densely rough-hirtellous; leaf blades with 3 or more longitudinal veins at the base, all or mostly folded inwards longitudinally . . . . .4. **S. riddellii**
        5. Pedicels smooth and glabrous or nearly so; leaf blades with one longitudinal vein (but often some principal lateral veins), flat . . . . .5. **S. ohioensis**
1. Heads in an elongate or pyramidal inflorescence or in axillary clusters
  6. Inflorescence terminal, often ± pyramidal (broadest toward base, about equally long, slightly nodding at top) but sometimes grading into axillary branches, *and* with curving, one-sided branches (the heads mostly directed upwards on well developed branches)
    7. Cauline leaves (at least the main ones) "triple-nerved," i.e., with a pair of elongate veins arising below the middle of the midrib and distinctly stronger than other lateral veins
      8. Leaves entire, succulent; plant of saline habitats (e.g., edges of heavily salted highways) . . . . .6. **S. sempervirens**
      8. Leaves with at least tiny and/or irregular teeth, of normal herbaceous texture; plants of widespread habitats
        9. Axis, pedicels, and branches of inflorescence glabrous; plant of prairie and dry prairie-like habitats, blooming *late* in the season; lower and rosette leaves linear-lanceolate . . . . .7. **S. missouriensis**
        9. Axis, pedicels, and branches of inflorescence at least sparsely but distinctly pubescent [use lens!]; or if glabrous (*S. juncea*), the lower and rosette leaves much larger than the mid-cauline leaves, ± elliptic, and the plant blooming *early* in the season in dry habitats

10. Stem glabrous all of its length below the inflorescence, rarely with a few scattered, spreading, short hairs [Note: Occasionally a plant of *S. uliginosa* may have obscurely triple-nerved leaves, but that species can usually be readily identified by the clasping base of the lowest leaves, which encircle half or more of the stem.]

    11. Basal (including rosette) and lower cauline leaves with oblanceolate to elliptic blades and long petioles, persistent; middle and upper cauline leaves remote (relatively few), distinctly smaller than basal leaves, and only weakly 3-nerved; plants blooming early (starting in July); branches of inflorescence glabrous or occasionally sparsely spreading-pubescent . . . . .8. **S. juncea**

    11. Basal leaves none; cauline leaves narrowly (rarely broadly) elliptic and the lowest withered by flowering time; middle and upper cauline leaves crowded (numerous), about the same size as the lowest leaves or larger, and distinctly 3-nerved; plants blooming late (starting Aug.–Sept.); branches of inflorescence ± densely pubescent . . . . .9. **S. gigantea**

10. Stem pubescent all or most of its length [Note: Occasionally a plant of *S. nemoralis* may have obscurely triple-nerved leaves, but that species can be easily distinguished by the dense, ashy-gray puberulence of the stem and leaves.]

    12. Involucres all or nearly all 2–3 mm long .10. **S. canadensis**

    12. Involucres all or mostly 3.1–4.6 (5) mm long . . . . .11. **S. altissima**

7. Cauline leaves with distinct midrib but the other (weaker) veins ± pinnate

13. Stems ± pubescent, at least on the upper half of the plant

    14. Cauline leaves entire or obscurely crenate-toothed; leaves and stems uniformly and densely puberulent throughout; lower and basal (including rosette) leaves oblanceolate, tapered into a winged petiole and larger than mid-cauline leaves; plant of sandy or rocky, open and usually very dry soil . . . . .12. **S. nemoralis**

    14. Cauline leaves sharply toothed; leaves beneath (at least on main veins) and stem with mostly spreading, longer hairs (over 0.5 mm); lower and basal leaves (none in rosettes) no larger than mid-cauline leaves (but usually absent at flowering time), all of them elliptic-lanceolate; plants of moist or shaded ground . . . . .13. **S. rugosa**

13. Stems glabrous (except sometimes just below and in the inflorescence)

    15. Lowest cauline leaves with tapering base clasping stem (encircling it for at least half its circumference); plant of wet habitats, with leaves nearly smooth above . . . . .20. **S. uliginosa**

    15. Lowest cauline leaves not clasping stem; plants of dry habitats or, if wet, the leaves very scabrous above

        16. Stem with strongly raised angles or ribs; upper leaf surface very scabrous, with dense, tiny, stiff conical projections (feeling like a cat's tongue and resisting rubbing toward the base with a finger); plant of swamps and other wet habitats . . . . .14. **S. patula**

        16. Stem terete (may be many-ridged); upper leaf surface smooth to slightly scabrous; plants of ± dry open or woodland habitats

            17. Basal (including rosette) and lower cauline leaves much larger than mid-cauline leaves, persistent (blades often 7–20 cm long on petioles half or more as long); branches of inflorescence glabrous or occasionally sparsely spreading-pubescent; leaves often tending to have prominent longitudinal veins, usually glabrous beneath but occasionally with some hairs on midrib; plant throughout the state, beginning to bloom in July (before other goldenrods) . . . . .8. **S. juncea**

17. Basal and lower leaves often withered by flowering time or, if present, not much larger than mid-cauline leaves; branches of inflorescence rather densely pubescent; leaves clearly pinnate-veined, with midrib and principal veins beneath spreading-pubescent (as in *S. rugosa*); plant in southern Lower Peninsula, blooming late . . . . .15. **S. ulmifolia**
6. Inflorescence axillary or terminal, but even if pyramidal the branches not one-sided and the top not nodding
  18. Leaves decreasing in size from middle of stem to the base, the mid to upper cauline leaves sharply toothed, much exceeding the distinctly axillary inflorescences (not necessarily any *branches*) they subtend; stems glabrous (except rarely on upper internodes), the lowest leaves usually withered by flowering time; achenes ± densely pubescent
    19. Leaf blades broadly ovate-elliptic, abruptly contracted to a winged petiole; stem ribbed or angled throughout, ± zigzag from node to node; leaves (at least the midrib beneath and petiole margins) ± sparsely pubescent . . . . . .16. **S. flexicaulis**
    19. Leaf blades narrowly elliptical, sessile; stem terete, glaucous when fresh, not (or scarcely) zigzag; leaves glabrous (except for short-ciliate margin) . . . .17. **S. caesia**
  18. Leaves increasing in size from middle of stem to the base, the mid-cauline leaves usually entire to crenate-toothed and usually not subtending inflorescences (these more clearly terminal); stems glabrous or pubescent, the lowest leaves usually persistent; achenes glabrous or glabrate except in *S. simplex*
    20. Stem ± densely pubescent its entire length and leaves pubescent on both sides
      21. Rays white or cream when fresh; involucres ca. 3–4 (4.5) mm long . . .18. **S. bicolor**
      21. Rays yellow; involucres 4–7 mm long
        22. Achenes nearly or quite glabrous; involucres 4–6 mm long . . . . .19. **S. hispida**
        22. Achenes ± densely antrorse-strigose; involucres 6–7.5 mm long . . . .21. **S. simplex**
    20. Stem glabrous or nearly so, at least below the middle and leaves glabrous or essentially so
      23. Lower cauline leaves ca. 6–18 times as long as broad, the petiole clasping the stem for half or more of its circumference; plant of wet habitats (including rock crevices on Lake Superior) . . . . .20. **S. uliginosa**
      23. Lower cauline leaves ca. 3–8 times as long as broad, not clasping (leaves of basal rosettes sometimes as much as 11 times as long as broad); plants mostly of dry habitats
        24. Involucres (4.5) 5–8 mm long; achenes ± densely antrorse-strigose; lower and basal leaves at least partly toothed or crenate-margined; involucres and leaves often resinous (easier determined when fresh, but appearing varnished, shiny, or glandular when dry) . . . . .21. **S. simplex**
        24. Involucres (2.5) 2.7–4.2 (5) mm long; achenes glabrous or nearly so; lower and usually basal leaves essentially entire; involucres resinous or not but leaves not resinous [Note: Plants tending to softer pubescence (if any) in the inflorescence, crenulate leaf margins, and slightly larger involucres are likely to be *S. hispida* var. *tonsa*.] . . . . .22. **S. speciosa**

1. **S. rigida** L.   Fig. 231   Stiff Goldenrod

Map 651. Prairies, dry fields and hillsides; may spread along roadsides and railroads.

This is a tall, handsome, large-headed goldenrod. The blades of the lower cauline leaves, like the basal ones, are much larger than the upper cauline leaves, and instead of being sessile or nearly so are on long petioles.

2. **S. ptarmicoides** (T. & G.) B. Boivin   Upland White or
                                              Sneezewort Goldenrod

Map 652. Local in jack pine plains and sandy prairie-like areas; calcareous flats along Lake Huron; limestone outcrops; fens and calcareous river
banks; rock crevices and ledges on Lake Superior.

The species name derives from an imagined resemblance to [*Achillea*]
Ptarmica, sneezewort, from a Greek word for sneeze. Sneezewort was once
used for a snuff. The name has nothing whatsoever to do with the common
name "ptarmigan" for a kind of grouse, derived from an old Gaelic name,
and attempts to link the two to invent a common name are ill-informed. If
recognized in the segregate genus *Oligoneuron*, the correct name for this
species is *O. album* (Nutt.) Nesom.

This species was long placed in the genus *Aster*, which it resembles in the
numerous, long, and white rays compared with those of the other flat-
topped goldenrods. However, genetically it hybridizes with typical goldenrods, not with true asters. Hybrids with *S. ohioensis* [*S.* ×*krotkovii* B.
Boivin] occur where the two species grow together, e.g., on Drummond
Island and along the Escanaba River. First-generation hybrids are recognizable in the field because of the pale yellow or cream-colored rays. Such
plants may resemble *S. houghtonii*; however, the rays are not only paler,
but also more numerous and/or longer; the pedicels may be smooth and
glabrous, scabrous, or rough-hirtellous. Some intermediates appear closer
to one presumed parent and some to the other; backcrossing can be suspected. Hybrids with *S. riddellii* and *S. rigida* have been reported elsewhere but not yet from Michigan, where their ranges (and habitats) overlap very little. All these are diploid species (2n = 18). See also comments
under the next species.

3. **S. houghtonii** A. Gray   Plate 7-C   Houghton's Goldenrod

Map 653. Primarily in damp interdunal flats and hollows, with some
plants on associated low dunes and beaches; nearby fens and swales; limestone crevices and pavements (alvar on Drummond Island); often with *S.
ohioensis* (which reaches peak flowering distinctly earlier), *Lobelia kalmii*,
and other calciphiles. This very local species, listed as threatened under
both Michigan and Federal law, occurs nowhere in the world except along
the northern shores of Lakes Michigan and Huron. It is very nearly endemic to Michigan.

Douglass Houghton, Michigan's first State Geologist, collected this species August 15, 1839, as he and his assistant (George Bull) were heading by
rowboat from Mackinac Island to Green Bay. On that date, they traveled
from a campsite east of Epoufette to their next one, west of Naubinway.
The type locality is therefore along the western shore of Mackinac County.
The great American botanist Asa Gray, who had spent the day August 15,
1838, visiting Houghton at his home in Detroit, named the new goldenrod

in his honor in the first edition of the celebrated Gray's *Manual* (in 1848 — 3 years after Houghton drowned in Lake Superior).

Solidago houghtonii is a hexaploid (2n = 54), apparently derived from hybridization between *S. ptarmicoides* and either *S. ohioensis* or *S. riddellii*. Morton has suggested backcrossing with *S. ohioensis* as the yellow-flowered parent, along with tripling of chromosomes. The rays are a strong yellow and conspicuous, making this a handsome plant in bloom. Reports of *S. houghtonii* from inland sites are based on diploid (and generally pale-rayed) hybrids or on a unique octoploid involving *S. ptarmicoides* as one ancestor (Crawford Co., Michigan) or on an unusual hexaploid (Genesee Co., New York) of similar ancestry. The Crawford County plants are especially robust, with rays and involucres at or exceeding the maximum length in *S. houghtonii*.

Those who are not familiar with *S. houghtonii* in the field are likely to understate the vigor of the species. Some keys have even stressed "few" heads (not over 15) as an identifying feature. The plant shown on plate 7 had over 200 heads, and while larger than average was by no means unique. While the number of heads on a plant may average fewer than in *S. ohioensis*, often an accompanying species, plants with over 100 heads are not unusual. The most vigorous plants seem to be associated with recent burial of their base, as at the lee margin of a dune encroaching on a wet interdunal hollow. John H. Ehlers, of the University of Michigan Department of Botany, was studying this species in the early 1930s, shortly before he retired, but he did not publish on it. From his correspondence, we know he was finding that plants "with 75 to 100 heads are common" and over 400 occurred. He referred to one plant in the Botanical Gardens with over 1200 heads in an inflorescence 8–10 inches broad.

4. **S. riddellii** Frank   Riddell's Goldenrod
    Map 654. Fens, wet prairies, shore meadows, low ground around lakes and along rivers.

    The distinctive narrow, curved (arcuate), infolded leaves are suggestive of *Helianthus maximilianii*. The leaves taper narrowly to an acute (but callus-tipped) apex, and the lower ones are often withered by flowering time.

5. **S. ohioensis** Riddell   Ohio Goldenrod
    Map 655. Damp sandy or rocky calcareous shores and interdunal flats; fens, wet prairies, sedge meadows, calcareous river banks.

    The leaves taper abruptly to a rounded or obtuse tip. The lower leaves, especially in rosettes, may have blades as broad as 6 cm and on very long petioles — a striking feature as they do not wither until late in the season. However, the leaves are sometimes about as narrow as in *S. houghtonii* and small plants are sometimes confused with that large-headed species. Hybrids with *S. ptarmicoides* are discussed under that species above.

6. **S. sempervirens** L. Seaside Goldenrod

Map 656. A native of the east coast, first reported from Michigan in 1978 by Reznicek (Michigan Bot. 19: 26) and apparently spreading along expressways (saline habitat) and on fly ash deposits.

7. **S. missouriensis** Nutt. Missouri or Prairie Goldenrod

Map 657. This western species barely reaches to the Great Lakes region and there may be some question whether it is native in Michigan, where it has been collected very rarely, first by S. H. and D. R. Camp in Jackson County in 1893.

Our plants are var. *fasciculata* Holz., once often segregated as *S. glaberrima* M. Martens. From the western region of Lake Superior westward, the species is not always clearly distinguished from *S. juncea*.

8. **S. juncea** Aiton Early Goldenrod

Map 658. Dry sandy fields, jack pine plains, oak and aspen woods (especially along borders); clearings, rocky openings, occasionally along roadsides and railroads.

This species is usually the first goldenrod to start blooming in an area (in July) but it has a long season. The tiny, nearly linear, entire leaves or

652. Solidago ptarmicoides

653. Solidago houghtonii

654. Solidago riddellii

655. Solidago ohioensis

656. Solidago sempervirens

657. Solidago missouriensis

bracts which are usually present on the uppermost portion of the stem are often conspicuous and helpful for identification, especially when the large basal leaves are missing. The later generally have blades ca. 7–20 cm long, on petioles half or more as long. Occasionally the branches of the inflorescence and, more rarely, the stem have rather sparse, spreading pubescence, but usually these parts are glabrous. Very rarely the branches of the inflorescence are not one-sided (see comments under *S. speciosa*, below).

9. **S. gigantea** Aiton   Late Goldenrod
Map 659. Dry or usually damp open ground, roadsides, fields, thickets; prairies, fens, wet meadows, tamarack swamps, marshes; swamps, along streams and rivers; deciduous woods and borders.
This late-blooming species resembles *S. canadensis* and *S. altissima* but has a glabrous stem, while the inflorescence is ± densely pubescent. The stem is often glaucous. The leaves range from glabrous to pubescent, and the uppermost ones tend to be more elliptical and toothed than in the preceding species. Furthermore, the lower leaves in *S. gigantea* are smaller than the middle ones, rather than distinctly larger as in *S. juncea*.

10. **S. canadensis** L.   Canada Goldenrod
Map 660. Fields, roadsides, railroads, fencerows, moist thickets; wet meadows, marshy ground, river banks, creek margins; fens, shores, floodplains, damp woodland, borders of conifer swamps and woods, dry rocky ground.
Concentrated in the southeastern portion of the state, but scattered throughout, are plants with the undersides of the leaves puberulent across the entire surface and with the stems tending to be pubescent all the way to the base. These have been called var. *hargeri* Fernald, although in the strictest sense that name applies to plants with relatively long hairs on the lower stem. Croat describes the stem as "densely spreading-puberulent." Typical var. *canadensis* has the undersides of the leaves glabrous or at most puberulent on the principal veins and is glabrous on the lower half

658. Solidago juncea   659. Solidago gigantea   660. Solidago canadensis

of the stem; it is found almost entirely in parts of the state other than the southeastern.

See also comments under the next species, which is often included in *S. canadensis* as var. *scabra* (Willd.) T. & G.

11. **S. altissima** L.   Plate 7-D   Tall Goldenrod

Map 661. In habitats similar to those of the preceding species: fields, prairie remnants, roadsides, fencerows, thickets; meadows, marshy ground, fens, damp shores, low woodlands; borders of rivers, woods, and swamps.

Involucre length has to be measured carefully on mature heads to be sure of the distinction from the small-headed *S. canadensis*. Measure several involucres to get a majority opinion if necessary. *S. gigantea* has involucres of similar size (or sometimes a trifle smaller), but the stem is glabrous beneath the inflorescence and the leaves are often glabrous beneath. In *S. altissima* the stem is pubescent to below the middle, usually to the base, but loss or abrasion of the hairs often makes it difficult to discern pubescence toward the base of the plant. The leaves generally are ± pubescent all across the lower surface, as in *S. canadensis* var. *hargeri*, and some of the main leaves are more often entire or nearly so than in the distinctly serrate-toothed *S. canadensis*.

One collection (*Richards 4227* in 1950, MICH) from south of Houghton appears to be the northern *S. lepida* DC.—whatever that is, at least as interpreted by Melville and Morton, who admitted that further work on this taxon is needed to determine its status. Others, too, have puzzled over the status of *S. lepida* and its circumscription, and the problems cannot be solved here, where I allow it to "run into" *S. altissima* just as it seems to do in nature. The Houghton specimen has distinctly pale, sharply serrate leaves with dense pubescence beneath, copious stem pubescence of spreading hairs [becoming sparser and absent (abraded?) below], and an upswept aspect to the inflorescence, of which the base is closely enveloped by leafy bracts (though not quite as much so as in the Melville & Morton photograph); the involucre is 3.5–4.2 mm long.

Two collections by Dodge in 1916 (MICH), referred by him to *S. lepida* (Univ. Michigan Mus. Zool. Misc. Publ. 4: 14. 1918), are certainly unusual. One is from Houghton and the other from L'Anse (Baraga County). The involucres hardly exceed 3 mm, the stems and leaves have tangled, whitish (i.e., ± tomentose) pubescence, and the leaves are nearly or quite entire. Another collection (*Hermann 679* in 1926, MICH) from Houghton County was determined by Shinners as *S. lepida*; it has leaves sharply toothed but glabrous except on the main veins below, a puberulent stem, and involucres barely 3 mm long. An 1888 Farwell collection from Keweenaw County (MSC) has the upswept panicle branches scarcely secund, resembling *S. lepida*, and the leaves and stems are nearly glabrous, but the involucres are only about 2.5 mm long (in *S. lepida* they are supposed to be 3–4 mm).

Other Farwell collections (in 1887, GH, BLH) have involucres of appropriate size, upswept bracted panicles, and hairy stems and leaves; these were identified by Fernald as *S. lepida* var. *fallax*. This group of goldenrods surely needs study on the Keweenaw Peninsula!

12. **S. nemoralis** Aiton   Gray Goldenrod
   Map 662. Dry fields, prairies, and open barren ground; open woods of aspen and oak, jack pine plains, sand dunes, rocky clearings and outcrops, including limestone pavements.
   The close, dense puberulence generally gives an ashy gray aspect to this species. If the branches of the inflorescence are not clearly one-sided (as sometimes happens, especially on pressed specimens), individuals might be confused with *S. hispida*, but differ in usually a nodding apex to the inflorescence, in the antrorse-strigose achenes, and in stiffer (less soft) pubescence.

13. **S. rugosa** Miller   Fig. 227   Rough-leaved Goldenrod
   Map 663. Moist woods and thickets; swamps (both coniferous and hardwood), especially along borders; peatlands; fields, fencerows, ditches; often in disturbed areas in woods and swamps, as along trails and in clearings. The Keweenaw County record is suspiciously disjunct. The specimen was collected in 1889 by Farwell (*732*, BLH), who said in his notes "Common." No one else has ever collected the species that far west in the Upper Peninsula.
   Variable in pubescence, leaf texture, and inflorescence. The latter is often large and elm-like in shape, with several main forks, or it can be narrow with the lower branches shorter than the subtending leafy bracts [var. *villosa* (Pursh) Fernald]. The leaves are strongly rugose (with veins sunken, as viewed from above, or raised, as viewed from beneath, the surface thus appearing pitted or like hammered metal) in var. *aspera* (Aiton) Fernald. Both of these variants occur throughout the range of the species in the state, as does var. *rugosa*, and, indeed, some specimens display key features of both vars. *aspera* and *villosa*, and both intergrade

661. Solidago altissima   662. Solidago nemoralis   663. Solidago rugosa

with var. *rugosa*. There seems to be some tendency for the "aspera" plants to tolerate drier, weedier sites.

Very rarely, the stems are completely glabrous below the inflorescence. Such plants would key to *S. ulmifolia*, but can generally be distinguished if they possess most or all of the following characters: lower leaves smaller than the middle ones, leaves scabrous-hispidulous above, achenes strongly antrorse-strigose, rays 6 or more per head, stems arising from elongate rhizomes. In the quite similar *S. ulmifolia* (which is also more of a woodland than wetland plant), lower leaves are somewhat larger than the middle ones and often coarsely toothed, leaves usually have long straight often appressed hairs on the upper surface, achenes are short- and/or sparsely pubescent, rays are 3–5 per head (fig. 229), and stems arise from a strong caudex without rhizomes.

An apparent hybrid between this species and *S. uliginosa* has been collected at Sault Ste. Marie, where it was growing with the putative parents (*Hiltunen 2258*, WUD).

14. **S. patula** Willd.   Fig. 228   Rough-leaved Goldenrod
    Map 664. Swamps (cedar and hardwoods), fens, sedge meadows, thickets.

This is one of the most distinctive goldenrods, with very scabrous leaves, the blades of the large lower ones often 5–8 or even 10 cm broad, tapering into winged petioles.

Plants of intermediate appearance from two Berrien County sites are presumably hybrids with *S. uliginosa*. The leaves are scabrous above, but narrower than in *S. patula*, and the stems are scarcely angled. At the site where I gathered specimens in 1958, about one mile west of Millburg, it was growing with both putative parents. At the other site, only *S. uliginosa* was reported (Michigan Bot. 28: 212. 1989). Similar plants have been collected in Kent, Macomb, Oakland, and Wayne counties.

15. **S. ulmifolia** Willd.   Fig. 229   Elm-leaved Goldenrod
    Map 665. Oak woods; shaded shores and banks.

Quite similar to *S. rugosa*, especially if one does not have complete

664. Solidago patula

665. Solidago ulmifolia

666. Solidago flexicaulis

specimens. A glabrous or glabrate stemmed goldenrod of this sort, with flat (not rugose) leaves and from the southern Lower Peninsula, may require careful comparison of the two species for identification. See comments under *S. rugosa* above.

16. **S. flexicaulis** L.   Fig. 230   Zigzag Goldenrod
Map 666. Low, rich woods, often in moist hollows, ravines, and banks; rocky (especially limestone) woods; shaded creek borders, swampy woods (cedar, ash).

Specimens of *S. patula* in which the one-sided nature of the inflorescence branches is obscure would key here (the leaf shape and angled stem are remarkably similar), but *S. patula* has very scabrous upper surfaces on the leaves, which are smooth in *S. flexicaulis*. The leaves are closely and sharply, even jagged, toothed in both species. See also comments on *S. macrophylla* just above the references for this genus.

The stem is rarely pubescent on the upper internodes. The zigzag aspect of the stem seems to be more striking in the field than on pressed specimens. Intermediate plants tending to more rounded stems, narrower leaves, and sometimes fully glabrous leaves appear to be hybrids, probably with *S. gigantea*.

17. **S. caesia** L.   Bluestem Goldenrod
Map 667. Usually in rich deciduous woods (e.g., beech-maple), swamp forest, or at least damp clearings and thickets; but also in dry woods (e.g., jack pine and aspen or oak-hickory) and on wooded dunes.

The rather delicate glaucous stem and mostly small axillary inflorescences make this an easily recognized species. The little inflorescences may be on elongate axillary branches or the plant may be unbranched. Both forms can occur together. The narrowly elliptic, usually toothed leaves might remind one of *S. canadensis*, but they are not at all triple-nerved.

18. **S. bicolor** L.   White Goldenrod; Silver-rod
Map 668. Dry sandy woods and banks.

Authors mostly agree that the white rays are the only sure character for

667. Solidago caesia   668. Solidago bicolor   669. Solidago hispida

227. *Solidago rugosa*, head ×2
228. *S. patula* ×½; head ×2
229. *S. ulmifolia*, head ×2
230. *S. flexicaulis* ×½; head ×2

distinguishing this species from *S. hispida*. Unfortunately, only one Michigan collector (D. Cooley) explicitly recorded the flower color on his labels. Since the color cannot be safely determined from long-dried specimens, I have mapped as *S. bicolor* only specimens for which the collector in publications or available field notes, if not on the labels, stated the color to be white or indicated that color by distinguishing between *S. bicolor* and *S. bicolor* var. *concolor* (= *S. hispida*). Even so, the species is clearly very rare and local, if not extirpated, in Michigan.

19. **S. hispida** Willd.   Plate 7-E   Hairy Goldenrod
     Map 669. Sandy woods (oak, hickory, jack pine, aspen) and clearings; dunes, sandy or rocky plains, rock outcrops and pavements of all kinds; rarely in damp ground.
     Ordinarily distinctive in its larger lower leaves, softly pubescent stems and leaves, and glabrous or glabrate achenes. However, a glabrous-stemmed form grows on rock outcrops of the Upper Peninsula: var. *tonsa* Fernald. It is easily confused with *S. simplex*, as mostly dwarfed individuals of the two species may grow together in rock crevices. Our records of var. *tonsa* are from the Keweenaw county mainland, Isle Royale, and Drummond Island.
     Most 19th century collectors included this species in *S. bicolor*, as var. *concolor* T. & G. if distinguished nomenclaturally; hence, old reports of *S. bicolor* should not be interpreted as referring to white-rayed plants without further evidence (see above).

20. **S. uliginosa** Nutt.   Bog Goldenrod
     Map 670. Bogs, fens, swamps of cedar and other conifers, marly places; marshes, wet meadows and prairie-like ground, shores and interdunal pools or hollows; rock crevices and pools on Lake Superior.
     This is one of the most variable species in a genus noted for such variability! The inflorescence ranges from a broad pyramidal one with one-sided branches resembling that of *S. juncea* to the more common shape: a long cylindrical inflorescence. The latter goes well with the streamlined lower and basal leaves (up to 0.5 m long), of which the petioles are very distinctive, nearly or quite encircling the stem at their very base. The stems and leaves are completely glabrous, except that the pedicels and branches of the inflorescence may be rough-pubescent or sparsely hirsute.
     Variability also results from hybridization (and presumed backcrossing), especially with neighboring species in wet habitats. Hybrids with *S. ptarmicoides* could resemble *S. houghtonii* (although with paler rays). Hybrids with *S. patula* (see also above) can be recognized by somewhat angled stems and strongly scabrous upper surfaces of the leaves (as well as a tendency to have more toothed leaves). The leaves of *S. uliginosa* are nearly or quite entire, at most with rather obscure crenulate teeth. A few plants from Delta and Washtenaw counties with sharp teeth but smooth surfaces on the leaves suggest hybridization with *S. gigantea*.

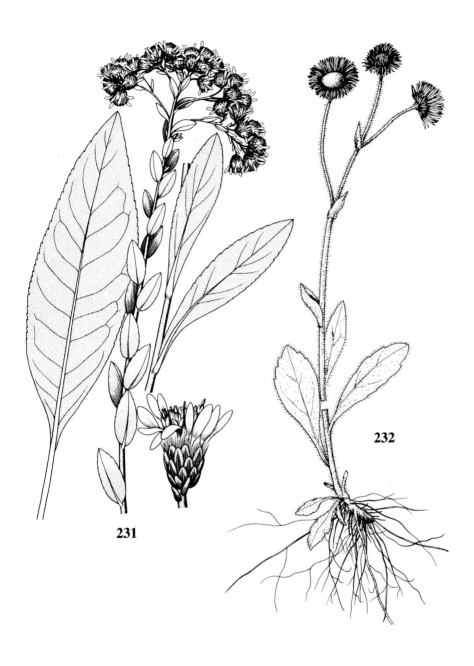

231. *Solidago rigida* ×½; head ×2
232. *Erigeron pulchellus* ×½

21. **S. simplex** Kunth   Plate 7-F   Gillman's Goldenrod

Map 671. Sand dunes and beaches along the shores of the Great Lakes; sandy plains and savanas with jack pine and oak; rock crevices and ledges along Lake Superior and inland on cliffs and outcrops.

There seems now to be widespread agreement that *S. simplex* is the oldest name applicable to this complex, which includes plants variously referred in the past to *S. spathulata* DC., *S. randii* (Porter) Britton, *S. racemosa* Greene, *S. glutinosa* Nutt., and *S. gillmanii* (A. Gray) E. S. Steele — among others. I use here a common name that appears in several works ("Dune Goldenrod" is also used), although strictly speaking it applies only to the most frequent phase of the species in this area, in order to give some deserved recognition to a neglected scientist and scholar in the early history of Michigan (see Voss 1978, pp. 47–50). Henry Gillman (1833–1915) was an assistant engineer with the United States Lake Survey 1851–1869 and an assistant superintendent of lighthouse construction on the Great Lakes 1870–1876; his plant-collecting was accomplished during those periods. He also achieved some fame as an archeologist and, later, as a diplomat and librarian (in charge of the Detroit Public Library 1880–1885).

The forms of this broadly defined species are as diverse as the interests of Mr. Gillman. Common and showy late in the summer on sand dunes along Lakes Michigan and Superior and northern Lake Huron is var. *gillmanii* (A. Gray) Ringius, which sometimes has pubescent stems (unlike most *S. simplex*). This taxon was originally described from the "shores of Lakes Superior and Michigan." The collection recently designated as lectotype (Taxon 36: 155. 1987) was grown in the Botanic Garden of Harvard University (by Asa Gray) from roots supplied by William Boott from "Upper Michigan." It is a particularly robust individual, as might be expected under conditions of cultivation. Field-collected specimens from Gillman exist from several sites on Lakes Michigan and Huron. This is a robust variety, even if not so much so as the specimen now serving as type of the name, the stems sometimes a meter tall, the inflorescences often branching and large-headed, and the leaves toothed.

670. Solidago uliginosa

671. Solidago simplex

672. Solidago speciosa

Plants of the jack pine plains, even large ones, tend to have slightly smaller involucres, sometimes only 4.5–5 mm long, but also to have clearly glutinous phyllaries and leaves. Whether these should be included in var. *simplex* (according to Ringius, reaching its eastern limit at Lake Superior) or recognized as var. *randii* (Porter) Kartesz & Gandhi is not clear to me. Plants (usually ± stunted) of Upper Peninsula rock outcrops are also sometimes small-headed and apparently run into *S. hispida* var. *tonsa* on the Lake Superior shores of Isle Royale and the Keweenaw Peninsula. There, plants with glabrous achenes *and* stems, or pubescent stems but densely pubescent achenes (and with varying densities of pubescence) grow together, along with more typical plants of both species. Immature specimens without achenes or full-grown involucres cannot be placed.

Reports of the western *S. decumbens* Greene from Michigan may be based on a Keweenaw County collection by Fernald (crest of Lookout Mountain in 1934), so-labeled but very immature, with heads still in bud but glutinous (as are the leaves). This and other collections resemble similar stunted forms of *S. simplex*.

Dwarf plants may bloom when as short as 5 cm, and such low individuals may bloom earlier, presumably encouraged by the warm influence of the hot sands on which they grow.

22. **S. speciosa** Nutt.   Showy Goldenrod

Map 672. Dry open sandy ground, including oak and jack pine savanas, prairies, fields, rarely dunes — and associated roadsides and railroads; thin soil on high rock mountains in the western Upper Peninsula (Marquette & Ontonagon cos.).

This is another variable species, but well named for its large showy inflorescences late in the season. In the southern part of the Lower Peninsula and apparently also in the western Upper Peninsula is var. *speciosa*, with basal leaves running 3–8 cm broad. Less robust plants, with fewer and narrower leaves, blooming earlier, are var. *jejunifolia* (E. S. Steele) Cronquist (TL: Indian River, Cheboygan Co.); these plants range south through the jack pine plains from Cheboygan County, whence they were

673. Chrysopsis villosa

674. Chrysopsis camporum

675. Haplopappus ciliatus

first described, into the range of var. *speciosa*. The separation of varieties is not always clear, especially since the basal leaves often wither by flowering time. Boivin has allied var. *jejunifolia*, a plant of pine and oak barrens, with the wetland *S. uliginosa*, an unlikely relationship.

Rarely a plant of *S. juncea* has branches of the inflorescence not one-sided, and such plants are likely to run here in the key because of their larger lower leaves, glabrous or glabrate stems, and small heads. However, at least the lower leaves are ± strongly toothed, the blooming season is much earlier, and the inflorescence is predominantly terminal (even vase-like).

60. Chrysopsis Golden-aster

Some authors include one or both of our species in the genus *Heterotheca*.

REFERENCES

Semple, John C., Vivian C. Blok, & Patricia Heiman. 1980. Morphological, Anatomical, Habit, and Habitat Differences Among the Goldenaster Genera Chrysopsis, Heterotheca, and Pityopsis (Compositae — Astereae). Canad. Jour. Bot. 58: 147–163.

Semple, John C. 1983. Range Expansion of Heterotheca camporum (Compositae: Astereae) in the Southeastern United States. Brittonia 35: 140–146.

KEY TO THE SPECIES

1. Plants low (up to ca. 0.5 m), ± spreading and several-stemmed or -branched; leaves mostly obtuse (but mucronate), eglandular .... 1. **C. villosa**
1. Plants tall (over 0.5 m), erect; leaves mostly acute, stipitate-glandular (at least beneath) .... 2. **C. camporum**

1. C. villosa (Pursh) DC.

Map 673. Native and widespread to the west, a waif eastward into Michigan. Collected in 1902 by Emma Cole at the Phillips farm, Grand Rapids, where many other western waifs occurred. Henson collected the species 80 years later along the Soo Line railroad in Schoolcraft County.

Sometimes placed as *Heterotheca villosa* (Pursh) Shinners.

2. C. camporum Greene

Map 674. A prairie plant, adventive eastward, where it is becoming well established (Swink & Wilhelm 1994). Collections from Ottawa County (*Gillis 13513* in 1976) and Washtenaw County (*Reznicek 7088* in 1982) have been identified by Semple as *Heterotheca camporum* var. *glandulissimum* Semple [= *C. camporum* var. *glandulissima* (Semple) Cronquist], which is the weedy variant expanding its range. An earlier collection from Wayne County (*Churchill* in 1953, MSC) appears to be the same.

Our specimens all have showy rays ca. 1.5 cm or a little longer—about 50% longer than usually stated for this species. The many short stipitate-glandular hairs on stems and leaves of var. *glandulissima* make it easier (than the typical variety) to distinguish from the preceding species. Furthermore, the largest hairs on the lower leaves of *C. camporum* (as in other *Heterotheca* sens. str.) are even more stiff and bulbous-based than the slender hairs characteristic of *C. villosa* (representing *Chrysopsis* sens. str.).

61. Haplopappus

Those who split up this genus — with considerable justification — place our species as *Prionopsis ciliata* (Nutt.) Nutt. or even transfer it to *Grindelia*, as *G. papposa* Nesom & Suh.

1. **H. ciliatus** (Nutt.) DC.　　　　　　　　　　　　　Goldenweed
Map 675. A native of the southern Great Plains, but spreading sparingly eastward along roadsides and railroads. First collected in Michigan in Berrien County in 1972 (*Schulenberg*, MOR). Found in Schoolcraft County in 1988 (*Henson 2687*, MICH).

Superficially resembles *Grindelia squarrosa*, of similar habitats (and to which in fact it is apparently closely related), but readily distinguished by the spiny tips of the leaf teeth, soft pappus hairs, and absence of both gum on the involucre and resinous dots on the leaves.

62. Inula

1. **I. helenium** L.　　　　　　　　　　　　　　　Elecampane
Map 676. Roadsides, fields, trails and clearings in woods; occurs in both dry and wet ground. Apparently native to central Eurasia, but widely spread as an escape from cultivation, as it is both ornamental and medicinal.

This is a robust, large-headed perennial, up to 2 m tall. The massive root has long been revered for its medicinal properties, and the enormous leaves may at first glance be mistaken for those of *Silphium terebinthinaceum*.

63. Callistephus

1. **C. chinensis** (L.) Nees　　　　　　　　　　　　Chinese Aster
Map 677. A native of China, widely cultivated, and only rarely escaped. Collected (*Farwell 8151*, MICH, BLH) in 1927 near Farmington, where it was said to be "Scattered over considerable territory" (Am. Midl. Nat. 11: 70. 1928).

64. Boltonia

1. **B. asteroides** (L.) L'Hér.　　　　　　　　　False Aster; Boltonia
Map 678. Native mostly south and west of Michigan, where it was first collected in 1905 in Cass County (*Pepoon 433*, MSC). It seems now to be concentrated in Monroe County, where it was first collected in 1910

(*Farwell 2184*, GH) and where its abundance fluctuates greatly from year to year. After exposure of moist ground following flooding, bushy plants as tall as 2 m may dominate for a year or two, soon to die out, presumably succumbing to competition and/or increasing dryness. It is likely to be found with *Penthorum sedoides, Echinochloa walteri, Ammannia robusta* (another recent arrival), *Bidens* spp., *Helenium autumnale, Juncus torreyi,* and other plants of marshy mudflats.

The yellow disk and pink rays make this an attractive species, and indeed it is sometimes cultivated, often as a form with white rays. In view of the relative recency of Michigan records, it is possible that the species is not native here. Our plants are var. *recognita* (Fernald & Griscom) Cronquist.

65. **Erigeron**   Fleabane

This genus is quite similar to *Aster*, but fleabanes tend to bloom in spring and early summer, while asters are typical of the late summer and fall landscape. In our species (not necessarily elsewhere), the rays of *Erigeron* are more numerous and narrower than is usual in the asters, while the plants tend to be lower and fewer-leaved than in most asters. Thus, in this region, the genus can be more easily recognized than might be supposed.

676. Inula helenium

677. Callistephus chinensis

678. Boltonia asteroides

679. Erigeron pulchellus

680. Erigeron
     philadelphicus

681. Erigeron flagellaris

To determine the width of the rays, it is necessary to examine *flat* ones (well pressed if dried), for the margins curl inward as they shrivel and such rays appear excessively narrow.

REFERENCES

Cronquist, Arthur. 1947. Revision of the North American Species of Erigeron, North of Mexico. Brittonia 6: 121–300.

Fernald, M. L., & K. M. Wiegand. 1913. A Northern Variety of Erigeron ramosus. Rhodora 15: 59–61.

Morton, J. K. 1988. Variation in Erigeron philadelphicus (Compositae). Canad. Jour. Bot. 66: 298–302.

KEY TO THE SPECIES

1. Cauline leaves broadly rounded at base, sessile and usually ± clasping; heads ca. 1.5–3.5 cm broad
  2. Rays (when not withered) ca. 0.8–1.7 mm broad; heads 1–4 per stem . . . 1. **E. pulchellus**
  2. Rays less than 0.5 mm broad; heads usually 3–30 (40) per stem . . . .2. **E. philadelphicus**
1. Cauline leaves tapered to a non-clasping base; heads (if radiate) mostly ca. (1) 1.5–2.2 cm broad
  3. Plants scapose (the heads solitary on stalks naked or nearly so) and becoming stoloniferous at the base; pubescence appressed on leaves and stems . . . .3. **E. flagellaris**
  3. Plants with leafy stems, not stoloniferous; pubescence various
    4. Rays inconspicuous, exceeding the pappus by less than 2 mm; pubescence of pedicels glandular, spreading . . . . .4. **E. acris**
    4. Rays showy, longer (heads rarely discoid in *E. strigosus*); pubescence of pedicels antrorse, eglandular
      5. Leaves very numerous, much overlapping, linear, less than 5 mm broad, often in axillary tufts besides the principal cauline leaves; heads 1–4 (5) per stem or main branch, on long nearly or quite naked pedicels . . . . .5. **E. hyssopifolius**
      5. Leaves not dense, all or mostly over 4 mm broad, with no axillary tufts; heads usually numerous (except on depauperate plants) — even as many as 100 or more on robust individuals
        6. Middle region of stem moderately to densely pubescent with only short (0.5 mm or less) mostly appressed-antrorse hairs; principal cauline leaves linear to oblanceolate, ca. 2.5–10 (15) mm wide, entire .6. **E. strigosus**
        6. Middle region of stem glabrate to pubescent with all or many of the hairs long (0.5–1.2 mm) and spreading; principal cauline leaves usually elliptic to ovate, ca. 10–35 (40) mm wide, with a few large teeth .7. **E. annuus**

1. **E. pulchellus** Michaux   Fig. 232   Robin's-plantain
    Map 679. Woodland of oak, jack pine, etc., doing well after fire; banks, often moist, as above fens, lakes, and other wet areas; meadows, deciduous woods.

The stems often tend to be even more pubescent at the base than in the next species. The heads average larger as well as fewer than in *E. philadelphicus* and are quite showy. The plants are stoloniferous and colonial.

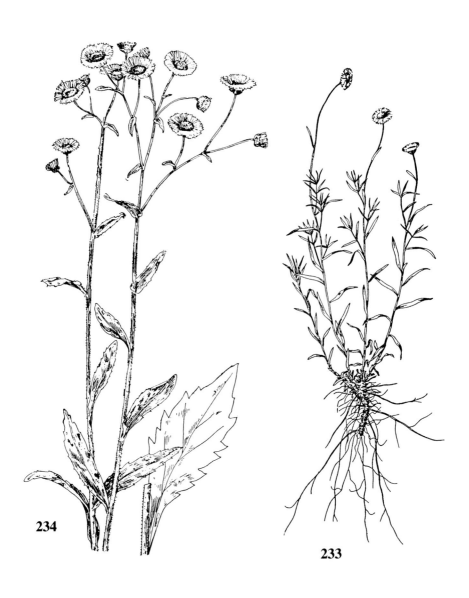

234

233

233. *Erigeron hyssopifolius* ×½
234. *E. annuus* ×½

2. **E. philadelphicus** L.   Common or Philadelphia Fleabane
    Map 680. Almost ubiquitous: roadsides, railroads, fields, meadows; roads, trails, and clearings (including logged areas) in cedar swamps and upland mesic hardwoods; river banks, shores.
    The heads in this species are smaller than in the preceding, averaging ca. 1.5–2 cm broad. A dwarf glabrate extreme with very small heads and white rays is var. *provancheri* (Vict. & J. Rousseau) B. Boivin. It has been found in the Manitoulin Island area and the Bruce Peninsula in Ontario, and should be sought on Drummond Island and perhaps elsewhere in crevices of limestone (dolomite). The rays of var. *philadelphicus* range from pale to deep pink (rarely white), and the plants are ± hairy.

3. **E. flagellaris** A. Gray
    Map 681. A native of western North America, collected in 1902 by Emma Cole from the same farm where refuse from New Mexico led to the temporary introduction of *Thelesperma megapotamicum* (see genus 19 above), *Verbesina encelioides* (genus 46), and other species.
    This is a distinctive low, small-headed and small-leaved plant.

4. **E. acris** L.   Kamchatka Fleabane
    Map 682. A circumpolar species, in eastern North America barely ranging south into the United States (farther south in the West). Our few specimens mostly lack habitat information (but must have come mostly from ± rocky areas). Clearings, open paper birch woods, and sandy shaded banks are cited on labels.

5. **E. hyssopifolius** Michaux   Fig. 233   Hyssop-leaved Fleabane
    Map 683. A boreal species of North America, ranging from New England to the Yukon, very local in Michigan. Locally frequent in rock crevices along the north shore of Lake Superior in Ontario, thence inland northward. Collected by Farwell as "rare" in Keweenaw County in 1890 (*827*, BLH); otherwise, known only from a marly fen and associated open tamarack-cedar swamp near Moran (see Michigan Bot. 23: 11–18. 1983).

682. Erigeron acris   683. Erigeron
                               hyssopifolius
                                                  684. Erigeron strigosus

Page 483

Originally described from Lake Mistassini, Quebec, whence *Primula mistassinica* received its name. This is an attractive, delicate-looking plant with white or pinkish rays, the color perhaps depending in part on the age of the flowers.

6. **E. strigosus** Willd.   Daisy Fleabane
Map 684. Characteristically in dry open sandy ground, including woodland of oak, aspen, and/or jack pine; clearings, rock outcrops, banks, prairie remnants, shores; fields, roadsides, railroads; sometimes in wet places. Considered indigenous in North America, but weedy in habit and an introduced weed in Europe.

Usually a shorter, more slender plant than the next species, up to 5 (8) dm tall. American authors have not, for the most part, followed the Europeans in merging this species with the next, although the two are sometimes difficult to distinguish. Rarely (in Michigan) plants with essentially entire, narrow, oblanceolate leaves as in *E. strigosus* may have the long spreading stem hairs of *E. annuus*, and I have associated those plants here with the latter. They may represent what has been called *E. strigosus* var. *septentrionalis* (Fernald & Wiegand) Fernald. Fernald and Wiegand (1913) observed that such transitional plants are unlikely to be hybrids because they occur in some places where *E. annuus* is unknown. On the other hand, perhaps they *are E. annuus*.

Very rarely the rays are much reduced or absent [f. *discoideus* (A. Gray) Fernald]. Such plants have been collected in Keweenaw and Lenawee counties. This species was once widely known as *E. ramosus* (Walter) BSP.

7. **E. annuus** (L.) Pers.   Fig. 234   Daisy Fleabane
Map 685. Usually in ± disturbed ground (dry or moist) including fields, roadsides, waste ground, clearings and trails in woods, cultivated ground; floodplains, wetlands, open mixed woods, shores. Like the preceding, a weedy native of North America, introduced in Europe.

Often a taller plant than *E. strigosus*, the lower and basal leaves generally toothed. The rays in both species are 0.6–1.1 mm broad, thus distinguishing them from *E. philadelphicus* (besides the rays' being usually white, though occasionally pale pink). See also comments under the preceding species, for ambiguous specimens.

66. **Aster**   Aster
This is a large and extraordinarily difficult genus, sometimes divided into two or more genera. A broad circumscription is followed here. Some authors place species 8, 12, and 20 in the genus *Virgulus*, with a base chromosome number of 5. Others segregate *A. brachyactis* as *Brachyactis ciliata* (Ledeb.) Ledeb., with 7 as the base number. Our other species (base number 8, rarely 9) have, less often, been placed in as many as five additional genera. Nesom (1995) divides our asters among seven genera, along

somewhat different lines, and still other schemes are under study by various workers. Furthermore, the plant long known as *Aster ptarmicoides* T. & G. is now generally accepted as a species of *Solidago*.

Polyploidy is frequent, hybridization (with subsequent backcrossing) seems more frequent than in most genera (although perhaps not as common as sometimes supposed), morphological variation is great within species (sometimes depending upon habitat). None of the wild plants have read their job descriptions, much less attempted to conform to them, and the student of *Aster* can expect exceptions to almost any statement in the key. As usual for such complex genera, one must balance the alternatives, try to gain experience with as many taxa as possible, and agree to give up on some individuals that are too depauperate, abnormal, incomplete, or otherwise perverse. Especially frustrating are many herbarium specimens that lack basal and underground parts, are pressed too badly to allow measurement of parts, or for which the label fails to state fresh flower colors, habitat, or other essentials. People who go looking for odd and unusual asters often get just what they are looking for: trouble. The percentage of strange variants and presumed (but frequently unidentifiable) hybrids seems much larger in herbaria than in nature.

Rays that are white when fresh may turn pink or blue when dried; others are colored from the beginning. Disk corollas often turn from yellow to rose or violet as they mature; the depths of their lobing in relation to the whole limb (the expanded, not tubular, portion of the corolla) is very important, being deep in *A. lateriflorus* and *A. ontarionis*. The large lobes of these species are best seen on mature (rosy) corollas with the lobes spreading. In measuring length of rays, it is of course necessary to find mature, flat (well pressed, if dry) ones; look for the *largest* ones on a specimen.

Very few of the herbarium specimens I have seen are associated with cytological data, but there is much information, with references, in Semple and Heard (1987) and some older data are in Van Faasen (1971). Whether we have in Michigan all the same ploidy levels within the same species as elsewhere (or what their distribution might be), I do not know, nor is there cytological evidence for the intermediate specimens that appear to be of hybrid origin. Species with a base chromosome number of 8 are generally known to hybridize with relative ease, and if different ploidy levels are involved, triploids and pentaploids may result. The text here mentions only some of the crosses for which botanists have guessed at a hybrid origin when annotating our specimens: usually the most obvious or frequent ones. In the 1950s, L. H. Shinners annotated a considerable number of Michigan specimens; more recently Almut Jones and her associates have looked at a great many more (and without being able to pigeonhole neatly many ambiguous sheets). Experts are not in full agreement on what the species are and where the lines should be drawn, but the detailed treatments by Jones (1989) and by Semple and Heard (1987) provide much helpful additional information. These and other principal references useful in elucidating key characters as well as recent name changes are listed below. Semple and Heard include all Michigan species except nos. 16 and 17.

In view of the long-chaotic state of both nomenclature and taxonomy in this genus, with names shifting in their application and misidentifications rampant, I have not tried to explain all previous reports from Michigan, some of which are included in Van Faasen (1963). In some cases specimens exist but seem unacceptable as authentic Michigan plants; probably I should have been stricter about some others as well. See note on *A. prenanthoides* under *A. laevis*. The range of *A. patens* Aiton is sometimes said to include southern Michigan, but the only specimen I have found is an old one (1887, MSC) with a printed label giving the locality as "Lansing" with no further details as to locality, habitat, or collector. It is possible that the plant was cultivated. Presumably it was accessible to Beal, who listed (1905) only Ann Arbor and Macomb counties as localities (and from which no specimens have materialized). The species otherwise ranges well to the south and east of Michigan (not known from Ontario or the northern portions of Ohio, Indiana, and Illinois). The leaves are relatively small, nearly or quite sessile and strongly cordate-clasping at the base. The species would run in the key below to *A. novae-angliae*, from which it differs in being much less hairy and glandular (only the tips of the phyllaries glandular, the phyllaries otherwise merely antrorse-strigose).

REFERENCES

Jones, Almut G. 1983. Nomenclatural Changes in Aster (Asteraceae). Bull. Torrey Bot. Club 110: 39–42.

Jones, Almut G. 1989. Aster and Brachyactis in Illinois. Illinois Nat. Hist. Surv. Bull. 34: 135–194.

Les, Donald H., James A. Reinartz, & Lawrence A. Leitner. 1993 ["1992"]. Distribution and Habitats of the Forked Aster (Aster furcatus: Asteraceae), a Threatened Wisconsin Plant. Michigan Bot. 31: 143–152.

Nesom, Guy L. 1995 ["1994"]. Review of the Taxonomy of Aster Sensu Lato (Asteraceae: Astereae), Emphasizing the New World Species. Phytologia 77: 141–297.

Pringle, James S. 1967. The Common Aster Species of Southern Ontario. Roy. Bot. Gard. (Hamilton) Tech. Bull. 2. 15 pp.

Semple, John C., & Ronald A. Brammall. 1982. Wild Aster lanceolatus × lateriflorus Hybrids in Ontario and Comments on the Origin of A. ontarionis (Compositae — Astereae). Canad. Jour. Bot. 60: 1895–1906.

Semple, John C., & Jerry G. Chmielewski. 1986. Revision of the Aster lanceolatus Complex, Including A. simplex and A. hesperius (Compositae: Astereae): A Multivariate Morphometric Study. Canad. Jour. Bot. 65: 1047–1062.

Semple, John C., & Stephen B. Heard. 1987. The Asters of Ontario: Aster L. and Virgulus Raf. (Compositae: Astereae). Univ. Waterloo Biol. Ser. 30. 88 pp. [Cites many previous papers.]

Shinners, L. H. 1941. The Genus Aster in Wisconsin. Am. Midl. Nat. 26: 398–420.

Van Faasen, Paul. 1963. Cytotaxonomic Studies in Michigan Asters. Michigan Bot. 2: 17–27.

Van Faasen, Paul. 1971. The Genus Aster in Michigan. I. Distribution of the Species. Michigan Bot. 10: 99–106.

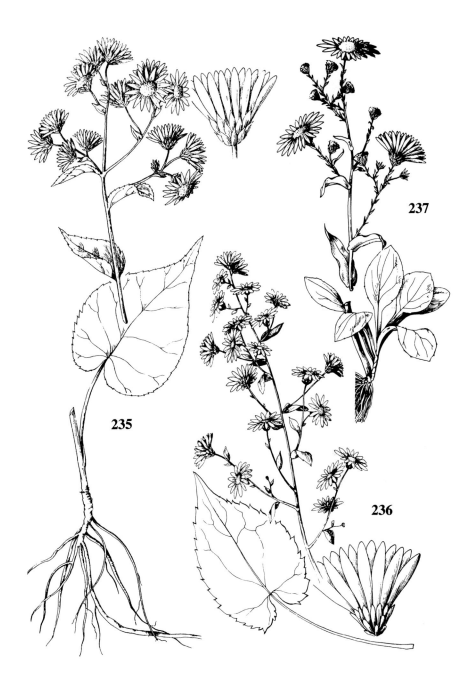

235. *Aster macrophyllus* ×½; head ×1
236. *A. cordifolius* ×½; head ×2
237. *A. laevis* ×½

KEY TO THE SPECIES

1. Blades of at least the lower and/or basal leaves with distinct petioles and cordate to broadly rounded or truncate bases
  2. Involucres mostly (6.5) 7–9 (10.5) mm long, with outer phyllaries 1–2.2 mm broad; inflorescence corymbiform
    3. Pedicels and involucres without glands; stems leafy up to the inflorescence . . . . .1. **A. furcatus**
    3. Pedicels and usually involucres with ± dense short-stalked glands (these sometimes overtopped by longer eglandular hairs); stems usually with leaves much reduced below the inflorescence . . . . .2. **A. macrophyllus**
  2. Involucres less than 6.5 (7.5) mm long, with no phyllaries more than 1 mm broad; inflorescence elongate (racemose or paniculate)
    4. Margins of leaves at middle of stem entire or at most with a few irregular teeth
      5. Leaves at middle of stem sessile and clasping, smooth and glabrous, ± glaucous. . . . . .9. **A. laevis**
      5. Leaves at middle of stem petioled or tapered to base, scabrous to the touch (at least above), not glaucous
        6. Upper part of stem and undersides of leaves evenly rough-hirtellous with tiny widely spreading hairs; phyllaries strigose on back, all distinctly less than 1 mm wide, with the diamond-shaped green tip mostly at least twice as long as wide . . . . .3. **A. shortii**
        6. Upper part of stem essentially glabrous (or unevenly pubescent, e.g., in lines or stripes) and undersides of leaves scabrous to sparsely pubescent (with some longer, softer hairs especially along the midrib); phyllaries glabrous, at least the widest usually (0.9) 1 mm broad, with the diamond-shaped green tip often only 1–2 times as long as wide . . . . .4. **A. oolentangiensis**
    4. Margins of leaves at middle of stem toothed along all or most of the blade
      7. Lower and basal leaves with blades deeply cordate (sinus 5–15 (30) mm deep) and prominently toothed (some teeth (1.5) 2–5 mm long on forward margin); petioles, especially on mid-cauline leaves, wingless or narrowly winged (1 mm or less on each side); phyllaries merely acute, the diamond-shaped green area usually ca. 3 times as long as broad or shorter (at least on outer phyllaries); inflorescence open, paniculiform, the heads with rays pale blue to purplish (occasionally white) . . . . .5. **A. cordifolius**
      7. Lower and basal leaves with blades truncate, broadly rounded, or subcordate at base (sinus rarely over 5 mm deep) and usually more shallowly toothed; petioles, especially on mid-cauline leaves, often winged 1–3.5 (6) mm on each side (especially near the blade); phyllaries acuminate or attenuate, the diamond-shaped green area prolonged or (especially in *A. sagittifolius*) obscure; inflorescence and rays various
        8. Rays 3.5–6 (6.7) mm long, white (rarely blue or pink); inflorescence racemiform, with ascending branches and all (or most) pedicels about equally short. . . . . .6. **A. sagittifolius**
        8. Rays (6.5) 7–10.5 (12) mm long, blue to purple; inflorescence open-paniculiform, with spreading branches and heads on pedicels of very uneven lengths . . . . .7. **A. ciliolatus**
1. Blades of leaves not cordate or if somewhat so, the leaves sessile, not distinctly petioled
  9. Leaves and phyllaries silvery-silky with long appressed hairs on both surfaces . . . . .8. **A. sericeus**
  9. Leaves and phyllaries with quite different sorts of pubescence or glabrous

Page 488

10. Upper cauline leaves sessile, ± clasping (auriculate) or the phyllaries with ± dense stalked glands (or both conditions present)
  11. Leaves, stems, involucres, and often pedicels glabrous, eglandular, and glaucous; outer phyllaries about half as long as the inner ones or even shorter . . . . .9. **A. laevis**
  11. Leaves, stems, involucres, and/or pedicels pubescent, in some species glandular, and not glaucous; outer phyllaries about equalling the inner ones or even longer
    12. Phyllaries glabrous (except sometimes for marginal cilia), eglandular; leaves (largest ones) often at least sparsely toothed . . . . .10. **A. puniceus**
    12. Phyllaries (and pedicels) partly to densely glandular; leaves entire
      13. Leaf blades only slightly auriculate; phyllaries usually green throughout; achenes sparsely strigose; stems arising singly from an elongate rhizome; plant of the Lake Superior region . . . . .11. **A. modestus**
      13. Leaf blades strongly auriculate; phyllaries usually flushed with purple, especially on the acuminate tip; achenes densely strigose; stems arising in clusters from a stout caudex; plant of southern range . . . . .12. **A. novae-angliae**
10. Upper cauline leaves without auriculate bases (may, if sessile, be slightly sheathing) *and* the phyllaries without glands
  14. Rays none or minute (less than 2 mm long, scarcely exceeding the involucre); plants annual, with small taproots
    15. Heads with tiny (blue–pink) rays; leaf margin essentially smooth; broadest leaves (except in the most depauperate individuals) (3.5) 6.5–10 (11) mm wide . . . . .13. **A. subulatus**
    15. Heads rayless; leaf margins scabrous-ciliate with stout but tiny prickles; broadest leaves less than 4 mm wide . . . . .14. **A. brachyactis**
  14. Rays conspicuous; plants perennial, usually rhizomatous
    16. Basal leaves with blades as broad as 8 cm or more, several times as large as the principal cauline leaves, on long winged petioles; plant a tall (to 2 m) escape from cultivation . . . . .15. **A. tataricus**
    16. Basal leaves (if any) much smaller, scarcely if at all larger than the principal cauline leaves (often withered at flowering time), scarcely if at all petioled; plant usually smaller, most species clearly native
      17. Phyllaries (at least the middle ones) tough and parchment-like, green at most at tip (the midvein not bordered to the base); pappus hairs slightly thickened toward apex; rays few (ca. 3–8); achenes densely silky
        18. Leaves obscurely to definitely toothed, broadly oblanceolate to elliptic; disk florets [9–20] . . . . .16. **A. paternus**
        18. Leaves entire, narrowly oblanceolate to linear; disk florets [5–10] . . . .
. . . . .17. **A. solidagineus**
      17. Phyllaries herbaceous in texture, with midvein bordered with color; pappus hairs thickened or (usually) not; rays usually more numerous (if rays few and/ or pappus hairs thickened, leaf venation distinctly areolate); achenes glabrous to moderately pubescent
        19. Middle phyllaries with broad dark (green or reddish) zone bordering midvein mostly ± uniform in width (at mid-length, nearly or quite as wide as the pale area on each side); heads solitary to numerous in a corymbiform inflorescence; plants further distinctive as described in couplet 20
          20. Leaves linear to very narrowly elliptic (less than 7 mm wide), glandular beneath but not strongly reticulate-veined, with revolute margins; heads ca. 2.5–4 cm across, solitary to few on long pedicels, with pink rays . . . . .
. . . . .18. **A. nemoralis**

20. Leaves elliptic, mostly (7) 10–30 mm wide, glabrous or pubescent but eglandular and reticulate with strongly contrasting dark veinlets and tiny pale areoles ca. 0.6 mm or less across, with flat margins; heads ca. 2 cm or less across, numerous in flat-topped corymbs, with white rays . . . .
. . . . .19. **A. umbellatus**
19. Middle phyllaries with dark hair-thin midvein ± abruptly expanded to a short or elongate broad diamond-shaped area; heads in an elongate paniculate inflorescence (not flat-topped)
21. Phyllaries pubescent (at least slightly strigose) on back
22. Tips of phyllaries (or most of them) and bracts in the inflorescence with a tiny white or colorless spinule . . . . .20. **A. ericoides**
22. Tips of phyllaries and bracts at most with a colored callus (not spinule)
. . . . .24. **A. ontarionis**
21. Phyllaries glabrous on back (may be ciliate)
23. Tips of phyllaries (or most of them) spinulose or with prolonged apex formed by inrolled margin; involucre slightly urn-shaped (constricted just above the middle); basal rosettes of ± oblanceolate leaves usually present at flowering time . . . . .21. **A. pilosus**
23. Tips of phyllaries merely acute or callused, not spinulose or prolonged; involucres top- or cup-shaped, not constricted; basal rosettes usually absent at flowering time (or with broad elliptic-obovate leaves)
24. Leaves beneath with distinct regular reticulate pattern formed by dark veinlets around paler green areoles ca. 0.6 mm in diameter (fig. 242); rays blue (very rarely white); plant of southern Lower Peninsula . . . . . .
. . . . .22. **A. praealtus**
24. Leaves beneath without distinct reticulation or the areoles clearly irregular or elongate; rays white (occasionally pinkish; blue in one species and one variety, both northern); plants of various range
25. Lobes of disk florets about half or more the total length of the limb (expanded corolla, not the slender tube); leaves normally pubescent on at least the midrib beneath
26. Leaves pubescent primarily on the midrib beneath (rarely glabrous); stems often clumped, from a sturdy caudex . . . . .23. **A. lateriflorus**
26. Leaves on upper stem evenly short-pubescent all across the lower surface (moderately to densely so — rarely nearly glabrous); stems solitary from slender rhizomes . . . . .24. **A. ontarionis**
25. Lobes of disk florets about a third or less of the total length of the corolla limb; leaves glabrous or at most scabrous beneath [if pubescent, cf. *A. oolentangiensis*]

685. Erigeron annuus   686. Aster furcatus   687. Aster macrophyllus

27. Heads on pedicels (or branchlets) mostly (0.7) 1–5 cm long with many of the bractlets less than 5 (7) mm long (fig. 243); heads small (less than 1.5 cm broad when fresh or well pressed) . . . . .25. **A. dumosus**
27. Heads on often shorter pedicels with the bractlets few (or none) and mostly not so reduced; heads various, larger (1.5–2.5 (3) cm broad) except in some forms of *A. lanceolatus*
  28. Cauline leaves ± clasping (most bases, especially lower on the stem, circling more than half the circumference of the stem); phyllaries nearly equal in length (or the outer ones over half as long as the inner ones); rays bright blue-violet; plants of the Upper Peninsula and northern Lower Peninsula . . . . .26. **A. longifolius**
  28. Cauline leaves not clasping (all bases circling half or less the circumference of the stem); phyllaries of different lengths, imbricate; rays usually white; plants throughout the state
    29. Rhizomes and stems less than 2 (2.2) mm in diameter; inflorescence usually with divaricate or spreading branches (if any) and 1–12 (20) heads; leaves less than 6 mm broad, the margins often revolute; plant of wet peatlands, occasionally shores . . . . .27. **A. borealis**
    29. Rhizomes and stems stouter (at least 2.5 mm); inflorescence with ± ascending branches and (except in the most depauperate specimens) more than 20 heads; leaves often 6–18 (33) mm broad (though sometimes narrower), with flat margins; plant of damp habitats generally . . . . .28. **A. lanceolatus**

1. **A. furcatus** E. S. Burgess   Forked Aster
   Map 686. This is a rare aster throughout its midwestern range, from southern Michigan and Wisconsin to Missouri, in moist woods, especially along rivers. It can form large colonies from rhizomes. (See Les et al. 1993.) Collected in 1906 (*S. Alexander*, MICH) on banks of the Raisin River in Washtenaw County and in 1934 (*Dreisbach 8359*, MICH) on the Chippewa River in Midland County.
   The leaves are distinctly scabrous above and the rays normally white.

2. **A. macrophyllus** L.   Fig. 235   Large- or Big-leaved Aster
   Map 687. In woods of all kinds: beech-maple, hemlock-northern hardwoods, drier sites (with oak, hickory, aspen, jack pine, or mixed conifer-hardwoods), less often in swamp forests and river banks; common on wooded dunes, also in northern rocky woods with spruce and fir. Often abundant after disturbance such as road construction, logging, or fire; spreading to roadsides, railroad embankments, pine plantations.
   Very rarely a specimen lacks glands on the pedicels and phyllaries, but it can be distinguished from the preceding by the strongly cordate bases of the lower (or even middle) leaf blades, although the middle and upper leaves tend to be much reduced in size. In *A. furcatus*, the base is at most truncate and the leaves are well developed at the upper and middle nodes. The upper leaf surface in *A. macrophyllus* is often smooth, but may be scabrous. The rays are sometimes white, but usually pink to blue. Sterile

heart-shaped leaves are generally abundant in woods, and noticeable because of their large size; in comparison, flowering stems are rather sparse in most years and favor borders, trails, and openings. Stems may flower, with only one or two heads, when as short as 15 cm.

3. **A. shortii** Lindley   Short's Aster

Map 688. This is a species of thin, sometimes rocky woods but the labels on only one Michigan collection mention any specific kind of woods (oak-hickory); also along roadsides.

4. **A. oolentangiensis** Riddell   Sky-blue Aster

Map 689. Thin woodland and savanas of oak (with hickory, sassafras, and/or pine — jack pine in Clare Co.); dry open sandy ground as well as wet prairies.

The relatively broad phyllaries and usually strongly scabrous upper surfaces of the leaves will distinguish this species from others with which it might be confused.

This was long known as *A. azureus* Lindley, but that proves (see Jones 1983) to have been published several months later than Riddell's name. Replacement by the latter, derived from the Olentangy River in central Ohio, is deplored by many botanists, but I personally do not mind, having been born in a city only a few miles upstream on the Olentangy from the type locality.

5. **A. cordifolius** L.   Fig. 236.   Heart-leaved or Blue Wood Aster

Map 690. Beech-maple, oak-hickory, and swampy woods (as on floodplains), especially at borders and clearings; thickets, sometimes weedy in habit in urban areas. The only collection from Cheboygan County (ex herb. Kofoid, in 1890, GH) is perfectly typical but unusually far north.

The rays are usually as short as in the next species but are almost always blue or purple, and the inflorescence is more open, the phyllaries merely acute (with rather clear diamond-shaped green tips on the midvein), and the leaves with more deeply toothed and cordate blades on petioles wingless or nearly so.

688. Aster shortii   689. Aster oolentangiensis   690. Aster cordifolius

Somewhat ambiguous specimens resembling this species or the next may be unusual (or poorly collected) individuals or hybrids (including backcrosses). Some specimens from the southern Lower Peninsula annotated by aster students as possibly showing introgression from *A. ciliolatus* more likely have some influence of one or the other of these two species, since *A. ciliolatus* is known in the state only from farther north. (See also comments below.)

6. **A. sagittifolius** Willd.   Arrow-leaved Aster
Map 691. Dry open sandy (sometimes rocky) woodland and savana, with oak, sassafras, aspen, and/or pine (especially jack pine); wooded banks and hillsides; stabilized dunes; fields, grassy roadsides, hedgerows; rarely in wet areas.

Most specimens can be readily placed by the short white rays, attenuate phyllaries, and conspicuously winged petioles. Rarely a plant will have the leaf blades as strongly toothed and cordate as in *A. cordifolius*, but even on these the blades tend to be decurrent into broad wings on the petiole. In all but unusually shaded or otherwise depauperate individuals, the inflorescence has a characteristic pyramidal shape (or diamond-shaped outline). The disk corollas, as in the preceding species and the next one, turn purple with age, but only rarely are the rays pink or blue (a result of hybridization?).

A few (mostly old) specimens from Ionia, Jackson, Kalamazoo, and Kent counties have been determined by Jones or Shinners as "near" *A. drummondii* Lindley but generally with some evidence of influence from *A. sagittifolius*. They do resemble the latter species, including the winged petioles, but are somewhat more pubescent on the upper stem (which is glabrous or glabrate in *A. sagittifolius*). In view of the weak endorsement of these specimens as *A. drummondii*, as well as the fact that the taxon is by some authors reduced to *A. sagittifolius* var. *drummondii* (Lindley) Shinners anyway, it seems premature to admit Drummond's aster as a species to be recognized in Michigan. Some additional specimens from the southwestern Lower Peninsula also seem to approach the "*drummondii*" variant.

Recently there has been an interpretation placing this name in the synonymy of *A. cordifolius* and calling the present species *A. urophyllus* Lindley. There is, however, disagreement regarding identity of the type of Willdenow's name, Jones claiming that it is a variant of *A. cordifolius* while Cronquist, the most recent botanist familiar with American asters to examine it, concluded that it does indeed represent what we have long called *A. sagittifolius*.

7. **A. ciliolatus** Lindley   Plate 8-B   Lindley's Aster
Map 692. Dry to moist deciduous, mixed, and coniferous woods, especially in clearings and along borders including adjacent roads, trails, and

fields; thickets, logged and burned areas, dune ridges, rocky ground. The range of this northern species does not (with barely an exception) appear quite to overlap that of the southern *A. cordifolius* in Michigan (although they do overlap in Ontario).

The long blue to purple rays and large heads will rather easily separate this species from *A. sagittifolius* even if the aspect of the inflorescence is not clear. The heads are 16–28 mm broad when fresh and fully expanded, while in the two preceding species they run less than 18 mm broad.

A few plants clearly intermediate with other species are presumably hybrids, e.g., *A. ciliolatus* × *A. lanceolatus* from Beaver Island and Cheboygan, Chippewa, and Mackinac counties and *A. ciliolatus* × *A. laevis* from Schoolcraft County. Jones named a few collections from southern Michigan as possible hybrids involving *A. ciliolatus* but as that species appears not to range so far south, I suspect some other parentage may be involved.

8. **A. sericeus** Vent.   Silky Aster
   Map 693. Prairies, dry banks and fields. The Keweenaw County record (*Farwell 242* in 1889, BLH, GH) looks quite out of line, but the species is known from far western Ontario and adjacent Minnesota (as well as southern Minnesota).

   An absolutely distinctive species, with rather uniformly small, oblong-elliptic, very silky leaves.

9. **A. laevis** L.   Fig. 237   Smooth Aster
   Map 694. Dry open woods (oak, aspen, jack pine) and thickets, especially at borders and clearings; fields and roadsides; sandy plains, bluffs, stabilized dunes; prairies, meadows, shores; rarely in wet places such as fens.

   In its relatively few, large (long-rayed) heads this species somewhat resembles *A. ciliolatus*; furthermore, sometimes broad leaves below the middle of the stem taper to the base, even appearing to have narrowly to broadly winged petioles. However, *A. ciliolatus* does not have the clasping

691. Aster sagittifolius   692. Aster ciliolatus   693. Aster sericeus

leaf bases so characteristic of *A. laevis*. The species is quite variable in leaf shape and number of heads. Apparently it hybridizes with other species occasionally, including *A. pilosus* and *A. ciliolatus*.

Aster prenanthoides Willd. resembles some forms of *A. laevis* and would be likely to key here should it occur in Michigan. One specimen labeled as from Keweenaw County in 1890 (*Farwell 754*, BLH) is so unlikely a find for this southern species that I am very dubious about accepting it. The stem is crooked (i.e., zigzag at the nodes) and finely pubescent in lines above; the leaves are glabrous to scabrous above, distinctly acuminate at the apex and tapered to a narrow winged base auriculate at the stem; the inflorescence is broadly corymbiform. The specimen does not appear to have been glaucous. Farwell's notes for this number add: "Borders of woods," but no additional locality besides the county.

10. **A. puniceus** L.   Fig. 238   Swamp or Purple-stemmed Aster
Map 695. Edges of moist woods, thickets, marshes; swamps (cedar and other conifers, hardwoods), wet hollows, fens, sedge meadows, fields; shores, swales, ditches, stream and river banks. *A. puniceus* var. *firmus* tends to form large showy colonies often conspicuous in open ground as along wet roadsides.

There are two varieties of this species throughout the state, easily recognized, especially in the field, in the extreme but tending too often to intergrade in most characters for acceptance as species (although some authors do so). In var. *puniceus*, the stem arises from a short, stout rhizome and is fairly uniformly pubescent its entire length (though the hairs on the lower internodes may be abraded), the leaves are more hispidulous, and the internodes are longer, giving a more open or lax appearance. In var. *firmus* (Nees) T. & G. there are extensive elongate rhizomes, the stem is glabrous or glabrate below, with the upper internodes glabrous or with pubescence uneven (in narrow stripes or lines), the leaves are glabrous (except sometimes on the midrib beneath), and the internodes are shorter, giving a very compact and leafy appearance to the top of the plant. This

694. Aster laevis   695. Aster puniceus   696. Aster modestus

taxon is treated as *A. firmus* Nees at the rank of species and was long called *A. lucidulus* (A. Gray) Wiegand. Both varieties may grow together in the same fens.

The large rays are blue to purple (or even rose). Plants with white rays occasionally occur. These may be called f. *albiligulatus* Pease & A. H. Moore or, if var. *firmus* is distinguished, the albino of it may be called f. *etiamalbus* Venard.

11. A. modestus Lindley

Map 696. This boreal species ranges from James Bay to the Yukon, south barely into the United States in the Lake Superior region and the Pacific Northwest. There are several collections from shores, alder thickets, and open fields at Isle Royale, along with one from a cedar swamp at Cat Harbor on the mainland (*Wells & Thompson 3724* in 1972, MICH, BLH).

12. A. novae-angliae L.   Fig. 239   New England Aster

Map 697. Open, usually moist to wet ground, including shores, meadows, fields, shrubby swamps, fens and wet prairies, edges of streams and rivers, thickets; dryish to moist woods (especially along banks and roadsides); somewhat weedy in habit. The Schoolcraft County specimen is said to have escaped from cultivation, and it is possible that some of the other northern records are similar, although from natural habitats.

Like many other asters, the ray color varies from blue to purple, occasionally rose. White-rayed plants have apparently not yet been documented from Michigan. Where this species grows with *A. ericoides*, an attractive hybrid, *A. ×amethystinus* Nutt., may be expected. It is quite intermediate between these very different parents, with the leaves resembling miniature ones of *A. novae-angliae* but the copiously flowering inflorescence with smaller heads resembling that of *A. ericoides*. The phyllaries mostly lack the tiny spinose tips of the latter, although the leaves may have such tips. This hybrid has been collected in Macomb, Monroe, Oakland, Washtenaw, and Wayne counties.

697. Aster novae-angliae   698. Aster subulatus   699. Aster brachyactis

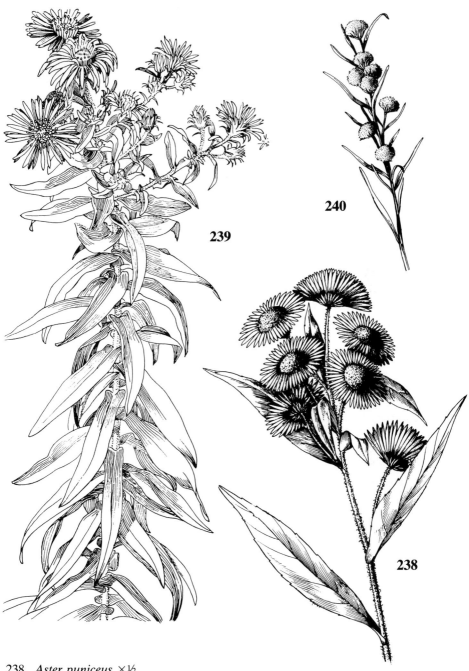

238. *Aster puniceus* ×½
239. *A. novae-angliae* ×½
240. *A. brachyactis* ×½

13.   **A. subulatus** Michaux   Saltmarsh Aster
   Map 698. First collected in Michigan in 1914 at the same salt mine area
where *Pluchea* was found. This is another Coastal halophyte and it, too,
probably spread inland and survives now along well salted highways, where
it will doubtless continue to expand its range.

14.   **A. brachyactis** S. F. Blake   Fig. 240   Rayless Aster
   Map 699. This is another halophyte, but somewhat more widespread
and considered native west of the Great Lakes. First collected in Michigan
by Dodge in St. Clair County (1900) and Huron County (1909). Now, local
on the shores of the Great Lakes and some inland lakes as well as in waste
ground along railroads and salted highways and by salt storage areas.
   Sometimes segregated in a separate genus as *Brachyactis ciliata* (Ledeb.)
Ledeb., long known as *B. angusta* (Lindley) Britton.

15.   **A. tataricus** L. f.   Tartarian Aster
   Map 700. A native of Asia, rarely escaped from cultivation. Collected in
1951 (*Rapp Cass-663*, MICH) at Jones, without further data.
   The rays are blue-purple and the heads large, in corymbs.

16.   **A. paternus** Cronquist   White-topped Aster
   Map 701. Native mostly east and south of Michigan. Collected but once
in this state and of dubious status: "dry ground," Galesburg (*Farwell 5097a*,
GH; Pap. Michigan Acad. 1: 100. 1923).
   Long known as *Sericocarpus asteroides* (L.) BSP.

17. **A. solidagineus** Michaux
   Map 702. Like the preceding, native mostly west and south of Michigan.
An unnumbered specimen purporting to have been collected in Keweenaw
County in 1886 (*Farwell*, BLH) is of very dubious status. Farwell's only
published report (as also in his field notes), from Keweenaw County in
1904, seems not to be supported by an extant specimen (Pap. Michigan
Acad. 1: 100. 1923).
   Long known as *Sericocarpus linifolius* (L.) BSP.

18.   **A. nemoralis** Aiton   Fig. 241   Bog Aster
   Map 703. This attractive little northern aster is quite unlike any other
and here at the western edge of its range is usually said to inhabit "acid
bogs." But that is far from a fair statement of its occurrence in Michigan.
Here it grows only in the eastern Upper Peninsula, which is underlain with
limestone, and it thrives in fens, especially patterned peatlands where there
may be expanses of it over many acres, sometimes with no sphagnum in
sight. It can be found with such typical fen species as *Cladium mariscoides*
and in thickets of *Myrica gale*. It is at times in sphagnum (with rhizomes

beneath), and at times on damp sandy shores, but it seems to avoid both the most acid and the most alkaline sites.

19. **A. umbellatus** Miller   Flat-topped Aster
   Map 704. This is one of our most distinctive asters and is common throughout the state, although in some areas such as the northernmost Lower Peninsula quite local. Borders and openings in moist woods, swamp forests (conifers, hardwoods), and thickets (willows, alders, etc.); oak-hickory, aspen-pine, and mixed woods; along streams and rivers, sedge meadows and fens, shores, ditches, marshy flats, swales; rock shores of Lake Superior.
   Pubescence varies a great deal in this species. Plants with leaves glabrous beneath (except on the midrib) are var. *umbellatus* and occur from the southern border of the state to Keweenaw County, apparently absent only from the westernmost Upper Peninsula and Isle Royale. Plants with leaves pubescent across the lower surface are var. *pubens* A. Gray and occur throughout the Upper Peninsula, with scattered locations as far south as Lansing. Sometimes this variant is recognized as a distinct species, *A. pubentior* Cronquist, but intermediates are not uncommon and neither Jones nor Semple believes that specific rank is justified.

700. Aster tataricus

701. Aster paternus

702. Aster solidagineus

703. Aster nemoralis

704. Aster umbellatus

705. Aster ericoides

20. **A. ericoides** L.   Heath, Wreath, or White Prairie Aster
Map 705. Prairies (dry to wet), marshy shores, meadows; often along roadsides, railroads, fencerows, fields; sometimes weedy in waste ground.

Distinctive in its numerous, crowded, small heads, often on long one-sided branches. The rays are normally white. The cauline leaves are very small, numerous, ± linear, rather rough-pubescent (or at least ciliate). Hybridizes with *A. novae-angliae* (see above). Sometimes may be mistaken for the next species (or *vice versa*) — to which it is not at all closely related. Even if the pubescence on the phyllaries of *A. ericoides* is not clear, the shorter, stubbier look to the phyllaries and leaves will help.

21. **A. pilosus** Willd.   Frost Aster
Map 706. We have two rather well marked varieties, although intergrades rarely occur. Along and near the shores of the Great Lakes from Berrien and Huron counties north is var. *pringlei* (A. Gray) S. F. Blake, with stems and leaves glabrous or nearly so. This is definitely a calciphile, on sandy and gravelly shores and beaches, interdunal flats and swales, limestone pavements and other outcrops, rarely in fens (inland in Lenawee Co.); fields and streamsides. The Ontonagon County collection (*Voss 13718*, MICH), from sandstone shore of Lake Superior, is not as far out of line as it appears, for this variety is along the Lake Superior shore in northern Wisconsin.

Chiefly inland in the southern Lower Peninsula, with occasional plants in disturbed ground northward, is var. *pilosus*, with spreading hairs on stems and leaves. This variety is in dry to wet fields and prairies, along roadsides and shores.

The bracts and leaves around the inflorescence often have spinulose tips; the awl-like appearance of these and the tips of the phyllaries make this a reasonably easy species to recognize. The rays are usually white, occasionally pinkish. The pappus of the disk florets is shorter than the corollas (as in most *A. lateriflorus* and *A. ontarionis*, but not in the last four species below).

706. Aster pilosus

707. Aster praealtus

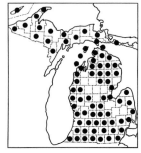
708. Aster lateriflorus

22. **A. praealtus** Poiret   Fig. 242   Willow Aster
Map 707. Moist fields (including recent clearings) and prairies.

The conspicuous reticulate venation of the undersides of the leaves suggests *A. umbellatus*, but that has much broader leaves. In *A. praealtus* the leaves are small, narrow, and crowded. A white-rayed form was collected in a hedgerow at Detroit Metropolitan Airport in 1992 (*Brodowicz 1089*, MICH).

23. **A. lateriflorus** (L.) Britton   Calico Aster
Map 708. Usually in ± shaded ground, including woods (beech-maple, oak-hickory, mixed hardwood and conifers), thickets, borders and clearings, swamp forests (conifers or hardwoods); floodplains, river banks, ravines, rocky banks and openings; also meadows, shores, roadsides, rarely peatlands.

Ordinarily well distinguished by the deeply lobed disk corollas, which all turn rosy early in anthesis (unlike most asters, *before* starting to wither); leaves with only the midrib hairy beneath; ± one-sided racemose branches of the inflorescence; and stems arising from a caudex rather than an elongate rhizome. The rather slender stems vary from glabrous to pubescent and are often branched or sprawling from the lower nodes. Leaves vary from linear and entire to elliptical and toothed. Pedicel length varies. Needless to say, names have been published for some of these intergrading variants. The basal leaves, when present, have relatively broad elliptic or rotund blades. The contrasting white rays, still fresh when the disks are rosy purple, are presumably the source of the common name.

24. **A. ontarionis** Wiegand   Lake Ontario Aster
Map 709. Wooded river banks and floodplains; moist thickets and woods, whether oak-hickory or beech-maple, rarely cedar or tamarack swamp; wet meadows and open sandy hollows; disturbed woods, roadsides, and fields (if labels be trusted).

If, as authors state, this species is distinguished by the deeply lobed disk corollas and evenly pubescent undersides of the upper stem leaves and bracts, considerable diversity of habit is encompassed. Some plants resemble *A. dumosus* in the numerous short bractlets on the pedicels.

25. **A. dumosus** L.   Fig. 243   Bushy Aster
Map 710. Sandy to mucky or marly shores of lakes and ponds (or even wet borrow pits), interdunal hollows; sedge meadows, wet prairies, fens; conifer thickets, sandy banks and clearings, sometimes associated with oaks and jack pines.

The numerous, rather uniform, blunt or acute and callus-tipped bractlets on elongate pedicels (or branchlets) are quite characteristic of this species—but do sometimes occur in *A. ontarionis* (see above). A few plants with deeply lobed disk corollas, many little bractlets on the pedicels, and leaves glabrous beneath I suspect to be hybrids of *A. dumosus* and *A. lateriflorus* (or *A. ontarionis*). Our specimens of *A. dumosus* are often quite harshly scabrous on one or both surfaces of the leaves, which in some specimens are smooth. The stems are sometimes short-pubescent. The pubescent-scabrous extreme of this species was named var. *dodgei* Fernald (TL: Harsen's Island, St. Clair Co.). The rays are usually white with us, rarely pink or purple; in some parts of the range, the rays are said to be usually colored. Besides approaching *A. lateriflorus* and *A. ontarionis* in one direction, this species seems to approach *A. pilosus* and *A. lanceolatus* in another, and not all of the collections mapped are guaranteed pure. Certain odd-looking plants of the *dumosus* complex from Berrien and Van Buren counties, with very small heads, short pedicels, and one-sided recurved branches in the inflorescence, may be the puzzling *A. racemosus* Elliott [also known, apparently incorrectly, as *A. fragilis* Willd. and *A. vimineus* Lam.].

26. **A. longifolius** Lam.   Long-leaved Blue Aster
    Map 711. Damp open sandy, gravelly, or rocky ground, including shores, limestone alvars, seasonally wet glades and swales as on outwash plains with jack pine; calcareous river banks, rock shores of Lake Superior.
    This northern species, ranging west only to Wisconsin, has not been widely (or accurately) recognized. It is sometimes included in *A. novi-belgii* L. of the East Coast. The leaves are indeed long, usually overtopping most or all of the inflorescence. A number of plants with the long, narrow, clasping leaves and conspicuous blue rays of this species have the phyllaries ± imbricate, the outer ones distinctly shorter than the inner, rather than subequal. These may represent hybrids or else simply variation in definition of "subequal." Authors are not agreed upon the key characters distinguishing this species from *A. lanceolatus* ssp. *hesperius*, even admitting that they run into one another where the ranges overlap. Our intermediate or ambiguous specimens are all from within the area mapped for more clearcut material of *A. longifolius*.

709. Aster ontarionis

710. Aster dumosus

711. Aster longifolius

Page 502

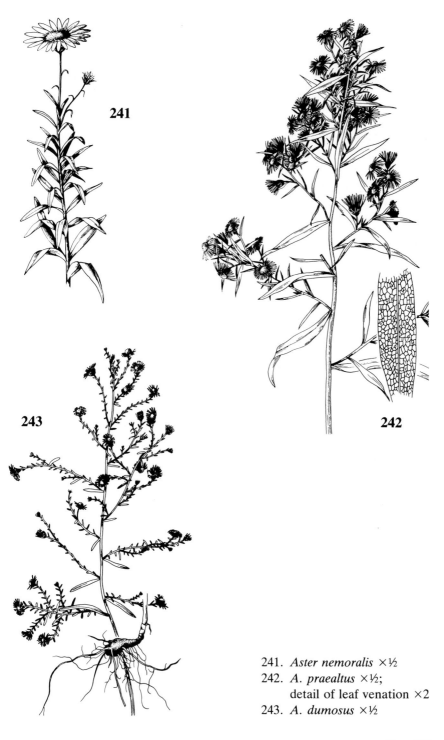

241. *Aster nemoralis* ×½
242. *A. praealtus* ×½;
  detail of leaf venation ×2
243. *A. dumosus* ×½

Sometimes *A. lanceolatus* ssp. *hesperius* is said to have a long peduncular bract subtending the involucre and equalling it in length, but in our specimens the presence of such bracts is not correlated with either strength of leaf clasping or phyllary length (subequal vs. imbricate). An Ontario collection from near Sault Ste. Marie and determined as *A. longifolius* by Semple has peduncular bracts and the outer phyllaries are distinctly shorter than the inner ones (but mostly more than half as long).

27. **A. borealis** (T. & G.) Prov. Rush Aster
Map 712. Fens, bogs, open conifer swamps (cedar, tamarack, spruce); wet, often sedgy, sand flats, shores, meadows, and swales.

The rays are usually white, sometimes pinkish or bluish. Occasional intermediate plants suggest hybridization with *A. lanceolatus* and *A. puniceus*.

Long known as *A. junciformis* Rydb. [and earlier misidentified as *A. junceus* Aiton], but the correct name was pointed out in 1968 (Nat. Canad. 95: 1511).

28. **A. lanceolatus** Willd. Fig. 244 Panicled Aster
Map 713. One of our commonest asters, especially in damp open ground including shores, river banks, edges of woods and swamps, meadows, ditches and swales, interdunal flats; wet prairies, marshes, fens; fields, along old railroads, roadsides.

Treated here in the broad sense adopted by Semple and his co-workers, *A. lanceolatus* includes tetraploids, pentaploids, hexaploids, and octoploids. Add to that variability in ploidy level a penchant for hybridization, and one can see why this species is one of our most diverse morphologically. *A. tradescantii* L. is perhaps a diploid ancester of *A. lanceolatus* although some consider the name to belong in the synonymy of *A. lateriflorus*; *A. tradescantii* is low in stature and occurs to the east of the Great Lakes.

Sometimes this species is confused with *A. pilosus*, and indeed these two species appear to hybridize, but besides its unique phyllaries and involucre shape, that species tends to have the pappus of the disk florets shorter than

712. Aster borealis 713. Aster lanceolatus 714. Arctium lappa

the corollas while in *A. lanceolatus* and its allies, the pappus equals or exceeds the disk corollas. Plants intermediate with *A. puniceus* (especially var. *firmus*) seem to be more frequent than some supposed hybrids. And a number of plants resembling *A. lanceolatus* but with deeply lobed disk corollas presumably have resulted from hybridization with *A. lateriflorus*, at least in the distant past, but no cytological data are available. Similarities between putative hybrids of such parentage and *A. ontarionis* suggested to Semple and Brammall (1982) that the latter arose out of a hybridization event.

This species has long been known under the names *A. simplex* Willd. and *A. paniculatus* Lam. Plants with large blue-rayed heads superficially resemble *A. longifolius* (see above), but differ in less clasping leaf bases, more imbricate phyllaries, and tendency to shorter leaves and more heads. Such plants represent *A. lanceolatus* ssp. *hesperius* (A. Gray) Semple & Chmielewski (see Semple & Chmielewski 1987) and are known from Delta, Schoolcraft, and perhaps Benzie and Manistee counties. All of our other plants are in ssp. *lanceolatus*. Plants with pubescence uniform (rather than in lines, or absent) on the middle and lower internodes have been named var. *hirsuticaulis* Semple & Chmielewski (TL: Oak Grove, Otsego Co.). Plants with quite small heads (involucre under 5 mm long), very numerous and crowded on branches of the inflorescence, have been called var. *interior* (Wiegand) Semple & Chmielewski.

67. **Arctium**   Burdock

The very large, coarse-looking leaves of burdock (blades sometimes half a meter long) might remind one of rhubarb, but the petioles would make a rather fibrous, tasteless pie. Each head is dispersed as a bur covered with hooked spines, readily embedded in the fur of animals (or clothes of humans) — nature's original Velcro. Persons who care for their dogs, sweaters, or butterfly nets will keep away from a patch of tall fruiting burdock plants!

Both of our species are Eurasian natives, the commoner one, at least, too familiar as a weed of waste places. *Flora Europaea* notes: "Specific limits within this genus cannot be clearly defined. . . . All taxa are interfertile [resulting] in innumerable intermediates which are fully fertile and breed true from seed."

REFERENCES

Gross, Ronald S., Patricia A. Werner, & Wayne R. Hawthorn. 1980. The Biology of Canadian Weeds. 38. Arctium minus (Hill) Bernh. and A. lappa L. Canad. Jour. Pl. Sci. 60: 621–634.
Moore, R. J., & C. Frankton. 1974. The Thistles of Canada. Canada Dep. Agr. Monogr. 10. 111 pp. [*Arctium*, pp. 12–18.]

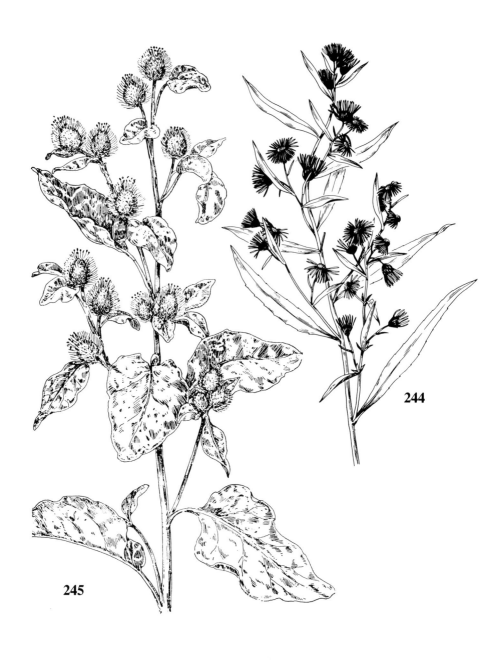

244. *Aster lanceolatus* ×½
245. *Arctium minus* ×½

KEY TO THE SPECIES

1. Mature heads ca. 2.5–4.5 cm broad (including spines), all or mostly on pedicels ca. 2.4–16 cm long, in terminal, corymbiform inflorescences; phyllaries smooth (at most weakly glandular) on margins and green (except for brownish spine-tips)....
....1. **A. lappa**
1. Mature heads ca. 1.5–2.8 (3.5) cm broad, all or mostly on pedicels ca. 0.5–3 (4) cm long, in elongate racemose or paniculiform inflorescences; phyllaries minutely serrulate below the middle and often purplish toward the apex ....2. **A. minus**

1. **A. lappa** L.   Great Burdock
    Map 714. Evidently not common (and perhaps no longer found) in Michigan. Collected along a roadside in Van Buren County in 1906 (*Gates 1490*, MICH, MSC) and near Monroe in 1923 (*Farwell 6720*, MICH, BLH).
    The open corymbiform inflorescence with long stalked large heads can be very distinctive in the field, but not always clear on fragmentary herbarium specimens. The petioles are solid, while in the next species they are hollow at the center. See also comments below.

2. **A. minus** Bernh.   Fig. 245   Common Burdock
    Map 715. Roadsides, railroads, fields, thickets, fencerows, farmyards, around old buildings, and waste ground everywhere (e.g., vacant lots, dumps, filled land); trails and clearings in upland woods and cedar swamps; river banks and other moist disturbed ground. Burdock does not endear itself to collectors and is doubtless under-represented in herbaria. Despite the poor showing on the map, it probably occurs in every county in the state.
    Intermediate plants, e.g., like *A. minus* but with heads as large as some *A. lappa* or somewhat corymbiform, or like *A. lappa* except for smaller heads or shorter pedicels, may result from hybridization but their origin cannot be accurately determined. Specimens ambiguous on these characters have been placed by the phyllary margins. In *A. minus*, at least some of the inner phyllaries are minutely serrulate below the middle. Most specimens are distinct enough and one need not resort to the 20× magnification, strong light, and sometimes dissection necessary to see clearly the firm sharp serrulations.
    A few old specimens have the deeply lobed or laciniate leaves that characterize a rare (probably diseased?) form of this species, with flowers poorly if at all developed [f. *laciniatum* Clute]. White-flowered plants have been named f. *pallidum* Farw. (TL: Detroit [Wayne Co.]).

68. **Centaurea**   Star-thistle; Knapweed
    This is a large and difficult genus, with over 200 species recognized in Europe alone. Our showiest ones are probably garden escapes and others are mostly unwelcome weeds. Hybridization helps to confuse the lines

between some species and the key cannot accommodate all intermediates. The classic work of Marsden-Jones and Turrill deals thoroughly (as far as our species are concerned) with *C. jacea* and *C. nigra*. Their extensive figures and photographs of plants and phyllaries illustrate and document the great variation that can result from hybridization and backcrossing (resulting in introgression of characters from one species into another).

REFERENCES

Marsden-Jones, E. M., & W. B. Turrill. 1954. British Knapweeds a Study in Synthetic Taxonomy. Ray Soc. Publ. 138. 201 pp. + 27 pl.

Moore, R. J., & C. Frankton. 1954. Cytotaxonomy of Three Species of Centaurea Adventive in Canada. Canad. Jour. Bot. 32: 182–186.

Moore, R. J. 1972. Distribution of Native and Introduced Knapweeds (Centaurea) in Canada and the United States. Rhodora 74: 331–346.

Moore, R. J., & C. Frankton. 1974. The Thistles of Canada. Canada Dep. Agr. Monogr. 10. 111 pp. [*Centaurea*, pp. 75–97.]

Watson, A. K., & A. J. Renney. 1974. The Biology of Canadian Weeds. 6. Centaurea diffusa and C. maculosa. Canad. Jour. Pl. Sci. 54: 687–701.

Watson, A. K. 1980. The Biology of Canadian Weeds. 43. Acroptilon (Centaurea) repens (L.) DC. Canad. Jour. Pl. Sci. 60: 993–1004.

KEY TO THE SPECIES

1. Involucres ca. 4.5–5.5 [10] cm in diameter; flowers yellow ....1. **C. macrocephala**
1. Involucres smaller (less than 3 cm broad); flowers white to pink, blue, or purple (yellow in *C. solstitialis*, with very long-spined involucres)
  2. Cauline leaves (except the uppermost and bracteal ones) deeply pinnatifid or bipinnatifid
    3. Phyllaries blackened at the tip, softly fringed with flattened projections (the terminal one obscure, no longer than the lateral ones; fig. 246); flowers usually pink-purple; pappus well developed (1–2.5 mm long) ....2. **C. maculosa**
    3. Phyllaries light yellow-brown at the tip, spinose-fringed with firm rounded projections (the terminal stiff spine longer than at least the lateral ones immediately below it; fig. 247); flowers usually white; pappus often none or vestigial
      4. Involucre ca. (7) 9–12 mm broad; plant a locally abundant weed ....3. **C. diffusa**
      4. Involucre less than 5 mm broad; plant a rare waif ....4. **C. virgata**
  2. Cauline leaves (except sometimes the basal ones) entire or with a few teeth or small lobes

715. Arctium minus   716. Centaurea   717. Centaurea maculosa
                 macrocephala

5. Phyllaries (at least the middle ones) tipped with a firm spine much longer than the body of the phyllary (usually at least twice as long); flowers yellow . . . .5. **C. solstitialis**
5. Phyllaries tipped by no spine or one much shorter than the body; flowers white to pink, blue, or purple
  6. Phyllaries entire, pilose but not fringed or lacerate; pappus ca. 6–9 mm long . . . . .6. **C. repens**
  6. Phyllaries lacerate-toothed or fringed; pappus shorter
    7. Middle (and all) phyllaries rounded or tapered to lacerate-toothed apex (with no expanded appendage; e.g., fig. 248)
      8. Enlarged marginal flowers exceeding involucre by ca. 2–3 cm; involucre ca. 2 cm (or more) long, the phyllaries with prominent blackish brown lacerate-toothed border; largest leaves ca. 2–3 cm broad . . . . .7. **C. montana**
      8. Enlarged marginal flowers exceeding involucre by ca. 0.9–1.5 cm; involucre ca. (1) 1.2–1.5 cm long, the phyllaries (at least the outermost) with pale (translucent to pinkish) border; largest leaves less than 1 cm broad . . . .8. **C. cyanus**
    7. Middle phyllaries with an abruptly expanded appendage at the apex (broader than the narrow neck below it; e.g., figs. 249–251)
      9. Appendage on middle phyllaries with body (not fringe) ± triangular, usually ca. 1.5 times as long as the base is broad, bordered with a prominent ciliate fringe; shaft of middle and inner phyllaries green (or purple-tinged), clearly visible between the tips and appendages of the overlying phyllaries . . . . .9. **C. nigrescens**
      9. Appendage on middle (not necessarily lower) phyllaries with body ± orbicular, the border various; shaft of middle and often inner phyllaries concealed by the large appendages of the overlying phyllaries
        10. Middle and outer phyllaries with appendages light brown, irregularly cleft, erose, or lacerate — or even entire, little if at all ciliate-fringed (fig. 249); pappus none; marginal flowers much larger than others in the head . . . . .10. **C. jacea**
        10. Middle and outer phyllaries with appendages blackened, all deeply pectinate-ciliate (fig. 250); pappus ca. 0.5–1.5 mm long; marginal flowers not larger than others in the head [See text and fig. 251 for hybrids with *C. jacea*.] . . . . .11. **C. nigra**

1. C. macrocephala Willd.

Map 716. This handsome ornamental is a subalpine native of the Caucasus region, rarely escaped to roadsides. First noted in 1962 and 1963 as established in two places near Bête Grise, Keweenaw County, and in 1985 and 1986 noted as well established near Wakefield, Gogebic County.

These tall plants (to ca. 1 m) with 1–4 enormous heads of yellow flowers are striking and unmistakable. The stems bear numerous unlobed leaves.

2. C. maculosa Lam. Fig. 246 Spotted Knapweed

Map 717. A Eurasian invader first collected in Michigan in 1911 in an alfalfa field at Hesperia (on the Oceana/Newaygo county line); no collector is recorded (MSC). In 1915 it was collected in Gladwin County, in 1917 in Oakland County, so it must have been established across a wide area well before 1920. Now, spread aggressively everywhere, the scourge of old fields on poor soil, roadsides, disturbed ground, and waste places, especially in the northern part of the state, where it takes over open sites. Although common in the Upper Peninsula in the 1960s, there is no record from Isle Royale until 1994.

Since the introduction of *C. diffusa*, these two species seem to have hybridized in some areas, blurring the distinctions stated in the key (see below). Moore and Frankton (1964) and Watson and Renney (1974) compare the two species quite thoroughly, though without any documentation of hybridization.

Some recent authors refer our material to *C. biebersteinii* DC., but I am unable to see consistent differences.

3. **C. diffusa** Lam.   Fig. 247   White-flowered or Tumble Knapweed
Map 718. This is another immigrant success story. The species was first collected in Michigan in Kalamazoo County in 1943, but it quickly spread northward (if not already there) to become locally common along with *C. maculosa* (see Brittonia 9: 96. 1957). Within a decade it was well established north to the Straits of Mackinac, often abundant over great areas of roadside and dry fields; also around railroads, sand dunes, vacant lots, dumps, gravel pits, parking areas, and other waste places.

The species is usually said to lack a pappus on the achene, but there is frequently a short one on plants otherwise resembling *C. diffusa*. Plants with black-tipped phyllaries having a spine at the tip are apparently hybrids. Other combinations of characters occur as well. *C. maculosa* has a white form (collected in Charlevoix, Iosco, and Kalamazoo counties), and *C. diffusa*, a pink one, so flower color alone will not discriminate although it may be a clue in a given population where hybridization is occurring. Hybrids seem especially rampant in Cheboygan, Emmet, and Kalkaska counties.

Occasionally the tips of the phyllaries are spreading (squarrose) as in the next species, but such plants have slightly larger involucres and lack a pappus.

4. **C. virgata** Lam.
Map 719. A native of dry ground in Turkey and perhaps neighboring areas. Immature specimens were collected in 1900 by Emma Cole (MICH, GH) in Paris Township. It is also reported from California and Utah.

The most comprehensive modern treatment of this species, sens. lat., is by Wagenitz (in P. H. Davis, *Flora of Turkey*, 5: 496–497. 1975). Our plant is apparently var. *squarrosa* Boiss. [sometimes called *C. squarrosa* Willd., an illegitimate name]. The species has a well developed pappus and the flowers are pink-purple, but the small heads and spine-tipped spreading phyllaries will easily distinguish it from *C. maculosa* (as well as from *C. diffusa*).

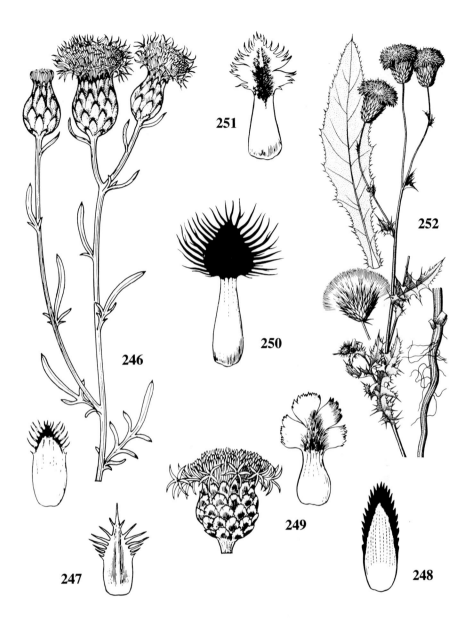

246. *Centaurea maculosa* ×1½; outer phyllary ×4
247. *C. diffusa*, outer phyllary ×4
248. *C. montana*, outer phyllary ×4
249. *C. jacea*, head ×1; outer phyllary ×4
250. *C. nigra*, outer phyllary ×4
251. *C. ×pratensis*, outer phyllary ×4
252. *Cirsium arvense* ×½

5. **C. solstitialis** L.   Yellow Star-thistle

Map 720. Native to the Mediterranean region; evidently a locally common weed in alfalfa fields (introduced with seed) in the first decade of the 20th century, but not collected in the state since.

6. **C. repens** L.   Russian Knapweed

Map 721. An Asian species collected in Michigan in 1928 (*Farwell 8225*, MICH, BLH, WUD, MSC) along railroad tracks near Ann Arbor (Am. Midl. Nat. 12: 114, 133. 1930). Since, found only by the Haneses (in 1936 and lasting for at least a decade) along a railroad in Kalamazoo County (Pap. Michigan Acad. 33: 139. 1938). It is a very serious weed in many parts of North America (Watson 1980).

C. picris Pallas is now included in this species, which is easily recognized by its entire phyllaries (scarious apically) and long (but early deciduous) pappus. Sometimes placed in a separate genus as *Acroptilon repens* (Pallas) DC.

7. **C. montana** L.   Fig. 248   Mountain-bluet

Map 722. A mountain species of central Europe, occasionally established as an escape from cultivation along roadsides, woodland borders, and waste places.

This is a very colorful species, with its large heads of deep blue flowers and phyllaries trimmed in black.

8. **C. cyanus** L.   Bachelor's-button; Cornflower

Map 723. A slender tap-rooted annual, originally native in the Mediterranean region and widely cultivated, sometimes escaping to roadsides, fields, shores, cultivated ground, dumps, and other waste places.

9. **C. nigrescens** Willd.   Short-fringed Knapweed

Map 724. A Eurasian native, occasionally established along roadsides, trails, and fields.

Included here are Michigan collections referred in the past to *C.*

718. Centaurea diffusa   719. Centaurea virgata   720. Centaurea solstitialis

vochinensis Bernh. and *C. dubia* Suter. The involucre (when pressed) tends to be almost as long as broad or a little longer, while in the next two species it tends to be distinctly shorter than broad. Plants superficially resemble the next two species in habit, but differ in the phyllaries as well as shape of the heads. A collection from Mackinac County (*Ehlers 4577* in 1929, MICH) is placed here, but differs from other specimens in the long, narrowly triangular (4–5 mm) phyllary appendages, which are long-ciliate; it might key to *C. pectinata* L. of Europe, but does not resemble that species at all, especially in foliage.

10. **C. jacea** L.   Fig. 249   Brown Knapweed
    Map 725. A native of Europe, local along gravelly roadsides, parking lots, fields, shores. Collected in waste places as early as 1895, in Bay County.
    Plants with enlarged marginal flowers and brown phyllary appendages, but with some of the middle and outer appendages strongly fringed, are hybrids with the next species, named *C. ×pratensis* Thuill. There is full gradation toward the parents, especially *C. jacea*. Detailed experimental work on this aggregation, including backcrosses, is reported by Marsden-Jones and Turrill (1954).

721. Centaurea repens

722. Centaurea montana

723. Centaurea cyanus

724. Centaurea nigrescens

725. Centaurea jacea

726. Centaurea nigra

11. **C. nigra** L. Fig. 250 Black Knapweed

Map 726. This is another immigrant from Europe, but apparently much rarer than *C. jacea*. Very local in waste places, e.g., dumps and roadsides.

C. nigra is considered to lack enlarged marginal flowers in the heads. Plants once referred to *C. nigra* var. *radiata* DC. are now thought to be the hybrid *C. ×pratensis* Thuill. (fig. 251), which is more common than *C. nigra*. Because this hybrid is so relatively widespread along roadsides and in fields, Map 727 is provided for documentation. Besides the radiate heads, it is distinguished by paler phyllary appendages (of which some middle ones are only split or lacerate, not fringed) and achenes without a pappus. In good *C. nigra*, only the innermost phyllaries lack a well developed ciliate fringe on the appendages.

69. Silybum

1. **S. marianum** (L.) Gaertner Milk-thistle

Map 728. Native to southern Europe, northern Africa, and the Near East, occasionally escaped from cultivation. Said by Dodge to be "plentiful" in yards and streets at Port Huron in the 1890s; collected by H. C. Beardslee [Jr.] in Cheboygan County in August of 1890 [no further data, MO]; collected on a rubbish heap in Grand Rapids in 1902 (specimen given to Emma Cole).

The large, auriculate, scarcely lobed leaves are spiny-margined and the strikingly large phyllaries are spine-tipped besides having spiny margins on the expanded bases.

70. Onopordon

REFERENCE

Dress, William J. 1966. Notes on the Cultivated Compositae 9. Onopordum. Baileya 14: 74–86.

727. Centaurea ×pratensis   728. Silybum marianum   729. Onopordon acanthium

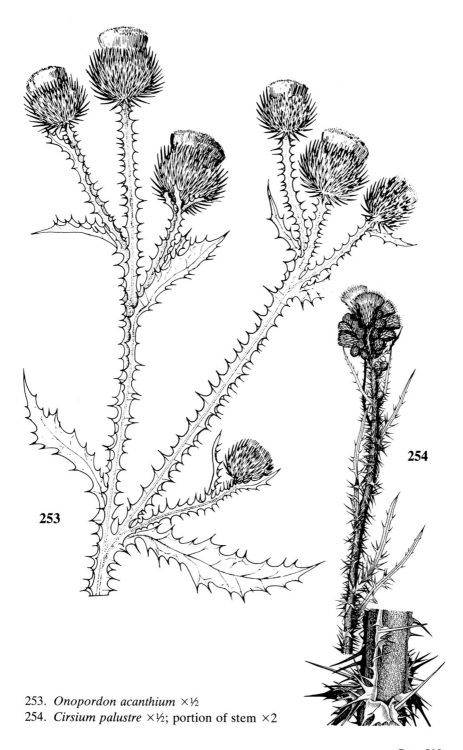

253. *Onopordon acanthium* ×½
254. *Cirsium palustre* ×½; portion of stem ×2

1. **O. acanthium** L.   Fig. 253   Cotton-thistle
Map 729. A species of Europe and western Asia, the most common one
of the genus in cultivation on this continent and the only one generally
escaping. Fields, thickets, dooryards, roadsides, and waste ground. Collected rarely in Michigan, 1894–1957.

The conspicuously winged stems and spiny-margined leaves are all covered with a cobwebby tomentum. The phyllaries are extremely numerous
and taper into a long apical spine.

71. **Carduus**   Plumeless Thistle

Hybrids between the two species treated here are well known where
they commonly grow together (Warwick et al. 1989) but have not yet been
observed in Michigan.

REFERENCES

Desrochers, A. M., J. F. Bain, & S. I. Warwick. 1988. The Biology of Canadian Weeds. 89.
    Carduus nutans L. and Carduus acanthoides L. Canad. Jour. Pl. Sci. 68: 1053–1068.
Moore, R. J., & C. Frankton. 1974. The Thistles of Canada. Canada Dep. Agr. Monogr. 10.
    111 pp. [*Carduus*, pp. 54–61.]
Stuckey, Ronald L., & Jane L. Forsyth. 1971. Distribution of Naturalized Carduus nutans
    (Compositae) Mapped in Relation to Geology in Northwestern Ohio. Ohio Jour. Sci. 71:
    1–15.
Warwick, Suzanne I., John F. Bain, Roger Wheatcroft, & Brian K. Thompson. 1989. Hybrid-
    ization and Introgression in Carduus nutans and C. acanthoides Reexamined. Syst. Bot.
    14: 476–494.

KEY TO THE SPECIES

1. Phyllaries (outer) 3–5 (6) mm wide below the middle; heads ca. 4–9 cm in diameter,
   nodding at maturity and solitary at the ends of branches . . . . .1. **C. nutans**
1. Phyllaries less than 2 mm wide; heads less than 4 cm in diameter, erect, solitary or
   clustered on branches . . . . .2. **C. acanthoides**

1. **C. nutans** L.   Nodding or Musk Thistle
Map 730. Roadsides, fields, and gravelly banks. The oldest Michigan
collection I have seen is from Eaton County in 1924. This Eurasian species
is locally introduced in North America. Stuckey and Forsyth found its
distribution in northwest Ohio to be closely correlated with the presence of
limestone or dolomite bedrock less than 6 feet below the surface. The
species is too sparse in Michigan to draw any conclusions here.

2. **C. acanthoides** L.
Map 731. Roadsides, railroads, fields, farmyards, and other disturbed
ground. First collected in Michigan in 1920 (Washtenaw Co.) and 1922
(Oakland Co.). Another Eurasian species.

All Michigan reports of *C. crispus* L. have turned out to be misidentifications of this species, which has at least a few crinkly septate hairs on the

veins beneath the leaves. In *C. crispus*, the hairs are unicellular and there is often some tomentum on the undersides of the leaves.

72. **Cirsium**   Thistle

Most of our thistles are tall (often 2 m), unpleasantly prickly plants, generally biennial or perennial. *C. hillii* is normally low and squat, with very large deep-colored heads besides, and hence one of the most attractive species. The tallest *C. pitcheri* I have seen was 1.6 m high, but most individuals are shorter. I have included the spines in measurements given for involucre widths, but it is hard to determine precise figures, especially for heads that have been flattened in pressing. All of our species except *C. pitcheri* normally have pink to purple flowers, but white-flowered forms may occur, apparently quite rarely. The phyllaries vary within the head, the outermost and usually the middle ones in most species spine-tipped, but the innermost ones spineless. Measurements of spines on leaves and phyllaries are of the pale or brown, non-green portion. The distinctive long-plumose pappus is readily deciduous, the hairs attached in a ring at the base—suggesting that wind dispersal may not be as effective as might otherwise be supposed.

The references cited below will help in providing additional characters, interpretation of possible hybrids, and understanding of the great variability in this genus. Johnson and Iltis (1964, cited under family references) usefully deal with the same 10 species as ours.

REFERENCES

Frankton, C., & R. J. Moore. 1961. Cytotaxonomy, Phylogeny, and Canadian Distribution of Cirsium undulatum and Cirsium flodmanii. Canad. Jour. Bot. 39: 21–33 + 1 pl.

Keddy, C. J., & P. A. Keddy. 1984. Reproductive Biology and Habitat of Cirsium pitcheri. Michigan Bot. 23: 57–67.

Loveless, M. D., & J. L. Hamrick. 1988. Genetic Organization and Evolutionary History in Two North American Species of Cirsium. Evolution 42: 254–265.

Moore, R. J., & C. Frankton. 1966. An Evaluation of the Status of Cirsium pumilum and Cirsium hillii. Canad. Jour. Bot. 44: 581–595.

730. Carduus nutans

731. Carduus acanthoides

732. Cirsium palustre

Moore, R. J., & C. Frankton. 1969. Cytotaxonomy of Some Cirsium Species of the Eastern United States, with a Key to Eastern Species. Canad. Jour. Bot. 47: 1257–1275.

Moore, R. J., & C. Frankton. 1974. The Thistles of Canada. Canada Dep. Agr. Monogr. 10. 111 pp. [*Cirsium*, pp. 19–53.]

Moore, R. J. 1975. The Biology of Canadian Weeds. 13. Cirsium arvense (L.) Scop. Canad. Jour. Pl. Sci. 55: 1033–1048.

Ownbey, Gerald B. 1951. Natural Hybridization in the Genus Cirsium — I. C. discolor (Muhl. ex Willd.) Spreng. × C. muticum Michx. Bull. Torrey Bot. Club 78: 233–253.

Ownbey, Gerald B. 1964. Natural Hybridization in the Genus Cirsium — II. C. altissimum × C. discolor. Michigan Bot. 3: 87–97.

KEY TO THE SPECIES

1. Internodes (especially upper ones) with continuous or intermittent spiny wings their full length (fig. 254); plants aggressive, introduced weeds
   2. Heads small (involucre ca. 1–1.5 cm broad), mostly in dense clusters; phyllaries at most with a spine tip ca. 0.5 (–1) mm long; upper surface of leaves with a few septate hairs . . . . .1. **C. palustre**
   2. Heads large (involucre ca. 3.5–6.5 cm broad), mostly solitary on short branches; phyllaries with a strong spine tip ca. 2–6 mm long; upper surface of leaves with ± appressed and dense spines mostly 0.5–1.5 mm long .2. **C. vulgare**
1. Internodes at most with leaf base slightly decurrent (less than halfway to the next lower node); plants native or sporadic waifs (except for *C. arvense*)
   3. Leaves green beneath (if loosely tomentose, the green or pale green surface not hidden)
      4. Heads large (involucre ca. 3.5–6 (7) cm broad), 1–3 (7) per stem; spines on outer phyllaries mostly 1.5–3.5 mm long . . . . .3. **C. hillii**
      4. Heads smaller (involucre ca. 1–3 (3.5) cm broad), often 5 or more per stem; spines on outer phyllaries none or at most ca. 1–1.5 mm long
         5. Spines on outer phyllaries none or less than 0.5 mm long; involucre ± cobwebby pubescent; largest spines on leaves less than 2.5 (3.5) mm long; plant producing rosettes (biennial or perennial) but not horizontal roots, native in moist habitats . . . . .4. **C. muticum**
         5. Spines on outer phyllaries ca. 0.5–1 (1.5) mm long; involucre at most cobwebby on phyllary margins; largest spines on leaves ca. 3.5–7 mm long (as short as 1.5–2 mm in forms with at least the upper leaves unlobed); plants perennial and colonial from horizontal roots, but not producing rosettes, introduced and mostly in dry ground . . . . .5. **C. arvense**
   3. Leaves densely tomentose beneath, the green surface fully hidden
      6. Upper internodes (if not all) densely white-tomentose

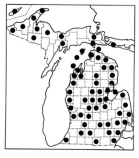

733. Cirsium vulgare   734. Cirsium hillii   735. Cirsium muticum

7. Cauline leaves deeply pinnate with linear lobes (entire or nearly so) about as wide as the rachis (less than 4 (8) mm); flowers white or cream (or somewhat pinkish from the style and anthers); plant native, on sand dunes along the Great Lakes . . . . .6. **C. pitcheri**
7. Cauline leaves with ± triangular lobes, the rachis over 5 mm broad; flowers purple (except in albino forms); plant seemingly a sporadic waif in waste places
   8. Lateral lobes of leaves not over 7 mm broad at base; involucre ca. 2–2.7 cm long; [achenes ca. 3–4 (5) mm long] . . . . .7. **C. flodmanii**
   8. Lateral lobes of leaves mostly over 7 mm broad at base; involucre ca. (2.7) 3–3.5 cm long; [achenes ca. 5–7 mm long] . . . . .8. **C. undulatum**
 6. Upper internodes not tomentose or only very lightly so
  9. Involucre ca. 1–1.8 (2.3) cm long or even shorter, the phyllaries with spines none or up to 1 (1.5) mm long . . . . .5. **C. arvense** var. **vestitum**
  9. Involucre ca. 2.1–3.3 cm long, the phyllaries with spines 3.5–6.5 mm long
   10. Middle and upper cauline leaves deeply pinnately lobed; scarious tips of innermost phyllaries all narrowly tapered to the apex, with no expansion at base . . . . .9. **C. discolor**
   10. Middle and upper cauline leaves unlobed to shallowly lobed; scarious tips on at least some innermost phyllaries usually expanded at the base (broader than the firm body of the phyllary tip and contracted before tapering to the apex) . . . . .10. **C. altissimum**

1. **C. palustre** (L.) Scop.   Fig. 254   Marsh or
                                                      European Swamp Thistle
   Map 732. A native of western Europe, collected in New Hampshire and Newfoundland early in the century and becoming locally abundant as it spread. First collected in Michigan in Marquette County in 1934 (Rhodora 37: 337. 1935) by Fernald, who thought it likely to be native but admitted that its behavior in the state "should be watched." Unfortunately, we have watched it spread throughout the Upper Peninsula, to islands in the Straits of Mackinac, and by 1959 to the mainland of the northern Lower Peninsula (Michigan Bot. 16: 139. 1977). It continues to move south. Tall plants form dense ungainly colonies for miles along roadside ditches and adjacent swamps, whence it spreads to shores and remoter wetlands. In cedar swamps and somewhat shaded fens it may try to masquerade as the native swamp thistle, *C. muticum*. Its large spiny rosettes and densely prickly stems appear very much out of place next to *Orchis rotundifolia* and other rarities. The tragic spread of this pest into natural wetlands is doubtless aided by logging roads and other human (or inhuman) disturbance.
   The crowded heads at the end of main stems and usually branches (which elongate as they mature) give this species a very distinctive appearance that makes it readily recognizable as one drives down the road. Spraying and mowing (e.g., along highway M-28) have not conquered it.

2. **C. vulgare** (Savi) Tenore   Bull Thistle
   Map 733. Fields, shores, ditches, vacant lots, dumps, and other waste ground; clearings and trails in woods; has a nice day after logging, fire, or

bulldozing; invades natural communities adjacent to the disturbed habitats. A native of Eurasia, long a weed of waste ground in North America. Collected in Michigan as early as the First Survey (1838). Doubtless throughout the state, though understandably collected less enthusiastically than most other plants. Towering, much-branched plants thriving in newly disturbed ground are truly horrendous.

3. **C. hillii** (Canby) Fernald   Hill's Thistle
    Map 734. Dry open places, especially with jack pine and oak, but also prairie-like ground, thin soil over limestone (Drummond Island), and sandy banks. This species ranges from the Great Lakes region to Iowa, and was originally described from the south end of Lake Michigan. The records from southern Michigan are all quite old (Oakland Co. 1849–1896; Washtenaw Co. in 1867; Jackson Co. 1894–1896; St. Clair Co. in 1904; Berrien Co. in 1932; and Kalamazoo Co. 1931–1939), and the species may well now be extinct in this part of the state. The rosettes of spiny leaves—the plants to bloom another year—are often conspicuous in grassy areas of jack pine savanas.
    A number of recent authors treat this taxon as a subspecies or variety of the eastern *C. pumilum* (Nutt.) Sprengel, i.e., as ssp. *hillii* (Canby) R. J. Moore & Frankton or var. *hillii* (Canby) B. Boivin (see Moore & Frankton 1966 for a full statement of the characters). All reports of *C. pumilum* from Michigan refer to this taxon, whatever rank one prefers for it, and it is probably the basis for some other old reports, such as *C. horridulum* Michaux, from the state (presumably misidentifications).
    The flowers are normally a rich rose-purple. I collected a white-flowered form in southern Cheboygan County in 1952, but it is not common: f. *albiflorum* (Scoggan) E. G. Voss.

4. **C. muticum** Michaux   Swamp Thistle
    Map 735. Calcareous shores, sedge meadows, fens; swamps of cedar, tamarack, and other conifers (less often hardwoods); low ground along rivers and streams.
    The flowers are a deep rose-purple; combined with the relatively unarmed stem, they make this species unusually attractive as thistles go. White-flowered plants [f. *lactiflorum* Fernald] have been collected in Keweenaw, Livingston, and Oakland counties. The phyllaries are less acute (even obtuse or blunt) compared with our other small-headed thistles (*C. arvense* and *C. palustre* — both introduced, weedy species), which have pale pink-purple flowers. Occasionally the leaves are lightly tomentose beneath, but not enough so to hide completely the green surface of the blade; none of our specimens with such leaves appear to be the hybrid with *C. discolor* described by Ownbey (1951).

5. **C. arvense** (L.) Scop.   Fig. 252   Canada or Field Thistle

Map 736. Widespread across Eurasia and northern Africa; introduced into North America before 1800 (already in 1795, legislation was passed in Vermont to control it, but aggressive plants, like aggressive people, can be no respecters of law). Dennis Cooley collected Canada thistle in Macomb County in 1851, noting "This plant has not yet become troublesome here — but is increasing." E. C. Allmendinger, in her 1876 Flora of Ann Arbor, wrote that it "will give trouble in the future if not soon exterminated." She had first collected it at Ann Arbor in 1862, and in 1875 A. B. Lyons collected it at Detroit. Soon it was throughout the state, although like many other immigrant plants, it is not adequately collected to document the details of its spread and occurrence. Its extensive and deep root system makes it very difficult to eradicate. Found at roadsides, ditches, railroads, fields, clearings, shores; disturbed swampy ground but not usually in very wet places; waste ground including filled land, parking areas, utility rights of way, dumps, neglected gardens, etc.

This is a variable species, although most Michigan specimens are the spiny, glabrous var. *horridum* Wimmer & Grab. as described in the key. A few collections are var. *arvense* [including var. *mite* Wimmer & Grab.], with the upper leaves unlobed and with small spines (up to 2 mm long). We have one collection, from Lenawee County in 1985 (*Smith 747*, MICH), with leaves densely tomentose beneath, at most shallowly lobed, and also with small spines; such plants have been called var. *vestitum* Wimmer & Grab. It is doubtful whether these varieties are very meaningful. White-flowered plants are rare, but have been collected in Baraga, Cheboygan, and Emmet counties [f. *albiflorum* E. L. Rand & Redfield]. Further contributing to the diversity of this species is the fact that it is dioecious. Plants with only staminate flowers will not produce viable seed (although they may have a reduced pappus).

6. **C. pitcheri** (Eaton) T. & G.   Plate 8-C   Dune or Pitcher's Thistle

Map 737. Endemic to the sand beaches and dunes of Lakes Michigan and Huron, and at two sites on Lake Superior. Listed as threatened under both Michigan and Federal law.

This thistle was first noted by Zina Pitcher "on the great sand banks of Lake Superior" — presumably the Grand Sable Dunes and during the 1826–1828 period when Pitcher was stationed as an Army surgeon at Fort Brady (Sault Ste. Marie). However, the species was first formally described as new to science in 1829 by Amos Eaton, who stated that his specimen had been collected by Edwin James (another Army surgeon) "at Lake Huron"— presumably on or near Mackinac Island, where Dr. James spent much of the summer of 1827, and which would therefore be the type locality. The Grand Sable Dunes remained the only known site for this thistle on Lake Superior for a century and a half, until it was discovered in 1977 almost due north, in the new Pukaskwa National Park in Ontario. Thorough searching of the continuously sandy shores between Whitefish Point and the Grand Sable Dunes has failed to reveal to me any specimens of Pitcher's thistle.

Dr. Pitcher, incidentally, resigned his commission while serving in Virginia in 1836 and settled in Michigan, practicing medicine in Detroit, where he was mayor for three years. He was an active regent of the University of Michigan 1837–1852, and is looked upon as the founder of what became the University's Medical School, on which faculty he served 1851–1872. In 1856 he was elected as the 10th president of the American Medical Association. He was a very active botanist, accumulating a large herbarium. Michigan should feel honored to have in its custody most of the world's population of the handsome thistle which grows in one of our most beautiful habitats and bears the name of one of our most prominent early citizens (see Voss 1978, pp. 12–15).

This is a relatively mild thistle, with only short spines at the tips of the leaf lobes and the phyllaries. The whitish foliage and flowers set this apart from all our other thistles, but they blend well with the dunes on which it grows. The species is apparently derived from the Great Plains *C. canescens* Nutt. (Loveless & Hamrick 1988). A first-year seedling consists of a very small rosette of leaves, but rosettes in subsequent years are more robust, from a very deep taproot, and flowering usually occurs in 2–8 years, after which the plant dies. (The species is not a biennial, as often described.) Most seeds travel relatively short distances from the parent plant, and entire heads are sometimes dispersed or buried, producing dense clusters of seedlings (Keddy & Keddy 1984). The fragrant flowers are popular with insect visitors and heads can be as broad as 5 cm at full bloom. Plants are occasionally well over 1 m tall, but are usually shorter and may be quite bushy-branched or single-stemmed. Consequently, the number of heads varies considerably. Plants with numerous heads may bloom over a long period (2 months), starting in late June with the earlier buds to open. A large bushy plant with heads as deep purple as in *C. vulgare* on a disturbed roadside through dunes in Mackinac County (*Voss 16509 & Hellquist* in 1995, MICH, UMBS) may represent an unusual color form (or unexpected hybrid); the foliage was just like that of normal white-flowered plants adjacent.

7. **C. flodmanii** (Rydb.) Arthur   Prairie Thistle

Map 738. Collected only once in Michigan, along a roadside in Cheboygan County (*Ehlers 3484*, MICH, UMBS). Both this species and the next are natives of the North American prairie and appear to be waifs in our area, although Frankton and Moore consider *C. flodmanii* to be a relic in the eastern part of the continent.

This species and the next have often been confused, but differ in chromosome number as well as other characters besides those in the key (see Frankton & Moore 1961). *C. flodmanii* tends to be a more delicate plant, with deeper-lobed, more pinnatifid leaves, while in the robust *C. undulatum* the leaves are sometimes only shallowly lobed. *C. flodmanii* has deeper purple heads.

8. **C. undulatum** (Nutt.) Sprengel   Wavy-leaved Thistle
    May 739. Collected in Michigan only as waifs: in a millyard in Emmet County (*Gleason* in 1917, UMBS; Pap. Mich. Acad. 4: 281. 1925) and along railroads in Washtenaw County (*Walpole* in 1919, BLH; *Farwell 7725* in 1926, MICH, BLH). The white-flowered form [f. *album* Farw.] was originally described from the railroad waif (TL: Ann Arbor).
    See also comments under the preceding species.

9. **C. discolor** (Willd.) Sprengel   Fig. 255   Pasture Thistle
    Map 740. Meadows, fields, clearings, hillsides, river banks, sparsely wooded sites; roadsides, vacant lots, pine plantations.
    This species is very similar to the next, especially in the heads, and the two do hybridize (although hybridization has not yet been documented in Michigan). The hybrids are best distinguished by intermediate leaf shape (Ownbey 1964), between the distinctive shapes of the two parents. They may be expected where there is some overlap in habitat — in time if not space, e.g., when the shaded habitat of *C. altissimum* has been cleared and *C. discolor* has spread into the newly open habitat that it prefers. The

736. Cirsium arvense

737. Cirsium pitcheri

738. Cirsium flodmanii

739. Cirsium undulatum

740. Cirsium discolor

741. Cirsium altissimum

heads of *C. discolor* run a little larger (involucre ca. 2.4–3.3 cm long in our material, compared with 2–2.7 cm in *C. altissimum*), but precise measurements are hard to determine. The longest spines on the phyllaries range from 3.5 to 6.5 (7) mm in both species.

The white-flowered form [f. *albiflorum* (Britton) House] has been collected in Calhoun, Lenawee, Oakland, and Washtenaw counties.

10. **C. altissimum** (L.) Sprengel   Tall Thistle
Map 741. Shaded river banks, floodplain and creek margin woods and thickets, oak-hickory and mixed hardwoods.

See comments under the preceding species.

73. **Antennaria**   Pussy-toes
The genus *Antennaria* shares with some of the troublesome genera of Rosaceae the problems of diploid species and apomictic polyploid derivatives; but it has an advantage in comprising fewer taxa. (See Beals & Peters 1967 for a general summary.) However, even with the evolutionary understanding and merging of named taxa accomplished by Bayer and Stebbins, the lines between taxa remain too often obscure.

Plants in this genus are dioecious and bloom in early spring (April–May). Involucral characters are based on pistillate plants (which are usually more common). Pappus hairs of staminate flowers are barbellate and/or thicker (± clavellate) than the very fine smooth hairs of pistillate flowers (but the style, if any, is undivided). Staminate flowers are broader than the filiform ones of pistillate flowers. The term "basal leaves" applies both to the rosettes at the base of stems and to those at the ends of stolons. Old basal leaves from the previous year are often best for determining the maximum leaf width and the number of prominent longitudinal veins, whereas young and older leaves of the current year are needed to determine the nature of pubescence.

The key to species employs characters used by Bayer and Stebbins (1994), and the nomenclature here is also theirs. I am, in addition, very grateful to R. J. Bayer for annotating in 1994 over 300 *Antennaria* specimens from the University of Michigan Herbarium and which support the great majority of the dots on the maps.

REFERENCES

Bayer, Randall J., & G. Ledyard Stebbins. 1982. A Revised Classification of Antennaria (Asteraceae: Inuleae) of the Eastern United States. Syst. Bot. 7: 300–313.

Bayer, Randall J. 1985. Investigations into the Evolutionary History of the Polyploid Complexes in Antennaria (Asteraceae: Inuleae). II. The A. parlinii Complex. Rhodora 87: 321–339.

Bayer, Randall J. 1989a. A Taxonomic Revision of the Antennaria rosea (Asteraceae: Inuleae: Gnaphaliinae) Polyploid Complex. Brittonia 41: 53–60.

Bayer, Randall J. 1989b. Nomenclatural Rearrangements in Antennaria neodioica and A. howellii (Asteraceae: Inuleae: Gnaphaliinae). Brittonia 41: 396–398.

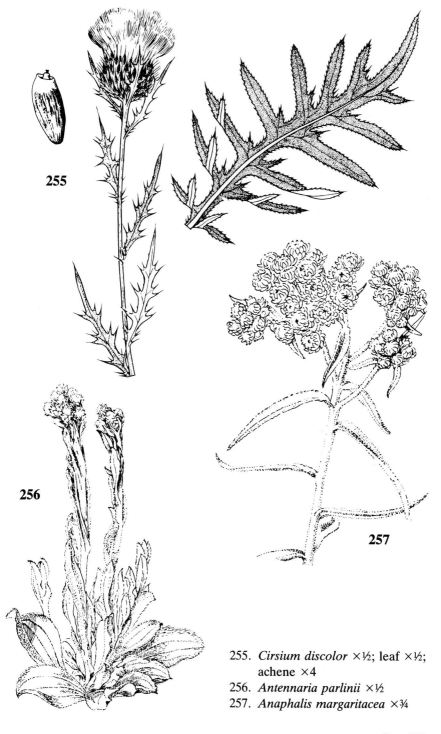

255. *Cirsium discolor* ×½; leaf ×½; achene ×4
256. *Antennaria parlinii* ×½
257. *Anaphalis margaritacea* ×¾

Bayer, Randall J., & G. Ledyard Stebbins. 1994 ["1993"]. A Synopsis with Keys for the Genus Antennaria (Asteraceae: Inuleae: Gnaphaliinae) of North America. Canad. Jour. Bot. 71: 1589–1604.

Beals, Edward W., & Ralph F. Peters. 1967 ["1966"]. Preliminary Reports on the Flora of Wisconsin No. 456. Compositae V—Composite Family V Tribe Inuleae. Trans. Wisconsin Acad. 55: 223–242.

Hyypio, Peter A. 1952. Antennaria rosea in the Lake Superior Region. Rhodora 54: 291.

KEY TO THE SPECIES

1. Blades of basal leaves (at least the largest) with 3–5 (7) longitudinal veins prominent (at least on lower surface) to beyond the broadest part of the leaf . . . . .1. **A. parlinii**
1. Blades of all leaves with only 1 prominent longitudinal vein
  2. Tips of phyllaries rose-pink; plant restricted to northwestern Lake Superior region . . . . .2. **A. rosea**
  2. Tips of phyllaries white; plants more widespread
    3. Basal leaves not over 6 mm wide, ± equally and densely pubescent on both surfaces; involucre of pistillate heads ca. 8–10 mm long . . . . .3. **A. parvifolia**
    3. Basal leaves at least 6 mm wide, or less pubescent (even glabrous) above, or usually both conditions present; involucre of pistillate heads ca. 5.5–8 (rarely 10) mm long
      4. Middle and upper cauline leaves merely acute or with firm subulate tip (bracts in inflorescence may have an appendage) . . . . .4. **A. howellii**
      4. Middle and upper cauline leaves with a flat (often ± curled or involute) scarious appendage at the tip
        5. New basal leaves of the season essentially glabrous above or very soon becoming so (may appear hairy along the margin from tomentum of underside) . . . .
        . . . . .4. **A. howellii**
        5. New basal leaves pubescent above when young (becoming glabrous only in age) . . . . .5. **A. neglecta**

1. **A. parlinii** Fernald   Fig. 256

Map 742. More often than our other species in shaded places, even woods, and rich soil, but usually in dry open sites, including rock outcrops, banks, grassy roadsides and hillsides; woodland of oak, hickory, sassafras, and/or jack pine, especially along borders and in clearings; but sometimes with beech and maple, low woods, or boggy sites.

This is one of our commonest species, as well as the most easily recognized one. It was long included by many authors in *A. plantaginifolia* (L.) Richardson, a species, as now interpreted, with smaller involucres and ranging primarily east of the Great Lakes; but also disjunct to the west of us. *A. parlinii* includes hybrids of multiple origin and therefore is maintained as distinct from the diploid relative, *A. plantaginifolia*. In *A. parlinii*, the involucres are ca. (7) 7.5–9.7 mm long.

Bayer and Stebbins recognize two subspecies: ssp. *fallax* (Greene) Bayer & Stebbins ranges throughout the state and has the basal leaves ± persistently pubescent above, often with bits of cobwebby tomentum adhering into late summer; ssp. *parlinii* has the basal leaves glabrous or nearly so above even when relatively young and also often has some tiny pink-purple hairs among the tomentum of the upper parts of the stem. However, these characters seem not always to hold on specimens determined by Bayer. The ssp. *parlinii* occurs only in the southern third of the Lower Peninsula—north to Kent County.

Occasionally (especially on the Keweenaw Peninsula) plants of *A. howellii* ssp. *neodioica* are found that approach *A. parlinii* in a tendency toward 3-nerved basal leaves; some of these have been called (by Greene and others) *A. farwellii* Greene, although Bayer refers the type of the latter name (TL: Clifton, Keweenaw Co.) to *A. parlinii* ssp. *fallax*.

2. **A. rosea** Greene

Map 743. This apomictic complex occurs in the Maritime Provinces of Canada, west to James Bay and Alaska, and south in the West nearly to Mexico; it is ± disjunct in the northwestern part of Lake Superior (on Caribou Island in the Isle Royale archipelago and also in nearby Ontario). Its Michigan occurrence (*H. A. Baggley* in 1940 & 1941, MSC, IRP, MICH) was early reported by Hyypio (1952).

3. **A. parvifolia** Nutt.

Map 744. Primarily a western species, collected in Michigan in 1991 (*D. Garlitz 1425*, MICH) in sandy soil on the west side of the village of Fife Lake.

4. **A. howellii** Greene

Map 745. Can be expected almost anywhere in dry open places. Frequent on rock ledges and outcrops of all kinds and in sandy or rocky woodland (especially in openings and clearings) with jack pine, oak, aspen, and/or red maple; sometimes on moist shores and in boggy ground; fields, roadsides, banks, even in lawns.

Our three subspecies have long been recognized at species rank (bearing the same epithets), although the characters emphasized to distinguish them have varied somewhat among authors — and many additional names have been published for trivial variants. Bayer and Stebbins earlier combined

742. Antennaria parlinii

743. Antennaria rosea

744. Antennaria parvifolia

them under the name *A. neodioica*; but with inclusion of the western *A. howellii* in the species, that earlier name must prevail for this broad complex of apomictic plants (Bayer 1989b). Staminate clones are very rare.

Cauline leaves of ssp. *canadensis* (Greene) Bayer share with *A. neglecta* (one of its ancestors) a distinctive little flat scarious tip or "flag," sometimes curled or sometimes involute (and then not always easy to distinguish from a merely subulate tip). The glabrous and bright green upper surfaces of the basal leaves are often quite striking compared to the relatively dull, gray, ± pubescent young leaves of the other subspecies. The ssp. *neodioica* (Greene) Bayer has rather short stolons, with all leaves nearly as large as the terminal rosette ones—and the basal leaves tend to have more distinct petioles as well as to be the most heavily pubescent above of all our subspecies. Thus, in this subspecies there are often dense mats of crowded rosettes of dull-looking leaves. The ssp. *petaloidea* (Fernald) Bayer tends to have elongate stolons bearing leaves along them distinctly smaller than their terminal rosette leaves. Cauline leaves of this subspecies apparently can rarely have the little "flag" tips that characterize ssp. *canadensis*. The basal leaves of ssp. *petaloidea*, unlike those of ssp. *neodioica*, tend to be obovate without distinct petioles. Both ssp. *petaloidea* and ssp. *neodioica* sometimes grade into *A. parlinii* in the possession of at least partly 3-nerved basal leaves, but they tend to shorter involucres (less than 7.5, or very rarely 8, mm) than in that species.

The distributions of ssp. *petaloidea* and ssp. *neodioica* are essentially throughout the range mapped for the whole complex; ssp. *canadensis* comes south only to Saginaw County, thus barely meeting the latitude in Gratiot County for the northernmost clearcut records of *A. neglecta*.

5. **A. neglecta** Greene
Map 746. Like other *Antennaria* species, primarily in dry open ground, including sandy woodland, fields, prairies; grassy hillsides.

As here defined, this is a sexually reproducing diploid species. Staminate clones are frequent. Cronquist has included the *A. howellii* complex in an even broader concept under the name *A. neglecta*. Indeed, the two are often very difficult to distinguish morphologically. However, the argument of Bayer (especially 1989b) is logical: "Including these apomictic hybrids within a single sexual diploid . . . distorts the relationship because many of the apomicts are genetically more closely related to one of the diploids other than *A. neglecta*."

This species is apparently [nearly ?] restricted in Michigan to the southern Lower Peninsula, although Bayer has identified as *A. neglecta* two staminate individuals from barrens in Crawford County (on a mixed sheet with *A. howellii* ssp. *petaloidea*). The cauline leaves on these plants lack "flags" and the distribution is suspicious. Likewise, a staminate collection (though with cauline leaves clearly "flagged") from Marquette County,

Page 528

labeled some 30 years after its 1894 collection date (no collector stated) is highly suspicious on geographic grounds and a mixing of labels (or specimens) is suspected.

74. Filago

1. **F. arvensis** L.
Map 747. Sandy roadsides, clearings, fields, shores, especially in disturbed areas of aspen and jackpine woodland. A European species, locally flourishing in northern Michigan in the mid 1950s (Brittonia 9: 97. 1957), but not collected in the state since 1977 (Otsego Co.).

The heads are ± racemose or paniculate, and the plants may be stiffly erect, barely 1 dm tall, or bushy-branched and much larger. Sometimes placed in a segregate genus as *Logfia arvensis* (L.) Holub.

75. Anaphalis

1. **A. margaritacea** (L.) Bentham   Fig. 257   Pearly Everlasting
Map 748. Locally very common northward, often in somewhat disturbed areas. While sometimes in moist ground, usually in dry sandy or rocky open places such as shores, dunes, fields, roadsides and railroads; in dry woodlands of aspens or mixed conifers and hardwoods, especially along borders and in clearings (following logging or fire) and on trails.

This is one of the few native American species that has become established as a weed of waste ground in Europe, where it is cultivated. It is also native in northeastern Asia. The gleaming white phyllaries spread in the aging heads, and plants make fine winter bouquets as the "everlasting" name suggests.

The leaves vary in density of white tomentum on the upper surface, which is sometimes fully green. That surface is usually smooth, although there are sometimes a few tiny gland-tipped hairs, usually well hidden in the tomentum.

745. Antennaria howellii   746. Antennaria neglecta   747. Filago arvensis

76. Gnaphalium   Cudweed

Some European workers have divided this genus into several (as they have *Filago*).

REFERENCE

Mahler, William F. 1975. Typification and Distribution of the Varieties of Gnaphalium helleri Britton (Compositae — Inuleae). Sida 6: 30–32.

KEY TO THE SPECIES

1. Inflorescence elongate (spicate or racemose); pappus bristles united in a ring at the base
  2. Plants annual or biennial, with sparse, slightly overlapping cauline leaves and *Antennaria*-like basal rosettes . . . . 1. **G. purpureum**
  2. Plants perennial, from a rhizome, with numerous overlapping cauline leaves and no basal rosettes . . . . 2. **G. sylvaticum**
1. Inflorescence corymbiform (or heads crowded at ends of branches, or sometimes spicate in *G. uliginosum* with involucres only 2–3 mm long); pappus bristles separate
  3. Involucre ca. 2–2.7 (3) mm long; plants (except the most depauperate) bushy-branched; heads in clusters overtopped by subtending leaves . . . . 3. **G. uliginosum**
  3. Involucre 4.5–6.5 (7) mm long; plants rarely branched (except at the top or in the inflorescence); heads not overtopped by subtending leaves
    4. Leaves decurrent (stem with 2 short wings of leaf tissue extending beneath each node; fig. 258); stems (at least on the less tomentose middle and lower internodes) with evident spreading glandular hairs all or mostly 0.3–0.7 mm long . . . . 4. **G. macounii**
    4. Leaves not decurrent; stems with glandular hairs, if any, at most ca. 0.2 mm long, usually not evident
      5. Stems with white tomentum, concealing any glands . . . . 5. **G. obtusifolium**
      5. Stems green, with evident but tiny (ca. 0.2 mm or less) gland-tipped hairs . . . . 6. **G. helleri**

1. **G. purpureum** L.   Purple Cudweed

Map 749. A southern species, rarely collected in Michigan (first in 1895 and 1896 in St. Clair Co.) and doubtless adventive here. Sandy fields, shores, excavations, and pastures.

2. **G. sylvaticum** L.

Map 750. A circumpolar species, ranging south in North America to New England and the northern Great Lakes region. Known thus far in Michigan only at Grand Island, where found by Henson in 1991 (*3413, 3443, & 3485*, MICH) along old trails in hardwoods.

3. **G. uliginosum** L.   Low Cudweed

Map 751. A Eurasian species, early introduced; collected as far back as 1837 by the First Survey in Kalamazoo County. Damp woods, clearings, shores; along stream beds and trail roads in swamps, often common in sticky clay soil; waste places, fields, roadsides, ditches; crevices of gull-distressed rock shores in the Isle Royale archipelago.

4. **G. macounii** Greene   Fig. 258   Clammy Cudweed

Map 752. Dry fields and barren ground, oak and jack pine woodlands, stabilized dunes, rocky ridges and outcrops; clearings, gravel pits, and other disturbed sites, including those logged or burned; occasionally in damp ground.

This species and the next two have glandular foliage and a prominent sweet balsamic odor. Some authors have included this species in the Mexican *G. viscosum* Kunth.

5. **G. obtusifolium** L.   Fig. 259   Fragrant Cudweed

Map 753. Dry open ground, fields, woods (especially open jack pine, oak, and aspen); disturbed ground, especially cleared or burned areas as well as gravel and borrow pits; occasionally in boggy ground.

6. **G. helleri** Britton

Map 754. Usually in dry sandy oak, aspen, and jack pine woodland.

Our plants are all var. *micradenium* (Weath.) Mahler, long treated as a variety of the preceding species (Mahler 1975).

748. Anaphalis
    margaritacea

749. Gnaphalium
    purpureum

750. Gnaphalium
    sylvaticum

751. Gnaphalium
    uliginosum

752. Gnaphalium macounii

753. Gnaphalium
    obtusifolium

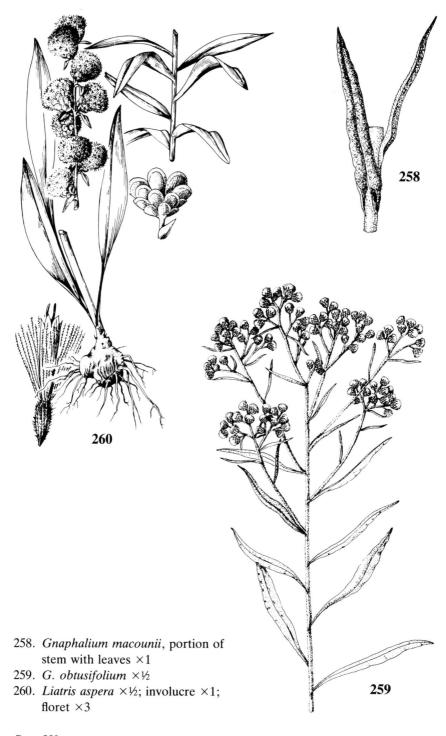

258. *Gnaphalium macounii*, portion of
     stem with leaves ×1
259. *G. obtusifolium* ×½
260. *Liatris aspera* ×½; involucre ×1;
     floret ×3

77. **Pluchea**

REFERENCE

Gillis, William T. 1977. Pluchea Revisited. Taxon 26: 587–591.

1. **P. odorata** (L.) Cass.   Saltmarsh-fleabane
   Map 755. A species of saline habitats, mostly along the coast, barely as far north or inland as Michigan. Collected in this state as early as 1912 by Farwell, near a salt mine in Wayne County (see also Rhodora 18: 243–244. 1916), but apparently not gathered anywhere in Michigan since 1932.
   This species was long known as *P. purpurascens* (Sw.) DC., and has also been confused with *P. camphorata* (L.) DC., which has at most a glandular but not pubescent involucre.

78. **Vernonia**   Ironweed
The line between our two species is not as clear as most people would like. Intermediate plants [called *V.* ×*illinoensis* Gleason by many authors] are frequent, and we have quite intermediate specimens from Allegan, Ingham, Kalamazoo, Kent, Lenawee, Livingston, Monroe, and Washtenaw counties. These mostly have either the small, few-flowered heads of *V. gigantea* with the leaf pubescence of *V. missurica*, or the larger heads of *V. missurica* but the very slightly pubescent or glabrate leaves of *V. gigantea*. Other plants that might be considered intermediate have been included here in *V. missurica*, to which they are closer and which seems to have a great range of pubescence length and density. Whether the variation results from introgression or not may be questioned. Jones and Faust (1978) only admit the existence of hybrids but do not describe or illustrate them nor account for the names applied to them. The latter, however, are provided in Faust (1972). Most Michigan specimens originally labeled as *V. illinoensis* appear easily accommodated in *V. missurica* as broadly defined, e.g., by Swink and Wilhelm for the Chicago area. Gleason himself later came to believe that his *V. illinoensis* should be included in *V. missurica*.

754. Gnaphalium helleri   755. Pluchea odorata   756. Vernonia gigantea

V. fasciculata Michaux has sometimes been reported from Michigan, but no Michigan specimens originally so labeled appear to be that species, which is glabrous or essentially so on the lower surface of the leaves, where the glandular dots are distinctly sunken in tiny pits. (In a few specimens of *V. missurica*, with leaves pubescent beneath, the dots are somewhat sunken.) *V. fasciculata* ranges mostly west of Michigan and has rather narrowly lanceolate leaves compared to our species.

REFERENCES

Faust, W. Zack. 1972. A Biosystematic Study of the Interiores Species Group of the Genus Vernonia (Compositae). Brittonia 24: 363–378.
Jones, Samuel B., & W. Zack Faust. 1978. Vernonieae. N. Am. Fl. II(10): 180–196.

KEY TO THE SPECIES

1. Flowers (and achenes) fewer than 30 per head; involucres ca. 5–7 mm broad; leaves nearly glabrous beneath, with minute, sparse, straight (often slenderly conical) hairs less than 0.3 mm long (sometimes slightly longer, denser, or crinkly hairs on midrib) . . . . .1. **V. gigantea**
1. Flowers (and achenes) 33–42 per head; involucres ca. (6) 6.5–9 (10) mm broad; leaves ± pubescent beneath, especially on main veins, with crinkly hairs, some at least 0.3 mm long . . . . .2. **V. missurica**

1. **V. gigantea** (Walter) Branner & Coville
Map 756. This ironweed is common immediately to the south of Michigan but barely enters the state. Meadows, floodplain woods, marshy thickets, roadsides.

The name *V. altissima* Nutt. was long used for this species.

Our plants have 22–27 florets per head, but a little larger range is to be expected.

2. **V. missurica** Raf.
Map 757. This species ranges from southern Michigan, where it is clearly the commoner one, southwest to eastern Texas. Riverbottom (rarely upland) woods; wet prairies, fens, sedge meadows; damp or dry open ground, river banks, fencerows, fields, roadsides.

Hybrids with the preceding species are considered frequent, or even common, especially by those who define the species more strictly (see comments above). Plants which appear to show some hybrid influence range north as far as Kent and Bay counties.

79. **Liatris**   Blazing-star; Gay-feather
This is a distinctive, strictly North American genus, with very showy heads. All of our species are in cultivation. The long, slender, dense spikes of *L. pycnostachya* and *L. spicata* are perhaps best known, but larger-headed species like *L. aspera* and *L. scariosa*, often abundant on jack pine

and oak plains, can be very showy, especially vigorous plants with 75–100 or more heads (as when well nourished or thriving in a newly disturbed spot). Some species occasionally display a white-flowered form, with pale pink intermediates to be found where normal rose-purple flowers also occur.

The underground parts of *L. punctata* are elongate and slender, like a taproot (or horizontal like a rhizome), but our other species all have one to many stems arising from a perennial, rounded, eventually rather large structure usually called a corm although it has no external evidence of leaves or nodes. Whatever it is anatomically, this is a good device to aid survival after fire (fig. 260). For these are largely plants of wet or dry grasslands, prairies, and savanas, maintained by fire. We have relatively few species compared with areas to the south and west of Michigan. (More have been described than are accepted here, where a moderately conservative approach leads to quite readily distinguishable entities without the plethora of alleged hybrids that plague the splitters.)

The terminal (or uppermost) head is the first to bloom in an inflorescence and hence tends to appear larger than the others at a given time. Characters of the phyllaries are best seen in fully mature heads. The distinctive mature phyllaries of *L. aspera* often have the broad scarious (but pinkish) apex neatly tucked beneath the firm part, making it appear to have a smooth, entire margin. At least, the margin is puckered and slightly tucked or folded.

REFERENCES

Dress, William J. 1959. Notes on the Cultivated Compositae 3. Liatris. Baileya 7: 23–32.
Gaiser, L. O. 1946. The Genus Liatris. Rhodora 48: 165–183; 216–263; 273–326; 331–382; 393–412.

KEY TO THE SPECIES

1. Pappus bristles clearly plumose, the lateral branches more than 8 times as long as the diameter of the main bristle
   2. Corolla lobes glabrous; heads mostly 4–6-flowered; leaves up to 4 mm wide . . . . .1. **L. punctata**
   2. Corolla lobes long-hairy within; heads at least 10-flowered; leaves (at least the widest lower ones) usually at least 4 (rarely 2.5) mm wide
      3. Phyllaries with spreading, prolonged, acuminate tips; stems at least sparsely pubescent . . . . .2. **L. squarrosa**
      3. Phyllaries with rounded or obtuse (rarely acute) tips, at most with a little apiculus, not spreading; stems glabrous . . . . .3. **L. cylindracea**
1. Pappus bristles strongly barbed, the lateral branches less than 5 times as long as the diameter of the main bristle
   4. Involucre ± cylindrical or conical, the heads with 5–10 flowers; inflorescence a long, cylindrical spike of dense sessile heads; habitat moist to wet places
      5. Axis of inflorescence moderately to densely pilose with spreading-crinkly hairs; tips of most phyllaries tapered to a narrowly acute, often outcurved apex . . . . .4. **L. pycnostachya**
      5. Axis of inflorescence glabrous; tips of phyllaries rounded to obtuse, neither acute nor outcurved . . . . .5. **L. spicata**

4. Involucre broadly rounded at the base, ± hemispheric, the heads with more than 10 flowers; inflorescence various; habitat mostly dry
   6. Middle phyllaries with a broad (ca. 1–2 mm) scarious margin apically, this often split or ± lacerate and usually puckered at the base or even tucked under the firmer portion of the phyllary; flowers ca. 14–28 (30) per head; inflorescence elongate, spike-like, the heads all or mostly sessile or on pedicels shorter than the involucre . . . . .6. **L. aspera**
   6. Middle phyllaries with a very narrow scarious margin, this slightly toothed but scarcely if at all puckered or tucked; flowers ca. 40 or more per head; inflorescence ± racemose, the heads all or mostly on pedicels or bracted branches equalling or longer than the involucre . . . . .7. **L. scariosa**

1. **L. punctata** Hooker

Map 758. Only from prairie soil in Kalamazoo County, where the Haneses first collected it in 1933 (Rhodora 38: 366. 1936). Only one or two multi-stemmed plants were known, but they were reported to have been there for over 25 years and they flourished for at least another decade. In 1941, one of the plants had over 50 blooming stems.

2. **L. squarrosa** (L.) Michaux

Map 759. Apparently the only Michigan collections were made by Farwell 1904–1914 at Palmer Park in Detroit. One specimen (*1853*, MICH) bears an old but undated identification label by B. L. Robinson; however, the species has never been explicitly attributed to Michigan in manuals and published records in local lists remain unverified. It is known immediately to the south in Ohio.

3. **L. cylindracea** Michaux   Fig. 261

Map 760. Dry sandy jack pine, oak, aspen woodland; fields, dunes (Huron Co.), and prairies. An old collection by Farwell (*626*, BLH) labeled as collected in Keweenaw County in 1888, is so far from other documented sites as to be very dubious and is not included on the map.

The involucres are ± cylindrical or conical, and heads are often on pedicels longer than the involucres. The white-flowered f. *bartelii* Steyerm. is not common. An uncommon hybrid with *L. aspera* has been named *L.*

757. Vernonia missurica   758. Liatris punctata   759. Liatris squarrosa

×*gladewitzii* (Farwell) Shinners (TL: Rochester [Oakland Co.] in 1925). This hybrid has also been collected by C. F. Wheeler in Ionia County in 1878 and at Whitmore Lake (Washtenaw & Livingston cos.) in 1857 by J. Q. A. Fritchey. It is a more robust plant than *L. cylindracea*, the inflorescence more spicate (heads short-stalked), and the stems usually at least sparsely pubescent. The heads are campanulate to hemispherical.

4. L. pycnostachya Michaux
Map 761. Widely cultivated, and our plants are doubtless escaped in marshy fields and ditches.

5. L. spicata Willd.
Map 762. Moist sandy plains and shores, marshy meadows; wet prairies, fens, tamarack swamps; mucky swales, marly shores, roadsides and fields; unlike other species, rarely in dry oak or jack pine woodland. Dodge wrote on an 1896 label: "Thousands of acres covered with it about Lake St. Clair."

This species hybridizes with the preceding. Rare individuals with scattered hairs in the inflorescence and other intermediate conditions indicate some introgression. White-flowered plants are f. *albiflora* Britton and have been found on Harsen's Island, St. Clair County.

6. L. aspera Michaux   Fig. 260
Map 763. Dry sandy prairies, fields, plains, clearings, and oak or jack pine woodland; associated roadsides and railroads.

Included here are Michigan plants referred (rightly or wrongly) to the mysterious *L. sphaeroidea* Michaux, sometimes treated as a hybrid closer to *L. aspera* than to another parent. Specimens so annotated by Gaiser do not appear significantly different from *L. aspera*. *L. sphaeroidea* has been variously distinguished by less puckered phyllaries (not true of our specimens), longer pedicels, or smoother leaves.

White-flowered plants are f. *benkei* (J. F. Macbr.) Fernald and have been recorded from Kalamazoo, Mecosta, and Newaygo counties.

760. Liatris cylindracea   761. Liatris pycnostachya   762. Liatris spicata

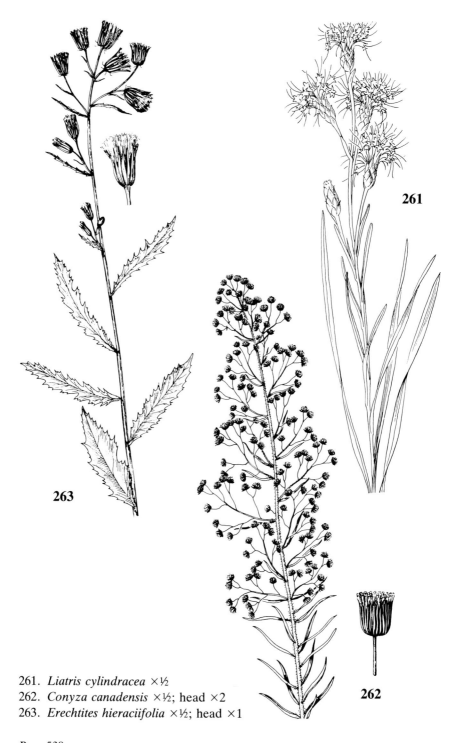

261. *Liatris cylindracea* ×½
262. *Conyza canadensis* ×½; head ×2
263. *Erechtites hieraciifolia* ×½; head ×1

7. **L. scariosa** (L.) Willd.   Plate 8-D
    Map 764. In the usual places for *Liatris*: dry sandy prairie remnants, fields, hillsides; woodlands and barrens with jack pine, oak, aspen; associated roadsides and railroads; survives fire, clearing, and mowing. A Farwell collection (*1826½*, BLH), labeled as collected in Keweenaw County in 1904, is, like his *L. cylindracea*, so far out of range as to be suspicious and is not here mapped.
    Our plants are all var. *nieuwlandii* (Lunell) E. G. Voss, which has been variously interpreted as a distinct species, as a hybrid, or as a variety of some other species. I here follow the disposition that seems agreed upon in most recent lists and manuals (although they attribute the combination to an author who seems never to have published it). Typical *L. scariosa* is more southeastern in range.

80. **Conyza**
This genus has often been included in *Erigeron*.

REFERENCES

Darbyshire, Stephen J. 1990. Conyza ramosissima, Another New Weed in the Ottawa District. Trail Landsc. 24: 94–98.
Shinners, Lloyd H. 1949. The Texas Species of Conyza. Field Lab. 17: 142–144.

1. **C. canadensis** (L.) Cronquist   Fig. 262   Horseweed
    Map 765. Now considered a nearly cosmopolitan weed, but apparently native to North America, widely naturalized in Europe and elsewhere. Especially characteristic of recently cleared, burned, abandoned, or otherwise disturbed ground, including rebuilt roadsides, railroads, fields, gravel pits, dumps, parking areas, shores, gardens; also in prairie-like areas.
    The rudimentary rays are white to pink. Stature of plants varies from a few centimeters (flowering as short as 3 cm, with heads solitary or few) to well over a meter tall (with hundreds of heads). The stem is usually erect and unbranched till near its summit. The prominent stiff hairs nearly or quite at right angles to the stem and leaf margins (which appear boldly

763. Liatris aspera

764. Liatris scariosa

765. Conyza canadensis

ciliate) usually make it an easy species to recognize, even vegetatively. However, some plants (particularly from the Upper Peninsula) are small, with ascending branches from near or below the middle. These and some others may have very sparse or more appressed pubescence, but most have the widely spreading setose hairs of *C. canadensis*. None thus far found, however, have the diffuse, decumbent branches at the base that characterize the more delicate and narrower-leaved *C. ramosissima* Cronquist which also has only antrorse or appressed hairs. That species has been collected in Ohio not too far from the Michigan border and should be sought here. Our small branched plants generally resemble those illustrated (but not in fact correctly) as *C. ramosissima* from Ontario by Darbyshire (1990). Whether these little plants result from hybridization, seasonal or genetic variation, or other factors deserves investigation.

81. **Petasites** Sweet-coltsfoot

These are all rather thick-stemmed plants that bloom in May — unlike most composites. The cauline leaves consist of broad multi-veined modified petioles with at most a rudimentary blade. The basal leaves arise from a rhizome, are long-petioled, with broad blades ± tomentose beneath, at least when young. They expand with or after the flowers. The basal leaves are conspicuous throughout the summer, long after the stems have withered. The copious silky white pappus makes a handsome sight when the plants are fruiting in late spring. The plants are functionally dioecious, the staminate flowers without rays but marginal pistillate ones often with conspicuous though relatively short rays. Pistillate heads may have numerous rays.

REFERENCES

Bogle, Alfred L. 1968. Evidence for the Hybrid Origin of Petasites warrenii and P. vitifolius. Rhodora 70: 533–551.
Cronquist, Arthur. 1978. Petasites. N. Am. Fl. II(10): 174–179.

766. Petasites hybridus   767. Petasites sagittatus   768. Petasites frigidus

KEY TO THE SPECIES

1. Flowers pink-purple, all without rays; plant locally established as an escape from cultivation . . . . .1. **P. hybridus**
1. Flowers creamy white, marginal pistillate ones often with rays; plants native
  2. Blades of basal leaves ± sagittate, pinnately veined, toothed but not lobed. . . . . .
  . . . . .2. **P. sagittatus**
  2. Blades of basal leaves ± orbicular and deeply palmately lobed . . . . .3. **P. frigidus**

1. **P. hybridus** (L.) Gaertner, Meyer, & Scherb.

Map 766. A native of Europe, locally well established in low ground at Marquette and in Eaton County.

Blades of the reniform to orbicular, closely toothed basal leaves are often as broad as 3.5 dm and are recorded as 1 m on plants in Eaton County.

2. **P. sagittatus** (Pursh) A. Gray   Plate 8-E

Map 767. Ranges from Labrador to Alaska, from the tundra south to the northern Great Lakes region. First discovered in Michigan by Henson in 1981. Swampy ground, ditches, swales between pine ridges, sedge fens.

Apparently hybridizes with *P. frigidus*, producing intermediate plants that may be called *P.* ×*vitifolius* Greene (as interpreted by Bogle). These are known from Alger and Schoolcraft counties.

3. **P. frigidus** (L.) Fries   Plate 8-F

Map 768. Low moist woods, coniferous and mixed, including cedar and tamarack swamps and some rocky sites; seems to bloom best along trails and after clearing.

This is our only common species, represented by var. *palmatus* (Aiton) Cronquist, long treated as *P. palmatus* (Aiton) A. Gray but later generally (except by Hultén) considered a distinct North American variety of the circumpolar arctic *P. frigidus*; typical *P. frigidus* of the tundra is smaller, with the basal leaves at most very shallowly lobed.

Large clones of leaves are often noted in the summer, but because of the early blooming this species is doubtless under-collected.

82. Erechtites

REFERENCE

Barkley, T. M., & Arthur Cronquist. 1978. Erechtites. N. Am. Fl. II(10): 139–142.

1. **E. hieraciifolia** (L.) DC.   Fig. 263   Fireweed

Map 769. This is one of the relatively few American plants of weedy habit to have become naturalized in parts of Europe. Often in ± moist ground and burned or cleared areas; shores, marshy places, floodplains, waste ground, fields.

The copious white pappus is as conspicuous as that of *Petasites* but comes late in the summer. At the base of the involucre are always some very slender, short, almost hair-like bractlets. Plants vary greatly in size and can be as tall as 2 m or as short as 6 cm when blooming.

83. **Cacalia**   Indian-plantain

REFERENCES

Athey, Joann T., & Richard W. Pippen. 1987. Pale Indian Plantain (Cacalia atriplicifolia, Asteraceae) a Little-known Plant in Michigan Part I — Observations on Growth Patterns/ Life History. Michigan Bot. 26: 3–15.
Athey, Joann T., & Richard W. Pippen. 1989. Pale Indian Plantain (Cacalia atriplicifolia, Asteraceae) a Little-known Plant in Michigan. Life History Part II. Seeds and Seed Dispersal. Michigan Bot. 28: 89–95.
Pippen, Richard W. 1978. Cacalia. N. Am. Fl. II(10): 151–159.

KEY TO THE SPECIES

1. Leaf blades entire or nearly so, unlobed, elliptical, with several prominent longitudinal veins; phyllaries prominently keeled . . . . .1. **C. plantaginea**
1. Leaf blades coarsely toothed or shallowly lobed, triangular-ovate, all but the upper ones cordate, with main veins ± palmate; phyllaries not keeled . . . . . . .2. **C. atriplicifolia**

1. **C. plantaginea** (Raf.) Shinners   Fig. 264
Map 770. Fens, prairies, sedge meadows, and calcareous shores. This calciphile ranges to the south and southwest from Lake Huron as far as eastern Texas. Bois Blanc Island is the northernmost known site in its entire range. Its main occurrence in Ontario is along the west shore of the Bruce Peninsula.

2. **C. atriplicifolia** L.   Fig. 265
Map 771. Dry (rarely moist) open or shaded ground, especially with oak, as on old dunes; prairies, creek banks, river floodplain forest.
  This is a tall, very glaucous, large-leaved, striking plant, apparently truly local, for it would be hard for collectors to overlook.

769. Erechtites hieraciifolia

770. Cacalia plantaginea

771. Cacalia atriplicifolia

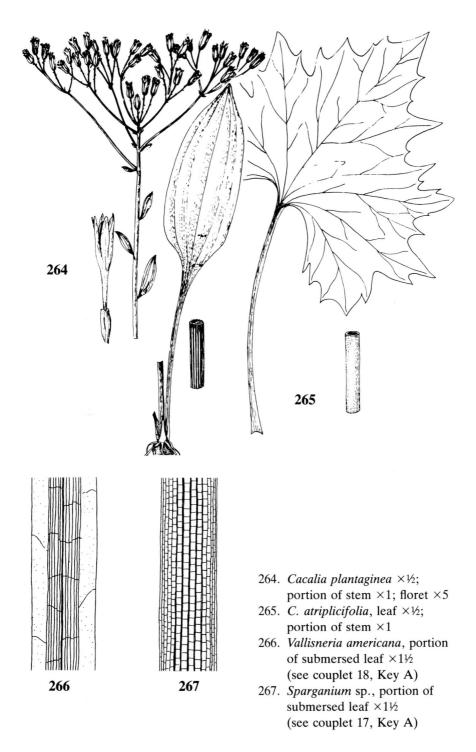

264

265

266   267

264. *Cacalia plantaginea* ×½;
    portion of stem ×1; floret ×5
265. *C. atriplicifolia*, leaf ×½;
    portion of stem ×1
266. *Vallisneria americana*, portion
    of submersed leaf ×1½
    (see couplet 18, Key A)
267. *Sparganium* sp., portion of
    submersed leaf ×1½
    (see couplet 17, Key A)

General Keys

GENERAL KEYS
(TO FAMILIES AND SPECIAL GROUPS)

INTRODUCTION

This series of keys is intended to lead one to the family, or sometimes directly to the genus or species, of an unknown plant. Therefore, the keys are based on the most reliable and easily observed characters for that purpose; only rarely (and incidentally) do they reflect the natural relationships and classification of the families. Diverse families (or their component genera and species) will usually key down at more than one place, so the keys often cannot be "worked backwards" to gain a concise description of the diagnostic features of a family. (Those features may, indeed, be too esoteric even to be mentioned for keying purposes. They can be learned by consulting a general text or more detailed manual.)

Furthermore, the keys accommodate a number of easy misinterpretations that a beginner might make (e.g., considering the 3 bracts of *Hepatica* to be sepals; or believing a series of perianth parts to be absent when in fact the calyx or corolla is present but either nearly obsolete or very early deciduous; or interpreting a *very* deeply lobed corolla as composed of separate petals). And they accommodate some ambiguities on the part of the plants themselves (e.g., if the ovary is only partly inferior, it should key under both "superior" and "inferior"; or if the perianth is only slightly bilateral it should key also under "regular"). However, plant taxa are composed of living organisms and hence are variable; there will always be some individuals so "far out" as to defy identification by any key.

These keys are designed to cover only seed-plants known from Michigan and may not work for other species in the same families or genera. Except for a few true aquatics, they will not work for ferns and their "allies" (horsetails, clubmosses, etc.), which never produce flowers or seeds. Ordinarily, flowers are required for use of these keys; the diversity of plants covered simply does not allow a key based on vegetative characters alone. Many aquatic plants are, however, seen but rarely in flower, and even then the flowers are often much reduced and obscure; but they *can* be identified vegetatively. Therefore, many of these plants are disposed of in Key A, the first of the individual keys.

The summary of characters at the beginning of each individual key is not always exhaustive; e.g., plants which will key as aquatics (Key A) or with neither green color nor developed leaves (Key C) may not run under later keys even though the later summary headings do not explicitly exclude such plants. Nevertheless, before forging ahead in any of the individual keys, the user should find it helpful to read the summary, as a reminder of how one arrived there and as an opportunity to confirm accurate reading thus far.

Unless explicitly stated to the contrary, references to "flowers" are to normal open flowers (showy or obscure) and not to cleistogamous flowers (fertilized and setting seed while remaining in "bud"). As always, keys should be read carefully. For instance, if the perianth (sepals and/or petals) is said to be regular (radially symmetrical), that does not necessarily mean that the reproductive parts (stamens and/or pistils) are also regular.

The introductory key below leads sometimes to a family but mostly to the 15 individual keys (A through O) which follow. When there are several genera in a family and only one of them will run at a given lead in a key, the name of the genus (or even of a species) is usually stated. Indicated in parentheses are the volume and page numbers in *Michigan Flora* where the stated family, genus, or species is treated. The organization of these keys owes very much to the general keys in the Gleason and Cronquist *Manual* (ed. 2), which themselves are based on decades of use and refinement. However, many differences exist as the result of omitting statements and taxa that do not apply to the Michigan flora; inclusion of some taxa not covered by the Gleason and Cronquist keys; the addition of some characters and even special groups that, especially in light of the smaller flora of this single state, will more easily serve to distinguish taxa; and the accommodation of some taxa that experience has shown would not run satisfactorily in the prior keys. The Gleason and Cronquist general keys can apparently trace their origin to Gleason's early (1918) *Plants of Michigan*, and it is a pleasure to present, with permission of the New York Botanical Garden, this more fully evolved and greatly altered descendent in *Michigan Flora*.

1. Plants strictly aquatic, the leaves or plant body *entirely* submersed or floating on the surface of the water (at most, the inflorescence and bracts, not leaves, held above the surface)[1] . . . . **KEY A** (p. 551)
1. Plants with at least some leaves (or stem if plant apparently leafless) above the water, or plants strictly terrestrial
  2. Plants woody (trees, shrubs, and woody vines), with erect, trailing, or viny aboveground stems living through the winter and continuing to grow the next season [hence, leaves may be evergreen or deciduous] . . . . **KEY B** (p. 558)
  2. Plants herbaceous, the perennial parts, if any, below or on the surface of the ground (to which the stems die back each year), not producing woody stems which survive the winter well above ground [hence, without evergreen leaves (although there may be basal winter-green leaves)]

3. Plant lacking green color (often wholly parasitic or saprophytic) and the leaves none or reduced to tiny scales)[2] .... **KEY C** (p. 565)
3. Plant with green color and the leaves usually developed (occasionally the stems photosynthetic, as in cacti)
    4. Inflorescences producing only small bulblets or tufts of little leaves (or modified floral parts), but no flowers or fruit[3] .... **KEY D** (p. 566)
    4. Inflorescences normal
        5. Perianth parts (2), 3, (4), or 6 (never 5) *and* leaves (or other green photosynthetic parts when leaves are absent or reduced) parallel-veined (the 3 or more main veins running from base of blade to apex and ± parallel—with or without minute cross-veins), entire, simple[4] .... **KEY E** (p. 567)
        5. Perianth parts various (often 5) but leaves netted-veined (or with only the midvein conspicuous), entire or toothed, simple or compound (the main veins, if more than 1, branching and ± reticulate)[5]

---

[1]In the field, plants are usually readily recognized as being aquatic if one is not misled by a rise in water level to assume that a temporarily inundated plant belongs to a normally aquatic species. In the herbarium, a proper label should record the habitat, but most of the larger true aquatics, even without complete data, can be recognized as such by the delicate structure of submersed stem and leaves, which are often extremely limp and flexible; hence when dry they still convey the impression of having been supported by water. The presence of algae, other aquatic organisms, or marl encrustations is also a handy clue to an underwater source. Rush-like plants (grasses, sedges, rushes) with erect stems extending above the water should *not* be sought under this part of the key unless they have definite limp aquatic foliage. Aquatics producing *only* floating leaves, their petioles extending to roots or rhizomes in the substrate, will key here as well as elsewhere in the General Keys on the basis of their floral characters. This key is slightly revised from its earlier incarnation in Michigan Bot. 6: 35–50 (1967).

[2]Included in Key C are a few leafless or apparently leafless herbaceous plants which are not parasitic or saprophytic but might be sought here because of decided yellowish (or at least non-green) color of at least some individuals as well as apparently leafless condition at flowering time.

[3]Key D is offered for certain plants that are obviously mature seed-plants (not ferns and their allies) but often produce sterile structures as described rather than flowers; specimens of these species, if they possess flowers, may also be run in the appropriate other portions of the General Keys.

[4]The plants in Key E are monocots (Liliopsida), which usually have floral parts in 3s (rarely 2s, never 5s) and parallel-veined leaves. The few monocots with apparently netted-veined leaves will also run in the alternative lead of this couplet, where all dicots (Magnoliopsida) belong, even those with apparently parallel-veined leaves (but not necessarily those included in Keys A, B, & C). All plants not covered in keys A–D and with flat ribbon-like, grass-like or sword-shaped, or ± terete linear-subulate leaves belong here, even if the veins are obscure, unless (1) perianth parts (sepals or petals or both) are present and 5, (2) the perianth is absent or chaffy with the stamens 1 or 4, or (3) the flowers are in an involucrate head (cf. Key F). All plants with rush-like stems *and* apparently leafless or the leaves bristle-like, 3-angled or involute, belong here.

[5]All plants not covered in keys A–D with compound, deeply lobed, or toothed leaves belong here, as do all others (even monocots) with distinctly netted-veined leaves, i.e. with some main veins diverging from the midvein rather than all running from base to apex of blade. Also included here are the 3 numbered exceptional "parallel-veined" options excluded (in footnote 4) under the alternative lead (to Key E), but generally recognizable as dicots by having perianth parts in 5's (commonly), a chaffy (or absent) perianth *and* 1 or 4 stamens, and/ or an involucrate head.

6. Inflorescence a dense "head" (either a true head or a spadix), consisting of few to many small sessile flowers on a common receptacle (not merely an elongate spike), subtended by 1 or more small or large bracts . . . . . . **KEY F** (p. 569)
6. Inflorescence not an involucrate head, or if head-like the individual flowers short-pediceled and/or the "head" not immediately subtended by 1 or more bracts
　7. Plants leafless but with thick, fleshy, green stem segments bearing strong spines . . . .**CACTACEAE** (ii: 603)
　7. Plants not both leafless and with spiny fleshy stems
　　8. Inflorescence of "false flowers" consisting of small cup-like structures (uniform in texture: not composed of separate parts like bracts or scales) each bearing 1–5 glands on its rim (sometimes with additional petaloid appendages) and including 2 or more stamens and 1 central stipitate 3-lobed pistil (which ripens into an exserted, 3-lobed capsule); sap milky[6] . . . .
　　　　. . . . .*Euphorbia* in **EUPHORBIACEAE** (ii: 520 & cf. figs. 254, 256, 260)
　　8. Inflorescence various, but not composed of such structures; pistil only rarely stipitate (and if so, not 3-lobed); sap various
　　　9. Anthers and stigma fused into a central structure obscuring the individual reproductive parts; ovaries 2, ripening into follicles, the seeds each with a tuft of hairs (except *Vinca*); sap milky
　　　　10. Inflorescence an umbel or (if plant a vine) a small cyme; filaments united into a tube around the pistil; anthers fused to stigma; carpels free (except for stigma); pollen grains cohering in pollinia (2 per anther) . . . . . .
　　　　　. . . . . **ASCLEPIADACEAE** (iii: 87)
　　　　10. Inflorescence a cyme (plant never a vine) or flowers solitary; filaments distinct; anthers ± adhering but not actually fused to stigma; carpels with styles (as well as stigmas) partly united; pollen grains separate . . .
　　　　　. . . . . **APOCYNACEAE** (iii: 84)
　　　9. Anthers and stigmas not fused to each other, of diverse but recognizable structure; ovaries, seeds, and sap various but not combined as above
　　　　11. Flowers unisexual, containing one or more stamens or pistils, but not both[7] . . . .**KEY G** (p. 570)
　　　　11. Flowers all or mostly perfect, containing both stamen(s) and pistil(s) (although these may not all be equally mature at the same time)
　　　　　12. Perianth none
　　　　　　13. Leaves deeply lobed or compound
　　　　　　　14. Pistil 1, with 2 stigmas; sap yellow . . . .
　　　　　　　　. . . . .*Macleaya* in **PAPAVERACEAE** (ii: 239)
　　　　　　　14. Pistils several, each with 1 stigma; sap watery . . . .
　　　　　　　　. . . . .**RANUNCULACEAE** (ii: 199)
　　　　　　13. Leaves unlobed (at most, blade cordate), entire
　　　　　　　15. Flowers axillary; stamen 1; leaves whorled, sessile, linear or nearly so . . . .**HIPPURIDACEAE** (ii: 640)
　　　　　　　15. Flowers in nodding spike-like racemes; stamens 6–8; leaves alternate, petiolate with cordate blade . . . .**SAURURACEAE** (ii: 29)

---

[6]Each stamen and pistil in *Euphorbia* is anatomically a very much reduced single flower, as evidenced by the stipe of the ovary and a joint or swelling on the stalk of a stamen, marking the junction of a pedicel and filament.

[7]If perfect flowers are also found on a plant, try also the alternative lead. Some plants with flowers in which either the stamens or the pistils mature distinctly earlier than the other are included here as well as under perfect flowers, but users should be cautious on this point.

12. Perianth present (but not always conspicuous)

KEY A

(Aquatic Plants with All Leaves Floating or Submersed)

1. Plants without distinct stem and leaves, free-floating at or below surface of water (except where stranded by drop in water level), the segments (internodes) small (up to 15 mm, but in most species much smaller), often remaining attached where budded from parent plant
  2. Plant body once to several times equally 2-lobed or 2-forked . . . .
. . . . Ricciaceae (a family of liverworts)
  2. Plant body not consistently dichotomous . . . .**LEMNACEAE** (i: 368)
1. Plants with distinct stem and/or leaves, usually anchored in substrate, mostly larger
  3. Plants with floating leaves present (blades, or at least their terminal portions, floating on the surface of the water, usually ± smooth and leathery in texture, especially compared with submersed leaves—or submersed leaves none)
    4. Blades of some or all floating leaves on a plant sagittate or deeply lobed at base, or compound, or peltate
      5. Floating blades compound (4-foliolate) . . . . .Marsileaceae (a family of ferns)
      5. Floating blades simple
        6. Floating blades (at least some of them) sagittate (the apex and lobes acute) [Caution: Plants with sagittate leaves extending *above* the surface of the water do not belong in this key.] . . . . *Sagittaria* in **ALISMATACEAE** (i: 101)
        6. Floating (and any other) blades circular to ± elliptical in outline, peltate or rounded at apex with deep sinus at base . . . .**NYMPHAEACEAE** (ii: 189)
    4. Blades of floating leaves all unlobed (at most subcordate at base), simple, the petiole marginal or (in ribbon-like leaves) none

---

[8]Several genera and families are included for convenience here that technically have both corolla and calyx but in which one of these series either is obsolete and hence usually overlooked or falls off very early in anthesis.

[9]A few plants are included here in which the ovary position is so obscure in a technically perigynous flower as to be easily misinterpreted as inferior.

7. Floating leaves small (less than 1 cm long), crowded in a terminal rosette; submersed leaves distinctly opposite; flowers solitary, axillary . . . .
. . . . **CALLITRICHACEAE** (ii: 528)
7. Floating leaves larger, not in a rosette; submersed leaves alternate, basal, or absent; flowers mostly in a terminal inflorescence
  8. Leaves narrow and ribbon-like, the blades many times as long as broad, without distinct petiole (though in some species a sheath surrounds the stem)
    9. Leaves ± rounded at tip (even if tapered), the floating portion smooth and shiny, somewhat yellow-green to bright green when fresh, occasionally keeled but midvein scarcely if at all more prominent than others; leaf not differentiated into blade and sheath, the submersed portion similar to the floating but more evidently with a fine closely checkered pattern (fig. 267); flowers and fruit in spherical heads . . . . **SPARGANIACEAE** (i: 71)
    9. Leaves sharply acute at tip, the floating portion rather dull, ± blue-green when fresh, with midrib; leaf including a sheath around stem and a membranous ligule at junction of sheath and blade; flowers and fruit in paniculate spikelets . . . . **GRAMINEAE** (i: 109 & 51)
  8. Leaves (at least floating ones) with ± elliptical blades and distinct petioles
    10. Leaves all basal; petals 3, white . . . . **ALISMATACEAE** (i: 100)
    10. Leaves alternate or opposite; petals 4–6, pink or dull and inconspicuous
      11. Venation netted; flowers bright pink, in dense ovoid to cylindrical spikes
      . . . . *Polygonum amphibium* in **POLYGONACEAE** (ii: 120)
      11. Venation parallel; flowers dull, in narrow cylindrical spikes . . . .
      . . . . **POTAMOGETONACEAE** (i: 75)
3. Plants without any floating leaves, entirely submersed (except sometimes for inflorescences and associated bracts)
  12. Leaves (or leaf-like structures) all basal and simple
    13. Leaves definitely flat, several times as broad as thick (widest about the middle or parallel-sided)
      14. Leaf blades not over twice as long as broad . . . .
      . . . . [juvenile] **NYMPHAEACEAE** (ii: 189)
      14. Leaf blades more than twice as long as broad
        15. Leaves stiff and erect or somewhat outcurved, less than 20 cm long
          16. Base of leaf somewhat sheathing, with a membranous ligule (as in a grass) at base of spreading blade . . . . *Pontederia* in **PONTEDERIACEAE** (i: 378)
          16. Base of leaf not sheathing and with no ligule . . . .
          . . . . *Sagittaria* in **ALISMATACEAE** (i: 101)
        15. Leaves limp, more than 20 times as long as broad
          17. Midvein not evident: all veins of essentially equal prominence, with the tiny cross-veins giving a checkered appearance to the leaf, which is thus uniformly marked with minute rectangular cells ca. 1–2 mm long or smaller (fig. 267) . . . . **SPARGANIACEAE** (i: 71)
          17. Midvein (and usually some additional longitudinal veins) evident, the veins not all of equal prominence, not dividing the leaf into minute rectangular cells
            18. Leaves with the central third (or more) of distinctly different pattern (more densely reticulate) than the two marginal zones (fig. 266); plants dioecious, the staminate flowers eventually liberated from a dense inflorescence submersed at base of plant, the pistillate solitary on a long ± spiraled stalk which reaches the surface of the water; plants without milky juice . . . .
            . . . . *Vallisneria* in **HYDROCHARITACEAE** (i: 105)
            18. Leaves ± uniform in venation, not 3-zoned; plants monoecious, with emergent inflorescence of white-petaled flowers (but these scarce on plants with

submersed tape-like leaves); plants often with milky juice . . . .
. . . . *Sagittaria* in **ALISMATACEAE** (i: 101)
13. Leaves (or similar sterile stems) filiform or terete or only slightly flattened (especially basally), elongate and limp to short and quill-like, less than twice as broad as thick
  19. Major erect structures solitary, spaced along a simple or branched delicate rhizome, consisting either of rather yellowish stems bearing minute alternate bumps as leaves or of filiform leaves mostly buried in the substrate and with a few minute bladder-like organs
    20. Leaves merely minute alternate bumps on stem; bladders not present; flowers sessile, inconspicuous, regular . . . .
    . . . . *Myriophyllum tenellum* in **HALORAGACEAE** (ii: 636)
    20. Leaves filiform, mostly buried in substrate (only the green tips, incurled when young, protruding); bladders (minute) usually present on the delicate branching rhizomes and buried leaf bases; flowers short-pediceled, showy (yellow or purple), bilaterally symmetrical . . . .
    . . . . *Utricularia* (couplet 2) in **LENTIBULARIACEAE** (iii: 259)
  19. Major erect structures solitary to densely tufted, consisting of filiform or quill-like leaves or stems, with neither alternate bumps or bladders
    21. Leaves very limp (retaining no stiffness when removed from water and hence irregularly sinuate, bent, or matted on herbarium specimens)—though a stiffer straight stem may also be present, mostly more than 20 cm long, ca. 0.2–1 mm in diameter
      22. Leaves (actually sterile stems) terete their entire length, not expanded basally nor sheathing each other, but each separate and closely surrounded at base for ca. (0.6) 1 cm or more by a very delicate membranous tubular sheath (this sometimes requiring careful dissection to distinguish); rhizome less than 2 mm in diameter; inflorescence (rare on plants otherwise entirely submersed) a single strictly terminal spikelet . . . .
      . . . . *Eleocharis* in **CYPERACEAE** (i: 336 & 54)
      22. Leaves slightly expanded basally for ca. (0.7) 2–10 cm, sheathing the next inner leaf at least dorsally (usually the sheath continued ventrally as an almost invisible membrane), with tiny ligule or pair of auricles at the summit; rhizome various; inflorescence a lateral spikelet or terminal cyme
        23. Leaf somewhat flattened or grooved ventrally for at least a few cm above the sheath (± crescent-shaped in section), with 1–5 longitudinal nerves evident, the tiny cross-veins connecting between nerves but not extending entirely across the leaf; sheath with a tiny ligule at summit; rhizome less than 2 mm in diameter; inflorescence a solitary lateral spikelet on a stiff wiry stem just above or near the surface of the water; flowers without petals and sepals; fruit an achene . . . .
        . . . . *Scirpus subterminalis* in **CYPERACEAE** (i: 354)
        23. Leaf terete above sheath, with no evident longitudinal veins, but numerous definite septa extending entirely across the blade (which shrinks between septa on drying); sheath with a minute pair of auricles at summit; rhizome ca. 2–5 mm thick; inflorescence an open cyme of many several-flowered heads on a very stout stem (several mm in diameter, over 50 cm tall); flowers with 6 tepals; fruit a capsule . . . *Juncus militaris* in **JUNCACEAE** (i: 389)
    21. Leaves usually firm (retaining stiffness when removed from water and hence straight or with an even curve in herbarium specimens), less (in most species much less) than 20 cm long, of various diameter
      24. Leaves filiform throughout, not broader basally nor sheathing each other, solitary (rarely) or in small tufts along a filiform whitish rhizome, each leaf

(actually a sterile stem) closely surrounded at its base for ca. 6 mm or more by a very delicate membranous tubular sheath (this sometimes requiring careful dissection to distinguish); inflorescence (rare on completely submersed plants) a single terminal spikelet . . . .
. . . . *Eleocharis acicularis* in **CYPERACEAE** (i: 340)

24. Leaves linear or tapered from base to apex, or if otherwise uniformly filiform then sheathing or expanded at base, without individual tubular sheaths as described above; inflorescence various

25. Leaf in section appearing composed of 2 hollow tubes, linear (± parallel-sided), broadly rounded at tip; flowers bilaterally symmetrical, in a few-flowered raceme . . . . *Lobelia dortmanna* in **CAMPANULACEAE** (iii: 331)

25. Leaf not (or rarely) of 2 hollow tubes, tapered and ± acute (or filiform); flowers regular and racemose, or solitary, or in a dense head or spike, or plant producing spores at base

26. Roots with prominent cross-septate appearance (checkered with fine transverse lines); inflorescence a small whitish or gray head (flowering in shallow or rarely deep water and on wet shores) . . . .
. . . .**ERIOCAULACEAE** (i: 374)

26. Roots not distinctly septate or cross-lined; inflorescence not as above

27. Leaves rather abruptly expanded at base to enclose sporangia, often dark green, composed of 4 hollow tubes (seen in cross-section), surrounding a hard corm-like stem; plant always submersed (unless stranded), non-flowering . . . . *Isoëtes* (a genus of pteridophytes)

27. Leaves gradually and slightly expanded or grooved on one side at a somewhat sheathing base but not composed of 4 tubes nor enclosing sporangia and no corm-like stem present; plants (except *Subularia*) not flowering when submersed but only on wet shores

28. Leaves somewhat flattened at least basally, widest at the base, gradually tapered to sharp apex; plants with buried rhizome or none

29. Plants connected by slender rhizomes (ca. 1 mm or narrower); sheathing basal portion of leaf (ca. 7 mm or more) with pale membranous borders abruptly terminating (or with minute auricles); leaves often 4 cm or more long, somewhat flattened laterally below, with 2–3 conspicuous hollow tubes evident in section; inflorescence (not on wholly submersed plants) a spreading cyme of solitary to paired 3-merous flowers . . . . *Juncus pelocarpus* in **JUNCACEAE** (i: 387)

29. Plants without rhizomes; sheath not abruptly auricled; leaves less than 4 cm long, somewhat flattened dorsoventrally (especially toward base), with numerous small hollow areas of irregular size; inflorescence (often submersed) a few- (often only 2-) flowered raceme of 4-merous flowers . . . . *Subularia aquatica* in **CRUCIFERAE** (ii: 307)

28. Leaves ± terete, scarcely or no wider at base than at middle, of ± uniform width at least to the middle (or even slightly thicker there before tapering to apex); plants with rhizomes or stolons at, near, or above surface of substrate

30. Plants with green stolons strongly arching above substrate; leaves filiform, ± uniform in diameter, ca. 0.5–1 mm thick, truncate at tip
. . . .*Ranunculus reptans* in **RANUNCULACEAE** (ii: 218)

30. Plants producing delicate horizontal white to green stolons at or near (above or below) surface of substrate (in addition to stouter short rhizome); leaves ca. 0.7–3 mm thick at middle, whence tapered to apex . . . . *Littorella* in **PLANTAGINACEAE** (iii: 271)

12. Leaves cauline, simple or compound (basal and dissected in one species)

31. Leaves compound, dissected, forked, or deeply lobed
  32. Leaves apparently in a basal rosette, few . . . .
  . . . . *Sium suave* in **UMBELLIFERAE** (ii: 663)
  32. Leaves definitely cauline: opposite, whorled, or alternate
    33. Leaves all or mostly opposite or whorled
      34. Leaves (or whorled branches) rolled inward at tip when young (circinate),
        bearing tiny stalked bladders; flowers emersed, bilaterally symmetrical,
        rose or purple . . . . *Utricularia purpurea* in **LENTIBULARIACEAE** (iii: 265)
      34. Leaves not inrolled at tip, without bladders; flowers various but not as
        above
        35. Petiole evident (5–15 mm long on well developed leaves), the blade fan-
          shaped and much dissected; flowers emergent, white . . . .
          . . . . *Cabomba* in **NYMPHAEACEAE** (ii: 190)
        35. Petiole absent or nearly so, the blade pectinate or much dissected or soon
          forking once or twice; flowers inconspicuous or yellow
          36. Leaves once or twice dichotomously forked, the segments usually
            sparsely toothed along one edge; flowers inconspicuous, axillary, sub-
            mersed . . . . **CERATOPHYLLACEAE** (ii: 186)
          36. Leaves not dichotomously forked, the segments entire; flowers emersed
            or (rarely) submersed
            37. Leaves pectinate (with straight central axis following midrib, once-
              pinnatifid or comb-like on both sides); flowers inconspicuous, in all but
              the rarest species emersed in terminal spike . . . .
              . . . . *Myriophyllum* in **HALORAGACEAE** (ii: 634)
            37. Leaves with no definite central axis, much dissected (fig. 189); flowers
              emersed in a showy yellow head (usually with at least one pair of merely
              serrate opposite leaves below it) . . . . *Megalodonta* in **COMPOSITAE** (iii: 379)
    33. Leaves definitely all alternate
      38. Leaves with a definite central axis (following midvein); flowers various
        39. Leaves pectinate (the lateral segments not again branched); flowers incon-
          spicuous, axillary; fruit a nutlet . . . . *Proserpinaca* in **HALORAGACEAE** (ii: 631)
        39. Leaves with lateral segments further narrowly divided; flowers with white
          corollas, in emersed raceme; fruit a silique . . . .
          . . . . *Armoracia aquatica* [*A. lacustris*] in **CRUCIFERAE** (ii: 282)
      38. Leaves with no definite central axis (except sometimes *after* initially forking
        at the stem); flowers emersed, with conspicuous corolla
        40. Petiole present (sometimes very short), ± adnate to a stipular sheath;
          plants without bladders; flowers regular, white or yellow, with numerous
          separate carpels forming achenes . . *Ranunculus* in **RANUNCULACEAE** (ii: 212)
        40. Petioles and stipular sheaths absent; plants with small stalked bladders on
          leaves (e.g., fig. 129) or on separate branches; flowers bilateral, yellow or
          purplish, with a single pistil producing a capsule . . . .
          . . . . *Utricularia* in **LENTIBULARIACEAE** (iii: 259)
31. Leaves simple, unlobed, usually entire (toothed in a few species)
  41. Leaves much reduced, ± scale-like, not over 7 mm long, never distinctly
    opposite or whorled
    42. Leaves minute, yellowish, merely widely spaced bumps or scales on stem . .
      . . . . *Myriophyllum tenellum* in **HALORAGACEAE** (ii: 636)
    42. Leaves up to 7 mm long, green or brownish, loosely overlapping . . . .
      . . . . (aquatic mosses & liverworts)
  41. Leaves much longer or distinctly opposite or whorled (or both conditions)
    43. Leaves alternate, with ligule-like stipules (these wholly adnate to leaves in
      Ruppia)

Page 555

44. Leaf blades ± filiform, terete or at least half as thick as broad, *and* the stipule adnate to leaf base for 10–30 mm or more, forming a sheath around the stem

   45. Sheath with no free stipular ligule at summit (the stipule wholly adnate to leaf blade, merely rounded at the summit); leaf blade terete; fruit stalked in an umbel-like arrangement on a ± spiraled and elongating limp peduncle . . . . *Ruppia* in **RUPPIACEAE** (i: 93)

   45. Sheath with a short ligule-like extension of free stipule at summit (the stipule only partly adnate); leaf blade often somewhat flattened; fruit sessile or subsessile in a spike with a straight ± stiff peduncle . . . .
. . . . *Potamogeton* in **POTAMOGETONACEAE** (i: 75)

44. Leaf blades definitely flattened and several times as broad as thick (even if narrow), *or* stipule little if at all adnate to blade (or both conditions)

   46. Blades flattened, ribbon-like (up to 5 or even 7.5 mm wide), with no definite midrib (no central vein more prominent than others except rarely toward base); flowers solitary, rare, cleistogamous in axils of submersed leaves or (these almost never on submersed plants) with 6 bright yellow tepals . . . . *Heteranthera* in **PONTEDERIACEAE** (i: 378)

   46. Blades flattened with a definite midrib or filiform; flowers in globose or cylindrical spikes, neither cleistogamous nor with showy yellow perianth
. . . . *Potamogeton* in **POTAMOGETONACEAE** (i: 75)

43. Leaves opposite or whorled, without stipules

   47. Leaves nearly filiform, not over 0.5 mm wide, very gradually tapered from base to apex but not abruptly expanded basally, perfectly smooth; plants perennial by slender rhizomes; flowers axillary, 1 staminate flower (a single stamen) and (1) 2–several carpels at a node; fruit slightly curved and minutely toothed on convex side . . . . **ZANNICHELLIACEAE** (i: 94)

   47. Leaves broader; or if filiform then abruptly expanded basally and with spiculate or toothed margins, the plants annual, and the fruit solitary and ellipsoid

      48. Leaves definitely whorled

         49. Whorled structures ("branches") cylindrical, elongate, usually stiff with calcium deposits; plants with distinctive musky odor . . . .
. . . . Characeae (a family of algae)

         49. Whorled structures (true leaves) flattened, short (not over 20 mm long) or elongate and very limp; plants without odor

            50. Leaves 6–12 (usually 9) in a whorl, not over 2.5 mm wide, ca. 12–25 times as long as wide; flowers perfect, apetalous, sessile in axils of emersed leaves or bracts . . . . **HIPPURIDACEAE** (ii: 640)

            50. Leaves mostly 3–4 (rarely 6) in a whorl, 0.8–5 mm wide, at most 10–13 times as long; flowers perfect or unisexual, but with petals

               51. Leaves mostly 3 (rarely 6) in a whorl, very thin (2 cell layers) and delicate; stem round (not angled), smooth; flowers unisexual, with 3 often pink petals, at least the pistillate long-stalked from entirely submersed stem . . . . *Elodea* in **HYDROCHARITACEAE** (i: 106)

               51. Leaves mostly 4 in a whorl, stiff and firm; stem 4-sided, often with minutely retrorse-scabrous angles; flowers perfect, with 3–4 white petals (usually not developed on wholly submersed plants) . . . .
. . . . *Galium* in **RUBIACEAE** (iii: 279)

      48. Leaves opposite (in some species, with bushy axillary tufts of leaves which may give a falsely whorled appearance)

         52. Largest leaves at least 1–4 cm long, with distinct petiole and expanded, entire blade

53. Leaf blades ± orbicular, with orange to black glandular dots especially beneath; flowers 5-merous with showy yellow petals and superior ovary............... *Lysimachia nummularia* in **PRIMULACEAE** (iii: 59)
53. Leaf blades ± diamond-shaped, without glandular dots; flowers 4-merous, inconspicuous, with inferior ovary.........................
..........................*Ludwigia palustris* in **ONAGRACEAE** (ii: 617)
52. Largest leaves smaller, or sessile, or toothed (or all of these)
54. Leaves large, 3–13 cm long, 5–20 mm wide
55. Leaves sessile and clasping, limp, at most obscurely and remotely toothed; flowers (rarely present on plants with all foliage submersed) in axillary racemes ....
..........*Veronica anagallis-aquatica* in **SCROPHULARIACEAE** (iii: 231)
55. Leaves sessile, clasping, tapered, or petioled, stiff, often regularly crenate or toothed; flowers various.........................(abnormally submersed individuals of various terrestrial or marsh species, chiefly Labiatae)
54. Leaves small (shorter or narrower than the above, or usually both)
56. Leaves linear and bidentate at apex when well submersed, often becoming obovate, ± weakly 3-nerved, and not necessarily bidentate toward summit of stem (or in floating rosettes); fruit solitary in axils, somewhat heart-shaped, of two 2-seeded segments .....**CALLITRICHACEAE** (ii: 528)
56. Leaves filiform to orbicular or tapered from base to apex, but essentially uniform on a plant and if linear not bidentate at apex; fruit various
57. Leaves at least 3 times as long as wide, broader at base than at middle; fruit absent or solitary in axils of leaves and ± ellipsoid
58. Leaves (especially lower ones) ± evenly tapered from broad base to minutely but bluntly bidentate apex, 3–10 times as long as wide, strictly entire, not subtending axillary tufts of leaves or fruit; plant often with a few scattered pale glandular dots on surface toward upper portion .... *Gratiola aurea* in **SCROPHULARIACEAE** (iii: 235)
58. Leaves filiform to linear-lanceolate, ± expanded at very base, acute or spiculate at apex, at least 6 times as long as wide, minutely spiculate to conspicuously toothed on margins, usually subtending axillary tufts of leaves and/or flowers or ellipsoid fruit; plant without glands on surface .... *Najas* in **NAJADACEAE** (i: 94)
57. Leaves less than 3 times as long as wide, often nearly round
59. Stems forming moss-like mats but the erect or ascending tips (above rooted nodes) less than 3 cm long; leaves with at most 1 weak nerve; stipules minute but usually evident with some leaves; flowers axillary, inconspicuous ....**ELATINACEAE** (ii: 582)
59. Stems greatly elongate (generally 10–30 cm); leaves more evidently veined; stipules none; flowers terminal, yellow (but usually absent on plants with all leaves submersed)
60. Stems stiffly erect; leaves weakly pinnately veined (with evident midvein), with reddish to blackish shiny dots or flecks (these often also on stem) .... *Lysimachia terrestris* in **PRIMULACEAE** (iii: 60)
60. Stems ± lax; leaves 3-nerved, without dark dots or flecks (though emersed leaves have translucent dots) ....
.......................*Hypericum boreale* in **GUTTIFERAE** (ii: 581)

KEY B

(Woody Plants)

1. Leaves scale-like (ca. 4 mm or less long and often appressed/imbricate) or needle-like (stiff and filiform to narrowly linear, less than 2.7 mm broad)
  2. Plant with leaves scale-like (or less than 3 mm long)
    3. Leaves alternate
      4. Flowers yellow; leaves ± densely pubescent ......*Hudsonia* in **CISTACEAE** (ii: 584)
      4. Flowers pink; leaves glabrous ....**TAMARICACEAE** (ii: 583)
    3. Leaves opposite or whorled
      5. Plants fragrant when crushed, producing small dry or berry-like cones but never flowers or true fruit ....**CUPRESSACEAE** (i: 66)
      5. Plants not fragrant, producing flowers and fruit in season
        6. Plant a parasite, less than 1.5 cm high, on branches of conifers, blooming in early May without showy perianth ....**VISCACEAE** (ii: 100)
        6. Plant a well developed terrestrial shrub, blooming in late summer with showy pink flowers .... *Calluna* in **ERICACEAE** (iii: 38)
  2. Plant with leaves needle-like or narrowly linear (over 3 mm long)
    7. Leaves opposite or whorled............................**CUPRESSACEAE** (i: 66)
    7. Leaves alternate or in clusters
      8. Plant a prostrate shrub; leaves less than 7 mm long; fruit a dark several-seeded berry (the seeds enclosed) ....**EMPETRACEAE** (ii: 530)
      8. Plant a bushy tree or shrub; leaves all or mostly 7 mm or longer; "fruit" a cone or a bright red 1-seeded fleshy structure (with seeds exposed)
        9. Seed solitary in a red, fleshy, cup-like aril; leaves flattened, with strongly decurrent base, persistent, appearing 2-ranked, all green on both sides (may be yellowish beneath) .... **TAXACEAE** (i: 58)
        9. Seeds borne on scales of a dry woody cone; leaves flattened or not (but if so, not decurrent, readily falling when dry, not 2-ranked, and/or with white lines) ....**PINACEAE** (i: 59)
1. Leaves with expanded (or dissected) blades, neither scale-like nor stiff and narrowly linear
  10. Leaves opposite or whorled or nearly so (evident from scars if not expanded at anthesis)
    11. Flowers appearing before leaves are expanded
    12. Perianth of both calyx and corolla
      13. Ovary superior; petals separate; flowers often unisexual ....**ACERACEAE** (ii: 545)
      13. Ovary inferior; petals united; flowers perfect .... **CAPRIFOLIACEAE** (iii: 295)
    12. Perianth of only one cycle of parts, or none
      14. Inflorescence an ament (catkin); bud scale 1 ....
        ....*Salix purpurea* in **SALICACEAE** (ii: 43)
      14. Inflorescence otherwise, of clustered or pediceled flowers but not an elongate ament; bud scales more than 1
        15. Flowers staminate or perfect
          16. Stamens 2 [–4] .... *Fraxinus* in **OLEACEAE** (iii: 65)
          16. Stamens 5 or more
            17. Calyx lobes 4; stamens 8; buds scurfy-pubescent ....
            ....*Shepherdia* in **ELAEAGNACEAE** (ii: 608)
            17. Calyx lobes 5; stamens ca. 5–10; buds not scurfy-pubescent ....
            ....**ACERACEAE** (ii: 545)
        15. Flowers pistillate
          18. Ovary with 2 divergent lobes ....**ACERACEAE** (ii: 545)

18. Ovary unlobed
    19. Flowers perigynous, the floral tube with a prominent disk at its summit; buds scurfy-pubescent; young fruit rotund ....
    ....*Shepherdia* in **ELAEAGNACEAE** (ii: 608)
    19. Flowers not perigynous, without a prominent disk; buds not scurfy; young fruit strongly flattened .... *Fraxinus* in **OLEACEAE** (iii: 65)
11. Flowers appearing after the leaves have expanded (i.e., leaves present)
  20. Leaves compound
    21. Plant a climbing or trailing vine
      22. Leaves pinnately compound; corolla well developed, showy, bilateral; flowers perfect; stamens 4; pistil 1........... *Campsis* in **BIGNONIACEAE** (iii: 252)
      22. Leaves all or mostly trifoliolate; corolla none (though calyx may be showy and regular); flowers unisexual; stamens and pistils numerous ....
      ....*Clematis* in **RANUNCULACEAE** (ii: 201)
    21. Plant erect, not a vine
      23. Petals none; fruit a samara (winged)
        24. Ovary 2-lobed; fruit united in pairs; stamens ca. 5–10; leaflets usually 3–5
        ....*Acer negundo* in **ACERACEAE** (ii: 545)
        24. Ovary not lobed; fruits not paired; stamens 2 [–4]; leaflets 5–11 ....
        ....*Fraxinus* in **OLEACEAE** (iii: 65)
      23. Petals well developed and conspicuous; fruit various but not a samara
        25. Petals united; leaves pinnately compound with 5 or more leaflets; fruit fleshy .... *Sambucus* in **CAPRIFOLIACEAE** (iii: 308)
        25. Petals separate; leaves trifoliolate or palmately compound; fruit dryish
          26. Leaflets 3; flowers in drooping panicles; petals and stamens 5; fruit an inflated, indehiscent, bladdery capsule ....**STAPHYLEACEAE** (ii: 545)
          26. Leaflets 5–7 (9); flowers in erect panicles; petals 4–5, stamens 6–8; fruit a leathery usually 1–2-seeded capsule ....**HIPPOCASTANACEAE** (ii: 551)
  20. Leaves simple
    27. Stamens more numerous than the petals or lobes of the corolla (or of the calyx if corolla is absent)—or flowers strictly pistillate
      28. Petals united ....**ERICACEAE** (iii: 34)
      28. Petals separate or none
        29. Stamens usually more than 10; corolla yellow
          30. Leaves well developed, punctate with translucent dots; styles 3–5 (but ± coherent); stamens very numerous (> 100) ....
          ....*Hypericum* in **GUTTIFERAE** (ii: 577)
          30. Leaves scale-like, not punctate; style 1; stamens fewer than 30.........
          ....*Hudsonia* in **CISTACEAE** (ii: 584)
        29. Stamens 10 or usually fewer—or flowers strictly pistillate; corolla various
          31. Leaves palmately lobed, toothed; fruit a samara, united in pairs ....
          ....**ACERACEAE** (ii: 545)
          31. Leaves unlobed, entire or toothed; fruit a berry or capsule, not paired
            32. Plant clearly a bushy shrub, with scurfy pubescence; flowers unisexual, inconspicuous, very early in spring; fruit a berry ....
            ....*Shepherdia* in **ELAEAGNACEAE** (ii: 608)
            32. Plant barely woody at base, glabrous to somewhat tomentose but not scurfy-pubescent; flowers perfect with showy pink (to white) petals; fruit a capsule
              33. Leaves evergreen, very shiny, toothed; stigma nearly sessile ....
              ....*Chimaphila* in **PYROLACEAE** (iii: 29)
              33. Leaves deciduous, ± dull, entire; stigma on an elongate style........
              ....**LYTHRACEAE** (ii: 609)

27. Stamens the same number as the lobes or petals of the corolla or fewer
  34. Petals separate
    35. Flowers in terminal inflorescences . . . .**CORNACEAE** (ii: 675)
    35. Flowers axillary
      36. Fruit a red to purple capsule, the seeds enclosed in a red or orange aril; styles unlobed; stamens alternating with the petals . . . .
        . . . .*Euonymus* in **CELASTRACEAE** (ii: 543)
      36. Fruit a dry inconspicuous capsule or fleshy and indehiscent, the seeds not arillate; styles often lobed; stamens opposite the petals . . . .
        . . . .**RHAMNACEAE** (ii: 556)
  34. Petals united
    37. Ovary inferior
      38. Flowers and fruits in dense spherical peduncled heads or paired at the ends of trailing branches; leaves entire, with broad stipules between the petiole bases . . . .**RUBIACEAE** (iii: 278)
      38. Flowers and fruits pediceled in small clusters or ± branched inflorescences; leaves entire or toothed, with stipules none or slender and partly adnate to petioles . . . .**CAPRIFOLIACEAE** (iii: 295)
    37. Ovary superior
      39. Corolla bilateral
        40. Plant a tree with large cordate ± whorled leaves. . . . .
          . . . .*Catalpa* in **BIGNONIACEAE** (iii: 252)
        40. Plant a mint with opposite leaves and scarcely woody at the base . . . . . .
          . . . . .some *Hyssopus* or *Thymus* in **LABIATAE** (iii: 138)
      39. Corolla regular
        41. Ovaries 2 (but styles and stigmas united); plant an evergreen creeper with blue flowers solitary in the axils . . . . *Vinca* in **APOCYNACEAE** (iii: 84)
        41. Ovary 1; plant erect, with flowers in inflorescences
          42. Stamens 2; leaves entire . . . .**OLEACEAE** (iii: 63)
          42. Stamens 4; leaves toothed . . . .**LOGANIACEAE** (iii: 71)
10. Leaves alternate
  43. Leaves deeply dissected into linear-filiform segments, aromatic . . . .
    . . . .*Artemisia* in **COMPOSITAE** (iii: 401)
  43. Leaves simple or compound but segments (leaflets), if any, broader, aromatic or not
    44. Plants dioecious
      45. Plant a climbing vine (or trailing in absence of support for tendrils or twining stem)
        46. Stems with tendrils
          47. Leaves entire; stems prickly (at least below); perianth of 6 tepals . . . .
            . . . .*Smilax* in **LILIACEAE** (i: 396)
          47. Leaves toothed; stems unarmed; perianth of 5 petals and 5 sometimes obsolete sepals . . . .**VITACEAE** (ii: 560)
        46. Stems without tendrils
          48. Leaves trifoliolate; plants climbing by adventitious roots (POISONOUS) .
            . . . .*Toxicodendron* in **ANACARDIACEAE** (ii: 532)
          48. Leaves simple or with more than 3 leaflets; plants climbing by twining stems
            49. Leaves pinnately veined, simple. . . . .*Celastrus* in **CELASTRACEAE** (ii: 541)
            49. Leaves palmately veined (or compound)
              50. Sepals and petals each 6; leaves ± peltate (petiole attached in from margin of the blade), at most somewhat lobed but not toothed . . . .
                . . . .**MENISPERMACEAE** (ii: 233)
              50. Sepals (often obsolete) and petals each 5; leaves with marginal petiole, toothed . . . .**VITACEAE** (ii: 560)

Page 560

45. Plant ± erect, not climbing
  51. Flowers in cylindrical to subglobose aments (catkins)
    52. Twigs and leaves with milky sap; calyx minute; leaves palmately or pinnately veined . . . . **MORACEAE** (ii: 88)
    52. Twigs and leaves with watery sap; calyx none; leaves pinnately veined
      53. Crushed foliage pungently aromatic; twigs resin-dotted; fruit an achene . . . . **MYRICACEAE** (ii: 54)
      53. Crushed foliage in most species not at all aromatic; twigs without resinous dots (may be generally shiny); fruit a capsule . . . . **SALICACEAE** (ii: 29)
  51. Flowers not in aments
    54. Leaves compound, present at anthesis
      55. Carpel 1, ripening into a large, flat pod; leaves even-pinnately compound or bipinnate . . . . *Gleditsia* in **LEGUMINOSAE** (ii: 472)
      55. Carpels more than 1, the fruit various (not a legume pod); leaves odd-pinnately compound or trifoliolate
        56. Leaves punctate with translucent oil glands; stems prickly in common species . . . . **RUTACEAE** (ii: 508)
        56. Leaves without translucent oil dots; stems unarmed
          57. Leaflets nearly or quite entire except for one or more coarse teeth near the base, each with a large gland beneath; fruit a samara; plant a weedy tree escaped from cultivation . . . . **SIMAROUBACEAE** (ii: 510)
          57. Leaflets ± regularly toothed (glandless) or entire; fruit a small smooth to glandular-pubescent drupe; plant a native shrub (SOME SPECIES POISONOUS) . . . . **ANACARDIACEAE** (ii: 531)
    54. Leaves simple, or unexpanded at anthesis
      58. Flowers pistillate
        59. Calyx and corolla both present (the former sometimes very small and inconspicuous)
          60. Inflorescences terminal, ± crowded . . . . *Rhus aromatica* in **ANACARDIACEAE** (ii: 536)
          60. Inflorescences axillary
            61. Stigma nearly or quite sessile . . . . **AQUIFOLIACEAE** (ii: 539)
            61. Stigma clearly on an elongate style
              62. Leaves entire; style simple . . . . **NYSSACEAE** (ii: 612)
              62. Leaves closely toothed; style divided above the middle . . . . **RHAMNACEAE** (ii: 556)
        59. Calyx and corolla not differentiated, or absent
          63. Perianth of 6 parts; bruised twigs and leaves spicy-aromatic . . . . **LAURACEAE** (ii: 236)
          63. Perianth of 3–5 parts or none; bruised parts not aromatic
            64. Style and stigma 1; leaves entire
              65. Stigma nearly or quite sessile . . . . *Nemopanthus* in **AQUIFOLIACEAE** (ii: 540)
              65. Stigma on an elongate style . . . . **NYSSACEAE** (ii: 612)
            64. Style divided above, stigmas 2–4; leaves toothed (at least on apical half)
              66. Leaf blades asymmetrical at base . . . . *Celtis* in **ULMACEAE** (ii: 87)
              66. Leaf blades ± symmetrical at base . . . . **RHAMNACEAE** (ii: 556)
      58. Flowers staminate
        67. Stamens more numerous than the sepals or the petals (or perianth none)
          68. Perianth none or minute; stamens usually 12; bruised parts not aromatic . . . . **NYSSACEAE** (ii: 612)

68. Perianth conspicuous (yellow, of 6 parts); stamens 9; bruised twigs and leaves spicy-aromatic ....**LAURACEAE** (ii: 236)
67. Stamens the same number as the perianth parts (same as the petals if both sepals and petals present)
  69. Inflorescences terminal, ± crowded ....
    ....*Rhus aromatica* in **ANACARDIACEAE** (ii: 536)
  69. Inflorescences axillary
    70. Stamens alternate with the sepals (opposite the petals if any) ....
      ....**RHAMNACEAE** (ii: 556)
    70. Stamens opposite the sepals (alternate with the petals if any)
      71. Leaf blades ± symmetrical at base, entire or toothed.............
        ....**AQUIFOLIACEAE** (ii: 539)
      71. Leaf blades asymmetrical at base, toothed (at least on apical half) ..
        ....*Celtis* in **ULMACEAE** (ii: 87)
44. Plants not dioecious, the flowers either perfect or unisexual (if the latter, then both sexes on the same individual)
 72. Flowers (at least the staminate) in aments or dense globose heads (always unisexual and individually inconspicuous)
  73. Staminate flowers in dense globose heads
    74. Leaves pinnately veined, with one midrib and lateral veins each ending in a single tooth; pistillate flowers in small clusters ....
      ....*Fagus* in **FAGACEAE** (ii: 84)
    74. Leaves palmately veined, with 3–5 principal veins; pistillate flowers in globose heads ....**PLATANACEAE** (ii: 336)
  73. Staminate flowers in cylindrical to ellipsoid aments
    75. Pistillate flowers solitary or in small clusters; styles 3 or leaves compound
      76. Leaves pinnately compound; styles 2 ....**JUGLANDACEAE** (ii: 55)
      76. Leaves simple (may be deeply lobed); styles usually 3 ....
        ....**FAGACEAE** (ii: 72)
    75. Pistillate flowers in aments, heads, or cone-like structures (in *Corylus*, the red styles protruding from an ament resembling a leaf bud); styles 2; leaves simple
      77. Twigs and leaves with milky sap; calyx minute, 4-parted; leaves palmately veined (or fruit in a large globose fleshy structure)...........
        ....**MORACEAE** (ii: 88)
      77. Twigs and leaves with watery sap; calyx usually none or 2-parted; leaves pinnately veined (fruit in aments or small clusters) ....
        ....**BETULACEAE** (ii: 61)
 72. Flowers not in aments or heads (often perfect and/or conspicuous)
  78. Perianth none or apparently of a single series of parts
    79. Leaves compound
      80. Plant a vine with twining stems ....**LARDIZABALACEAE** (ii: 233)
      80. Plant erect, not twining
        81. Stem prickly; plant at most shrub-like; fruit fleshy, in umbels.......
        ....*Aralia* in **ARALIACEAE** (ii: 644)
        81. Stem unarmed; plant a weedy tree; fruit a samara, in large panicles .
          ....**SIMAROUBACEAE** (ii: 510)
    79. Leaves simple (or absent at anthesis)
      82. Stamens more numerous than the segments or lobes (if any) of the perianth (or perianth none)
        83. Plant a twining vine; perianth bilateral ....
        ....*Aristolochia* in **ARISTOLOCHIACEAE** (ii: 102)

83. Plant erect, not a vine; perianth regular
  84. Perianth minute . . . . *Nyssa* in **NYSSACEAE** (ii: 612)
  84. Perianth well developed
    85. Stamens 8; perianth lobes 4 (or essentially none) . . . .
      . . . .**THYMELAEACEAE** (ii: 605)
    85. Stamens 5–7 or 9; perianth lobes or segments 5 or 6
      86. Leaves densely tomentose beneath, revolute; perianth segments
        5 . . . . *Ledum* in **ERICACEAE** (iii: 40)
      86. Leaves slightly if at all pubescent, not revolute; perianth segments
        6 . . . .**LAURACEAE** (ii: 236)
82. Stamens the same number as the lobes or segments of the perianth
  87. Styles 2
    88. Stems very prickly; fruit fleshy, in umbels . . . .
      . . . . *Oplopanax* in **ARALIACEAE** (ii: 641)
    88. Stems unarmed; fruit a samara or drupe, in small clusters . . . .
      . . . .**ULMACEAE** (ii: 85)
  87. Style 1 (may be branched above)
    89. Plant a vine, climbing or trailing by tendrils . . . .**VITACEAE** (ii: 560)
    89. Plant an erect shrub or tree
      90. Inflorescences terminal . . . .**CORNACEAE** (ii: 675)
      90. Inflorescences lateral
        91. Leaves beneath and branchlets silvery-scurfy; stamens 4 . . . .
          . . . . *Elaeagnus* in **ELAEAGNACEAE** (ii: 606)
        91. Leaves and branchlets glabrous or nearly so, not scurfy; stamens
          4–6
          92. Stamens alternating with the sepals . . . .**RHAMNACEAE** (ii: 556)
          92. Stamens opposite the sepals . . . .
            . . . . *Nemopanthus* in **AQUIFOLIACEAE** (ii: 540)
78. Perianth clearly differentiated into calyx and corolla
  93. Ovaries at least 3, distinct; stamens more than 10
    94. Sepals and petals each 5; leaves simple and toothed or compound; bud
      scales several . . . .**ROSACEAE** (ii: 336)
    94. Sepals 3, petals 6; leaves simple, entire; bud scales none or 2
      95. Leaves unlobed, estipulate; flowers dark purplish; fruit fleshy; buds
        naked . . . .**ANNONACEAE** (ii: 235)
      95. Leaves 4-lobed, stipulate (the stipules leaving a scar surrounding each
        node); flowers greenish yellow; fruit dry, in a cone-like structure; bud
        scales 2 . . . .**MAGNOLIACEAE** (ii: 235)
  93. Ovary 1; stamens 10 or fewer
    96. Corolla bilateral (or petal only 1); stamens 10 (usually with some of the
      filaments connate) . . . .**LEGUMINOSAE** (ii: 444)
    96. Corolla essentially regular; stamens various
      97. Petals united
        98. Stamens more numerous than the corolla lobes
          99. Leaves with stellate pubescence beneath and young twigs stellate;
            pith of older twigs chambered; fruit large (over 2 cm long), dry, 4-
            winged . . . .**STYRACACEAE** (iii: 63)
          99. Leaves and twigs glabrous or pubescent but not stellate; pith solid;
            fruit small, dry or fleshy but not winged . . . .**ERICACEAE** (iii: 34)
        98. Stamens the same number as the corolla lobes
          100. Stamens adnate to the corolla; plant a ± sprawling or climbing
            vine; fruit a berry . . . .**SOLANACEAE** (iii: 183)
          100. Stamens free from the corolla; plant an erect or trailing shrub (not
            climbing); fruit various

101. Stigma on a well developed style; fruit a capsule . . . .
. . . . **ERICACEAE** (iii: 34)
101. Stigma nearly sessile; fruit a red drupe . . . . **AQUIFOLIACEAE** (ii: 539)
97. Petals separate
  102. Ovary at least partly inferior
    103. Stamens at least twice as many as the petals
      104. Style 1 . . . . *Vaccinium* in **ERICACEAE** (iii: 42)
      104. Styles 2–5 . . . . **ROSACEAE** (ii: 336)
    103. Stamens the same number as the petals
      105. Petals 4
        106. Flowers white, in terminal cymes, blooming in early to mid
           summer; fruit fleshy; leaves entire . . . . **CORNACEAE** (ii: 675)
        106. Flowers yellow, in small axillary clusters, blooming in late fall;
           fruit a capsule; leaves with rounded teeth . . . .
           . . . . **HAMAMELIDACEAE** (ii: 334)
      105. Petals 5
        107. Flowers in racemes or small axillary clusters . . . .
           . . . . **GROSSULARIACEAE** (ii: 327)
        107. Flowers in numerous umbels . . . . **ARALIACEAE** (ii: 640)
  102. Ovary entirely superior
    108. Stamens more than twice as many as the petals; style 1
      109. Corolla yellow; fruit a capsule . . . . **CISTACEAE** (ii: 584)
      109. Corolla white to pink; fruit indehiscent
        110. Inflorescence apparently borne at the middle of a tongue-shaped
           bract; leaves palmately veined . . . . **TILIACEAE** (ii: 566)
        110. Inflorescence borne normally; leaves pinnately veined . . . .
           . . . . *Prunus* in **ROSACEAE** (ii: 365)
    108. Stamens twice as many as the petals or fewer
      111. Leaves compound
        112. Leaves even-pinnate or even-bipinnate; fruit a large woody le-
           gume (pod splitting on 2 sutures) . . . .
           . . . . **LEGUMINOSAE** (ii: 444, couplet 26)
        112. Leaves odd-pinnate, trifoliolate, or palmate
          113. Leaflets nearly or quite entire except for one or more coarse
             teeth near the base, each with a large gland beneath; fruit a
             samara . . . . **SIMAROUBACEAE** (ii: 511)
          113. Leaflets toothed or entire and without large glands; fruit vari-
             ous (a samara only in trifoliolate *Ptelea*)
            114. Inflorescences terminal
              115. Fruit a samara, in loose open cymes; leaves trifoliolate, punc-
                tate with translucent oil glands . . . . *Ptelea* in **RUTACEAE** (ii: 509)
              115. Fruit a glandular-pubescent drupe, in dense panicles; leaves
                pinnately compound, without translucent glands . . . .
                . . . . **ANACARDIACEAE** (ii: 531)
            114. Inflorescences lateral or axillary
              116. Leaflets strongly spiny-toothed; flowers yellow . . . .
              . . . . *Mahonia* in **BERBERIDACEAE** (ii: 231)
              116. Leaflets without spines; flowers greenish yellow
                117. Leaves palmately compound with mostly 5–7 sharply
                  toothed leaflets; plant a vine with tendrils; stamens oppo-
                  site the petals (i.e., alternate with the sepals) . . . .
                  . . . . *Parthenocissus* in **VITACEAE** (ii: 560)

117. Leaves trifoliolate or pinnately compound with entire or nearly entire leaflets (POISONOUS); plant a shrub or vine with adventitious roots (not tendrils); stamens alternating with the petals (i.e., opposite the sepals) . . . .*Toxicodendron* in **ANACARDIACEAE** (ii: 532)

111. Leaves simple

118. Styles 2, separate to the base; petals 4, yellow, linear . . . . **HAMAMELIDACEAE** (ii: 334)

118. Style 1 (may be lobed or cleft at summit); petals various

119. Stems spiny; flowers yellow, 6-merous . . . .*Berberis* in **BERBERIDACEAE** (ii: 230)

119. Stems unarmed; flowers white, pink, or greenish, 4–5-merous

120. Stamens more numerous than the petals; inflorescence an umbel or raceme; plant a low subshrub . . . . **PYROLACEAE** (iii: 26)

120. Stamens the same number as the petals; inflorescence various; plant a bushy shrub

121. Leaves evergreen, densely white- or brown-tomentose beneath, revolute . . . . *Ledum* in **ERICACEAE** (iii: 40)

121. Leaves deciduous, glabrous or nearly so, with flat margins

122. Stamens alternating with the sepals (i.e., opposite the petals); style 3-lobed . . . . **RHAMNACEAE** (ii: 556)

122. Stamens opposite the sepals (i.e., alternating with the petals); style nearly or quite obsolete . . .**AQUIFOLIACEAE** (ii: 539)

KEY C

(Herbaceous Plants Lacking Both Green Color and Developed Leaves at Flowering Time)

1. Plants not anchored in the ground, solely parasitic on and attached to stems of other plants at maturity

2. Stem up to 15 mm long, with minute opposite leaves (scale-like); flowers in May, unisexual (plants dioecious), the staminate with stamens adnate to calyx lobes, the pistillate with inferior ovary; parasites on conifers . . . . **VISCACEAE** (ii: 100)

2. Stem elongate, with minute alternate leaves; flowers in late summer, perfect, the stamens partly adnate to corolla and the ovary superior; parasites on flowering plants . . . . **CUSCUTACEAE** (iii: 100)

1. Plants clearly anchored in the ground, not attached to other above-ground plants

3. Stem buried in ground; flowers in late winter or earliest spring, crowded in a spadix with a nearly or partly buried hood-like brownish or mottled spathe (green leaves from rhizome appearing after flowering); stamens 4; plant with skunk-like odor . . . . *Symplocarpus* in **ARACEAE** (i: 367)

3. Stem or flower stalk above ground; flowers later, solitary or in a few- to many-flowered raceme or umbel; stamens various; plant with odor, if any, not skunk-like

4. Flowers completely 3-merous and regular, in an umbel on a naked peduncle arising from an underground, onion-smelling bulb . . . .*Allium tricoccum* in **LILIACEAE** (i: 410)

4. Flowers not completely 3-merous, regular or bilateral, not in an umbel, on aerial stems
  5. Scale-like leaves (and branches if any) opposite; flowers less than 5 mm long
    6. Stem thick and fleshy, appearing jointed, the flowers deeply embedded in it . . . . *Salicornia* in **CHENOPODIACEAE** (ii: 127)
    6. Stem normal, slender and wiry, the flowers not at all embedded in it
      7. Sepals and petals each 5, separate; stamens 5–10; styles 3 . . . . *Hypericum gentianoides* in **GUTTIFERAE** (ii: 579)
      7. Sepals and petals each 4 and each series connate basally; stamens 4; style 1 . . . . *Bartonia* in **GENTIANACEAE** (iii: 73)
  5. Scale-like leaves alternate (or apparently none); flowers of various size
    8. Petals 5, mostly united in a tube, the flower slightly to distinctly bilateral, not spurred; stamens 4 . . . . **OROBANCHACEAE** (iii: 254)
    8. Petals 3–5 but not united in a tubular corolla, the flower regular or strongly bilateral (sometimes spurred); stamens various
      9. Perianth strongly bilateral; stamens 1–2
        10. Sepals and petals 3, the lower petal a definite lip, the others little modified; ovary inferior; plants of various habitat but not aquatic; perianth of various color . . . . **ORCHIDACEAE** (i: 433)
        10. Sepals apparently 2 and petals 5, but corolla basically 2-lipped; ovary superior; plants of wet shores and bog pools, with perianth yellow or purple . . . . *Utricularia* in **LENTIBULARIACEAE** (iii: 259)
      9. Perianth regular; stamens 4–10
        11. Corolla less than 5.5 mm long; stamens 4
          12. Flowers sessile; plant of wet lakeshores, nearly or quite aquatic; stigmas 4, conspicuously exposed (corolla barely 2 mm long); fruit an indehiscent nutlet . . . . *Myriophyllum tenellum* in **HALORAGACEAE** (ii: 636)
          12. Flowers long-pediceled; plant of peaty habitats but not aquatic; stigma inconspicuous (corolla longer); fruit a capsule . . . . *Bartonia* in **GENTIANACEAE** (iii: 73)
        11. Corolla at least 5 mm long; stamens 8–10
          13. Stems usually 4–6 mm thick, from a thick ball of mycorrhizal roots; style straight; corolla bell- or flask-shaped; anthers opening by longitudinal slits . . . . **MONOTROPACEAE** (iii: 32)
          13. Stems and roots slender; style strongly declined; corolla of wide-spreading petals; anthers opening by terminal pores . . . . *Pyrola chlorantha* in **PYROLACEAE** (iii: 28)

KEY D

(Inflorescence Apparently Converted to Bulblets, Tufts of Leaves, etc.)

1. Leaves with flat, net-veined (or dissected) blades
  2. Leaves with narrow, sparsely toothed leaflets or further dissected; stem hollow; bulblets produced in the axils of broad-based acuminate bracts or leaves, not transversely segmented . . . . *Cicuta bulbifera* in **UMBELLIFERAE** (ii: 673)
  2. Leaves simple and entire; stem solid; bulblets otherwise

3. Bulblets unsegmented, in a terminal spike; well developed leaves mostly basal (or alternate) .... *Polygonum viviparum* in **POLYGONACEAE** (ii: 120)
3. Bulblets transversely segmented, in axils of normal leaves, which are cauline and opposite .... *Lysimachia terrestris* in **PRIMULACEAE** (iii: 60)
1. Leaves terete or slender and parallel-veined
  4. Bulblets in a ± spherical head or umbel; plants with odor of onion or garlic .....
    ....*Allium* in **LILIACEAE** (i: 410)
  4. Bulblets not in a distinct umbel or spherical head; plants without strong odor
    5. Leaves terete, septate (with hard cross-partitions, easily seen on dry specimens or felt by gently pinching a leaf and drawing it between the fingers).............
    ....*Juncus* in **JUNCACEAE** (i: 381)
    5. Leaves flat, neither terete nor septate
      6. Stem ± triangular and solid .... *Scirpus atrovirens* in **CYPERACEAE** (i: 357)
      6. Stem terete, with hollow internodes ....**GRAMINEAE** (i: 109)[†]

KEY E

(Monocots, with Parallel-Veined Leaves)

1. Plant a climbing or twining vine, in most species with tendrils; flowers unisexual; leaves net-veined
  2. Inflorescence an umbel; plants with tendrils; ovary superior; fruit a berry ....
    ....*Smilax* in **LILIACEAE** (i: 396)
  2. Inflorescence spicate to paniculate; plant without tendrils; ovary inferior; fruit a capsule ....**DIOSCOREACEAE** (i: 423)
1. Plant not a vine and without tendrils; flowers perfect or unisexual; leaves parallel- or net-veined
  3. Inflorescence a spadix, subtended by a spathe which may be broad and hood-like or elongate; leaves in some species compound or net-veined ....**ARACEAE** (i: 365)
  3. Inflorescence not a spadix (if flowers in a head, this with neither an elongate fleshy axis nor a conspicuous subtending spathe); leaves simple, rarely net-veined (in *Smilax ecirrata, Trillium,* and some Alismataceae)
    4. Perianth much reduced: absent, or composed solely of bristles (these small and stiff or elongate and cottony), or of chaffy or scale-like parts—never conspicuously petaloid
      5. Individual flowers subtended by 1 or 2 scales; leaves ± elongate, grass-like, usually with a sheath at the base surrounding the stem; inflorescence various but never a single terminal globose head on a scape; fruit a 1-seeded grain or nutlet (achene)
        6. Each fertile flower subtended by a *single* scale (others may be at base of spikelet); sheaths of leaves closed (margins connate); stems frequently triangular (but 4–several-angled or terete in many species), usually solid; leaves usually 3-ranked (especially in a species with terete hollow stem); stamens with filament attached to end of anther; fruit a definitely 2- or 3-sided (rarely nearly terete) nutlet....................................... **CYPERACEAE** (i: 244)

[†]Various plants might be sought here, including species of *Poa* and *Festuca* normally with some bulblets and others with deformed (diseased?) inflorescences.

6. Each flower subtended by 2 scales (almost opposite each other, one rarely absent); sheaths often open; stems ± terete (sometimes flattened), never triangular; leaves not clearly 3-ranked (basically 2-ranked); stamens with filament attached near middle of anther (or apparently so because of sagittate anthers); fruit usually a grain neither flattened (2-sided) nor triangular . . . .
. . . . **GRAMINEAE** (i: 109)
5. Individual flowers subtended by no scales or only by bristles, or with a *regular* perianth of chaffy scales (or tepals); leaves, inflorescence, and fruit various
  7. Inflorescence a single, very compact, almost spherical head (terminating an erect scape), less than 12 mm across
    8. Surface of head (tips of receptacular bracts) white-woolly; flowers chaffy, not concealed by involucral bracts; roots with abundant conspicuous transverse markings . . . . **ERIOCAULACEAE** (i: 374)
    8. Surface of head (bracts) glabrous; flowers yellow (properly running in alternative lead 4 below) or largely concealed by bracts; roots without transverse markings . . . . **XYRIDACEAE** (i: 372)
  7. Inflorescence not a single terminal head and/or exceeding 12 mm
    9. Inflorescence composed of separate staminate and pistillate portions, the former consisting of conspicuous stamens, sooner or later withering, leaving only the pistillate portion conspicuous
      10. Pistillate flowers in (1–) several globose heads; perianth of greenish sepals; leaves strongly keeled (3-angled in section) . . . . **SPARGANIACEAE** (i: 71)
      10. Pistillate flowers in an elongate densely flowered spike; perianth of white hairs; leaves flat-elliptical in section . . . . **TYPHACEAE** (i: 69)
    9. Inflorescence composed of perfect flowers, without conspicuously separate staminate and pistillate portions
      11. Flowers in a branched or umbellate inflorescence, solitary or, more often, clustered into small heads of 2 or more; fruit a 3- to many-seeded capsule
        . . . . **JUNCACEAE** (i: 379)
      11. Flowers in a single elongate spike or zigzag raceme; fruit indehiscent or a 1– 2-seeded follicle
        12. Spike (truly a spadix) apparently lateral; fruit of each flower indehiscent .
          . . . . *Acorus* in **ARACEAE** (i: 366)
        12. Spike or raceme terminal; fruit of each flower consisting of 3 or 6 1–2- seeded follicles . . . . **JUNCAGINACEAE** (i: 98)
4. Perianth at least in part of ± conspicuous white or colored petals
  13. Flowers bilaterally symmetrical
    14. Ovary inferior; fertile stamens 1 or 2, united with the pistil; flowers not blue (almost any other color) . . . . **ORCHIDACEAE** (i: 433)
    14. Ovary superior; fertile stamens 3 or 6, free; flowers blue (except albinos), at least in part
      15. Sepals colored like the petals; stamens 6, all fertile; flowers in a dense elongate inflorescence . . . . *Pontederia* in **PONTEDERIACEAE** (i: 378)
      15. Sepals greenish, unlike the petals; stamens 6, 3 with imperfect anthers; flowers few . . . . *Commelina* in **COMMELINACEAE** (i: 375)
  13. Flowers regular (radially symmetrical)
    16. Sepals and petals of quite different color and/or texture, the former green or brownish
      17. Leaves in a single whorl of 3 on the stem . . . . *Trillium* in **LILIACEAE** (i: 400)
      17. Leaves all basal or nearly so
        18. Petals yellow; flowers in a single compact head less than 12 mm across . . .
          . . . . **XYRIDACEAE** (i: 372)
        18. Petals blue, white, or pink; flowers in a more open or larger inflorescence

19. Pistils several in each flower, each developing into an achene; stamens 6–many; flowers unisexual or perfect; petals white or pinkish; leaves often broadly elliptical or sagittate, usually ± net-veined . **ALISMATACEAE** (i: 100)

19. Pistil 1 in each flower, developing into a capsule; stamens 6; flowers perfect; petals blue, purple, or rose (except in occasional albinos); leaves elongate, clearly parallel-veined . . *Tradescantia* in **COMMELINACEAE** (i: 376)

16. Sepals and petals both colored and petaloid, usually similar in shape (tepals) or the sepals (in *Iris*) of different size and shape

20. Ovary superior (or flowers unisexual)

21. Pistils 6, united only at the very base, ripening into follicles; stamens 9; flowers pink, in an umbel terminating a long scape . . . . . **BUTOMACEAE** (i: 105)

21. Pistil 1, sometimes the carpels slightly separate near the summit; stamens 3–6; flowers and inflorescence various

22. Stamens 3; tepals 6, yellow; plants creeping on wet shores . . . . *Heteranthera* in **PONTEDERIACEAE** (i: 378)

22. Stamens and tepals 4 or 6, the latter yellow or not; plants erect, of various habitats . . . . **LILIACEAE** (i: 392)

20. Ovary inferior (flowers perfect)

23. Stamens 3; leaves equitant . . . . **IRIDACEAE** (i: 424)

23. Stamens 6; leaves not equitant

24. Ovary clearly inferior, hairy . . . . **AMARYLLIDACEAE** (i: 423)

24. Ovary only half-inferior, part of it adnate to the perianth, glabrous (at most granular-roughened) . . . . **LILIACEAE** (i: 392)

KEY F

(Flowers in an Involucrate Head)

1. Flowers on a thick fleshy axis (inflorescence a spadix) subtended by a single large overtopping bract (spathe); perianth none or of 4 tepals . . . . **ARACEAE** (i: 365)

1. Flowers not in a spadix overtopped by a spathe; perianth various

2. Leaves parallel-veined, all basal, and less than 5 mm broad

3. Flowers yellow, mostly concealed by bracts; roots without transverse markings; surface of head (bracts) glabrous . . . . **XYRIDACEAE** (i: 372)

3. Flowers chaffy, not concealed by involucral bracts; roots with abundant conspicuous transverse markings; surface of head white-woolly (tips of receptacular bracts) . . . . **ERIOCAULACEAE** (i: 374)

2. Leaves net-veined or if parallel-veined then cauline and more than 5 mm broad

4. Ovary inferior

5. Leaves opposite (very rarely whorled), toothed or pinnatifid (entire in a very local waif, *Succisella*); corolla 4-lobed, lilac–purple (sometimes pale); stamens 4, separate . . . . **DIPSACACEAE** (iii: 321)

5. Leaves and corolla not combined as above (e.g., leaves alternate and/or entire or corolla 5-lobed and/or some other color)

6. Margins of cauline leaves and inflorescence bracts with stiff spines; corolla of separate petals; calyx present (no pappus); stamens separate . . . . *Eryngium* in **UMBELLIFERAE** (ii: 651)

6. Margins of cauline leaves and bracts various (spiny in a few species); corolla of united petals; calyx none (but a pappus of scales, awns, or bristles often present); stamens almost always fused in a ring around the style . . . .
. . . .**COMPOSITAE** (iii: 340)
4. Ovary superior
  7. Leaves alternate, compound; involucral bract 3-foliolate; flowers strongly bilateral, papilionaceous (as in other legumes) . . . . *Trifolium* in **LEGUMINOSAE** (ii: 451)
  7. Leaves opposite, simple; flowers often nearly or quite regular
    8. Plant with minty odor; ovary deeply 4-lobed, with 1 style; petals united . . . . .
    . . . . .some genera in **LABIATAE** (iii: 138)
    8. Plant without minty odor; ovary not lobed, with 2 styles; petals separate . . . .
    . . . . .some *Petrorhagia* and *Dianthus* in **CARYOPHYLLACEAE** (ii: 172 & 174)

KEY G

(Herbaceous Plants with Unisexual Flowers)

1. Leaves compound
  2. Leaves palmately compound (or 3-foliolate)
    3. Flowers in umbels
      4. Leaves cauline, in a single whorl . . . . *Panax* in **ARALIACEAE** (ii: 642)
      4. Leaves alternate and basal . . . . *Sanicula* in **UMBELLIFERAE** (ii: 654)
    3. Flowers in spikes or panicles
      5. Margins of leaflets entire; flowers at the base of a prolonged fleshy spadix subtended by a single large bract (spathe) . . . . *Arisaema* in **ARACEAE** (i: 366)
      5. Margins of leaflets toothed; flowers on normal herbaceous (but not fleshy) pedicels or axes
        6. Leaves all opposite; plant a vine; perianth showy . . . .
        . . . .*Clematis* in **RANUNCULACEAE** (ii: 201)
        6. Leaves alternate on upper part of stem; plant erect; perianth minute and inconspicuous . . . . *Cannabis* in **CANNABACEAE** (ii: 92)
  2. Leaves pinnately compound or more than once compound
    7. Plant a vine, climbing by tendrils; perianth 4-merous; fruit a large 3-lobed bladdery capsule . . . .**SAPINDACEAE** (ii: 554)
    7. Plant erect, without tendrils; perianth 4- or 5-merous; fruit not lobed, an achene, berry, or follicle
      8. Flowers in tight ovoid heads or umbels
        9. Leaves once pinnately compound; flowers in tight heads; perianth 4-merous . .
        . . . .*Sanguisorba* in **ROSACEAE** (ii: 440)
        9. Leaves 2–3 times compound; flowers pediceled, in umbels; perianth 5-merous . . . . *Aralia* in **ARALIACEAE** (ii: 644)
      8. Flowers in panicles
        10. Leaflets with only 3–9 (12) mostly rounded or obtuse teeth; perianth absent or of one series of parts; fruit an achene . .*Thalictrum* in **RANUNCULACEAE** (ii: 204)
        10. Leaflets sharply and closely toothed throughout; perianth with both calyx and corolla (but these small); fruit a follicle . . . . *Aruncus* in **ROSACEAE** (ii: 433)
1. Leaves simple

11. Plant with leaves all basal
  12. Flowers in dense spikes (or 1–3 at base in *Littorella*)...**PLANTAGINACEAE** (iii: 271)
  12. Flowers pediceled in panicles................ *Rumex* in **POLYGONACEAE** (ii: 105)
11. Plant with leaves all or mostly cauline
  13. Leaves peltate or pubescent with forked/stellate hairs ...**EUPHORBIACEAE** (ii: 516)
  13. Leaves neither peltate nor with forked/stellate hairs
    14. Leaves opposite or whorled
      15. Flowers solitary in axils of leaves; perianth none; stamen 1
        16. Leaves whorled; ovary and fruit terete; style 1 ....**HIPPURIDACEAE** (ii: 640)
        16. Leaves opposite; ovary and fruit ± flattened; styles 2 ....
          ....**CALLITRICHACEAE** (ii: 528)
      15. Flowers in axillary or terminal inflorescences
        17. Leaves hastate, otherwise unlobed but entire to coarsely or irregularly
          toothed; pistillate flowers and fruit mostly concealed by a pair of bracts with
          margins ± united at base.............*Atriplex* in **CHENOPODIACEAE** (ii: 129)
        17. Leaves not hastate, in some species deeply lobed, in some closely toothed;
          pistillate flowers without 2 basal bracts
          18. Inflorescence terminal; corolla white or colored
            19. Cauline leaves deeply pinnately lobed; style 1; stamens 3–4 ....
              ....*Valeriana* in **VALERIANACEAE** (iii: 318)
            19. Cauline leaves unlobed; styles 3–7; stamens 10 ....
              ....*Silene* in **CARYOPHYLLACEAE** (ii: 178)
          18. Inflorescence axillary; corolla none or of reduced scales
            20. Leaves entire; inflorescence (spike) shorter than the peduncle ....
              ....*Plantago arenaria* in **PLANTAGINACEAE** (iii: 273)
            20. Leaves toothed; inflorescence longer than the peduncle
              21. Plant a vine; leaves deeply 3–7-lobed....*Humulus* in **CANNABACEAE** (ii: 92)
              21. Plant erect, not a vine; leaves unlobed ....**URTICACEAE** (ii: 93)
    14. Leaves alternate (at least at upper nodes)
      22. Flowers with 6 petaloid tepals and 6 stamens or 3 carpels (dioecious);
        inflorescences on long peduncles from the nodes (not terminal); leaves with
        several prominent longitudinal veins (including midrib)
        23. Plant a vine with twining stems (no tendrils); inflorescence a spike, raceme,
          or panicle; ovary inferior, ripening into a winged capsule ....
          ....**DIOSCOREACEAE** (i: 423)
        23. Plant erect or a vine with tendrils; inflorescence an umbel; ovary superior,
          ripening into a berry......................... *Smilax* in **LILIACEAE** (i: 396)
      22. Flowers either with other numbers of tepals, stamens, and carpels or the
        inflorescence terminal (on main stem or branches); leaves various but without
        several prominent long veins
        24. Perianth with both calyx and corolla (sometimes very inconspicuous); plants
          climbing or trailing, with tendrils; stamens 3; pistil 1, with inferior ovary ...
          ....**CUCURBITACEAE** (iii: 325)
        24. Perianth absent or of 1 series of parts (tepals); plants erect or prostrate,
          without tendrils; stamens and pistils various
          25. Flowers very small, in axillary clusters [plants monoecious; look for pis-
            tillate ones for keying]
            26. Style 1; stamens 4 or 5............................**URTICACEAE** (ii: 93)
            26. Styles (or sessile stigmas) 2–3; stamens various
              27. Styles 3, branched; bracts in inflorescence well developed and at least 5–
                10-lobed...................... *Acalypha* in **EUPHORBIACEAE** (ii: 519)
              27. Styles 2–3, unbranched; bracts in inflorescence unlobed (may be
                toothed)

Page 571

28. Tepals and bracts acute, narrow, scarious; stamens 3 ....
...........................*Amaranthus* in **AMARANTHACEAE** (ii: 143)
28. Tepals and bracts (the latter only beneath pistillate flowers, broad and usually tuberculate and toothed with margins partly fused) obtuse to acute but herbaceous in texture (or even indurated in one species); stamens 5 .... *Atriplex* in **CHENOPODIACEAE** (ii: 129)
25. Flowers small or not, in chiefly terminal inflorescences (spikes, panicles, or racemes on main stem and/or branches)
29. Tepals petaloid (white to pink); flowers in large open racemes ....
....**PHYTOLACCACEAE** (ii: 150)
29. Tepals none or minute, not petaloid (often scarious); inflorescence various
30. Flowers consistently 3-merous (tepals 6, stamens 6, carpels 3); stipules united into a sheath (ocrea) surrounding the stem above each node ....
...................................*Rumex* in **POLYGONACEAE** (ii: 105)
30. Flowers not consistently 3-merous (tepals 5 or fewer, stamens usually 5, styles often 2); stipules none
31. Tepals acute, narrow, scarious; bracts of similar appearance throughout the inflorescence .... *Amaranthus* in **AMARANTHACEAE** (ii: 143)
31. Tepals various (usually not as above) but bracts herbaceous or indurated, not scarious ....**CHENOPODIACEAE** (ii: 125)

KEY H

(Herbaceous Dicots with Perfect Flowers, Perianth in One Series, & Inferior Ovary)

1. Stamens more numerous than the 1–4 perianth lobes or parts
2. Perianth with 1–3 (rarely 4) lobes; stamens 6 or 12 ...**ARISTOLOCHIACEAE** (ii: 102)
2. Perianth 4-parted; stamens 8 or numerous
3. Leaves pinnately compound; stamens numerous ..*Sanguisorba* in **ROSACEAE** (ii: 440)
3. Leaves simple; stamens normally 8 ....*Chrysosplenium* in **SAXIFRAGACEAE** (ii: 320)
1. Stamens the same number as or fewer than the perianth lobes or parts, or perianth 5-merous (or both conditions)
4. Leaves all or mostly opposite or whorled
5. Inflorescence a dense terminal cluster of flowers (sessile or nearly so)
6. Leaves apparently whorled; bracts below the inflorescence large and white....
...................................*Cornus canadensis* in **CORNACEAE** (ii: 677)
6. Leaves clearly opposite; bracts below the inflorescence greenish ....
....**DIPSACACEAE** (iii: 321)
5. Inflorescence of solitary, axillary, or clearly pediceled flowers
7. Leaves compound, in a single whorl .... *Panax* in **ARALIACEAE** (2: 642)
7. Leaves simple or deeply lobed, opposite or in several whorls
8. Leaves in whorls .... *Galium* in **RUBIACEAE** (iii: 279)
8. Leaves opposite
9. Plant low and densely matted, with linear leaves; perianth 5-merous ....
..............................*Scleranthus* in **CARYOPHYLLACEAE** (ii: 160)

9. Plant prostrate or erect, but with broader leaves; perianth 5- or 4-merous
  10. Flowers in rather dense terminal inflorescences (at ends of stem and branches); stamens 3 (occasionally 4) . . . . **VALERIANACEAE** (iii: 318)
  10. Flowers 1–few in axils or solitary at ends of branches; stamens 4
    11. Styles 2; flowers solitary at ends of branches; plant flowering in May . . . .
      . . . . *Chrysosplenium* in **SAXIFRAGACEAE** (ii: 320)
    11. Style 1; flowers sessile, axillary; plant flowering in summer . . . .
      . . . . *Ludwigia palustris* in **ONAGRACEAE** (ii: 617)
4. Leaves alternate or basal
  12. Tepals and stamens each 3 or 4
    13. Stamens and tepals 3 . . . . *Proserpinaca* in **HALORAGACEAE** (ii: 631)
    13. Stamens and tepals 4
      14. Leaves pinnately compound, with conspicuous stipules . . . .
        . . . . *Sanguisorba* in **ROSACEAE** (ii: 440)
      14. Leaves simple, without stipules . . . . [go to couplet 10]
  12. Tepals and stamens each 5
    15. Leaves entire, simple and unlobed; flowers in cymes or few-flowered cymules; style 1 . . . . **SANTALACEAE** (ii: 97)
    15. Leaves (at least the cauline ones) toothed or crenulate (entire only in *Bupleurum*, with upper leaves perfoliate) and usually deeply lobed or compound; flowers in umbels; styles 2 or 5
      16. Styles 5; fruit fleshy, berry-like . . . . *Aralia* in **ARALIACEAE** (ii: 644)
      16. Styles 2; fruit dry, splitting into 2 achene-like indehiscent parts (mericarps) . .
        . . . . **UMBELLIFERAE** (ii: 645)

KEY I

(Herbaceous Dicots with Perfect Flowers, Perianth in One Series, & Superior Ovary)

1. Ovaries more than 1 in each flower, the carpels separate at least above the middle of the ovaries
  2. Stipules conspicuous; leaves pinnately compound . . *Sanguisorba* in **ROSACEAE** (ii: 440)
  2. Stipules none or leaves simple
    3. Ovaries united for most of lower half; leaves simple, unlobed . . . .
      . . . . **PENTHORACEAE** (ii: 318)
    3. Ovaries distinct; leaves of most species lobed or compound . . . .
      . . . . **RANUNCULACEAE** (ii: 199)
1. Ovary 1 in each flower (bearing 1 or more styles), the carpels united at least below the styles
  4. Leaves bipinnately compound, prickly; fruit a legume, also prickly . . . .
    . . . . *Schrankia* in **LEGUMINOSAE** (ii: 477)
  4. Leaves simple or compound (but not bipinnate and not prickly); fruit not a legume
    5. Stamens more than twice as many as the perianth lobes or parts
      6. Leaves tubular, open at apex and hence pitcher-like . . . .
        . . . . **SARRACENIACEAE** (ii: 310)

6. Leaves flat, of normal structure, simple or compound but not hollow
  7. Perianth small and inconspicuous (stamens more showy); leaves compound with definite flat broad leaflets . . . . **RANUNCULACEAE** (ii: 199)
  7. Perianth well developed, showy; leaves simple or dissected into very narrowly linear segments
    8. Leaf blades entire, unlobed except for deeply cordate base; plants aquatic . .
    . . . . *Nuphar* in **NYMPHAEACEAE** (ii: 195)
    8. Leaf blades deeply lobed or dissected; plants terrestrial
      9. Perianth parts 5; leaves pinnately dissected; sap watery . . . .
      . . . . *Nigella* in **RANUNCULACEAE** (ii: 224)
      9. Perianth parts 4 or 8; leaves ternately dissected (with watery sap) or otherwise toothed, spiny-margined, or lobed (with milky or colored sap) . . . .
      . . . . **PAPAVERACEAE** (ii: 237)
5. Stamens only twice as many as the perianth lobes or parts, or fewer
  10. Style 1 or none (stigmas may be 2 or more)
    11. Stamens more numerous than the perianth divisions
      12. Flowers bilateral; perianth colorful (white, yellow, or pink) . . . .
      . . . . **FUMARIACEAE** (ii: 245)
      12. Flowers regular; perianth dull, greenish
        13. Leaves opposite; flowers mostly axillary . . *Ammannia* in **LYTHRACEAE** (ii: 610)
        13. Leaves alternate or basal; flowers mostly terminal . . . . . . **CRUCIFERAE** (ii: 251)
    11. Stamens the same number as or fewer than the perianth lobes or parts
      14. Leaves alternate or basal
        15. Perianth parts (and stamens) 6, 8, or 9 . . . . **BERBERIDACEAE** (ii: 230)
        15. Perianth parts (and usually stamens) 4
          16. Leaves simple, entire . . . . *Parietaria* in **URTICACEAE** (ii: 94)
          16. Leaves pinnately compound with toothed leaflets . . . .
          . . . . *Sanguisorba* in **ROSACEAE** (ii: 440)
      14. Leaves opposite
        17. Flowers solitary or few in axils of leaves, sessile or nearly so
          18. Perianth with 4 teeth or shallow lobes alternating with 4 appendages; style 1, with capitate stigma . . . . **LYTHRACEAE** (ii: 609)
          18. Perianth with 5 divisions; style none (stigmas 2, sessile) . . . .
          . . . . *Herniaria* in **CARYOPHYLLACEAE** (ii: 157)
        17. Flowers in terminal inforescences (on stems and branches)
          19. Perianth showy, pink to purple; inflorescences each subtended by a conspicuous petaloid or papery 5-lobed involucre which enlarges as fruit matures . . . . **NYCTAGINACEAE** (ii: 149)
          19. Perianth reduced, inconspicuous, whitish or scarious; inflorescences subtended at most by very small bracts
            20. Flowers in cymes; perianth glabrous; style none (sessile stigmas 2); filaments separate . . . . *Herniaria* in **CARYOPHYLLACEAE** (ii: 157)
            20. Flowers in dense heads or spikes; perianth woolly; style 1; filaments united into a tube (simulating a corolla except for its bearing anthers!) .
            . . . . *Froelichia* in **AMARANTHACEAE** (ii: 142)
  10. Styles 2 or more
    21. Flowers embedded in a succulent segmented stem; leaves reduced to tiny opposite scales . . . . *Salicornia* in **CHENOPODIACEAE** (ii: 127)
    21. Flowers not embedded in a succulent stem; leaves not scale-like
      22. Leaves opposite or whorled
        23. Margins of leaves crenate; stamens normally 8 . . . .
        . . . . *Chrysosplenium* in **SAXIFRAGACEAE** (ii: 320)
        23. Margins of leaves entire

24. Leaves opposite; styles 2 . . . .**CARYOPHYLLACEAE** (ii: 155)
24. Leaves whorled; styles 3 . . . .**MOLLUGINACEAE** (ii: 151)
22. Leaves alternate
25. Stamens and styles each mostly 10; flowers in distinct simple racemes . . . . .
. . . .**PHYTOLACCACEAE** (ii: 150)
25. Stamens and (especially) styles fewer than 10; flowers not in distinct racemes
26. Plant with a ± membranous stipular sheath (ocrea) surrounding the stem above each node . . . .**POLYGONACEAE** (ii: 104)
26. Plant lacking stipules of any kind . . . .**CHENOPODIACEAE** (ii: 125)

KEY J

(Herbaceous Dicots with Perfect Flowers, Perianth in Two Series, & Ovaries Two or More in Each Flower)

1. Style and/or stigmas united (i.e., 1 in each flower, but style may be branched)
2. Ovaries 2; corolla regular, of united petals; stamens 5; sap in most species milky
3. Inflorescence an umbel or (if plant a vine) a small cyme; filaments united into a tube around the pistil; anthers fused to stigma; carpels free (except for stigma); pollen cohering in pollinia (2 per anther) . . . .**ASCLEPIADACEAE** (iii: 87)
3. Inflorescence a cyme (plant never a vine) or solitary; filaments distinct; anthers ± adhering but not actually fused to stigma; carpels with styles (as well as stigmas) partly united; pollen grains separate . . . .**APOCYNACEAE** (iii: 84)
2. Ovaries 4 or more; corolla regular or bilateral, of united or separate petals; stamens 2, 4, 5, or numerous; sap not milky
4. Petals separate; ovaries apparently 5 or more; stamens numerous, their filaments connate, at least for much of their length, into a tube around the style; leaves palmately veined (may be deeply lobed) . . . .**MALVACEAE** (ii: 567)
4. Petals united; ovaries apparently 4; stamens 2, 4, or 5, their filaments not connate (but ± adnate to corolla); leaves mostly pinnately veined
5. Leaves alternate (except at lower nodes in the rare *Plagiobothrys*); stamens 5; corolla regular (bilateral only in the very bristly *Echium*); stems not angled (rarely winged) and foliage not aromatic . . . .**BORAGINACEAE** (iii: 115)
5. Leaves opposite; stamens 2 or 4; corolla bilateral or in a few genera essentially regular; stems usually 4-angled ("square") and foliage often aromatic when bruised ("minty" or citrus-like) . . . .**LABIATAE** (iii: 138)
1. Style and stigmas separate (1 on each ovary—or scarcely developed)
6. Perianth bilateral
7. Leaves deeply cleft; stamens numerous; perianth with spurs; fruit a follicle (3 per flower) . . . .*Delphinium* (cf. also *Aconitum*) in **RANUNCULACEAE** (ii: 202–203)
7. Leaves shallowly lobed; stamens 5; perianth without spurs; fruit a capsule . . . . .
. . . .*Heuchera* in **SAXIFRAGACEAE** (ii: 321)
6. Perianth regular
8. Sepals (or sepal-like bracts) 3

9. Plant aquatic, with peltate (often floating) round or shield-shaped alternate floating leaf blades or palmately dissected opposite submersed leaves . . . .
. . . .**NYMPHAEACEAE** (ii: 189)
9. Plant terrestrial, with leaves neither peltate nor palmately dissected
  10. Petals and usually carpels 3; stamens 3 or 6; leaves cauline, deeply and narrowly pinnate-lobed . . . .**LIMNANTHACEAE** (ii: 531)
  10. Petals 5 or more, carpels and stamens numerous; leaves basal, with 3 (–7) broad lobes . . . . *Hepatica* in **RANUNCULACEAE** (ii: 209)
8. Sepals 4 or more
  11. Leaves peltate, the blades round, often floating, and mostly more than 15 cm broad; flowers mostly over 10 cm broad, with carpels embedded in a top-shaped receptacle . . . . *Nelumbo* in **NYMPHAEACEAE** (ii: 192)
  11. Leaves not peltate, the blades (of leaflets, if leaves compound) neither round nor as broad as 15 cm; flowers smaller and carpels not embedded in receptacle
    12. Sepals separate to the base; stamens and petals individually falling from the receptacle after anthesis . . . .**RANUNCULACEAE** (ii: 199)
    12. Sepals, petals, and stamens united to form a saucer- or cup-like floral tube ("hypanthium") at the margin of which the stamens and petals are borne
      13. Carpels as many as, or more than, the petals
        14. Leaves succulent, simple, entire, estipulate . . . .**CRASSULACEAE** (ii: 315)
        14. Leaves not succulent, deeply lobed or compound (simple in *Dalibarda*), toothed, stipulate . . . .**ROSACEAE** (ii: 336)
      13. Carpels fewer than the petals
        15. Leaves simple, at most shallowly lobed . . . .**SAXIFRAGACEAE** (ii: 319)
        15. Leaves clearly compound . . . .**ROSACEAE** (ii: 336)

KEY K

(Herbaceous Dicots, with Perfect Flowers,
Perianth in Two Series,
& Ovary Inferior)

1. Stamens twice as many as the petals (or approximately so)
  2. Style 1 (sometimes very short) . . . .**ONAGRACEAE** (ii: 614)
  2. Styles 2 or more
    3. Sepals 2; leaves succulent; styles 3 . . . . *Portulaca* in **PORTULACACEAE** (ii: 153)
    3. Sepals (4–) 5; leaves not succulent; styles 2 . . . .**SAXIFRAGACEAE** (ii: 614)
1. Stamens of the same number as the petals or corolla lobes, or fewer
  4. Petals united
    5. Cauline leaves alternate
      6. Corolla bilateral . . . . *Lobelia* in **CAMPANULACEAE** (iii: 330)
      6. Corolla regular
        7. Flowers less than 3 mm broad . . . . *Samolus* in **PRIMULACEAE** (iii: 62)
        7. Flowers much larger . . . .**CAMPANULACEAE** (iii: 329)
    5. Cauline leaves opposite or whorled
      8. Leaves whorled . . . .**RUBIACEAE** (iii: 278)
      8. Leaves opposite
        9. Stipules present (connate around the stem) . . . .**RUBIACEAE** (iii: 278)

Page 576

9. Stipules absent
  10. Flowers sessile in densely packed terminal heads . . . .**DIPSACACEAE** (iii: 321)
  10. Flowers visibly pediceled (even if crowded) or axillary
    11. Flowers numerous, in rather dense terminal inflorescences (at ends of stem
      and branches) . . . .**VALERIANACEAE** (iii: 318)
    11. Flowers axillary or on paired pedicels on a peduncle . . . .
    . . . .**CAPRIFOLIACEAE** (iii: 295)
4. Petals separate
 12. Stamens and petals each 2. . . . . *Circaea* in **ONAGRACEAE** (ii: 615)
 12. Stamens (fertile) and petals each 4 or 5 (stamens sometimes alternating with
   staminodia, which may have gland-tipped divisions)
  13. Petals 4
    14. Principal leaves apparently whorled; flowers in a dense headlike terminal
      cluster subtended by 4 large white bracts . . . .
    . . . .*Cornus canadensis* in **CORNACEAE** (ii: 677)
    14. Principal leaves alternate; flowers neither in a head-like terminal cluster nor
      subtended by 4 large white bracts . . . . *Ludwigia* in **ONAGRACEAE** (ii: 617)
  13. Petals 5
   15. Leaves simple; styles 2 or stigmas 4 and sessile; inflorescence various
    16. Flowers in panicles, in cymes, or solitary. . . . .**SAXIFRAGACEAE** (ii: 319)
    16. Flowers in umbels . . . .**UMBELLIFERAE** (ii: 645)
   15. Leaves compound; inflorescence an umbel
    17. Styles 5; fruit berry-like. . . . . *Aralia* in **ARALIACEAE** (ii: 694)
    17. Styles 2–3; fruit various
     18. Leaves alternate or basal; fruit dry, splitting into 2 achene-like indehiscent
       parts (mericarps) . . . .**UMBELLIFERAE** (ii: 645)
     18. Leaves in a single whorl; fruit berry-like . . . .*Panax* in **ARALIACEAE** (ii: 642)

KEY L

(Herbaceous Dicots with Perfect Flowers,
Perianth in Two Series, Ovary One and Superior,
& Stamens More Numerous than the Petals)

1. Corolla bilaterally symmetrical
 2. Sepals all or partly petal-like in appearance or prolonged into a spur
  3. Spur none; stamens 6, 7, or 8; leaves entire[†] . . . .**POLYGALACEAE** (ii: 512)
  3. Spur present on one of the sepals
   4. Leaves merely crenate or toothed; stamens 5[††] . . . .**BALSAMINACEAE** (ii: 554)

[†]The flowers in *Polygala* superficially resemble those of the Leguminosae ("papilionaceous") but the perianth parts are not parallel. Of the 5 sepals, the 2 lateral ones are large and petaloid ("wings"); the petals are 3, the lower one forming a "keel" (usually with a fringe or appendage near the end).

[††]The flowers in *Impatiens* include 3 sepals, of which the lower one is large, petaloid, sac-like, and slender-spurred; the petals are apparently 3, each of the 2 lateral ones with a lobe.

4. Leaves palmately cleft; stamens numerous . . . .
. . . . *Consolida* in **RANUNCULACEAE** (ii: 202)
2. Sepals not petal-like in form or appearance, usually green
  5. Sepals 2, separate, usually deciduous early in anthesis; leaves dissected or twice-compound . . . . **FUMARIACEAE** (ii: 245)
  5. Sepals 4 or more, usually ± connate
    6. Lower 2 petals forming a laterally compressed "keel" that encloses the stamens; leaves once-compound (simple only in *Crotalaria*) . . . . **LEGUMINOSAE** (ii: 444)
    6. Lower petals not forming a keel nor enclosing the stamens
      7. Flowers completely 5-merous (sepals & petals 5; stamens 5 or usually 10; pistil long-beaked, the 5 carpels evident in a 5-fid apex and 5 mericarps to the fruit); corolla pink or purple, only slightly bilateral; leaves deeply lobed or cleft or compound (the principal cauline ones opposite and prominently toothed or cleft) . . . . **GERANIACEAE** (ii: 502)
      7. Flowers with at least the carpels (often one or more other cycles as well) fewer than 5; corolla and leaves various (carpels 5 only in *Dictamnus*, with alternate leaves and nearly entire leaflets)
        8. Leaves simple, deeply lobed to entire
          9. Styles 3 (or more); upper petal larger than the others . . . . **RESEDACEAE** (ii: 310)
          9. Style 1; upper petal(s) no larger than the others
            10. Sepals and petals each 4; stamens 6 . . . . **CRUCIFERAE** (ii: 251)
            10. Sepals and petals each (4) 5–7; stamens twice as many . . . .
. . . . **LYTHRACEAE** (ii: 609)
        8. Leaves compound
          11. Flowers in dense terminal spikes, each flower very small with 5 anthers, apparently 1 larger petal, and 4 smaller petals arising from a column of stamens . . . . *Dalea purpurea* in **LEGUMINOSAE** (ii: 491)
          11. Flowers not in dense spikes and not so modified
            12. Petals and sepals each 4; fruit a capsule
              13. Plant a vine, climbing by tendrils; capsule 3-lobed, large and bladdery . .
. . . . **SAPINDACEAE** (ii: 554)
              13. Plant erect, without tendrils; capsule elongate and unlobed . . . .
. . . . *Polanisia* in **CAPPARACEAE** (ii: 250)
            12. Petals and sepals each 5; fruit a legume (splitting on 2 sutures or indehiscent) or a woody capsule
              14. Corolla yellow; leaves pinnately compound with terminal leaflet none or represented by a bristle; fruit a legume . . . . *Cassia* in **LEGUMINOSAE** (ii: 479)
              14. Corolla pink or purple; leaves pinnately compound with a terminal leaflet; fruit a woody capsule . . . . *Dictamnus* in **RUTACEAE** (ii: 509)
1. Corolla regular (radially symmetrical)
  15. Leaves tubular, open at apex and hence pitcher-like; style greatly expanded, large and umbrella-shaped . . . . **SARRACENIACEAE** (ii: 310)
  15. Leaves flat or at most succulent, of usual shapes; style not unusually expanded
    16. Sepals 2
      17. Leaves lobed, compound, or coarsely toothed, not succulent; sap in most species colored (yellow–orange) . . . . **PAPAVERACEAE** (ii: 237)
      17. Leaves unlobed, entire, succulent; sap watery . . . .
. . . . *Portulaca* in **PORTULACACEAE** (ii: 153)
    16. Sepals 3 or more
      18. Stamens more than twice as many as the petals
        19. Leaves compound
          20. Plant clammy-pubescent; leaves palmately compound with 3 entire leaflets .
. . . . *Polanisia* in **CAPPARACEAE** (ii: 250)

Page 578

20. Plant glabrous or with a little non-glandular pubescence; leaves twice-compound with numerous sharply toothed leaflets...**RANUNCULACEAE** (ii: 199)
19. Leaves simple
  21. Plant truly aquatic, with all leaves basal, the petioles all arising from a rhizome buried under water (except when stranded) ....**NYMPHAEACEAE** (ii: 189)
  21. Plant terrestrial, with at least some leaves cauline
    22. Style 1 (or none, with 3 sessile stigmas) ....**CISTACEAE** (ii: 584)
    22. Styles 2 or more, evident
      23. Leaves opposite, with translucent dots; petals yellow ....
      ....*Hypericum* in **GUTTIFERAE** (ii: 577)
      23. Leaves alternate, without translucent dots; petals of various color ....
      ....**MALVACEAE** (ii: 567)
18. Stamens twice as many as the petals or fewer
  24. Stamens fewer than twice as many as the petals
    25. Styles 2–5; leaves opposite or whorled, simple and entire
      26. Petals yellow .... *Hypericum* in **GUTTIFERAE** (ii: 577)
      26. Petals white, pink, or red
        27. Stamens 9, in 3 distinct groups of 3 each, with 3 conspicuous glands alternating with the groups .... *Triadenum* in **GUTTIFERAE** (ii: 575)
        27. Stamens various but neither 9 nor in groups........................
        ....**CARYOPHYLLACEAE** (ii: 155)
    25. Style 1 or none; leaves usually alternate, simple or compound, entire or toothed
      28. Sepals 5 (of which the 2 outer ones may be much reduced); petals 3, minute (shorter than the calyx), reddish .... *Lechea* in **CISTACEAE** (ii: 585)
      28. Sepals and petals each 4; petals usually ± showy, of various color
        29. Leaves palmately compound, with entire leaflets; stamens 6 or more, but not with 2 distinctly shorter; pedicels subtended by bracts............
        ....**CAPPARACEAE** (ii: 249)
        29. Leaves simple or if palmately compound the leaflets coarsely toothed; stamens 6, of which 2 are distinctly shorter; pedicels usually bractless ..
        ....**CRUCIFERAE** (ii: 251)
  24. Stamens exactly twice as many as the petals
    30. Petals 3
      31. Leaves alternate, compound or deeply pinnately lobed ....
      ....**LIMNANTHACEAE** (ii: 531)
      31. Leaves in a single whorl, simple and unlobed.......................
      ....*Trillium* in **LILIACEAE** (i: 400)
    30. Petals 4 or more
      32. Sepals and petals each 6 or more
        33. Cauline leaves a single opposite pair, eccentrically peltate and deeply lobed (with one flower between them) ....
        ....*Podophyllum* in **BERBERIDACEAE** (ii: 232)
        33. Cauline leaves more than 2, neither peltate nor lobed (with several to many flowers on a plant) .... *Lythrum* in **LYTHRACEAE** (ii: 611)
      32. Sepals and petals each 4 or 5
        34. Leaves compound or deeply divided nearly to base of blade
          35. Leaves opposite
            36. Petals yellow; leaves pinnately compound; plant prostrate and hairy ..
            ....**ZYGOPHYLLACEAE** (ii: 508)
           36. Petals pink to purple or red; leaves palmately compound or lobed; plant ± erect, hairy or not .... *Geranium* in **GERANIACEAE** (ii: 502)
          35. Leaves alternate

37. Styles 5; leaves with 3 obcordate leaflets.........**OXALIDACEAE** (ii: 498)
37. Style 1; leaves various, but if 3-foliolate, the leaflets not obcordate
   38. Leaves palmately compound ....**CAPPARACEAE** (ii: 249)
   38. Leaves pinnately compound .... *Cassia* in **LEGUMINOSAE** (ii: 479)
34. Leaves simple and entire, toothed, or shallowly lobed
  39. Style 1
    40. Floral tube ("hypanthium") present, with petals and sepals borne at its margin; petals white to (usually) pink-purple
     41. Anthers opening by terminal pores, very showy (curved, yellow, appearing set at 90° on the filament) and stamens becoming skewed toward one side of the flower ....**MELASTOMATACEAE** (ii: 614)
     41. Anthers opening by longitudinal slits, not especially showy; stamens not skewed................................**LYTHRACEAE** (ii: 609)
    40. Floral tube none, all parts arising directly from the receptacle; petal color various
     42. Sepals of 2 sizes, the 2 outer ones very much narrower and often shorter than the 3 inner ones (appearing as mere appendages on them); petals yellow....................................**CISTACEAE** (ii: 584)
     42. Sepals all of nearly the same size and shape; petals white, greenish, or pink ....**PYROLACEAE** (iii: 26)
  39. Styles 2 or more
    43. Ovary lobed, with a style on each lobe
     44. Leaves not succulent, all or mostly basal, (cauline leaves, if any, few and small or a single pair); lobes of ovary 2 ....**SAXIFRAGACEAE** (ii: 319)
     44. Leaves succulent, all or mostly cauline; lobes of ovary 4 or 5 ....**CRASSULACEAE** (ii: 315)
    43. Ovary unlobed, the styles all arising together
     45. Petals yellow; leaves with translucent dots ....*Hypericum* in **GUTTIFERAE** (ii: 577)
     45. Petals white to pink or red (never yellow); leaves without translucent dots ....**CARYOPHYLLACEAE** (ii: 155)

KEY M

(Herbaceous Dicots with Perfect Flowers, Perianth in Two Series, Ovary One and Superior, Stamens the Same Number as the Petals or Fewer, & Petals Separate)

1. Leaves compound or dissected
  2. Flowers solitary on leafless peduncles arising from the ground
   3. Corolla bilateral, spurred; petals 5 ....**VIOLACEAE** (ii: 589)
   3. Corolla regular, without spur; petals 8 .....*Jeffersonia* in **BERBERIDACEAE** (ii: 232)
  2. Flowers on leafy stems
    4. Leaves dissected or 2–3-times compound; inflorescence open, peduncled
     5. Petals and stamens each 6; leaves 2–3-times compound, with flat, broad leaflets .... *Caulophyllum* in **BERBERIDACEAE** (ii: 231)

      5. Petals and stamens each 5; leaves dissected . . . .**GERANIACEAE** (ii: 502)
   4. Leaves pinnately compound; inflorescence of nearly sessile and axillary flowers
      or a dense terminal spike
      6. Flowers axillary or supra-axillary and sessile or almost so; corolla yellow; peti-
         oles with a conspicuous discoid short-stalked gland; leaves without a terminal
         leaflet . . . . *Cassia nictitans* in **LEGUMINOSAE** (ii: 480)
      6. Flowers in a dense terminal spike; corolla purple; petioles without a gland;
         leaves with a terminal leaflet . . . . *Dalea purpurea* in **LEGUMINOSAE** (ii: 491)
1. Leaves entire or toothed to deeply lobed
   7. Leaves opposite, entire
      8. Sepals 2 or 3; petals 2–6
         9. Plant aquatic; sepals and petals each of the same number (2 or 3) . . . .
            . . . . **ELATINACEAE** (ii: 582)
         9. Plant terrestrial; sepals 2; petals 5 or 6 . . . .**PORTULACACEAE** (ii: 153)
      8. Sepals and petals each 4–6 (or more)
         10. Leaves deeply palmately lobed . . . .**GERANIACEAE** (ii: 502)
         10. Leaves entire or merely toothed
            11. Style 1
               12. Floral tube ("hypanthium") well developed, with sepals and petals borne at
                  its margin . . . .**LYTHRACEAE** (ii: 609)
               12. Floral tube none
                  13. Stamens opposite the petals (i.e., each stamen oriented above the middle of
                     a petal) . . . .**PRIMULACEAE** (iii: 54)
                  13. Stamens alternating with the petals . . . .**GENTIANACEAE** (iii: 71)
            11. Styles 2–5
               14. Flowers completely 5-merous, including 5 styles; stamens with filaments
                  connate at the base around the ovary; ovary 5- (or 10-) locular . . . .
                  . . . .**LINACEAE** (ii: 496)
               14. Flowers with styles usually fewer than 5 (and petals sometimes 4); stamens
                  not connate; ovary with 1 locule
                  15. Petals yellow; leaves with translucent dots . . . .
                     . . . .*Hypericum* in **GUTTIFERAE** (ii: 577)
                  15. Petals white to pink or red; leaves without translucent dots
                     16. Petals distinctly separate all the way to the base; stem not 4-angled . . . . . .
                        . . . .**CARYOPHYLLACEAE** (ii: 155)
                     16. Petals slightly connate at base; stem sharply 4-angled (even narrowly
                        winged) . . . . *Sabatia* in **GENTIANACEAE** (iii: 74)
   7. Leaves alternate or basal, entire or toothed
      17. Leaves shallowly to deeply palmately lobed
         18. Corolla bilateral, spurred; style 1 . . . .**VIOLACEAE** (ii: 589)
         18. Corolla regular or nearly so, not spurred; styles 2 . . . .
            . . . .*Heuchera* in **SAXIFRAGACEAE** (ii: 321)
      17. Leaves unlobed or pinnately lobed
         19. Styles 2 or more
            20. Leaves all basal, with conspicuous stipitate glands . . . .**DROSERACEAE** (ii: 312)
            20. Leaves cauline, without glands . . . .**LINACEAE** (ii: 496)
         19. Style 1 or none
            21. Floral tube well developed and prolonged, with sepals and petals borne at its
               margin . . . .**LYTHRACEAE** (ii: 609)
            21. Floral tube none or very little developed
               22. Corolla bilateral, saccate or spurred at the base . . . .**VIOLACEAE** (ii: 589)
               22. Corolla regular, without a spur
                  23. Petals and sepals each 4 . . . .**CRUCIFERAE** (ii: 251)

23. Petals and sepals each 5
   24. Leaves pinnately lobed or dissected . . . . . . *Erodium* in **GERANIACEAE** (ii: 502)
   24. Leaves entire or merely toothed
      25. Flowers solitary, terminal; styles essentially none (stigmas 4, nearly sessile); stamens alternating with cleft, gland-tipped staminodia . . . .
       . . . . *Parnassia* in **SAXIFRAGACEAE** (ii: 32l)
      25. Flowers in a terminal umbel or raceme; style present; staminodia none .
       . . . . **PRIMULACEAE** (iii: 54)

KEY N

(Herbaceous Dicots with Perfect Flowers,
Perianth in Two Series, Ovary One and Superior,
Corolla Regular and Stamens the Same Number
as Its Lobes, & Petals United)

1. Leaves all basal; flowers solitary on scapes
  2. Leaves covered with conspicuous stipitate glands . . . . **DROSERACEAE** (ii: 312)
  2. Leaves without stipitate glands
    3. Perianth 4-merous; flowers in spikes or heads; corolla scarious . . . .
     . . . . **PLANTAGINACEAE** (iii: 271)
    3. Perianth 5-merous; flowers in umbels; corolla petaloid . . . . . . **PRIMULACEAE** (iii: 54)
1. Leaves all or mostly cauline
  4. Ovary deeply 4-lobed, appearing like 4 separate ovaries [and also keyed thusly] but with one style arising deep in the midst of the lobes
    5. Leaves opposite; stamens 2 or 4; stems 4-angled ("square") and foliage aromatic (minty or citrus-like) . . . . **LABIATAE** (iii: 138)
    5. Leaves alternate (except at lower nodes in the very rare *Plagiobothrys*); stamens 5; stem not angled (rarely winged) and foliage not aromatic . . . .
     . . . . **BORAGINACEAE** (iii: 115)
  4. Ovary not conspicuously lobed (may be slightly 4- or 2-lobed or notched at apex, where style arises
    6. Leaves opposite (or whorled), at least below the inflorescence
      7. Flowers in dense heads or short spikes; corolla 4-lobed
        8. Corolla scarious; leaves linear, entire . . . .
         . . . . *Plantago arenaria* in **PLANTAGINACEAE** (iii: 273)
        8. Corolla petaloid; leaves lance-elliptic, toothed . . . .
         . . . . *Phyla* in **VERBENACEAE** (iii: 134)
      7. Flowers in crowded or more open racemes or other inflorescences; corolla lobes 4–7
        9. Stamens opposite the corolla lobes (i.e., each stamen arising and oriented above the *middle* of a lobe) and readily visible . . . . **PRIMULACEAE** (iii: 54)
        9. Stamens alternating with the corolla lobes (sometimes hidden in a corolla tube or closed corolla)
          10. Lobes of corolla 4 . . . . **GENTIANACEAE** (iii: 71)
          10. Lobes of corolla 5

11. Stigmas 3; ovary with 3 locules . . . .**POLEMONIACEAE** (iii: 104)
11. Stigma 1 (may be 2-lobed); ovary with 2 (or 4 or 5) locules
  12. Leaves glabrous; ovary 1-locular; fruit a 2-valved capsule . . . .
    . . . .**GENTIANACEAE** (iii: 71)
  12. Leaves strongly clammy-pubescent; ovary 2-locular; fruit a berry or a 2-
    valved capsule . . . *Leucophysalis* and *Petunia* in **SOLANACEAE** (iii: 186 & 198)
6. Leaves alternate
 13. Blades of leaves deeply lobed, dissected, or compound
  14. Plant a twining or trailing vine
   15. Corolla deeply funnel-shaped (or even trumpet-shaped) . . . .
    . . . .**CONVOLVULACEAE** (iii: 95)
   15. Corolla ± flat (rotate) . . . . *Solanum* in **SOLANACEAE** (iii: 187)
  14. Plant, whether erect or prostrate, not a vine
   16. Anthers forming a cone around the pistil . . . .**SOLANACEAE** (iii: 183)
   16. Anthers clearly separate
    17. Leaves 3-foliolate . . . . **MENYANTHACEAE** (iii: 82)
    17. Leaves otherwise lobed, compound, or dissected
     18. Leaves pinnately compound or pinnately dissected into entire filiform
      lobes (pectinate); ovary 3-locular; stigmas or style branches 3; capsule 3-
      valved . . . .**POLEMONIACEAE** (iii: 104)
     18. Leaves not compound: pinnatifid or bipinnatifid, the segments not both
      entire and filiform; ovary 1-locular; stigmas or style branches 2; capsule 2-
      valved . . . .**HYDROPHYLLACEAE** (iii: 112)
 13. Blades of leaves entire, toothed, or at most shallowly lobed (or merely cordate)
  19. Leaves reduced to small scales; flowers 4-merous . . . .
   . . . . *Bartonia* in **GENTIANACEAE** (iii: 73)
  19. Leaves developed; flowers mostly 5-merous
   20. Flowers or inflorescences axillary
    21. Fruit a 4-seeded capsule; corolla large, funnel-shaped; stigmas clearly 2,
     separate (except in *Ipomoea*, where at most 2–3-lobed) . . . .
     . . . .**CONVOLVULACEAE** (iii: 95)
    21. Fruit a many-seeded berry or capsule; corolla large and funnel-shaped (in
     Datura, Petunia) or ± flat (rotate) or campanulate; stigma 1 . . . .
     . . . .**SOLANACEAE** (iii: 183)
   20. Flowers or inflorescences terminal
    22. Flowers solitary; corolla over 7 cm long . . . . . . *Datura* in **SOLANACEAE** (iii: 192)
    22. Flowers in clusters; corolla smaller
     23. Inflorescence branched (panicle or cyme)
      24. Leaves linear to narrowly lanceolate, less than 8 cm long . . . .
       . . . . *Collomia* in **POLEMONIACEAE** (iii: 105)
      24. Leaves broad and at least 10 cm long . . . .*Nicotiana* in **SOLANACEAE** (iii: 185)
     23. Inflorescence simple (spike, raceme, or umbel)
      25. Corolla campanulate
       26. Plant glabrous; flowers less than 3 mm broad, in open racemes; corolla
        white . . . . *Samolus* in **PRIMULACEAE** (iii: 74)
       26. Plant clammy-pubescent; flowers very much larger, bracted in spikelike
        racemes; corolla greenish yellow with purple veins . . . .
        . . . .*Hyoscyamus* in **SOLANACEAE** (iii: 192)
      25. Corolla flat or saucer-shaped
       27. Anthers separate, at least some of them on hairy filaments; fruit a cap-
        sule . . . . *Verbascum* in **SCROPHULARIACEAE** (iii: 204)
       27. Anthers in a cone around the pistil, on glabrous filaments; fruit a
        berry . . . . *Solanum* in **SOLANACEAE** (iii: 187)

KEY O

(Herbaceous Dicots with Perfect Flowers, Perianth in Two Series, Ovary One and Superior, Corolla Either Bilateral or Stamens Fewer than Its Lobes—or Both, & Petals United)

1. Fertile (anther-bearing) stamens 5
  2. Ovary deeply 4-lobed; plant strongly bristly-hairy; fruit (1–) 4 nutlets ....
     .... *Echium* in **BORAGINACEAE** (iii: 117)
  2. Ovary not lobed; plants glabrous or with dense stellate or clammy pubescence; fruit a capsule
    3. Corolla flat or saucer-shaped, white or yellow; filaments (or some of them) hairy .... *Verbascum* in **SCROPHULARIACEAE** (iii: 204)
    3. Corolla campanulate, greenish yellow with purple veins; filaments glabrous ....
     .... *Hyoscyamus* in **SOLANACEAE** (iii: 192)
1. Fertile stamens 2 or 4
  4. Corolla with a spur or sac at the base
    5. Calyx 2-parted .... *Utricularia* in **LENTIBULARIACEAE** (iii: 259)
    5. Calyx 5-parted
      6. Leaves all basal, glandular-sticky above; flowers solitary on scapes ....
       .... *Pinguicula* in **LENTIBULARIACEAE** (iii: 259)
      6. Leaves all or mostly cauline and not sticky (stipitate-glandular in *Chaenorhinum*); flowers not solitary .... **SCROPHULARIACEAE** (iii: 200)
  4. Corolla not prolonged into a spur or sac at the base
    7. Leaves all alternate .... **SCROPHULARIACEAE** (iii: 200)
    7. Leaves all or mostly opposite or whorled
      8. Ovary deeply 4-lobed, appearing like 4 separate ovaries around the base of the single style [and also keyed thusly], the fruit (1–) 4 nutlets; plants usually with a 4-angled ("square") stem and often a minty or citrus-like aroma when bruised
       .... **LABIATAE** (iii: 138)
      8. Ovary not 4-lobed (at most, somewhat 2-lobed), the fruit a capsule; stem in only a few species 4-angled or with aroma when bruised
        9. Fertile stamens 2
          10. Flowers in terminal racemes or spikes, or solitary or paired in the axils of the leaves .... **SCROPHULARIACEAE** (iii: 200)
          10. Flowers in axillary racemes or spikes
            11. Corolla almost regular, with a 4-lobed limb .... **SCROPHULARIACEAE** (iii: 200)
            11. Corolla distinctly bilateral, 2-lipped .... *Justicia* in **ACANTHACEAE** (iii: 270)
        9. Fertile stamens 4
          12. Corolla nearly regular, the lobes about equal
            13. Corolla salverform (trumpet-shaped, with a slender tube of almost uniform diameter)
              14. Corolla 15–22 mm long; leaves sessile, lance-ovate with few (3–6 per side) short teeth; plant a root-parasite (blackening in drying), considered extinct in Michigan .... *Buchnera* in **SCROPHULARIACEAE** (iii: 237)
              14. Corolla shorter or leaves more prominently toothed or even lobed; plants (some species) widespread .... *Verbena* in **VERBENACEAE** (iii: 134)
            13. Corolla funnel-shaped or campanulate, with a tube broad toward its summit
              15. Calyx 5-lobed, bilateral, split nearly to the base on the lower side; plant clammy-pubescent; fruit a drupe, the fleshy exocarp falling away to leave the persistent elongate woody endocarp bearing 2 prominent recurved beaks longer than the body .... **PEDALIACEAE** (iii: 253 & fig. 119)

15. Calyx 4-lobed or if 5-lobed not split beneath and the lobes ± equal; plant glabrous or pubescent but not glandular/clammy; fruit a typical capsule
  16. Calyx lobes 4, or short and relatively broad and 0–4; or corolla yellow
  . . . .**SCROPHULARIACEAE** (iii: 200)
  16. Calyx lobes 5, lanceolate to bristle-like, longer than the calyx tube; corolla pink to purple
    17. Stem glabrous to ± hirsute; leaves unlobed, entire, similarly glabrous to hairy . . . . *Ruellia* in **ACANTHACEAE** (iii: 270)
    17. Stem rough with stiff retrorse hairs; leaves (at least the uppermost) often with a pair of basal lobes, scabrous above . . . .
    . . . .*Tomanthera* in **SCROPHULARIACEAE** (iii: 241)
12. Corolla strongly bilateral
  18. Mature flowers and fruit strongly reflexed, nearly sessile and in remote pairs on opposite sides of a spike-like terminal raceme; calyx with 3 upper teeth bristle-like and 2 lower teeth deltoid . . . . . .*Phryma* in **VERBENACEAE** (iii: 133)
  18. Mature flowers and fruit not strongly reflexed, and otherwise not as above (long-pediceled, alternate, and/or crowded); calyx with teeth equal or subequal (never bristle-like)
    19. Upper lip of corolla well developed, of 2 lobes (or these ± fused into one), often nearly or quite as long as the lower lip . . . .
    . . . .**SCROPHULARIACEAE** (iii: 200)
    19. Upper lip of corolla apparently absent (the corolla split lengthwise above) or much shorter than the lower lip and (except in *Phyla*) 4-lobed
      20. Flowers in dense short spikes or heads on long axillary peduncles . . . . . .
      . . . .*Phyla* in **VERBENACEAE** (iii: 134)
      20. Flowers not in dense heads or spikes . . . .**LABIATAE**[†] (iii: 138)

[†]Our genera of Labiatae having an ovary not *deeply* 4-lobed and which therefore will key here are *Ajuga*, *Teucrium*, and *Trichostema*.

Glossary

When this glossary lists a word, usually a noun, derivative words are generally not listed separately; e.g., *whorled* means "in a whorl"; *petioled* means "with a petiole"; *mucronate* means "with a mucro"; *papillose* (or *papillate*) means "with papillae"; *auriculate* means "with auricles"; *umbelliform* means "in general form of an umbel"; *stipitate* means "having a stipe"; and so forth.

Likewise, negatives need not be separately defined: *apetalous* means "without petals"; *eciliate* and *eglandular* mean "without cilia" and "without glands," respectively.

Since nature does not always follow the book, many conditions intermediate between two defined ones occur, and honesty compels the botanist to use a lot of hyphens for such intermediacy (linear-lanceolate, ovate-orbicular, auriculate-clasping) as well as for combined conditions (glandular-pubescent).

Some specialized terms used in certain families and genera are explained in the remarks for the group concerned and are not necessarily repeated here. And certain other terms are explained at the time of use (introduced so as to help readers interpret other works, which may use them more often).

Some frequently used terms for habitats are included here; for a more full discussion, see Part I of this Flora, pp. 17–23. Other terms not listed here are probably not specialized botanical ones; if they are unfamiliar, consult a good dictionary. Note that this glossary does not apply to all three parts of the Flora, but only to Part III (including the General Keys).

Abaxial. Away from the axis; e.g., the "lower" or dorsal surface of a leaf, the "outer" side of a flower or group of nutlets. Cf. adaxial.

Achene. A dry indehiscent fruit, strictly speaking one derived from a single superior carpel, but broadly used for similar fruits ("nutlets") derived from more than one carpel or (as in Compositae) from an inferior ovary.

Acuminate. Prolonged into a very acute point (and often slightly concave below the point).

Acute. With the sides or margins converging at less than a 90° angle.

Adaxial. Toward the axis; e.g., the "upper" or ventral surface of a leaf, the "inner" side of a flower part or group of nutlets. Cf. abaxial.

Adherent. Sticking (but not fused) to parts of a different kind. Cf. adnate.

Adnate. United (fused) to parts of a different kind; e.g., stamens to petals, stipule to blade. Cf. connate.

Adventive. Spreading from a native or naturalized source but not [yet] well established.

Albino. Lacking normal color; i.e., white — usually in reference to flowers, at least the corolla, for which another color is usual.

Allo-. (See n.)

Alternate. Arranged singly at the nodes, as leaves on a stem or branches in an inflorescence; neither opposite nor whorled.

Alvar. Flat limestone rock ("pavement") with thin (if any) soil and usually graminoid vegetation.

Ament. A spike or spike-like inflorescence consisting of reduced (usually apetalous and unisexual) flowers and deciduous as a unit; also called a "catkin."

Amphidiploid. A taxon of hybrid origin including chromosomes from both parents and fertile as a result of their doubling. Cf. also *n.*

Anastomose. To merge (as veins in a leaf blade) so as to form a network.

Annual. Living for one year; i.e., germinating, flowering, and setting seed in a single growing season (lacking perennial roots, rhizomes, or other such parts). A winter annual begins its year in the fall and completes its cycle after winter.

Anther. The pollen-bearing part of a stamen.

Anthesis. The time at which a flower is fully expanded and functional.

Antrorse. Directed toward the apex or upward, e.g., barbs on a bristle or awn. Cf. retrorse.

Apiculus. A very small sharp beak-like tip.

Apomixis. As used here, reproduction by seed without fertilization—a form of asexual reproduction.

Appressed. Oriented in a parallel or nearly parallel manner to the surface or axis to which attached.

Aril. An appendage arising from or near the scar (hilum) on a seed marking its point of attachment; an aril may be quite small or may enclose the seed.

Ascending. Directed strongly upward or forward (in relation to the point of attachment), but not fully erect or at right angles.

Attenuate. Drawn out gradually to a slender tapering apex or base.

Auricle. A lobe or appendage, often small and ear-like, typically projecting at the base of an organ (as on a leaf blade); i.e., more than clasping but extending beyond the point of attachment.

Autonym. A scientific name "automatically created" by publication of another name; e.g., the publication of a name for a variety automatically creates an autonym that repeats the species epithet at varietal rank.

Awn. A terminal appendage or elongation, usually bristle-like.

Axil. The angle where a leaf or branch joins a stem or main axis, or where a lateral vein joins the midrib of a leaf.

Barb. A small sharp projection, usually retrorse, as on a fish-hook.

Barbellate. With little barbs.

Basal. At the base; i.e., unless the context indicates otherwise, at the base of the plant, or at ground level.

Basionym. A name that provides the epithet for a combination at a different rank or position; e.g., *Seymeria macrophylla* is the basionym for the later *Dasistoma macrophylla*.

Beak. A comparatively slender prolongation (sometimes of firmer texture) on a broader organ.

Beard. A concentration or tuft of hairs.

Berry. A fleshy indehiscent several–many-seeded fruit derived from a single ovary; sometimes loosely used for similar fruits that lack one or more of these criteria.

Bi-. A prefix meaning two or twice.

Biennial. Living for two years. Such plants often produce a rosette of leaves the first year and a flowering stem the second year.

Bifid. Cleft in two.

Bilateral; bilaterally symmetrical. Capable of division into similar (mirror-image) halves on only one plane (= "zygomorphic" of many works). Cf. regular.

Binomial. A scientific name consisting of two words — for a species, the name of the genus followed by the specific epithet.

Blade. The expanded portion of a leaf or other flat structure.

Bog. An acid peatland, rich in sphagnum; its water typically comes from the atmosphere (not groundwater) and ideally it displays ± concentric zones of vegetation from open water of a small lake or pond to surrounding swamp forest or upland. Cf. fen.

Bract. A reduced leaf-like, sometimes scale-like, structure, often subtending a flower, pedicel, branch, etc. Distinctions between a bract and a leaf are often obscure and dependent on context.

Bractlet; bracteole. A secondary bract, e.g., one of smaller size than others or on a branch rather than main axis.

Bulb. A short underground shoot which bears fleshy overlapping leaves (as in an onion).

Bulblet. A vegetative propagule often shed from the parent plant and suggesting a tiny bulb in presence of modified leaves or bracts (though rarely onion-like).

Bulbous. With a bulb-shaped thickening.

Calcareous. Limy – rich in calcium carbonate as from limestone (or dolomite) or marl.

Calciphile. Favoring alkaline (calcareous) habitats.

Callus. A firm thickening or protuberance.

Calyx. The outer series of perianth parts (or the only one); the sepals, collectively.

Campanulate. Bell-shaped.

Capillary. Hair-like (i.e., extremely slender).

Capitate. Like a pin-head (as certain stigmas on the style).

Capsule. A fruit that dehisces along two or more sutures (derived from two or more carpels), usually several- or many-seeded.

Carpel. The basic female structural unit of the flower, homologous to a sepal, petal, or stamen; in a compound pistil, the carpels are united (connate), but the number can often (but not invariably) be determined from the number of styles, stigmas, sutures, or locules.

Caudate. With a well-defined prolonged (tail-like) appendage or abruptly acuminate apex.

Caudex. A stout perennial (sometimes almost woody) base of some herbaceous plants.

Cauline. On or pertaining to the stem — often in contrast to basal.

Chaff. (Receptacular chaff.) The scales, bracts, or bristles on the receptacle of a head in the Compositae, ordinarily subtending the ovaries (later, achenes).

Chasmogamous. (Of a flower) open and showy (as opposed to cleistogamous).

Cilia. Hairs along a margin or edge.

Ciliolate. Minutely ciliate.

Circinate. Inrolled from the tip downwards (as in the fiddlehead of a fern).

Circumscissile. Dehiscing by a circular line around the fruit.

Clasping. Sessile and at least slightly surrounding the stalk to which attached.

Clavate. Club-shaped; i.e., with a ± prolonged and narrow base.

Clavellate. Shaped as in a very slender club.

Cleistogamous. (Of a flower) fertilized and setting seed without opening.

Cline. A "character gradient"; i.e., ± regular or continuous change in a character across a geographic area.

Code. The International Code of Botanical Nomenclature (see References), which contains the rules and recommendations regarding the formation and selection of scientific names for plants.

Coherent. Sticking together in a group but not actually fused (as individual pollen grains).

Compound. Composed of more than one part, or branched; e.g., a leaf with two or more blades (leaflets), a pistil with more than one carpel, a branched inflorescence. Cf. simple.

Connivent. Coming into close contact but not actually fused (as in some anthers, e.g., in Solanaceae).

Connate. United (fused) to other parts of the same kind; e.g., petals to petals, leaves (opposite) to leaves. Cf. adnate.

Conserve. To retain (by official action of an International Botanical Congress) a scientific name that otherwise would not be usable under the Code.

Cordate. Broadly two-lobed; heart-shaped.

Coriaceous. Leathery in texture; firm.

Corm. A short thick underground stem lacking the fleshy leaves that characterize a bulb.

Corolla. The inner series of perianth parts (when there are two series); the petals, collectively.

Corymb. An inflorescence of the racemose or paniculate type, flowering from the margins inward, but with the lower pedicels or branches longer than the upper so that the inflorescence is relatively short, broad, and flat-topped.

Crenate. With very rounded teeth; scalloped.

Crenulate. Finely crenate.

Crisped. More or less puckered.

Cultivar. A named horticultural variety; it may be designated either by placing the cultivar epithet in single quotation marks or by preceding it with the abbreviation *cv*.

Cuneate. Wedge-shaped; i.e., with straight but not parallel sides.

Cuspidate. With a firm, sharp point.

Cyme. A type of inflorescence in which the terminal (rather than lower) flower matures first.

Cytological. Pertaining to cells, specifically to chromosomes.

Deciduous. (Of leaves) falling off naturally at the end of the growing season; (of floral parts) shed readily, often ephemeral.

Declined. Bent or turned slightly downwards.

Decumbent. Prostrate basally but ascending toward the tip.

Decurrent. Extending downward and along, as leaf blade tissue along a petiole or stem.

Deflexed. Reflexed.

Dehiscent. Splitting open naturally at maturity at one or more definite points.

Deltoid. Broadly triangular.

Dentate. With ± outward-pointing (often coarse and/or obtuse) marginal teeth.

Denticulate. With minute, usually ± remote, marginal teeth.

Depauperate. Stunted or otherwise poorly developed.

Dichotomous. Forking into two ± equal branches.

Dioecious. Having the sexes on separate plants; i.e., all flowers on one plant either staminate or pistillate. Cf. monoecious.

Diploid. (See *n*.)

Discoid. Consisting only of disk flowers (Compositae).

Disk. In the Compositae, the portion of the head consisting of radially symmetrical flowers. A ring of tissue around the base of the ovary (an enlargement on the receptacle) or at the margin of a floral tube.

Dissected. (Of a leaf) so finely divided that the blade tissue is nearly restricted to bordering the main veins (definite leaflets not evident).

Distal. At or toward the apex; i.e., toward the opposite end from that at which a structure is attached.

Distinct. Not connate. In general use, easy to see.

Divaricate. Strongly divergent; spreading or forking at about a 90° angle or more.

Divergent. Spreading away from the surface or axis to which attached.

Dorsal. Pertaining to the surface (e.g., of a leaf, sepal, or nutlet) away from the axis to which a structure is attached; abaxial. Cf. ventral.

Double. (Of a flower) with extra cycles of perianth parts (morphologically derived from stamens and carpels converted to petals); (of a serrate margin) with primary teeth again toothed.

Drupe. A fleshy indehiscent fruit with the seed (or seeds) enclosed in a hard tissue (endocarp) forming one (usually) or more central pits (or "stones").

Elaiosome. An appendage (modification of seed coat) on a seed, containing lipids attractive to ants (which are dispersal agents).

Elliptic; elliptical. Longer than wide, broadest at the middle, and tapering ± equally toward both ends.

Ellipsoid. = elliptical, but applied to a 3-dimensional object rather than to a plane surface.

Endocarp. Inner portion (surrounding the seeds) of the wall of a fruit.

Emersed. Normally extending above the water. Cf. submersed.

Entire. Without teeth; with a continuous margin.

Ephemeral. Lasting for a short time (of flower parts, less than a day).

Equitant. Folded lengthwise and straddling the structure beneath, as in the leaves of *Iris*.

Erose. Irregular (of a margin), as if chewed or gnawed.

Excurrent. Running beyond, as a vein prolonged beyond the margin of a leaf or sepal.

Exserted. Protruding beyond the surrounding structure(s), as stamens beyond a corolla. Cf. included.

Farinose. Covered with a pale powdery (mealy) substance.

Fascicle. A close tuft or cluster.

Fen. An "alkaline bog": a peatland so nourished by calcareous groundwater as to have more sedges and little if any sphagnum compared to the typical acid bog. A fen may resemble a bog in many ways, indeed, may ultimately become a bog through succession, and some peatlands may be transitional or a mosaic. Other fens are more like marshes.

Fertile. Normally reproductive; e.g., a fertile stamen produces pollen, a fertile flower bears seed (or at least reproductive parts), a fertile plant bears flowers. Cf. sterile.

Filament. The stalk of a stamen, usually thread-like but sometimes flattened or expanded.

Filiform. Thread-like; very slender and approximately as broad as thick.

Fimbriate. Fringed (with somewhat more substantial structures than mere cilia).

Floating. On the *surface* of the water (floating leaves neither rise above the surface nor live entirely under the surface).

Floral tube. The usually saucer- or cup-shaped structure formed by the adnate portions of perianth and stamens, on which the free portions of these organs are inserted. (In some works, = "hypanthium.")

Foliaceous. Leaf-like (in color, texture, or size).

Follicle. A fruit that dehisces along a single suture (derived from a single carpel).

Forest. Vegetation dominated by trees closely enough spaced to provide a ± continuous or closed canopy.

Form. A taxonomic rank below that of variety, usually used for minor, sporadic variants involving such features as flower color or pubescence, without any geographic coherence.

Free. Not adnate.

Fruit. A ripened ovary and any closely associated structures.

Glabrate. Nearly glabrous.

Glabrescent. Glabrate; becoming glabrous.

Glabrous. Without pubescence of any kind.

Gland. A secretory structure; any small protuberance (often of different texture, e.g., shiny or sticky in appearance) resembling such a structure.

Glaucous. Covered with a pale (gray to blue-green) waxy coating or "bloom."

Globose. Spherical.

Halophyte. Plant of saline habitats.

Hastate. Shaped like an arrowhead but with basal lobes diverging. Cf. sagittate.

Haustoria. The structures on a parasitic plant by which it is attached to its host.

Head. A compact inflorescence of sessile flowers or fruits crowded on a receptacle. Loosely used for compact clusters of fruits from a single flower.

Hemiparasitic. Partly parasitic, i.e., attached to a host plant but also with capacity for photosynthesis.

Heterostylous. With styles (and generally stamens) of different lengths in different flowers.

Hirsute. With rather coarse or stiff hairs.

Hirtellous. Minutely hirsute [essentially the same as hispidulous].

Hispid. With stiff hairs or bristles.

Hispidulous. Minutely hispid.

Hoary. With fine gray or whitish pubescence.

Homonym. Any of two or more identical scientific names. Ordinarily, only the first one to be published is available for use; later homonyms are illegitimate under the Code.

Hyaline. Thin and translucent.

Hypanthium. = floral tube.

Illegitimate. Contrary to one or more Articles of the Code, under which names, even though validly published, in violation of certain rules are not available for use (unless conserved). Note that the concept of illegitimacy has to do only with conditions of a *name*, and has nothing to do with whether or not the plant to which it was applied merits naming or recognizing (those are taxonomic, not nomenclatural, judgments). Cf. valid.

Imbricate. Overlapping, like shingles on a roof.

Impressed. Slightly sunken, as the veins on the surfaces of some leaves.

Incised. Cut ± deeply (but not as deeply as in pinnatifid or dissected).

Included. Not protruding beyond the surrounding structure(s). Cf. exserted.

Indehiscent. Not splitting open naturally. Cf. dehiscent.

Indurated. Firm and hardened.

Inferior. (Of an ovary) below the perianth. Cf. superior.

Inflorescence. An entire flower cluster, including pedicels and bracts; often used to cover clusters of fruit as well.

Infructescence. A fruiting inflorescence.

Inserted. Attached to or on; appearing to arise from (as stamens from a corolla to which they are adnate).

Internode. The portion of a stem or axis between nodes.

Introgression. The gradual infiltration of genes from one taxon into another, as the result of hybridization and back-crossing with the parents.

Involucre. The bract or bracts (or even leaves) at the base of an inflorescence. Cf. spathe.

Involute. With the margins rolled in (i.e., adaxially). Cf. revolute.

Keel. A ridge ± centrally located on the long axis of a structure, such as a sepal or an achene.

Lacerate. Ragged, irregularly cleft, appearing as if torn.

Lanceolate. Narrow and elongate, broadest below the middle.

Leaflet. One of the blades of a compound leaf.

Ligule. An appendage (e.g., membranous collar or fringe of hairs) at the base of a leaf or summit of a leaf sheath and on its adaxial side. In Compositae, the corolla of a bilaterally symmetrical (petal-like) flower; cf. ray.

Limb. The expanded part of an organ; in this volume, generally the expanded part of a corolla in contrast to the narrow tubular portion.

Linear. Narrow and elongate with ± parallel sides.

Lip. In the Orchidaceae, the one odd petal which is specially modified, usually the lowest (through twisting of the ovary 180°); in many other bilaterally symmetrical flowers, one of a set of lobes (e.g., 3 lobes in one lip and 2 lobes in another, representing a total of 5 corolla lobes).

Lobe. A projecting portion or segment, generally set off by an indentation (sinus).

Locule. A compartment or cavity, as in an anther or ovary (sometimes termed a "cell").

Marl. A deposit of calcium carbonate resulting from the activity of photosynthetic plants in altering the carbonate/bicarbonate balance in a lake or pond.

Marsh. Treeless vegetation at least seasonally wet.

Meadow. A loosely defined term for a treeless area (including one with many introduced species), often less level and hence less wet than a marsh though usually with many grasses or sedges.

Mericarp. One of the (usually indehiscent, 1-seeded) parts into which certain fruits separate.

-merous. -parted; i.e., with parts in the number cited or a multiple thereof.

Midrib. The prominent central vein of many leaves (often best seen from the lower side).

Monoecious. Having the sexes in separate flowers but on the same individual. Cf. dioecious.

Monotypic. (Of a family or genus) containing only one species.

Mucro. A short, sharp, slender point.

Mycorrhiza. A mutually beneficial combination of fungus and plant root.

n. The haploid or gametophytic number of chromosomes; ordinary cells of a seed plant have this number of *pairs* of chromosomes. Many plants have more than the basic two sets or complements of chromosomes (diploid) and the number of these is indicated with the suffix *-ploid*: triploid (3n) = 3 sets; tetraploid (4n) = 4 sets; pentaploid (5n) = 5 sets; octoploid (8n) = 8 sets; etc. An alloploid is of hybrid origin, including sets of chromosomes from different parents (rather than mere multiplication of chromosomes from one species).

Nectary. An organ that produces nectar.

Nerve. A vein or ridge, usually a relatively weak or less strong one.

Net- (or *netted-*) *veined.* With main veins (if more than one) branched (other veins diverging from the main veins, ± anastomosing or reticulate). Cf. parallel-veined.

Node. The point on a stem (extended to include the axis of an inflorescence) at which a leaf or branch arises.

Nothovariety. A taxon at the rank of variety within a hybrid treated at the rank of species.

Nutlet. An achene or similar tiny 1-seeded indehiscent fruit.

Oblique. Asymmetrical, unequal, or slanting.

Ob-. A prefix signifying inversion, usually with adjectives indicating shape; e.g., obovate, obconic, or obcordate (with the small end basal).

Oblong. Longer than wide and ± parallel-sided (but not as elongate as linear).

Obtuse. With the sides or margins converging at more than a 90° angle.

Opposite. Two at a node (and ± 180° apart), as in some leaves. Centered upon rather than alternating with, as stamens opposite the petals (cf. figs. 18 & 99).

Orbicular. Circular in outline or nearly so.

Ovary. The lower portion of a pistil, usually ± expanded, in which the seed or seeds are produced; ripens into a fruit.

Ovate. Shaped in general outline like a longitudinal section of an egg; i.e., broadest below the middle (but broader than lanceolate).

Ovoid. Egg-shaped.

Ovule. The immature seed within an ovary.

Palate. In a 2-lipped corolla, a projection or hump on the lower lip that closes the throat.

Palmate. Radiating from a common point, as veins or leaflets in a leaf.

Panicle. A "branched raceme"; i.e., an inflorescence in which the pedicels arise from a branched axis rather than a simple central axis. The lowermost flowers mature first, although inflorescences technically cymose are often said to be paniculate if they resemble panicles in their branching.

Papilla. A minute blunt or rounded projection on a surface.

Pappus. The bristles, hairs, scales, or other structures on the summit of the ovary (or achene) in the Compositae, i.e., at the base of the corolla (where one would expect a calyx).

Parallel-veined. With three or more main veins (± parallel) running from the base of the blade to the apex of the leaf (with or without minute cross-veins). Cf. net-veined.

Parasitic. Dependent upon (and attached to) another plant for nutrition.

Pectinate. Very deeply pinnatifid, with central axis and unbranched lateral segments like teeth on a (double) comb.

Pedicel. The stalk of an individual flower, spikelet, or head.

Peduncle. The stalk of an entire inflorescence (or of a solitary flower when there is only one).

Peltate. With the stalk attached to the mid-surface of a blade-like structure (rather than at the margin).

Pendent. Hanging or drooping.

Perennial. Living three or more years.

Perfect. (Of a flower) containing both stamen(s) and pistil(s); bisexual; hermaphrodite.

Perfoliate. With the stem (or other stalk) appearing to pass through the leaf (or other blade); i.e., the blades sessile (or two opposite blades connate) and their basal tissue surrounding the stem.

Perianth. All of the calyx and corolla collectively insofar as they are present, in contrast to the reproductive organs of the flower.

Perigynous. Surrounding the ovary or ovaries (but not adnate); possessing a floral tube (or hypanthium).

Petal. One of the divisions of the corolla.

Petiole. The stalk portion of a leaf.

Petiolule. The stalk of a leaflet in a compound leaf.

Phyllary. One of the bracts in the involucre of the Compositae.

Pilose. With soft, usually long and ± straight hairs.

Pinnate. Arranged in two rows, one on each side of a common axis, as veins in a leaf or leaflets in a compound leaf. In an odd-pinnate leaf, there is a terminal leaflet; in an even-pinnate one, there is no terminal leaflet. Twice-pinnate: with the primary divisions again pinnate.

Pinnatifid. Deeply lobed or cleft in a pinnate pattern. Bipinnatifid (or twice pinnatifid): with the primary divisions again pinnatifid.

Pinnatisect. Very deeply cleft in a pinnate pattern (often to the midrib).

Pistil. One of the female or seed-producing structures of a flower, whether composed of a single carpel or two or more carpels; usually consisting of one ovary and one or more styles and stigmas.

Pit. The seed and its stony covering of endocarp in a drupe. A tiny but often relatively deep depression on a surface.

Pith. The spongy center of a stem (consisting of thin-walled cells).

Placentation. The arrangement of ovules in an ovary.

-ploid. (See *n*.)

Plumose. (Of hairs) with lateral branches, like a feather — a pectinate hair (but usually 3-dimensional, not flat).

Pollen. The grains (microspores, containing male gametes) produced in the anther.

Polygamous. Bearing perfect and unisexual flowers on the same individual.

Pore. A small ± round natural opening through which pollen or seeds can escape.

Prairie. A naturally treeless area (but drier than a marsh, into which "wet prairies" grade).

Puberulent. Minutely or finely pubescent.

Pubescent. With hairs (of whatever size or texture).

Punctate. Dotted with tiny pits or glands or spots.

Pustulate. With blister-like swellings.

Raceme. A type of inflorescence in which each flower is on an unbranched pedicel attached to an unbranched ± elongate central axis; the flowering sequence is from the base to the apex.

Rachis. The central axis of an inflorescence or a compound leaf.

Radiate. With ray flowers in the head (Compositae).

-ranked. -rowed; 2-ranked structures are in two rows on opposite sides of an axis and 3-ranked structures are in 3 rows (best seen by examining from above the end of the axis).

Ray. In Compositae, the expanded portion (limb) of a petal-like (bilaterally symmetrical) flower or ligule.

Receptacle. The surface on which the parts of a flower are inserted or on which the flowers in a head or other dense inflorescence are inserted.

Receptacular chaff. (See chaff.)

Reflexed. Bent back or downwards.

Regular. With radial symmetry; capable of division into similar halves on more than one plane (= "actinomorphic" of many works). Cf. bilaterally symmetrical.

Remote. Relatively far apart.

Reniform. Shaped in general outline like a longitudinal section of a kidney, i.e., broader than long, ± shallowly cordate at base, and otherwise ± rounded.

Reticulate. Having the appearance of a net.

Retrorse. Directed toward the base or downward. Cf. antrorse.

Revolute. With the margins rolled back or under (i.e., abaxially). Cf. involute.

Rhizome. An underground stem, usually ± elongate and growing horizontally (distinguishable from a root by the presence of nodes).

Rosette. A ± dense and circular cluster of leaves.

Rotate. (Of a corolla) having a broad, flat limb and a very short tube.

Rugose. Wrinkled or puckered in appearance.

Sagittate. Arrowhead-shaped, with basal lobes pointing downward (not divergent, but often ± parallel). Cf. hastate.

Salverform. Having a slender tube and abruptly expanded flat limb.

Samara. A dry indehiscent nut-like fruit with a well developed wing.

Saprophyte. A plant incapable of photosynthesis, but not directly parasitic on any green plant (usually on a fungus).

Savana. Vegetation consisting mostly of grassland with trees scattered (not forming a closed canopy) or in scattered clumps.

Scabrous, scabrid. Rough (to the touch).

Scale. Any small thin bract, such as subtends an individual flower in Gramineae, Cyperaceae, or Salicaceae or covers a bud.

Scape. A peduncle arising from the base of a plant (directly from root, rhizome, etc.); a "leafless stem."

Scarious. Thin and dry, papery in texture.

Sepal. One of the divisions of the calyx.

Septate. With cross-partitions; jointed.

Serrate. With sharp, ± forward-pointing, marginal teeth. In a doubly serrate margin, there are teeth on the primary teeth.

Serrulate. Minutely serrate.

Sessile. Attached without a stalk.

Seta. A bristle-like hair.

Setaceous. Bristle-like.

Silique. A 2-carpellate fruit which dehisces from the base upward, leaving a septum between the locules, characteristic of the Cruciferae.

Simple. Composed of a single or unbranched part; e.g., a leaf with one blade, a pistil of one carpel, an unbranched inflorescence; an unbranched hair.

Sinuate. Broadly scalloped, with ± open sinuses and low teeth; coarsely dentate or wavy-margined.

Sinus. The space or cleft between two lobes.

Spadix. An inflorescence consisting of small sessile flowers on a ± elongate fleshy axis.

Spathe. A single bract (occasionally more) at the base of an inflorescence (equivalent to an involucre, but used only in the monocots).

Spatulate. Broad and flat distally, contracted or tapered toward the base; spoon-shaped.

Spicule. A minute sharp slender point, as on the margins of some leaves.

Spike. An elongate unbranched inflorescence in which the flowers are sessile; loosely, a dense elongate spike-like inflorescence with crowded flowers.

Spikelet. The unit of the inflorescence in a grass or sedge (i.e., a small spike, with reduced flowers on a central axis).

Spinulose. With minute spines.

Stamen. One of the male or pollen-producing structures of a flower, usually consisting of a filament and an anther.

Staminodium. A sterile stamen (so determined by its location), without an anther, but sometimes cleft, hairy, or otherwise considerably modified.

Stellate. (Of a hair) ± radially branched.

Sterile. Lacking flowers or fruit (= vegetative); not fertile.

Stigma. The part of a pistil that is receptive to pollen, usually distinguished by a sticky, papillose, or hairy surface.

Stipe. A stalk (generally used when no precise term such as petiole or pedicel is applicable); e.g., the short stalk on which some ovaries are elevated above the receptacle.

Stipule. An appendage on the stem at the base of a leaf (sometimes connate and sometimes partly or wholly adnate to the petiole).

Stolon. An elongate stem, on or near ground-level, growing ± horizontally and rooting at the nodes and/or apex.

Striate. With slender lines or stripes or low ridges.

Strigose. With short, straight, strongly appressed hairs.

Style. The portion of a pistil between the ovary and the stigma—often narrow and elongate.

Sub-. A prefix meaning almost, not quite, just below; e.g., subterminal, just below the end; subglobose, almost spherical.

Submersed. Normally occurring under water and so adapted (not merely flooded = submerged). Cf. emersed.

Subspecies. A subdivision of a species. This rank is higher than the rank of variety; i.e., a subspecies may include two or more varieties, but not vice versa. (Most taxonomists prefer to use one of these ranks or the other, although some use both.)

Subtend. Occur immediately below, as a bract below a flower or pedicel.

Subulate. Awl-shaped: very slender, firm, and sharp-pointed.

Succulent. Fleshy, juicy.

Superior. (Of an ovary) with the perianth and stamens inserted beneath it. Cf. inferior.

Suture. The line or joint along which two parts are fused (and along which they may separate).

Swale. A natural (unlike a ditch) ± elongate depression, at least seasonally wet.

Swamp. A wet (at least seasonally) and wooded area.

Sympetalous. With the petals united to each other (connate) for at least part of their length.

Taproot. A central primary root continuing the axis of the stem into the ground.

Tautonym. A binomial scientific name in which the generic name and the specific epithet are identical. Tautonyms (e.g., *Linaria linaria*) are not allowed under the Code for names of plants.

Taxon. Any taxonomically recognized unit, regardless of rank; e.g., genus, species, variety, form.

Tendril. A slender coiling or twining organ, as on some vines.

Tepal. One of the divisions of the perianth when the sepals and petals are similar in color, texture, and (usually) size (though usually distinguishable by position, the sepals being the outer series and the petals the inner one).

Terete. Round in cross-section.

Thicket. A loosely defined term for usually small areas (or narrow ones, as along a stream) with ± dense shrubs or small trees.

Throat. In a calyx or corolla of united parts, the region where the tube and the limb join — the entrance to the tube.

Tomentum. More or less dense, curly, matted hairs.

Tribe. A subdivision of a family, ranking above genus.

Trifid. Cleft into three.

Trifoliolate. With three leaflets.

Truncate. Ending abruptly (at base or apex), as if cut off squarely (neither tapered nor lobed).

Tube. The fused portion of a cycle of perianth parts, beyond which the calyx lobes or corolla lobes extend.

Tuber. A thickened portion of rhizome or root, usually a starch-storing organ.

Tubercle. A small ± knobby projection.

Turion. A "winter-bud" of an aquatic plant, consisting of a modified branch or bud (e.g., very short compact internodes and reduced leaves).

Type. The specimen that fixes application of a name; a name of a species, e.g., applies at least to its type — and to all other specimens deemed to belong to the same species. If a taxon is divided into two or more, the name remains with that element which includes its type, and a new name is required for the other element(s). Use of name in a sense that would exclude its type is termed a misapplication of the name, and such use must be abandoned.

Umbel. An inflorescence in which the pedicels arise from the same point or nearly so; in a compound umbel, each primary ray bears an umbellet.

Undulate. Wavy, sinuate.

Unisexual. (Of a flower) containing only stamen(s) or pistil(s); imperfect. Cf. perfect.

Valid. (Of publication of a name) meeting the requirements for publication stated in the Code, such as distribution, form of name, necessary inclusions (e.g., description, Latin, type designation). Note that validity has to do only with conditions of *publication* of a name, and has nothing to do with whether or not the plant to which it was applied merits naming or recognizing (those are taxonomic, not nomenclatural, judgments). Cf. illegitimate.

Valve. One of the parts into which a dehiscent fruit (or other structure) splits.

Variant. Any deviation from "normal" in a plant (without necessarily taxonomic recognition at any given rank).

Variety. A subdivision of a species; this rank is below the rank of subspecies. Cf. subspecies.

Vein. A bundle of vascular tissue; the external ridge marking the location of an underlying vein.

Ventral. Pertaining to the surface (e.g., of a leaf, nutlet, etc.) toward the axis to which a structure is attached; adaxial. Cf. dorsal.

Villous. With soft, not necessarily straight, hairs — practically synonymous with pilose.

Viscid. Sticky, glutinous.

Waste places, waste ground. Seriously disturbed areas that are not cultivated, such as roadsides, vacant lots, dumps, and construction sites.

Whorl. A ring of 3 or more similar structures around a stem or other axis (i.e., at the same node).

Wing. A flat, ± thin extension or appendage on the edge or surface of an organ (as on a seed or stem).

Woodland. Vegetation with trees less dense than in a forest (often a savana).

Woods. Forest; a "mixed woods" is one of both coniferous and deciduous trees

Index

Names of species followed by a map number (M-000) in parentheses are accepted ones for species mapped as part of our flora. An asterisk indicates that the species is illustrated. References to the color plates (pages 15–22) and to the figures (numbered consecutively throughout the volume) may be found at the main entry for each species—which is the first page cited for adopted names of accepted species. Additional pages cited refer to discussion, comparison, or significant observations elsewhere in the volume; page numbers where species are compared within the same genus are not indexed, nor are pages where taxa appear in the keys. Species and genera, with their descriptive key characters, can generally be found easily in the keys as they are numbered in sequence.

Names of species not followed by a map number represent synonyms of accepted names, most hybrids, and species mentioned incidentally (including those not considered established and those erroneously reported from the state, as well as familiar species cited only as cultivated).

Varieties and other infraspecific taxa are not included in the index. When only one species is mentioned in a genus, the name of the genus is not separately indexed. Common names consisting of two elements are ordinarily indexed only under the second one (e.g., Spotted *Wintergreen*, Small *Cranberry*), especially if the elements are not hyphenated—unless the second element is a broad, unseparated, or meaningless word like "berry," "plant," or "weed" (e.g., Cran*berry*, Blue*weed*, Trail *Plant*, Blue-*curls*). Hyphens (or spaces) and multiplication signs (designating hybrids) are ignored in alphabetizing.

Chamaemelum nobile, 392
Chamaesaracha, 187
Chamomile, 392
  Corn, 393
  False, 392
  Scentless, 392
  Stinking, 393
  Yellow, 393
Chamomilla, 391
  recutita, 392
  suaveolens, 392
Charlie, Creeping, 180
Chelone, 225
  glabra (M-305), 226
  obliqua (M-304), 225*
Cherry, Ground-, 195
Chicory, 345
Chimaphila, 29
  maculata (M-6), 29*
  umbellata (M-7), 29
Chinese-lantern-plant, 195
Chiogenes, 51
Chondrilla juncea (M-511), 369*
Chrysanthemum, 395, 400
  balsamita (M-551), 396*
  bipinnatum, 401
  cinerariifolium, 396
  leucanthemum (M-549), 395*, 413
  majus, 396
  parthenium (M-550), 396
  serotinum (M-552), 396
  ×superbum (M-553), 397
  uliginosum, 396
Chrysochus auratus, 84
Chrysopsis, 478
  camporum (M-674), 478
  villosa (M-673), 478
Cichorium, 345
  endiva, 345
  intybus (M-476), 345*
Cigar-tree, 252
Cinchona, 278
Cirsium, 517
  altissimum (M-741), 524
  arvense (M-736), 521*
  canescens, 522
  discolor (M-740), 523*
  flodmanii (M-738), 522
  hillii (M-734), 520
  horridulum, 520
  muticum (M-735), 520
  palustre (M-732), 519*
  pitcheri (M-737), 521*

[Cirsium]
  pumilum, 520
  undulatum (M-739), 523
  vulgare (M-733), 519
Citations of specimens, 23–24
Citrullus, 326
  lanatus (M-450), 326
  vulgaris, 326
Clary, Meadow, 166
Cleavers, 283
Clinopodium, 168, 178
  vulgare (M-220), 168*, 165
Cocklebur, 407
Code (of nomenclature), 7
Coffea, 278
Coffee, 278
Collinsia, 248
  parviflora (M-343), 248
  verna (M-342), 248
Collinsonia canadensis (M-209), 162*
Collomia linearis (M-115), 105
Coltsfoot, 457
  Sweet-, 540
Columbo, American, 74
Comfrey, 120
  Common, 120
  Northern Wild, 130
  Prickly, 120
Common names, 8
Compass Plant, 418
COMPOSITAE, 340
Coneflower, 437
  Cutleaf, 437
  Prairie, 435
  Purple, 413
  Showy, 440
  Sweet, 438
  Tall, 437
  Yellow, 435
Conobea, 222
Conoclinium coelestinum, 450
Conopholis americana (M-348), 256*
Constance, Lincoln, xiv
CONVOLVULACEAE, 95, 100
Convolvulus, 97
  arvensis (M-102), 97*
  japonicus, 99
  pellitus, 99
Conyza, 539
  canadensis (M-765), 539*
  ramosissima, 540
Cooley, Dennis, 4, 5, 521
Cooperrider, Tom S., xiv, 201

LOGANIACEAE, 71
Logfia arvensis, 529
Long, Robert W., xiv
Lonicera, 300
  ×bella (M-424), 307
  caerulea, 304
  canadensis (M-419), 304*
  caprifolium (M-411), 302
  dioica (M-414), 303*
  hirsuta (M-413), 302
  involucrata (M-417), 304
  japonica (M-415), 303
  maackii (M-420), 305*
  morrowii (M-423), 307
  oblongifolia (M-418), 304*
  prolifera, 302
  reticulata (M-412), 302
  ruprechtiana, 308
  sempervirens (M-410), 302
  tatarica (M-421), 305
  villosa (M-416), 304
  xylosteum (M-422), 307
Loosestrife, 58
  Fringed, 61
  Garden, 60
  Gooseneck, 60
  Tufted, 61
  White, 60
  Whorled, 61
Lopseed, 133
Lousewort, 212
Lungwort, 119
  Tall, 120
Lycium, 185
  barbarum (M-245), 185*
  chinense (M-244), 185
  halimifolium, 185
Lycopersicon, 187
  esculentum (M-248), 187*
  lycopersicum, 187
Lycopus, 147
  americanus (M-184), 150*
  amplectens, 150
  asper (M-182), 149
  europaeus (M-183), 150
  rubellus (M-181), 149
  ×sherardii, 148
  uniflorus (M-179), 148
  virginicus (M-180), 149
Lyonia ligustrina, 36
Lysimachia, 58
  ciliata (M-46), 61
  clethroides (M-41), 60

[Lysimachia]
  ×commixta, 60
  hybrida (M-49), 61
  lanceolata (M-50), 62
  nummularia (M-40), 59*
  ×producta, 60
  punctata (M-42), 60
  quadriflora (M-47), 61
  quadrifolia (M-48), 61*
  terrestris (M-44), 60
  thyrsiflora (M-45), 61*
  vulgaris (M-43), 60

Madder, 278
  Family, 278
  Field-, 279
Madia glomerata (M-593), 419
Maleberry, 36
Man-of-the-earth, 96
Map, Michigan, 6
Maps, explained, 4
Marbleweed, 119
Marguerite, Golden, 393
Marigold,
  Pot-, 433
  Stinking-, 446
  Water-, 380
Marjoram, Wild-, 162
Marrubium vulgare (M-228), 175
Marsh-elder, 408
Martyniaceae, 253
Matricaria, 391, 393
  chamomilla, 392
  discoidea (M-543), 391*
  inodora, 392
  maritima, 392
  matricarioides, 392
  perforata (M-544), 392
  recutita (M-545), 392
Matrimony Vine, 185
Mayweed, 393
Mazus, 237
  japonicus, 237
  pumilus, 237
  reptans (M-326), 237
McLouth, C. D., 446
McVaugh, Rogers, xiv
Measurements, 24
Megalodonta, 379
  beckii (M-523), 380*
Melampyrum lineare (M-344), 250*
Melissa officinalis (M-234), 178
Melon, 328

ONT.

ONTARIO

LAKE SUPERIOR

ISLE ROYALE

KEWEENAW

HOUGHTON

ONTONAGON

BARAGA

GOGEBIC

MARQUETTE

A L G E R

LUCE

CHIPPEWA

IRON

DICKINSON

SCHOOLCRAFT

DELTA

MACKINAC

MENOMINEE

MACKINAC IS

BOIS BLANC IS

DRUMMOND IS

FOX IS

BEAVER IS

EMMET

CHEBOYGAN

PRESQUE ISLE

LAKE HURON

MANITOU IS

CHARLEVOIX

LEELANAU

ANTRIM

OTSEGO

MONT-MORENCY

ALPENA

WISCONSIN

BENZIE

GRAND TRAVERSE

KALKASKA

CRAWFORD

OSCODA

ALCONA

MANISTEE

WEXFORD

MISSAUKEE

ROS-COMMON

OGEMAW

IOSCO

LAKE MICHIGAN

MASON

LAKE

OSCEOLA

CLARE

GLADWIN

ARENAC

CHARITY IS

HURON

OCEANA

NEWAYGO

MECOSTA

ISABELLA

MIDLAND

BAY

TUSCOLA

SANILAC

MUSKEGON

MONTCALM

GRATIOT

SAGINAW

KENT

IONIA

CLINTON

SHIA-WASSEE

GENESEE

LAPEER

ST. CLAIR

OTTAWA

ALLEGAN

BARRY

EATON

INGHAM

LIVINGSTON

OAKLAND

MACOMB

VAN BUREN

KALAMAZOO

CALHOUN

JACKSON

WASHTENAW

WAYNE

ONT.

BERRIEN

CASS

ST. JOSEPH

BRANCH

HILLS-DALE

LENAWEE

MONROE

LAKE ERIE

INDIANA

OHIO

MICHIGAN

0   10   50
Scale of Miles
UMMZ—1957—WLB